DISCRETE DYNAMICS AND DIFFERENCE EQUATIONS

DISCRETE DYNAMICS AND DIFFERENCE EQUATIONS

Proceedings of the Twelfth International Conference on Difference Equations and Applications

Lisbon, Portugal, 23 – 27 July 2007

Saber N Elaydi

Trinity University, USA

Henrique Oliveira • José Manuel Ferreira • João F Alves

Instituto Superior Técnico, Portugal

Editors

World Scientific

NEW JERSEY • LONDON • SINGAPORE • BEIJING • SHANGHAI • HONG KONG • TAIPEI • CHENNAI

Published by

World Scientific Publishing Co. Pte. Ltd.
5 Toh Tuck Link, Singapore 596224
USA office: 27 Warren Street, Suite 401-402, Hackensack, NJ 07601
UK office: 57 Shelton Street, Covent Garden, London WC2H 9HE

British Library Cataloguing-in-Publication Data
A catalogue record for this book is available from the British Library.

DISCRETE DYNAMICS AND DIFFERENCE EQUATIONS
Proceedings of the Twelfth International Conference on Difference Equations and Applications

ISBN-13 978-981-4287-64-7
ISBN-10 981-4287-64-4

Printed in Singapore by World Scientific Printers.

José de Sousa Ramos
(1948–2007)

José Sousa Ramos was the main supporter of the idea of organizing the International Conference on Difference Equations and Applications 2007 in Lisbon, at the Instituto Superior Técnico. The ICDEA, jointly with ECIT (European Conference on Iteration Theory) were the conferences that he most enjoyed to attend. Unfortunately he deceased January 1, 2007, and therefore could not play the role that he had planned in the organization of this Conference.

Sousa Ramos had a first degree in Physics and obtained his Ph.D. in Mathematics. In these sciences he showed a huge knowledge and a particular capacity of analyzing physical phenomena into the mathematical framework. He had a very creative and autonomous mind. This often led him to rather original ideas and very different and interesting ways of looking at several aspects of either Mathematics or Physics, which he enjoyed to discuss with everyone around.

He left us a strong legacy in the domain of Dynamical Systems, namely in the so called Symbolic Dynamics. A large work that he would have hated to build up only by his own, in complete solitude. Indeed, his extreme simplicity and modesty jointly with a rare joy of sharing explain the reason why it is so difficult for one to find a simple paper signed by him only.

An important aspect of his daily work, were the appointments he continuously had with his students and collaborators. In those meetings the advances in several problems were discussed and understood not only theoretically, but also experienced with a computer aid, a method that he thought essential and that largely increased as an "experimental mathematics".

Little by little, Sousa Ramos set up to build a large group of around twenty people from several universities, spread out over the country from north to south, that worked with him in almost permanence. In this way, he was able to achieve something very uncommon in our country: being the founder of a mathematical school. A true school in Dynamical Systems which is already continuing his legacy through the participation and organization of this conference.

Preface

These proceedings include articles based on talks presented at the 12th International Conference on difference equations and applications (ICDEA07). The conference was held in Lisbon, July 23–27, 2007, under the auspices of the International Society of Difference Equations (ISDE).

This volume encompasses articles on a variety of current topics such as stability, bifurcation, functional equations, chaos theory, mathematical biology, mathematical economics, boundary value problems, neural networks, cellular automata, combinatorics, and numerical methods. There are articles on hyperbolic dynamics in Nash maps (Misiurewicz *et al.*), discrete versions of the Lyapunov-Schmidt method (Vanderbauwhede), hyperbolic and minimal sets (Pinto *et al.*), difference equations with continuous time (Sharkovsky *et al.*), interval maps on cellular automata (Ramos *et al.*), and chaotic synchronization (Caneco). On mathematical biology, we include articles on stability through migration (Fujimoto *et al.*), a two-dimensional competing species model (Johnson *et al.*), and a multiplicative delay population dynamics model (Braverman *et al.*). In addition, we have papers on Stochastic difference equations (Appleby *et al.*), combinatorics of Newton maps (Balibrea), fuzzy dynamical systems (Rodriguez-Lopez), discretized panograph equations, nonlinear boundary problems (Sharkovsky *et al.*), and neural networks (Cheng *et al.*) are also included. Due to our limited space, we have not mentioned many papers in these proceedings. The interested reader may find a complete list of the titles in this volume under the Contents section.

The editors sincerely appreciate the excellent job of our referees, without which this volume would never have materialized. We present this volume to the mathematical community and hope that it will be a stimulus to research in discrete time problems that arise in a plethora of scientific disciplines as well as those that are related to various fields within mathematics.

Saber N. Elaydi
Henrique Oliveira
José Manuel Ferreira
João F. Alves

Contents

Part 1
Papers by Main Speakers

Combinatorics of Newton Maps on Quintic Equations

Francisco Balibrea

Facultad de Matemáticas, Universidad de Murcia,
Campus de Espinardo, 30100 Murcia, Spain
E-mail: balibrea@um.es

We study some dynamical properties of the Newton maps associated to real quintic polynomial equations of one variable. Using Tschirnhaus transformations and topological conjugation, we suppose the equation has been reduced to the form $p_c = x^5 - cx + 1 = 0$ where $c \in \mathbb{R}$. Then we use symbolic dynamics and in particular the construction of kneading sequences which allow to know the dynamical behavior of the Newton map $N_{p_c}(x) = x - \frac{p_c(x)}{p'_c(x}$ associated to the map p_c.

Keywords: Newton maps; difference equation; symbolic dynamics; Markov partitions; quintic equation.

Dedicated to the memory of Professor José de Sousa Ramos

1. Introduction and motivation

Finding roots of polynomials is an old problem in mathematics and consideration of dynamics of Newton maps as applied to real polynomials has a long research history even in one variable. In this setting numerical Newton method supplies one of the simplest and most widely-used root-finding methods constructing sequences approaching the roots of the equation.

The fundamental property of Newton method for a general map f on $\mathbb{R}$ is that it transforms the problem of finding roots of $f(x) = 0$ when f is differentiable, into the problem of finding attracting fixed points of the associate Newton map $N_f(x)$.

We bring here some interesting problems, not yet completely solved even for real cubic polynomial equations. For example, there may have open sets of initial points which do not lead to any root but instead to an attracting cycle of period grater than one. Boundaries of the basin of attraction of attracting cycles have a fractal structure which it is not yet well understood (see [7] and [8]).

Another type of results was given by Barna [1] saying that for real polynomials in one variable of degree at least 4 with distinct simple real roots, the set of initial points not converging to a root is homeomorphic to a Cantor set.

Under standard assumptions, the Newton method is locally convergent in a suitable set surrounding the solution. The possibility that a small change in x_0 can cause a drastic change in convergence indicates the nasty nature of the convergence problem.

A detailed treatment of the cubic polynomial case can be seen in [4] and a complete description of its combinatorics is given in [11].

In this paper we are dealing with the general quintic equation

$$x^5 + c_1x^4 + c_2x^3 + c_3x^2 + c_4x + c_5 = 0$$

which has been largely considered in the literature trying to construct convergent sequences to its roots.

To this aim, we reduce the number of parameters until 2 using the Tschirhaus transformation [3], obtaining the equation

$$x^5 + ax + b = 0$$

and then by topological conjugation we transform our problem in that of solving the equation

$$p_c(x) = x^5 - cx + 1 = 0$$

depending only of the real parameter c.

In section 3, using standard symbolic dynamics, we introduce the admissibility rules of the sequences associated to Newton map N_{p_c} and study their structure. The techniques of symbolic dynamics are based on notions of the kneading theory for one-dimensional multimodal maps, (see Milnor and Thurston [9]). It allows us to construct a tree of kneading sequences for N_{p_c}.

It is also possible to see a connection between kneading theory and subshifts of finite type is shown by using a commutative diagram derived from the topological configurations associated with m-modal maps which can be appreciated in [6].

2. Newton maps for quintics

We transform the polynomial equation $x^5 + ax + b = 0$ using topological conjugacy. It is well known that f and g are topologically conjugate provided there is an homeomorphisms τ such that $f \circ \tau = \tau \circ g$. In such case for f^n and g^n we have the same dynamical properties.

Proposition 2.1. *Let $g(x) = x^5 + ax + b^5$ where $b \neq 0$ and $f(x) = x^5 + cx + 1$ where $c = a/b^4$. Then N_g and N_f are topologically conjugate via the homeomorphism $\tau(x) = x/b$.*

Let us see what happens when $b = 0$.

Proposition 2.2. *Let $g(x) = x^5 + a^4x$, $\tau(x) = x/a$, with $a \neq 0$, and $p_+(x) = x^5 + x$. Then N_g and N_{p_+} are topologically conjugate by τ. By other hand, if $g(x) = x^5 - a^4x$ and $p_-(x) = x^5 - x$ then N_g and N_{p_-} are topologically conjugate by τ.*

When $a = 0$ we have $f(x) = x^5$.

Last two propositions imply that the dynamics of Newton map for quintic $g(x) = x^5 + ax + b$ is equivalent to the dynamics of Newton map for the polynomial family $f_c(x) = x^5 + cx + 1$ or to $g_a(x) = x^5 + ax$. Moreover, the Newton map for function $g_a(x)$ is topologically conjugate to Newton map for one of the three polynomials:

$$p_-(x) = x(x^4 - 1), \quad p_+(x) = x(x^4 + 1), \quad \text{or } p_0(x) = x^5.$$

Therefore the study of Newton map for quintic polynomials is reduced to the case $f_c(x) = x^5 - cx + 1$, where $c \in \mathbb{R}$.

- When $c < 0$, it is easy to verify that N_{p_c} has exactly one real root and that its *stable* set (the set of points which are forward asymptotic to it) is $\mathbb{R}$.
- When $c = 0$, there is also one real root and its stable set contains all real numbers except 0.
- When $c > 0$ we have three interesting cases which will be considered later.

The polynomial $p_c(x)$ has a maximum at $d_1 = -\sqrt[4]{c/5}$ and a minimum at $d_3 = \sqrt[4]{c/5}$.

Note that when c increases, the minimum of f decreases and the maximum increases. When $c = c_0 = 5 \times 2^{-8/5} = 1.64938...$ the minimum is 0.

Note also that when c is bigger than c_0, f has three real roots.

In last case, has sense to use the following result due to Rényi

Theorem 2.1. *Let $f : \mathbb{R} \to \mathbb{R}$. Let us suppose that $f''(x)$ is monotone increasing for all $x \in \mathbb{R}$ and that f has exactly three real roots a_i $(i = 1, 2, 3)$. Then the sequence $x_{n+1} = x_n - f(x_n)/f'(x_n)$ converges to one of the roots for every choice of x_0 except for x_0 belonging to a countable set E of singular points. For any $\varepsilon > 0$ there exists an interval $(x, x + \varepsilon)$ containing three points y_i $(x < y_i < x + \varepsilon,\ i = 1, 2, 3)$ having the property that if $x_0 = y_i$, then $(x_n)_{n=0}^{\infty}$ converges to a_i $(i = 1, 2, 3)$.*

The polynomial $p_c(x) = x^5 - c\,x + 1$ has three real roots when $c > c_0$ and $p_c''(x) = 20x^3$ is monotone increasing for all $x \in \mathbb{R}$, so we are in the conditions of last theorem.

Finally the most interesting situation happens when c is between 0 and c_0, in this case $p_c(x)$ has only one real root.

Now we consider $p_c(x) = x^5 - c\,x + 1$, so

$$N_{p_c}'(x) = \frac{p_c''(x)p_c(x)}{(p_c'(x))^2} = \frac{20\,x^3 p_c(x)}{(p_c'(x))^2}.$$

If $N_{p_c}'(x) = 0$ we have $x = 0$ or $p_c(x) = 0$.

As the roots of $p_c(x)$ are super-stable fixed points $(p'(x) = 0)$, the only interesting critical point of N_{p_c} is 0 denoted by d_2, so for the study of the iteration of N_{p_c} we will start at $x_0 = d_2$.

Let us now describe the numerical experiments which can be performed in the (x, c) - plane, computing the bifurcation diagram for N_{p_c} with $p_c(x) = x^5 - c\,x + 1$, where $c \in (0, c_0)$. There is a sequence of windows where $N_{p_c}(d_2)$ converges to a stable periodic orbit with period $(n \in \mathbb{N})$, intercalated with intervals where the critical point d_2 converges to the fixed point d_0 which can be seen in Fig 1.

Until now we have studied the case with only two coefficients in the quintic polynomials (in such a case we have at most three roots).

But there are other possibilities for a general quintic equation,

(1) the quintic equation has four distinct real roots, one of them double;
(2) the quintic equation has five distinct real roots.

Now it applies the follow result:

Theorem 2.2 (Barna). *If f is a real polynomial having all real roots and at least four distinct ones, then the set of initial values for which Newton method does not yield to a root of f is homeomorphic to a Cantor set. The set of exceptional initial values $J(f)$ is of Lebesgue measure zero.*

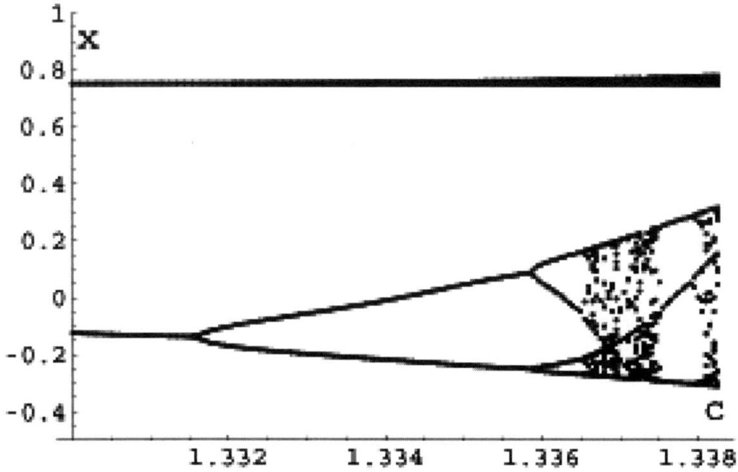

Fig. 1.

Proof. For a proof we refer to Hurley and Martin [5]. They all give modern proof of Barna result [1]. The underlying idea is to show that the set $J(f)$ arises in a way which is very similar to the usual Cantor set construction. Wong proves this result using symbolic dynamics [13]. □

Therefore in next section we will concentrate in the most interesting case $p_c(x) = x^5 - cx + 1$ for $c \in (0, c_0)$.

3. Symbolic dynamics

The dynamical system associated to $N_{p_c}(x)$ can be globally studied using the kneading theory which is an appropriate tool to classify topologically the dynamics of maps. In what follows we will use an extension of the Milnor-Thurston's theory to discontinuous maps defined in not necessarily compact subsets of $\mathbb{R}$ adapted to maps $N_{p_c}(x)$ where $p_c(x) = x^5 - cx + 1$, and $0 < c < c_0$ which allows us to compute in some cases the topological entropy of such maps.

To this aim, let us consider the alphabet $\mathcal{A}_c = \{A, A_0, B, L, C, M, R\}$ and the set $\Omega_c = \mathcal{A}_c^{\mathbb{N}_0}$ ($\mathbb{N}_0 = \mathbb{N} \cup \{0\}$), the space of all sequences composed of elements of $\mathcal{A}_c$. In order to avoid all preimages of $\{d_1, d_3\}$, instead of in $\mathbb{R}$ we will work in the subset

$$\Lambda_c = \mathbb{R} \backslash \bigcup_{n \in \mathbb{N}_0} N_{p_c}^{-n}\left(\{d_1, d_3\}\right) = \mathbb{R} \backslash A_c$$

where A_c is a countable set of points. Given $x \in \Lambda_c$, its itinerary $i_c(x) = (i_c(x)_0, i_c(x)_1, ..., i_c(x)_n, ...)$ is defined by

$$i_c(x)_m = \begin{cases} A \ \text{ if } N_{p_c}^m(x) \in (-\infty, d_0(c)) = I_0(c) \\ \\ A_0 \ \text{ if } N_{p_c}^m(x) = d_0(c) \\ \\ B \ \text{ if } N_{p_c}^m(x) \in (d_0(c), d_1(c)) = I_1(c) \\ \\ L \ \text{ if } N_{p_c}^m(x) \in (d_1(c), d_2) = I_2(c) \\ \\ C \ \text{ if } N_{p_c}^m(x) = d_2 \\ \\ M \ \text{ if } N_{p_c}^m(x) \in (d_2, d_3(c)) = I_3(c) \\ \\ R \ \text{ if } N_{p_c}^m(x) \in (d_3(c), \infty) = I_4(c) \end{cases}$$

observe that $d_2 = 0$ and does not depend on c and that the equation $N_{pc}(x) = 0$ has as unique solution $x = 4^{\frac{-1}{5}}$ which does not depend on c.

If we now consider the shift operator $\sigma : \Omega_c \to \Omega_c$, $\sigma(X_0X_1X_2...) = (X_1X_2X_3...)$, then it is immediate the commutativity of the diagram

$$\begin{array}{ccc} & N_{p_c} & \\ \Lambda_c & \longrightarrow & \Lambda_c \\ i_c \downarrow & & \downarrow i_c \\ \Omega_c & \longrightarrow & \Omega_c \\ & \sigma & \end{array}$$

We introduce in Ω_c the lexicographic ordering, derived from the natural ordering of symbols in $\mathcal{A}_c$:

$$A < A_0 < B < L < C < M < R$$

and state also that

$$(-1)R < (-1)M < C < (-1)L < (-1)B < A_0 < (-1)A$$

After this we state in Ω_c the following ordering

Definition 3.1. We say that $X \prec Y$ (or $Y \succ X$) for $X, Y \in \Omega_c$, if there exists $k \geq 0$ such that for $0 \leq i \leq k-1$ is $X_i = Y_i$ and $(-1)^{n_{BL}(X_0...X_{k-1})}X_k < (-1)^{n_{BL}(X_0...X_{k-1})}Y_k$ where $n_{BL}(X_0...X_{k-1})$ is equal to the number of symbols B and L which appear in $X_0...X_{k-1}$, that is, the number of times that the iterates of the point belong to decreasing intervals of the map $N_{p_c}(x)$.

Definition 3.2. We say $X \preceq Y$ (respectively $Y \succeq X$) if $X \prec Y$ or $X = Y$ (respectively $Y \succ X$ or $X = Y$).

Example 3.1. Let $MRRM...$ and $MRRR...$ be two sequences with a common block MRR. Then $n_{BL}(MRR)$ is even and therefore $MRRM... \prec MRRR...$. For the sequences $RLRA..., RLRR..., n_{BL}(RLR)$ is odd and then $RLRA... \succ RLRR...$

The lexicographic ordering is total in Ω_c and it is compatible with the usual ordering on Λ_c. The relationship between the two orderings is stated in the following result whose proof is similar to that of Proposition 3 from [9].

Proposition 3.1. *Let $x, y \in \Lambda_c$*

(1) If $x < y$ then $i_c(x) \preceq i_c(y)$
(2) If $i_c(x) \prec i_c(y)$, then $x < y$

Our task now is to introduce a criterium of admissibility of itineraries in Ω_c. Of special interest are the itineraries of d_2 and d_0. In the case d_0 the unique possible itinerary is $(A_0)^\infty$.

To this end, consider the subset $J \subset (0, c_0)$ given by

$$J = \{c \in (0, c_0) : \bigcup_{n \in \mathbb{N}_0} N^n_{p_c}(d_2) \cap \{d_1, d_3\} = \emptyset\} = J_1 \cup J_2.$$

where $J_1 = (0, c*) \cap J$, $J_2 = (c*, c_0) \cap J$ and $c* = \sqrt[5]{5}$ which corresponds to the solution of $N_{p_c}(0) = \frac{1}{c} = d_3$.

$J_1 = (0, c*) \backslash H_1$ (respectively $J_2 = (c*, c_0) \backslash H_2$) where H_1, H_2 are some subsets of parameters.

We distinguish between this two subsets because the corresponding maps of the family have different dynamics. In the first case, is $N_{p_c}(0) = \frac{1}{c} > d_3$ and $N_{p_c}(0) = \frac{1}{c} < d_3$ in the second but the shape of the graphs of the maps are similar.

For each $c \in J$ we will denote the itinerary of the orbit of the critical point d_2 by Y^c.

We describe the rules for admissible sequences looking at the graph of $N_{p_c}(x)$ and also construct the admissible matrix

$$T_c^1 = \begin{bmatrix} 1\,0\,0\,0\,0\,0\,0 \\ 0\,1\,0\,0\,0\,0\,0 \\ 1\,0\,0\,0\,0\,0\,0 \\ 0\,0\,0\,0\,0\,0\,1 \\ 0\,0\,0\,0\,0\,0\,1 \\ 0\,0\,0\,0\,0\,0\,1 \\ 1\,1\,1\,1\,1\,1\,1 \end{bmatrix}$$

for $c \in J_1$ and alphabet $\mathcal{A}_c$

If $c \in J_2$ the alphabet is similar, but not the admissible matrix which is

$$T_c^2 = \begin{bmatrix} 1\,0\,0\,0\,0\,0\,0 \\ 0\,1\,0\,0\,0\,0\,0 \\ 1\,0\,0\,0\,0\,0\,0 \\ 0\,0\,0\,0\,0\,1\,1 \\ 0\,0\,0\,0\,0\,1\,0 \\ 0\,0\,0\,0\,0\,1\,1 \\ 1\,1\,1\,1\,1\,1\,1 \end{bmatrix}$$

where in both matrix, $T_{ij}^k = 1$ $(k = 1, 2)$ whenever if $x \in I_i(c)$ then $N_{p_c}(x)$ can belong to $I_j(c)$ where the interval $I_j(c)$ were introduced in the definition of $i_c(x)$ for $j = 0, .., 4$, and $T_{ij}^k = 0$ otherwise.

We introduce the following admissibility rules:

$$\begin{cases} \sigma^i(Y^c)_0 = A \Rightarrow \sigma^{i+1}(Y^c) = (A)^\infty \\ \sigma^i(Y^c)_0 = B \Rightarrow \sigma^{i+1}(Y^c) = (A)^\infty \\ \sigma^i(Y^c)_0 = M \Rightarrow \sigma^{i+1}(Y^c) \succ Y^c \\ \sigma^i(Y^c)_0 = L \Rightarrow \sigma^{i+1}(Y^c) \succ Y^c \\ CH(M)^\infty \quad \text{and} \quad CH(R)^\infty \quad \text{are not admissible} \end{cases} . \tag{1}$$

If

$$\Omega_c^+ = \{Y \in \Omega_c : T_{Y_i, Y_{i+1}} = 1 \;\text{ and holding }\; (1)\}$$

then

$$\Omega^+ = \bigcup_{c \in J} \Omega_c^+$$

contains all admissible sequences for all $c \in J$ and will be called the set of admissible sequences. For example, the sequence $(CMMRR)^\infty$ is admissible.

It is clear that Ω_c^+ is totally ordered by $\succeq$.

Example 3.2. To see the admissibility we must pay attention to the fact that the critical point d_2 is a local minimum and, in such case, if we have $\sigma^i(Y)_1 = L$ or $\sigma^i(Y)_1 = M$ (where Y is a periodic sequence with period n of the critical point d_2, $1 \le i < n$) then we must have $\sigma^i(Y) > Y$. So the sequence $(RLRC)^\infty$ is admissible and appears near $c = 1.3346...$, while the sequences $(LMAC)^\infty$ and $(RMRC)^\infty$ are not admissible for the same reason.

Again, in a similar way to what is made in [9] we define the kneading increments associated to kneading data by

$$\nu_{d_i} = \theta(d_i^+) - \theta(d_i^-) \text{ with } i = 0, 1, 2, 3$$

where $\theta(x)$ is the invariant coordinate of each symbolic sequence associated to the itinerary of each point d_i, see [9].

Using this we define the kneading matrix $N(t)$ and the kneading determinant

$$\begin{aligned} D(t) &= \frac{(-1)^{i+1} D_i(t)}{(1 - \varepsilon_i t)} \\ &= \frac{d_Y(t)}{(1-t)(1-t^k)} \end{aligned}$$

where $D_i(t)$ is the determinant of $N(t)$ without the column i and the cyclotomic polynomials in the denominator correspond to the stable periodic orbits of d_0 and d_2, see [9].

Example 3.3. If we compute the kneading increment for the sequence $RLRC$, we obtain the kneading matrix

$$N(t) = \begin{bmatrix} -\frac{1+t}{1-t} & 1 & 0 & 0 & 0 \\ -\frac{t}{1-t} & -1 & 1 & 0 & -\frac{t}{1-t} \\ 0 & 0 & -1 + \frac{2t^2 - 2t^4}{1-t^4} & 1 & \frac{2t - 2t^3}{1-t^4} \\ \frac{t}{1-t} & 0 & 0 & -1 & 1 - \frac{t}{t-1} \end{bmatrix}.$$

With $i = 2$ we have $\varepsilon_2 = -1$ (because $N'_{f_c}(x)|_{[d_0, d_1]} < 0$), from which we get

$$D(t) = \frac{(-1)\ D_2(t)}{1+t} = \frac{(1+t)(1-t-t^2-t^3)}{(1-t)(1-t^4)}.$$

Next we denote by $d_Y(t)$ the numerator of $D(t)$ given by $D(t)\ (1-t)\ (1-t^k)$, where k is the period of the critical point d_2 and Y is the kneading sequence associated to d_2. Each kneading data determines a kneading determinant but the most significant factor of the numerator is determined by the kneading sequence Y.

It is easy to see the following result

Proposition 3.2. *To the set Ω^+ of the ordered kneading sequences can be associated the tree $\mathcal{T}_Y$, where in each k-level of the tree are localized kneading sequence of k length.*

Corollary 3.1. *To $\mathcal{T}_Y$ we associate a tree $\mathcal{T}_{d_Y(t)}$ of the numerators of kneading determinant.*

To proof this corollary we need the following result

Lemma 3.1. *Let Y be an admissible periodic sequence corresponding to orbit of the critical point d_2 of period k whose the numerator of the kneading determinant $d_Y(t)$, now we designate by $d_k(t)$. Then $d_k(t)$ has degree $n = k$ and the polynomials correspondent to the periodic sequences of period $k+1$ ($k+1$ - level of the tree) follow the rule of construction:*

$$(1-t)d_k(t) = \underbrace{1 - t + a_2t^2 + a_3t^3 + \ldots + a_kt^{k-1}}_{p(t)} \underbrace{-\delta t^k - \delta t^{k+1}}_{q(t)} = p(t) + q(t)$$

$$\swarrow \qquad \downarrow \qquad \downarrow \qquad \searrow$$

$$A \qquad L \qquad M \qquad R$$

$$(1-t)d_{k+1}(t) = p(t) - 2\delta t^k;\quad p(t) + 2\delta t^k;\quad p(t) - 2\delta t^{k+1};\quad p(t) + tq(t)$$

with $a_k \in \{-2, 0, 2\}$ and $\delta = (-1)^{n_L}$ where n_L is equal to the number of times that the symbol L appears in the symbolic sequence Y.

Proof. Computing the kneading determinants of the sequences in each level k of the set Ω^+ of the ordered kneading sequences and analyzing the passage from level k to level $k+1$ and using the induction it goes out the rule of the indicated construction. □

If we want to see the symbolic dynamics in the tree for the converging points in the Newton method, it is of ramifications ending with A^∞. In this case $d_Y(t)$ stays constant after reaching the first symbol.

Theorem 3.1. *Let P and Q be kneading sequences in Ω^+ with the lexicographic order $\prec$. If $P \prec Q$ then $c_P > c_Q$, where c_P (respectively c_Q) is the parameter value corresponding to the kneading sequence P (respectively Q). If $c_1 > c_2$ then there are $P, Q \in \Omega^+$ with $P \preceq Q$, where P (respectively Q) is the kneading sequence realized by the parameter value c_1 (respectively c_2). Moreover, for the Newton map N_{p_c} associated to quintic map p_c, the topological entropy is a non-decreasing function of c on* $(0, c_0)$

Proof. We are given only some ideas of the proof which will appear elsewhere. For we extend the Tsujii results on the quadratic map to the Newton map [12]. Then we prove that the kneading sequence P associated to the orbit of the critical point d_2 is monotone decreasing with respect to parameter c on the range $(0, c_0)$. □

Acknowledgements

The author has been partially supported by Ministerio de Ciencia e Innovación, Grant CGL2008-05688-C02-02/CLI.

References

1. B. Barna, Über die Divergenzpunkte des Newtons Verfahrens zur Bestimmung von Wurzeln algebraischer Gleichungen, I, *Pub. Math. Debrecen* 3 (1951) 109-118; II, 4 (1956) 384-397; III, 8 (1961) 193-207; IV, 14 (1967) 91-97.
2. P. Collet and J.-P. Eckman, *Iterated maps on the interval as dynamical systems*, Birkhäuser, 1980.
3. D. S. Dummit, Solving solvable quintics, *Math. Comp.*, 57 (1991), 387-401.
4. A. R. Holmgren, *A first course in discrete dynamical systems*, Universitext, Springer-Verlag (2000).
5. M. Hurley and C. Martin, Newton's Algorithm and Chaotic Dynamical Systems, *SIAM J. Math. Anal*, 15 (1984), 238-252.
6. J. P. Lampreia and J. Sousa Ramos, Trimodal maps, *Int. J. Bifurcation and Chaos*, 3 (1993), 1607-1617.
7. Curt McMullen, Families of rational maps and iterative root-finding algorithms, *Ann. Math.* (2) **125**(3) (1987), 467-493.
8. Curt McMullen, Braiding of the attractor and the failure of iterative algorithms, *Invent. Math.* **91**(2) (1988), 259-272.

9. J. Milnor and W. Thurston, *On iterated maps of the interval*, In *Dynamical systems: proceedings* 1986-87, pp. 465-563, Lecture Notes in Mathematics 1342, Springer-Verlag, 1988.
10. A. Rényi, On Newton method of approximation (Hungarian), *Mat. Lapok*, 1 (1950) 278-293.
11. L. Silva and J. Sousa Ramos, Renormalization of Newton maps. Iteration theory (ECIT '96) (Urbino), *Grazer Math. Ber.*, 339 (1999), 313–328, Karl-Franzens-Univ. Graz, Graz.
12. M. Tsujii, A simple proof for monotonocity of entropy in the quadratic family, *Ergod. Th. & Dynam. Sys.* 20, (2000), 925-933.
13. Sherman Wong, Newton's method and symbolic dynamics, *Proc. Amer. Math. Soc.* **91**, 2 (1984), 245-253.

Hyperbolic Dynamics in Nash Maps

R. A. Becker

Department of Economics, Indiana University,
Bloomington, IN 47405, USA
E-mail: becker@indiana.edu

S. K. Chakrabarti

Department of Economics,
Indiana University — Purdue University Indianapolis,
425 University Blvd., Indianapolis, IN 46202, USA
E-mail: imxl100@iupui.edu

W. Geller, B. Kitchens and M. Misiurewicz

Department of Mathematical Sciences,
Indiana University — Purdue University Indianapolis,
402 N. Blackford Street, Indianapolis, IN 46202, USA
E-mails: wgeller@math.iupui.edu,
bkitchens@math.iupui.edu,
mmisiure@math.iupui.edu

We investigate the dynamics of the Nash better response map for a family of games with two players and two strategies. This family contains the games of Coordination, Stag Hunt and Chicken. Each map is a piecewise rational map of the unit square to itself. We describe completely the dynamics for all maps from the family. All trajectories converge to fixed points or period 2 orbits. We create tools that should be applicable to other systems with similar behavior.

Keywords: Nash maps; Hyperbolic dynamics; Periodic points.

1. Introduction

1.1. *Economics introduction*

Nash[1–3] produced three proofs of the existence of an equilibrium point for n-person games. Each proof applied a fixed point theorem to a mapping from the Cartesian product of the players' mixed strategy sets to itself. Iteration of such a map leads to a discrete time dynamical system on the product space of the players' mixed strategies.

Nash's[3] proof focused on a *better response* map with the property that each fixed point is an equilibrium point, and vice versa. Becker and Chakrabarti[4] generalized Nash's better response map to extend the existence theorem to incorporate some forms of nonexpected utility theories as well as to games with a continuum of pure strategies.

In Ref. 5 we analyzed the dynamics of the Nash map for Matching Pennies and showed that the game's equilibrium point was unstable – the players' mixed strategies converge to an orbit of period eight from any initial starting point except the mixed strategy equilibrium point. The Nash dynamics that one observes is quite different from what one gets if the players use their best responses. When the players use the best response map then the play simply cycles over the pure strategies.

The fact that the Nash map has interesting dynamics in the game of Matching Pennies raises questions about the dynamics of such behavior in other 2×2 games. Thus one would be curious about what would happen, say, in the case of the games of Coordination, Stag Hunt, or Chicken.

The purpose of this paper is to present a thorough analysis of the Nash better response dynamics for a one-parameter family of 2×2 games that includes versions of the above games. We focus on the *essential Nash map* defined on the unit square. This map is derived from Nash's better response map by recognizing that it is sufficient to study the evolution of each players' probability of playing one of the available two pure strategies.

The payoff matrices that describe 2×2 games form an eight parameter family of matrices. Some of these matrices describe equivalent games under various notions of equivalence. The payoff matrices we investigate here form a five parameter family and are described as follows.

strategy	Left	Right
Top	$(c+\alpha, b'+\alpha)$	(b, b')
Bottom	(c, c')	$(b+\alpha, c'+\alpha)$

where $\alpha \geq 0$. All these games have the common feature that there are two pure strategy Nash equilibrium points, namely, (Top, Left) and (Bottom, Right) and a mixed strategy equilibrium. Some Coordination games belong to this class of games. Let $b = 0 = b'$ and $c = 0 = c'$. Then the payoff matrices become

strategy	Left	Right
Top	(α, α)	$(0, 0)$
Bottom	$(0, 0)$	(α, α)

where $\alpha \geq 0$. Every game described by the five parameter family of matrices above has the same essential Nash map as a game described by a payoff matrix in this one parameter family (see Sec. 2, Eqs. (1) and (2)).

The following version of the Stag-Hunt game also fits our model (see Refs. 6 and 7).

strategy	Hunt Stag	Chase Rabbit
Hunt Stag	$(4, 4)$	$(1, 3)$
Chase Rabbit	$(3, 1)$	$(2, 2)$

The class of games also includes realizations of Chicken, as given below:

strategy	Not Blink	Blink
Not Blink	$(-10, -10)$	$(5, -5)$
Blink	$(-5, 5)$	$(0, 0)$

If we interchange the strategy of one of the players then this game also fits our model.

In our analysis of the Nash dynamics of this class of games we find that the nature of the dynamics related to the mixed strategy equilibrium point is sensitive to the payoffs of the game. For example, the mixed strategy equilibrium point can change from being an orientation preserving saddle point to an orientation reversing repelling point. This latter observation is especially interesting since, as far as we know, the existing literature on evolutionary games has not explicitly noted this phenomenon. This behavior cannot occur in the case of evolutionary dynamics in continuous time. The observations about the local instability of the pure strategy equilibrium is, however, quite consistent with results obtained in the literature on evolutionary dynamics in continuous time, as for example in Ref. 8.

1.2. *Mathematics introduction*

From the mathematical point of view, the problem we solve in the paper is to describe completely maps from a certain concrete one-parameter family $\mathcal{H}$ of maps of the unit square to itself. Problems of this type are often the most difficult ones in the whole theory of dynamical systems. While a lot of abstract theory exists, only a small part of this theory can be used in a concrete situation. Quite often the assumptions of general theorems are impossible to check. Sometimes the reasons why we cannot decide what is going on are deep. This applies especially to systems with complicated dynamics. How can one practically distinguish between a dense orbit and a periodic orbit of period $10^{100000000}$? However, for systems with relatively simple behavior a full description should be possible.

The systems considered in this paper fall into this category of "systems with relatively simple behavior," where after making some computations and drawing several computer pictures one can guess what is going on, but there is no obvious way to confirm it rigorously.

To resolve this problem, we create tools that allow us to prove exactly what we need. They are described in Sec. 4. They basically consist (after the standard preliminary use of symmetry to simplify the map) of several steps. The first one is to show that in the interesting region our map is an orientation preserving homeomorphism. The second step is to identify regions where the points are moved by the map in specific directions (cones of directions). This is basically treating our system as a difference equation. The third step is to see where the points from those regions can be mapped. Here we use the fact that under a homeomorphism the regions cannot "jump" over each other; the cyclic order around a point where several regions meet must stay the same.

Although this looks very similar to the classical nullcline method, here we deal with a system with discrete instead of continuous time (in other words, with difference equations, rather than differential ones). This makes a substantial difference in considering what can happen to the trajectories of the system. We do not formalize our methods, although we believe that they can be used in many similar situations. However, each specific situation may require minor modifications, so it may be difficult to pinpoint exact assumptions that have to be made.

The computationally most difficult situations in the family $\mathcal{H}$ arise for parameter values for which bifurcations occur and the system is not hyperbolic. We do not consider them very important, and even omit them in the statement of Theorem 2.1, although we include them in Sec. 4 for completeness. Therefore the reader should not feel disturbed by formulas like those from the proof of Lemma 4.9 (starting with Eq. (19) this proof ceases to be computational).

2. Description of the family $\mathcal{H}$ and results

Let

$$\begin{bmatrix} (a,a') & (b,b') \\ (c,c') & (d,d') \end{bmatrix}$$

define a two person, two pure strategy game with X the row player and Y the column player. The two players have strategies $\bar{x} = (x, 1-x)$ and $\bar{y} = (y, 1-y)$, respectively. The payoff matrices for X and Y are

$$R_x = \begin{bmatrix} a & b \\ c & d \end{bmatrix} \quad \text{and} \quad R_y = \begin{bmatrix} a' & b' \\ c' & d' \end{bmatrix}.$$

The expected payoff for X is $\bar{x}R_x\bar{y}^T$ and the expected payoff for Y is $\bar{x}R_y\bar{y}^T$.

Define

$$\begin{aligned} t_x &= \quad x \quad + \max\{0, (e_1 - \bar{x})R_x\bar{y}^T\}, \\ t_{1-x} &= (1-x) + \max\{0, (e_2 - \bar{x})R_x\bar{y}^T\}, \\ t_y &= \quad y \quad + \max\{0, \bar{x}R_y(e_1 - \bar{y})^T\}, \\ t_{1-y} &= (1-y) + \max\{0, \bar{x}R_y(e_2 - \bar{y})^T\}, \end{aligned}$$

with e_1, e_2 the standard basis vectors.

The *Nash map* on the pair of probability vectors is

$$(\bar{x}, \bar{y}) \mapsto \left(\left(\frac{t_x}{t_x + t_{1-x}}, \frac{t_{1-x}}{t_x + t_{1-x}} \right), \left(\frac{t_y}{t_y + t_{1-y}}, \frac{t_{1-y}}{t_y + t_{1-y}} \right) \right).$$

All information is contained in the *essential Nash map* on the unit square

$$n = (n_1, n_2) : [0,1]^2 \to [0,1]^2,$$

defined by

$$n_1(x,y) = \frac{t_x}{t_x + t_{1-x}} = \frac{x + \max\{0, (e_1 - \bar{x})R_x\bar{y}^T\}}{1 + \max\{0, (e_1 - \bar{x})R_x\bar{y}^T\} + \max\{0, (e_2 - \bar{x})R_x\bar{y}^T\}},$$

$$n_2(x,y) = \frac{t_y}{t_y + t_{1-y}} = \frac{y + \max\{0, \bar{x}R_y(e_1 - \bar{y})^T\}}{1 + \max\{0, \bar{x}R_x(e_1 - \bar{y})^T\} + \max\{0, \bar{x}R_y(e_2 - \bar{y})^T\}},$$

with $\bar{x}$ and $\bar{y}$ as before. It is clear from the definition that in this setting the essential Nash map is a continuous map of the unit square into itself.

If we let $[r]^+ = \max\{0, r\}$, $[r]^- = \max\{0, -r\}$ and $\alpha = a - c$, $\beta = b - d$, $\gamma = a' - b'$, $\delta = c' - d'$, the essential Nash map reduces to

$$n_1(x,y) = \frac{x + (1-x)[\alpha y + \beta(1-y)]^+}{1 + (1-x)[\alpha y + \beta(1-y)]^+ + x[\alpha y + \beta(1-y)]^-}, \tag{1}$$

$$n_2(x,y) = \frac{y + (1-y)[\gamma x + \delta(1-x)]^+}{1 + (1-y)[\gamma x + \delta(1-x)]^+ + y[\gamma x + \delta(1-x)]^-}. \tag{2}$$

Consequently, essential Nash maps arising from the 2 by 2 games form a four parameter family of piecewise rational, continuous maps of the unit square to itself. In what follows we examine the dynamics of a one parameter family $\mathcal{H}$ of essential Nash maps. It is given by $\alpha = \gamma > 0$ and $\beta = \delta = -\alpha$.

The game of Coordination is defined by

$$\begin{bmatrix} (1,1) & (-1,-1) \\ (-1,-1) & (1,1) \end{bmatrix}.$$

The essential Nash map for Coordination occurs in $\mathcal{H}$ with $\alpha = 2$.

The game of Chicken is defined by

$$\begin{bmatrix} (-10,-10) & (5,-5) \\ (-5,5) & (0,0) \end{bmatrix}.$$

Interchanging the ordering for X's choices one observes that the essential Nash map for Chicken occurs in $\mathcal{H}$ with $\alpha = 5$.

An essential Nash map $n = (n_1, n_2) \in \mathcal{H}$ reduces to

$$n_1(x,y) = \frac{x + \alpha(1-x)[2y-1]^+}{1 + \alpha(1-x)[2y-1]^+ + \alpha x[2y-1]^-},$$
$$n_2(x,y) = \frac{y + \alpha(1-y)[2x-1]^+}{1 + \alpha(1-y)[2x-1]^+ + \alpha y[2x-1]^-}.$$

The *diagonal* is $\{(x,x) : 0 \leq x \leq 1\}$. The region above the diagonal is $\{(x,y) : 0 \leq x < y \leq 1\}$ and the region below the diagonal is $\{(x,y) : 0 \leq y < x \leq 1\}$. Reflection about the diagonal is given by the map $r_d(x,y) = (y,x)$. The *anti-diagonal* is $\{(x, 1-x) : 0 \leq x \leq 1\}$. The region above the anti-diagonal is $\{(x,y) : x + y > 1\}$ and the region below the anti-diagonal is $\{(x,y) : x + y < 1\}$. Reflection about the anti-diagonal is given by $r_a(x,y) = (1-y, 1-x)$. A set A is called *invariant* for a map f if $f(A) \subseteq A$.

Let $f : U \to V$, where $U, V \subseteq \mathbb{R}^2$, be a local diffeomorphism. The map f is *orientation preserving* at $p \in U$ if the Jacobian at p is positive and *orientation reversing* at p if the Jacobian is negative at p.

The main results about the one parameter family $\mathcal{H}$ of essential Nash maps are summarized in the following theorem. The reader can get slightly more information from Secs. 3 and 4.

We will say that *a point x is attracted to A* (where A can be a point or a compact set) if the trajectory of x converges to A, that is, the distance from the k-th image of x to A converges to 0 as $k \to \infty$. When we say that a fixed point is orientation preserving or reversing, we mean the behavior of the map in a neighborhood of that point.

Theorem 2.1. *Let $n \in \mathcal{H}$ be the essential Nash map described above.*

(a) For all α, $(0,0)$ and $(1,1)$ are orientation preserving, topologically attracting fixed points and $(1/2,1/2)$ is a fixed point. There are no other fixed points.
(b) When $0 < \alpha \leq 1/2$, then n is a homeomorphism onto its image. When $1/2 < \alpha$, then n is not one-to-one.
(c) For $0 < \alpha < 2$, $(1/2,1/2)$ is an orientation preserving saddle fixed point whose stable manifold is the anti-diagonal and unstable manifold is the diagonal (without $(0,0)$ and $(1,1)$).
(d) For $2 < \alpha < 4$, $(1/2,1/2)$ is an orientation reversing saddle fixed point whose stable manifold is the anti-diagonal and unstable manifold is the diagonal (without $(0,0)$ and $(1,1)$).
(e) For $\alpha \leq 4$, all points below the anti-diagonal are attracted to $(0,0)$ and all points above the anti-diagonal are attracted to $(1,1)$.
(f) For $4 < \alpha < 2(1+\sqrt{2})$, $(1/2,1/2)$ is an orientation reversing repelling fixed point. There is a saddle period two orbit on the anti-diagonal. One point of this orbit lies below $(1/2,1/2)$ and its stable manifold is the part of the anti-diagonal with $x < 1/2$. The other point lies above $(1/2,1/2)$ and its stable manifold is the part of the anti-diagonal with $x > 1/2$. All points below the anti-diagonal are attracted to $(0,0)$ and all points above the anti-diagonal are attracted to $(1,1)$.
(g) For $\alpha > 2(1+\sqrt{2})$, $(1/2,1/2)$ is an orientation reversing repelling fixed point. There is an attracting orbit of period two on the anti-diagonal. There are two saddle period two orbits that follow the orbit of period two on the anti-diagonal. One saddle orbit lies below the anti-diagonal and one above. There are no other periodic points. Every other point is attracted to one of the periodic orbits mentioned (including $(0,0)$ and $(1,1)$). For illustration, see Fig. 1.

The set of nonwandering points of each map in $\mathcal{H}$ consists of a finite number of periodic points. The points $(0,0)$ and $(1,1)$ are always topolgically attracting, but the derivative of the essential Nash map at these points is the identity. Except for the special values of α: 2, 4 and $2(1+\sqrt{2})$, and except for the points $(0,0)$ and $(1,1)$, all nonwandering points are hyperbolic periodic points. For this reason we refer to the dynamics as *hyperbolic*.

In Fig. 1 we have marked all periodic points of n for $\alpha = 6$. Additionally we have marked the anti-diagonal and the stable and unstable manifolds of the period 2 saddles. Note that the stable manifolds serve as a boundary between the basins of attraction of the points $(0,0)$ and $(1,1)$ and the basin

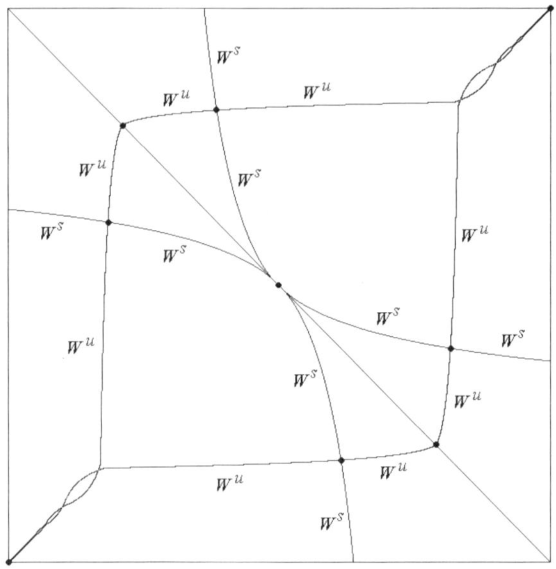

Fig. 1. Periodic points for $\alpha = 6$.

of attraction of the period 2 attracting orbit. The intersections between unstable manifolds are possible because the map n is not a homeomorphism.

The four functions that occur in the essential Nash map simplify. We denote them as follows.

$$n_1^+(x,y) = \frac{x + \alpha(1-x)(2y-1)}{1+\alpha(1-x)(2y-1)} \qquad \text{if } y \geq 1/2,$$

$$n_1^-(x,y) = \frac{x}{1+\alpha x(1-2y)} \qquad \text{if } y \leq 1/2,$$

$$n_2^+(x,y) = \frac{y + \alpha(1-y)(2x-1)}{1+\alpha(1-y)(2x-1)} \qquad \text{if } x \geq 1/2,$$

$$n_2^-(x,y) = \frac{y}{1+\alpha y(1-2x)} \qquad \text{if } x \leq 1/2.$$

The two lines $x = 1/2$ and $y = 1/2$ divide the square into four quadrants. We refer to the quadrants by compass points. The *northeast quadrant* is $NE = \{(x,y) : 1/2 \leq x \leq 1, 1/2 \leq y \leq 1\}$, and similar descriptions hold for the SE, SW and NW quadrants. In the NE quadrant the essential Nash map is $n = (n_1^+, n_2^+)$, in the SE quadrant it is $n = (n_1^-, n_2^+)$, in the SW quadrant $n = (n_1^-, n_2^-)$ and in the in the NW quadrant $n = (n_1^+, n_2^-)$. We will refer to the lines separating the quadrants as borders. Also there are the *North*, *South*, *East* and *West* regions (see Fig. 2).

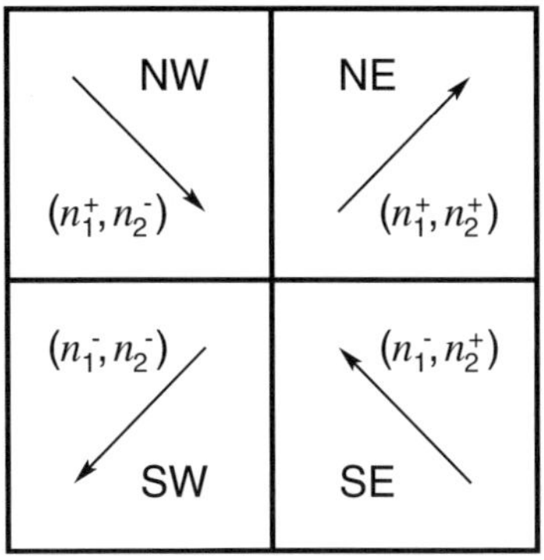

Fig. 2. Quadrants, functions and directions.

3. Essential Nash map

In this section we will investigate the essential Nash map n in the whole square $[0, 1]^2$.

Lemma 3.1. *The essential Nash map n commutes with r_d and with r_a. Therefore both the diagonal and the anti-diagonal are invariant for n.*

Proof. The first statement follows from a simple computation. The second part follows from the first one and the fact that the diagonal is the set of the fixed points of r_d and the anti-diagonal is the set of the fixed points of r_a. □

Lemma 3.2. *Consider the essential Nash map acting on the unit square.*

(a) *The number $n_1(x, y) - x$ is positive if $x \neq 1$ and $y > 1/2$, negative if $x \neq 0$ and $y < 1/2$, and zero otherwise (including $y = 1/2$). Similarly, the number $n_2(x, y) - y$ is positive if $y \neq 1$ and $x > 1/2$, negative if $y \neq 0$ and $x < 1/2$, and zero otherwise (including $x = 1/2$). Loosely speaking, this means that under the action of the essential Nash map, all points in the open NE quadrant move to the northeast, all points in the open SE quadrant move to the northwest, all points in the open SW quadrant move to the southwest, and all points in the open NW quadrant move to the southeast (see Fig. 2).*

(b) The essential Nash map has three fixed points. They are $(0,0)$, $(1,1)$ *and* $(1/2,1/2)$.

(c) The northeast and southwest quadrants are invariant for n. *Every point in the southeast quadrant, except* $(1/2,1/2)$, *is attracted to the point* $(0,0)$ *and every point in the northeast quadrant, except* $(1/2,1/2)$, *is attracted to the point* $(1,1)$.

Proof. Statement (a) follows from simple computations. Statements (b) and (c) follow from (a). □

Lemma 3.3. *The anti-diagonal and the regions above and below the anti-diagonal are invariant for* n. *Moreover, there exists a neighborhood* U *of* $(1/2,1/2)$ *such that for every point* $p \in U$ *which does not lie on the anti-diagonal the distance of* $n(p)$ *from the anti-diagonal is larger than the distance of* p *from the anti-diagonal.*

Proof. We have

$$n_1^+(x,y)+n_2^-(x,y)-1=(x+y-1)\frac{1+2\alpha y(1-x)}{[1+\alpha(1-x)(2y-1)][1+\alpha y(1-2x)]}. \tag{3}$$

Since the fraction on the right is always positive, we see that if a point p is above the anti-diagonal and in the NW quadrant then $n(p)$ is above the anti-diagonal. Since n commutes with r_d, we get the same result when we replace NW by SE. The NE quadrant is invariant by Lemma 3.2 (c). Thus, the region above the anti-diagonal is invariant for n. Since n commutes with r_a, the region below the anti-diagonal is also invariant for n. By continuity, the anti-diagonal is also invariant.

Using (3) and assuming that $x+y-1>0$, one can compute easily that

$$n_1^+(x,y)+n_2^-(x,y)-1>x+y-1$$

is equivalent to

$$1-x-y+2xy>\alpha(1-x)(1-2x)(2y-1)y.$$

At $x=y=1/2$ the left-hand side of this inequality is equal to $1/2$, while the right-hand side is 0. Therefore the inequality holds in some neighborhood of $(1/2,1/2)$. Thus, if p belongs to this neighborhood and lies above the anti-diagonal in the NW quadrant, the distance of $n(p)$ from the anti-diagonal is larger than the distance of p from the anti-diagonal. Using reflections about the anti-diagonal and diagonal and Lemma 3.2 (a), we get a whole neighborhood U of $(1/2,1/2)$ such that for every point $p \in U$ which does not lie on the anti-diagonal the distance of $n(p)$ from the anti-diagonal is larger than the distance of p from the anti-diagonal. $\square$

We will need all eight partial derivatives of the four functions that define the essential Nash map for the discussion that follows. They are

$$\begin{aligned}
\frac{\partial n_1^+}{\partial x} &= \frac{1}{[1+\alpha(1-x)(2y-1)]^2}, \\
\frac{\partial n_1^+}{\partial y} &= \frac{2\alpha(1-x)^2}{[1+\alpha(1-x)(2y-1)]^2}, \\
\frac{\partial n_1^-}{\partial x} &= \frac{1}{[(1+\alpha x(1-2y))]^2}, \\
\frac{\partial n_1^-}{\partial y} &= \frac{2\alpha x^2}{[(1+\alpha x(1-2y))]^2}, \\
\frac{\partial n_2^+}{\partial x} &= \frac{2\alpha(1-y)^2}{[1+\alpha(1-y)(2x-1)]^2}, \\
\frac{\partial n_2^+}{\partial y} &= \frac{1}{[1+\alpha(1-y)(2x-1)]^2}, \\
\frac{\partial n_2^-}{\partial x} &= \frac{2\alpha y^2}{[1+\alpha y(1-2x)]^2}, \\
\frac{\partial n_2^-}{\partial y} &= \frac{1}{[1+\alpha y(1-2x)]^2}.
\end{aligned} \tag{4}$$

We know from Lemma 3.2 (c) that the fixed points $(0,0)$ and $(1,1)$ are topological attractors (although from the differentiable point of view they are neutral; it is easy to check that the derivative at them is the identity). Let us investigate the nature of the third fixed point, $(1/2,1/2)$.

Lemma 3.4.

(a) All four derivatives of the essential Nash map at $(1/2,1/2)$ *are*

$$\begin{bmatrix} 1 & \frac{\alpha}{2} \\ \frac{\alpha}{2} & 1 \end{bmatrix}.$$

(b) The above matrix has two eigenvalues. The larger in modulus is $1+\frac{\alpha}{2}$ *with eigenvector* $(1,1)^T$ *and the smaller in modulus is* $1-\frac{\alpha}{2}$ *with eigenvector* $(1,-1)^T$.

(c) When $0<\alpha<2$, $(1/2,1/2)$ *is an orientation preserving saddle fixed point. When* $2<\alpha<4$, $(1/2,1/2)$ *is an orientation reversing saddle fixed point. When* $4<\alpha$, $(1/2,1/2)$ *is an orientation reversing repelling fixed point.*

Proof. Statements (a) and (b) are computations. Statement (c) follows immediately from the first two statements. □

Let us turn to the investigation of the behavior of n on the diagonal and anti-diagonal.

Lemma 3.5.

(a) The diagonal is mapped by n homeomorphically onto itself.
(b) The anti-diagonal is mapped by n homeomorphically into itself if and only if $\alpha \leq 1/2$ (then orientation is preserved) or if $\alpha \geq 2$ (then orientation is reversed).
(c) The upper and lower halves of the diagonal are invariant for n.
(d) The upper and lower halves of the anti-diagonal are invariant for n if and only if $\alpha \leq 1$.

Proof. The essential Nash map restricted to the diagonal in the NE quadrant is given by

$$x \mapsto \frac{x + \alpha(1-x)(2x-1)}{1 + \alpha(1-x)(2x-1)}.$$

Its derivative is

$$\frac{1 + 2\alpha(1-x)^2}{[1 + \alpha(1-x)(2x-1)]^2} > 0$$

and both endpoints of this segment are fixed points of n. In the SW quadrant the situation is similar because of the symmetry with respect to the anti-diagonal. This proves statements (a) and (c).

The essential Nash map restricted to the anti-diagonal in the NW quadrant is given by

$$y \mapsto \frac{y}{1 + \alpha y(2y-1)}.$$

Its derivative is

$$\frac{1 - 2\alpha y^2}{[1 + \alpha y(2y-1)]^2}.$$

If $\alpha \leq 1/2$ then the derivative above is positive for all $y \in [1/2, 1)$, if $\alpha \geq 2$ then it is negative for all $y \in (1/2, 1]$, and if $1/2 < \alpha < 2$ then it changes sign in $(1/2, 1)$. In the SE quadrant the situation is similar because of the symmetry with respect to the diagonal. This proves statement (b).

To prove statement (d), note that for $y > 1/2$

$$\frac{y}{1+\alpha y(2y-1)} \geq \frac{1}{2}$$

is equivalent to $\alpha y \leq 1$, which holds for all $y \in [1/2, 1]$ if and only if $\alpha \leq 1$ (and the situation is similar in the SE quadrant). □

Let us return to the global view on the essential Nash map. We will use the following lemma (see Ref. 9, cf. Ref. 5).

Lemma 3.6. *Suppose that C is a Jordan curve in the plane and D is the closure of the region bounded by C. Let f be a map from D into $\mathbb{R}^2$, which is a local homeomorphism on the interior D° of D and a homeomorphism from C onto its image. Then:*

(a) $f(C)$ is a Jordan curve and $f(D^\circ)$ is the region bounded by $f(C)$;
(b) f is a homeomorphism of D onto its image.

Lemma 3.7. *If $\alpha \leq 1/2$ then n is a homeomorphism onto its image.*

Proof. Assume that $\alpha \leq 1/2$. By Lemma 3.1 it follows that it is enough to prove that n restricted to the triangle T with the vertices $(0,1)$, $(1,1)$ and $(1/2, 1/2)$ is a homeomorphism onto its image and that $n(T) \subseteq T$. From the formulas for the partial derivatives of n we see that the Jacobian of n in the NW quadrant is a fraction with a positive denominator and the numerator equal to $1-4\alpha^2(1-x)^2y^2$. Thus, this Jacobian is positive except at $(0,1)$, where it is zero. Similarly, the Jacobian of n in the NE quadrant is a fraction with a positive denominator and the numerator equal to $1 - 4\alpha^2(1-x)^2(1-y)^2$, so it is positive everywhere. Therefore by Lemma 3.6 it remains to check that the image of the boundary of T is contained in T and that n restricted to this boundary is a homeomorphism.

Let us figure out how n acts on the three segments that comprise the boundary of T. By Lemma 3.5, n restricted to the segment joining $(0,1)$ with $(1/2, 1/2)$ is a homeomorphism onto its subsegment, and n maps the segment joining $(1/2, 1/2)$ with $(1,1)$ homeomorphically onto itself. In the NW quadrant we have

$$n(x,1) = \left(\frac{x+\alpha(1-x)}{1+\alpha(1-x)}, \frac{1}{1+\alpha(1-2x)}\right).$$

The second coordinate of the image above is a 1-to-1 function of x, so n restricted to the segment I joining $(0,1)$ with $(1/2,1)$ is a homeomorphism.

By Lemma 3.3 the image lies above the anti-diagonal (except the point $n(0,1)$).

To check that it lies above the diagonal, note that it is equivalent to

$$\frac{x+\alpha(1-x)}{1+\alpha(1-x)} < \frac{1}{1+\alpha(1-2x)},$$

which is in turn equivalent to

$$\alpha^2(1-2x)(1-x)+\alpha x(1-2x)+(x-1)<0.$$

If $x=1/2$, this is true. Fix $x\in[0,1/2)$ and look at the expression above as a quadratic polynomial of α. The coefficients by α^2 and α are positive, so the maximal value (for $\alpha\in(0,1/2]$) is attained at $\alpha=1/2$. Then the value of this polynomial is $(-2x^2+3x-3)/4<0$, so $n(I)$ lies above the diagonal.

On the segment J joining $(1/2,1)$ with $(1,1)$ we have $n_2^+(x,1)=1$, so the image is contained in $I\cup J$. In particular, one eigenvector of the derivative of n is horizontal. The Jacobian there is positive, so the corresponding eigenvalue is non-zero. Therefore n restricted to J is a homeomorphism onto its image. The set $n(J)$ intersects the diagonal only at $(1,1)$. Moreover, the image of I intersects the line $y=1$ only at $n(1/2,1)$. This completes the proof. □

Lemma 3.2 (c) tells us what happens to the trajectories of points that fall into the NE or SW quadrant. Let us consider other possibilities.

Lemma 3.8. *If the whole trajectory of a point p stays in the NW quadrant then p belongs to the anti-diagonal and is attracted to $(1/2,1/2)$. Similarly, if the whole trajectory of a point p stays in the SE quadrant then p belongs to the anti-diagonal and is attracted to $(1/2,1/2)$.*

Proof. Assume that the whole trajectory of a point p stays in the NW quadrant. By Lemma 3.2 (a), along the trajectory both coordinates change in a monotone way. Therefore the trajectory converges to a fixed point. The only fixed point in the NW quadrant is $(1/2,1/2)$, so the trajectory converges to this point. By Lemma 3.3, when the points on the trajectory are sufficiently close to $(1/2,1/2)$, they have to lie on the anti-diagonal. Since by Lemma 3.3 both the regions below and above the anti-diagonal are invariant, already p had to lie on the anti-diagonal.

The similar statement for the SW quadrant follows from the symmetry of the map with respect to the diagonal. □

4. Making use of the symmetry

By the results of the preceding section, the only trajectories with unknown behavior are those that go through both NW and SE quadrants. In order to investigate them, we consider $r_d \circ n$ in the NW and SE quadrants. This map commutes with r_d because n does. Therefore, again looking at the NW and SE quadrants will be equivalent. In order to simplify further computations, we conjugate this map via the affine map $s(x, y) = (1-2x, 2y-1)$. It maps the NW quadrant to the whole square $[0,1]^2$, the fixed point $(1/2, 1/2)$ is mapped to $(0,0)$, and the anti-diagonal is mapped to the diagonal. The formula for our new map $f = s \circ r_d \circ n \circ s^{-1}$ in $[0,1]^2$, that we have to investigate, is

$$f(x,y) = \left(\frac{-2y + \alpha x(1+y)}{2 + \alpha x(1+y)}, \frac{-2x + \alpha y(1+x)}{2 + \alpha y(1+x)} \right).$$

We will use notation $f(x, y) = (x', y')$.

Lemma 4.1. *If $\alpha \leq 2$ then for every $(x, y) \in [0,1]^2$, except the fixed point $(0,0)$, we have $x' < x$ and $y' < y$.*

Proof. The inequality $x' < x$ can be written as

$$\frac{-2y + \alpha x(1+y)}{2 + \alpha x(1+y)} < x$$

and this is equivalent to

$$(1+y)(\alpha x^2 - \alpha x + 2) > 2(1-x). \tag{5}$$

Since $\alpha \leq 2$, we have for $x > 0$

$$\alpha x^2 > 0 \geq (\alpha - 2)x$$

and if $x = 0$, there is an equality. Therefore

$$\alpha x^2 - \alpha x + 2 > 2(1-x)$$

with equality for $x = 0$, and (5) follows for $(x, y) \neq (0,0)$.

The same computations with x, y switched give us $y' < y$. □

Let us return to the map n. The next lemma reduces the set of trajectories with unknown behavior to the ones that alternate (regularly) between the NW and SE quadrants.

Lemma 4.2. *Assume that $\alpha > 2$. If $(x, y) \neq (1/2, 1/2)$ belongs to the NW quadrant then $n(x, y)$ does not belong to the NW quadrant. Similarly, if*

$(x, y) \neq (1/2, 1/2)$ belongs to the SE quadrant then $n(x, y)$ does not belong to the SE quadrant.

Proof. Suppose that the point $(x, y) \neq (1/2, 1/2)$ and its image both belong to the NW quadrant. Then

$$0 \leq x \leq 1/2 \leq y \leq 1 \tag{6}$$

and

$$\frac{x + \alpha(1-x)(2y-1)}{1 + \alpha(1-x)(2y-1)} \leq 1/2 \leq \frac{y}{1 - \alpha y(2x-1)}. \tag{7}$$

The first inequality of (7) is equivalent to $\alpha(1-x)(2y-1) \leq 1 - 2x$, and the second one to $\alpha y(1-2x) \leq 2y - 1$. From this and (6) we get

$$\alpha(1-x) \leq \frac{1-2x}{2y-1} \leq \frac{1}{\alpha y}.$$

Thus, $(1-x)y \leq 1/\alpha^2$. Together with (6), $\alpha > 2$ and $(x, y) \neq (1/2, 1/2)$, this gives us a contradiction. □

Lemma 4.3. *If $2 < \alpha \leq 3 + 2\sqrt{2}$ then the region R of the square $[0,1]^2$ where $x' > x$ is bounded from below by the segment of the lower side of the square from $x = 0$ to $x = (\alpha - 2)/\alpha$ and from above by the graph of*

$$y = \frac{2(1-x)}{\alpha x^2 - \alpha x + 2} - 1. \tag{8}$$

If $\alpha > 3 + 2\sqrt{2}$ then R is bounded from below by the segment of the lower side of the square from $x = 0$ to $x = (\alpha - 2)/\alpha$, from the left and right by the pieces of the graph of (8), and from above by the segment of the upper side of the square from

$$x = \frac{\alpha - 1 - \sqrt{1 - 6\alpha + \alpha^2}}{2\alpha}$$

to

$$x = \frac{\alpha - 1 + \sqrt{1 - 6\alpha + \alpha^2}}{2\alpha}.$$

At the points of the boundary of R where (8) holds we have $x' = x$. In the rest of $[0,1]^2$ we have $x' < x$.

The same statements are true if we switch x with y and x' with y'.

Proof. As we already noticed in the proof of Lemma 4.1, the inequality $x' > x$ is equivalent to

$$(1+y)(\alpha x^2 - \alpha x + 2) < 2(1-x). \tag{9}$$

Moreover, $x' = x$ is equivalent to (8). If $\alpha x^2 - \alpha x + 2 > 0$, then (9) is equivalent to

$$y < \frac{2(1-x)}{\alpha x^2 - \alpha x + 2} - 1. \tag{10}$$

Let us look closer at (8). We have $y = 0$ when $x = 0$ or $x = (\alpha - 2)/\alpha$. If $\alpha < 8$ then $\alpha x^2 - \alpha x + 2 > 0$, so $y > 0$ is equivalent to $0 < x < (\alpha - 2)/\alpha$. If $\alpha \geq 8$ then the solutions to $\alpha x^2 - \alpha x + 2 = 0$ are

$$x = \frac{1}{2}\left(1 \pm \sqrt{1 - \frac{8}{\alpha}}\right).$$

We have

$$0 < \frac{1}{2}\left(1 - \sqrt{1 - \frac{8}{\alpha}}\right) < \frac{1}{2}\left(1 + \sqrt{1 - \frac{8}{\alpha}}\right) < \frac{\alpha - 2}{\alpha}.$$

The last inequality above follows from the inequality $(\alpha - 2)/\alpha > 1/2$ and the fact that the value of the polynomial $\alpha x^2 - \alpha x + 2$ at $x = (\alpha - 2)/\alpha$ is $4/\alpha > 0$.

Next we differentiate the function given by (8). We get

$$\frac{dy}{dx} = \frac{2(\alpha x^2 - 2\alpha x + \alpha - 2)}{(\alpha x^2 - \alpha x + 2)^2}.$$

Thus, if $0 \leq x < 1 - \sqrt{2/\alpha}$ then $dy/dx > 0$ and if $1 - \sqrt{2/\alpha} < x \leq 1$ then $dy/dx < 0$ (except at the points where the denominator is 0). Note that if $\alpha \geq 8$ then

$$\frac{1}{2}\left(1 - \sqrt{1 - \frac{8}{\alpha}}\right) \leq 1 - \sqrt{\frac{2}{\alpha}} \leq \frac{1}{2}\left(1 + \sqrt{1 - \frac{8}{\alpha}}\right).$$

Putting all this together we get the following description of the region R. It is bounded from below by the segment of the lower side of the square from $x = 0$ to $x = (\alpha - 2)/\alpha$. If the graph of (8) on this interval fits in the square, it bounds R from above. This happens when (8) with $y = 1$ has no solutions or one solution. A simple computation shows that this is when $\alpha \leq 3 + 2\sqrt{2}$. For larger values of α, up to 8, the graph of (8) goes out of the square through the upper side, and we get a description of R as in the statement of the lemma (simple computations give the values of x for which this graph intersects the upper side of the square). If $\alpha \geq 8$ then this description is still valid, since as the denominator of the right-hand side of (8) goes to 0, y goes to infinity; the part of the graph of (8) between the values of x where the denominator is infinity does not count because there $\alpha x^2 - \alpha x + 2 < 0$.

The last statement of the lemma follows from the symmetry of f with respect to the diagonal. □

Lemma 4.4. *If $\alpha \leq 4$ then for the map n all points below the anti-diagonal are attracted to $(0, 0)$ and all points above the anti-diagonal are attracted to $(1, 1)$.*

Proof. By Lemma 3.2 (c) and Lemma 3.1, it is sufficient to prove that the trajectories of all points of the NW quadrant above the anti-diagonal converge to $(1, 1)$.

If $\alpha \leq 2$ then by Lemmas 3.2 (a) and 4.1 as we move along such a trajectory then the distance from the diagonal decreases. Thus, either the trajectory enters the NE quadrant (and then it converges to $(1, 1)$) or it converges to $(1/2, 1/2)$. However, the latter is impossible by Lemma 3.3.

If $2 < \alpha \leq 4$ then by Lemma 4.2 in order to find out how the trajectory of a point p behaves it is enough to analyze the trajectory of $s(p)$ under f. To do this we will show that the region R, described in Lemma 4.3, lies below the diagonal. Then its reflection from the diagonal lies above the diagonal, and since $s(p)$ also lies above the diagonal, on the trajectory of $s(p)$ for f the first coordinate decreases. Thus, this trajectory either gets out of the square or it converges to $(0, 0)$. Therefore the trajectory of p for n either enters the NE quadrant (and then it converges to $(1, 1)$) or it converges to $(1/2, 1/2)$. Again, the latter is impossible by Lemma 3.3.

To show that R lies below the diagonal, we note that if this is not the case then the graph of (8) intersects the diagonal at some point other than $(0, 0)$. However, if the right-hand side of (8) is equal to x and $x \neq 0$ then $\alpha x^2 = \alpha - 4$, and this has no solution (with $x \neq 0$) if $\alpha \leq 4$. This completes the proof. □

Let us illustrate the dynamics of f. We mark the region R and the region $\widetilde{R}$ symmetric to it with respect to the diagonal, the regions which are mapped outside the square, the fixed points inside the square, and approximate directions in which the points move (that is, approximate directions of vectors $f(p) - p$). The formulas for the regions which are mapped outside the square are simple. Since the denominators in the formula for f are positive, $x' < 0$ is equivalent to

$$x < \frac{2y}{\alpha(1+y)}.$$

To get the inequality equivalent to $y' < 0$ we have to switch x and y. Figure 3 illustrates the case $2 < \alpha \leq 4$.

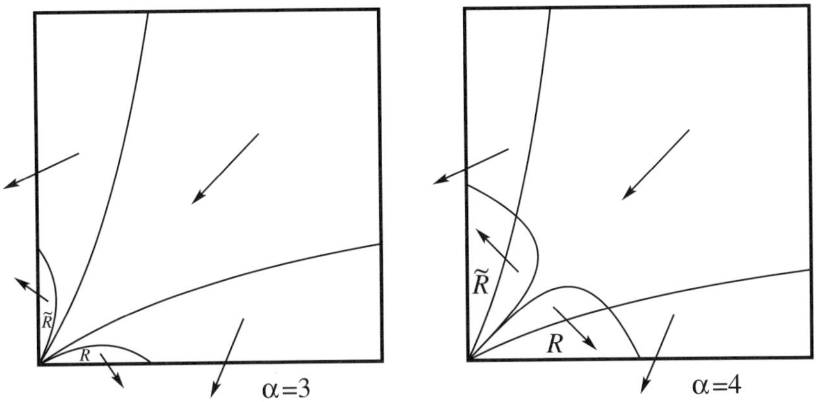

Fig. 3. Map f for $\alpha = 3$ and $\alpha = 4$.

Let us now investigate the dynamics of n for $\alpha > 4$. As we already noticed, Lemma 3.2 (c) tells us everything about the behavior of the trajectories that enter the NE or SW quadrants, and by Lemma 4.2 in the remaining two quadrants in order to understand the dynamics of n it is enough to understand the dynamics of f in $[0,1]^2$. In particular, except the point $(0,0)$, all fixed points of f correspond to period 2 orbits of n. Therefore, let us start with investigating the fixed points of f.

We will need the following simple lemma.

Lemma 4.5. *Let T be the trace and D the determinant of a 2 by 2 matrix A. Then A has two real eigenvalues, one larger than 1 and another smaller than 1, if and only if $T > D + 1$.*

Proof. The equation for the eigenvalues of A is $x^2 - Tx + D = 0$. Its discriminant is $\Delta = T^2 - 4D$, and the solutions are $(T \pm \sqrt{\Delta})/2$. They are both real, one larger than 1 and another smaller than 1, if and only if $\Delta > 0$

and

$$\frac{T-\sqrt{\Delta}}{2} < 1 < \frac{T+\sqrt{\Delta}}{2},$$

which is equivalent to $\sqrt{\Delta} > |T-2|$, that is, $\Delta > (T-2)^2$. Note that this implies $\Delta > 0$. Since $\Delta > (T-2)^2$ is equivalent to $T > D+1$, the proof is complete. □

Lemma 4.6.

(a) If $\alpha > 4$ then $(0,0)$ and

$$q = \left(\sqrt{1-\frac{4}{\alpha}}, \sqrt{1-\frac{4}{\alpha}}\right)$$

are fixed points of f. If $\alpha > 2(1+\sqrt{2})$ then additionally

$$q_1 = \left(\frac{\alpha+\sqrt{\alpha^2-4\alpha-4}}{2(\alpha+1)}, \frac{\alpha-\sqrt{\alpha^2-4\alpha-4}}{2(\alpha+1)}\right)$$

and

$$q_2 = \left(\frac{\alpha-\sqrt{\alpha^2-4\alpha-4}}{2(\alpha+1)}, \frac{\alpha+\sqrt{\alpha^2-4\alpha-4}}{2(\alpha+1)}\right)$$

are fixed points of f. These are all fixed points of f.

(b) If $4 < \alpha < 2(1+\sqrt{2})$ then q is a saddle; if $\alpha > 2(1+\sqrt{2})$ then q is attracting.

(c) Points q_1 and q_2 are saddles.

Proof. Let us find all fixed points of f. For this we have to solve the system of equations

$$x = \frac{-2y+\alpha x(1+y)}{2+\alpha x(1+y)},$$
$$y = \frac{-2x+\alpha y(1+x)}{2+\alpha y(1+x)}.$$

It can be rewritten as

$$2x+2y = \alpha x(1+y)(1-x), \tag{11}$$
$$2x+2y = \alpha y(1+x)(1-y). \tag{12}$$

Therefore we get

$$x(1+y)(1-x) = y(1+x)(1-y),$$

which is equivalent to

$$(y-x)(xy+x+y-1)=0.$$

This means that either $x=y$ or $xy=1-x-y$.

If $x=y$ then (11) becomes $4x=\alpha x(1-x^2)$ which has non-negative solutions $x=0$ and $x=\sqrt{1-\frac{4}{\alpha}}$. This proves that if $\alpha>4$ then $(0,0)$ and q are fixed points of f.

If $xy=1-x-y$ then (11) is equivalent to $x+y=\alpha xy$. Plugging this value of $x+y$ back to the equation $xy=1-x-y$ we get $xy=1/(\alpha+1)$ and $x+y=\alpha/(\alpha+1)$. This means that x and y are the two roots of the equation

$$x^2-\frac{\alpha}{\alpha+1}x+\frac{1}{\alpha+1}=0. \tag{13}$$

This equation has real roots if $\alpha\geq 2(1+\sqrt{2})$ (where for $\alpha=2(1+\sqrt{2})$ both roots are $\sqrt{1-4/\alpha}$), and this gives us additional fixed points q_1 and q_2 for $\alpha>2(1+\sqrt{2})$.

The method we used gave us all solutions of our system of equations, so statement (a) is proved.

To prove statements (b) and (c), we need the partial derivatives of f. They are:

$$\frac{\partial x'}{\partial x}=\frac{2\alpha(1+y)^2}{[2+\alpha x(1+y)]^2},$$
$$\frac{\partial x'}{\partial y}=\frac{-4}{[2+\alpha x(1+y)]^2},$$
$$\frac{\partial y'}{\partial x}=\frac{-4}{[2+\alpha y(1+x)]^2},$$
$$\frac{\partial y'}{\partial y}=\frac{2\alpha(1+x)^2}{[2+\alpha y(1+x)]^2}.$$

Thus, the eigenvalues of the derivative of f at q are

$$\frac{2\alpha(1+x)^2-4}{[2+\alpha x(1+x)]^2},$$

corresponding to the eigenvector $(1,1)^T$, and

$$\frac{2\alpha(1+x)^2+4}{[2+\alpha x(1+x)]^2},$$

corresponding to the eigenvector $(1,-1)^T$, where $x=\sqrt{1-4/\alpha}$. Note that both eigenvalues are positive. Simple computations (where we replace x^2

by $1 - 4/\alpha$ whenever it appears) show that the first eigenvalue is smaller than 1 if and only if

$$\alpha^2 - 6\alpha + 8 > (4\alpha - \alpha^2)x.$$

The left-hand side is equal to $(\alpha - 4)(\alpha - 2)$, so for $\alpha > 4$ it is positive, while the right-hand side is negative. This shows that the first eigenvalue is smaller than 1. Similarly, the second eigenvalue is smaller than 1 if and only if

$$\alpha^2 - 6\alpha + 4 > (4\alpha - \alpha^2)x.$$

If $\alpha \geq 3 + \sqrt{5}$ then the left-hand side is non-negative, while the right-hand side is negative, so the second eigenvalue is smaller than 1. If $4 < \alpha < 3 + \sqrt{5}$ then both sides are negative, so we take their squares and change the direction of the inequality. We get after a short computation $\alpha^2 - 4\alpha - 4 > 0$, which is true for $\alpha > 2(1 + \sqrt{2})$. Thus, the second eigenvalue is smaller than 1 if $\alpha > 2(1 + \sqrt{2})$ and greater than 1 if $4 < \alpha < 2(1 + \sqrt{2})$. This proves statement (b).

Consider now the fixed point $q_1 = (x, y)$ (or $q_2 = (x, y)$; the computations are the same). As we established in the proof of statement (a),

$$xy + x + y = 1 \tag{14}$$

and x, y are the two roots of the equation (13). Therefore,

$$xy = \frac{1}{\alpha + 1}, \qquad x + y = \frac{\alpha}{\alpha + 1} \tag{15}$$

and

$$x^2 = \frac{\alpha x - 1}{\alpha + 1}, \qquad y^2 = \frac{\alpha y - 1}{\alpha + 1}. \tag{16}$$

In view of (14), we have $x(1 + y) = 1 - y$ and $y(1 + x) = 1 - x$, so the derivative of f at q_1 is

$$\begin{bmatrix} \frac{2\alpha(1+y)^2}{[2+\alpha(1-y)]^2} & \frac{-4}{[2+\alpha(1-y)]^2} \\ \frac{-4}{[2+\alpha(1-x)]^2} & \frac{2\alpha(1+x)^2}{[2+\alpha(1-x)]^2} \end{bmatrix}.$$

Its determinant is positive, so by Lemma 4.5 in order to prove statement (c) it remains to check that its trace is larger than its determinant plus 1. This is equivalent to

$$\left[2\alpha(1 + y)^2 - (2 + \alpha(1 - y))^2\right]\left[2\alpha(1 + x)^2 - (2 + \alpha(1 - x))^2\right] < 16. \tag{17}$$

Using (16) we get

$$2\alpha(1+y)^2 - (2+\alpha(1-y))^2 = \frac{(\alpha^3+12\alpha^2+8\alpha)y - (\alpha^3+2\alpha^2+8\alpha+4)}{\alpha+1}$$

and

$$2\alpha(1+x)^2 - (2+\alpha(1-x))^2 = \frac{(\alpha^3+12\alpha^2+8\alpha)x - (\alpha^3+2\alpha^2+8\alpha+4)}{\alpha+1}.$$

Thus, (17) is equivalent to

$$(sy-t)(sx-t) < 16(\alpha+1)^2, \tag{18}$$

where

$$s = \alpha^3+12\alpha^2+8\alpha \quad \text{and} \quad t = \alpha^3+2\alpha^2+8\alpha+4.$$

By (15), we have

$$(sy-t)(sx-t) = s^2\frac{1}{\alpha+1} - st\frac{\alpha}{\alpha+1} + t^2,$$

so (18) is equivalent to

$$s^2 - \alpha st + (\alpha+1)t^2 < 16(\alpha+1)^3.$$

This in turn is equivalent to

$$8\alpha^6 - 8\alpha^5 - 104\alpha^4 - 184\alpha^3 - 128\alpha^2 - 32\alpha > 0.$$

The left-hand side of this inequality can be factorized as

$$8\alpha(\alpha+1)^3(\alpha^2-4\alpha-4),$$

so it is positive for $\alpha > 2(1+\sqrt{2})$. This completes the proof. □

Remark 4.7. A quick look at the formulas for the partial derivatives of f tells us that the Jacobian of f is positive in the whole square for all $\alpha > 2$. Moreover, by Lemmas 3.1 and 3.3 the diagonal, the region above the diagonal and the region below the diagonal are mapped by f respectively to the diagonal, above the diagonal and below the diagonal.

Lemma 4.8. *The map f is a diffeomorphism from $[0,1]^2$ onto its image.*

Proof. In view of Lemma 3.6, Remark 4.7 and the symmetry of f with respect to the diagonal, it is enough to show that f restricted to the union of the segments $I = [0,1]\times\{0\}$ and $J = \{1\}\times[0,1]$ is a homeomorphism onto its image. On I we have $y' = -x$ and on J we have $y' = 1 - 2/(1+\alpha y)$. In both expressions the right-hand side is a strictly monotone function of the variable, so f is one-to-one, and therefore a homeomorphism onto image, on both I and J. It remains to check that $f(I)$ and $f(J)$ are disjoint except for the common endpoint. Suppose that $f(x,0) = f(1,y)$. Then by the formulas above, $-x = 1 - 2/(1+\alpha y)$. Comparing x' we get $1/(2+\alpha x) = (1+y)/(2+\alpha(1+y))$. Eliminating x from this system of two equations we get

$$\frac{1+\alpha y}{2+\alpha+2\alpha y-\alpha^2 y} = \frac{1+y}{2+\alpha+\alpha y}.$$

This is equivalent to $y(\alpha-1)(\alpha y+\alpha+1) = 0$. Therefore the only solution is $y = 0$, which gives us $x = 1$. This completes the proof. □

According to Lemma 4.6, we have to distinguish between the case $4 < \alpha \le 2(1+\sqrt{2})$, when f has only one fixed point other than $(0,0)$, and the case $\alpha > 2(1+\sqrt{2})$, when f has three such points. Let us consider the first case (see Fig. 4). For $\alpha = 2(1+\sqrt{2})$ the derivative of f at the fixed point q has one eigenvalue equal to 1. Therefore we need a special lemma about the nature of this point.

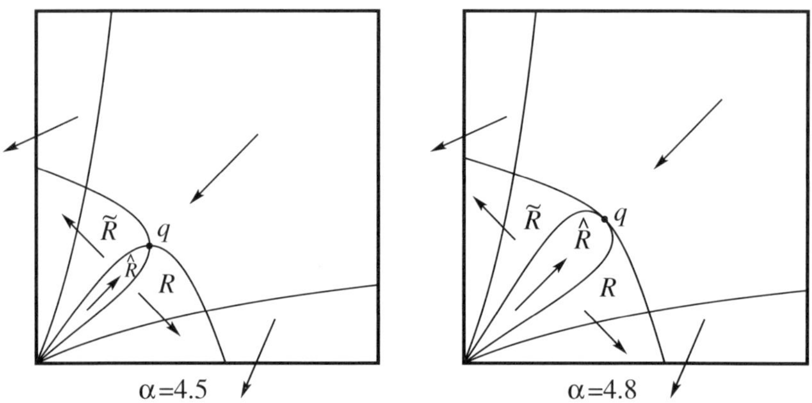

Fig. 4. Map f for $\alpha = 4.5$ and $\alpha = 4.8$.

Lemma 4.9. *Assume that $\alpha = 2(1+\sqrt{2})$. Let $p \in [0,1]^2$ be a point that does not lie on the diagonal. Then the trajectory of p does not converge to the fixed point q.*

Proof. According to Lemma 4.6 (a), $q = (\sqrt{2}-1, \sqrt{2}-1)$. Let us change the variables, so that q becomes the origin, one axis is along the diagonal, and the other perpendicular to it. That is,

$$u = (x - \sqrt{2} + 1) + (y - \sqrt{2} + 1), \qquad v = x - y.$$

If the image of (u, v) is (u', v'), then

$$u' = \frac{2[(2+\sqrt{2})(u^4+v^4)-(4+2\sqrt{2})u^2v^2+(6+10\sqrt{2})u(u^2-v^2)+(36+4\sqrt{2})u^2-(20+12\sqrt{2})v^2+(8+16\sqrt{2})u]}{[(1+\sqrt{2})(u^2-v^2)+(6+2\sqrt{2})u+(2+2\sqrt{2})v+(4+4\sqrt{2})][(1+\sqrt{2})(u^2-v^2)+(6+2\sqrt{2})u-(2+2\sqrt{2})v+(4+4\sqrt{2})]},$$

$$v' = \frac{4v[(1+\sqrt{2})(u^2-v^2)+(8+4\sqrt{2})u+(12+8\sqrt{2})]}{[(1+\sqrt{2})(u^2-v^2)+(6+2\sqrt{2})u+(2+2\sqrt{2})v+(4+4\sqrt{2})][(1+\sqrt{2})(u^2-v^2)+(6+2\sqrt{2})u-(2+2\sqrt{2})v+(4+4\sqrt{2})]}.$$

Taking the second order approximation to u' and v'/v we get

$$\begin{aligned} u' &= (4\sqrt{2}-5)u - \frac{19-13\sqrt{2}}{2}u^2 - \frac{3-\sqrt{2}}{2}v^2 + h_1(u,v), \\ v' &= v\left[1 - (3\sqrt{2}-3)u + \frac{48-31\sqrt{2}}{4}u^2 + \frac{4-\sqrt{2}}{4}v^2 + h_2(u,v)\right], \end{aligned}$$

where

$$\lim_{u,v\to 0} \frac{h_1(u,v)}{u^2+v^2} = \lim_{u,v\to 0} \frac{h_2(u,v)}{u^2+v^2} = 0.$$

Since the numbers $19-13\sqrt{2}$, $3-\sqrt{2}$, $48-31\sqrt{2}$ and $4-\sqrt{2}$ are positive, there exists a neighborhood U of $(0,0)$ such that if $(u,v) \in U$ then

$$u' \le \lambda u \quad \text{and} \quad |v'| \ge |v|(1-\mu u), \tag{19}$$

where

$$\lambda = 4\sqrt{2}-5 \quad \text{and} \quad \mu = 3\sqrt{2}-3.$$

Note that $0 < \lambda < 1$ and $\mu > 0$.

Suppose that the trajectory of a point $p \in [0,1]^2$ that does not lie on the diagonal converges to q. We are working in the coordinates u, v, so $q = (0,0)$. By replacing p by its image under a large iterate of the map we may assume that the whole trajectory is contained in U. Denote the n-th point of this trajectory by (u_n, v_n). By Remark 4.7, $v_n \ne 0$ for all n. If for some n we have $u_n \le 0$, then by (19) $u_m \le 0$ for all $m \ge n$, so, again by (19), $|v_{m+1}| \ge |v_m|$ for all $m \ge n$. Therefore v_m does not converge to 0, a contradiction. Thus, $u_n > 0$ for all n. By (19) we get $u_n \le u_0\lambda^n$ for all n, so, once more by (19),

$$|v_n| \ge |v_0| \prod_{i=0}^{n-1} (1-\mu u_0 \lambda^i) > |v_0| \prod_{i=0}^{\infty} (1-\mu u_0 \lambda^i)$$

(of course, we may assume that $\mu u_0 < 1$). The infinite product above is convergent to a positive number, so $|v_n|$ is bounded away from 0, a contradiction. This completes the proof. □

Lemma 4.10. *Assume that $4 < \alpha \le 2(1+\sqrt{2})$. Then for f the trajectories of all points on the diagonal except $(0,0)$ converge to the fixed point q. The trajectories of all points not on the diagonal leave the square $[0,1]^2$.*

Proof. The reader is advised to consult Fig. 4 when reading this proof.

By Remark 4.7, the diagonal is invariant. By Lemma 4.3, all points of the diagonal except $(0,0)$ and q are mapped by f in the direction of q. This proves that the trajectories of all points on the diagonal except $(0,0)$ converge to q.

Consider the set $\widehat{R} = \overline{R \cap \widetilde{R}}$ and a point $p = (x,y)$ from the part of its boundary above the diagonal. Then p belongs to the boundary of R, so $x' = x$. It also belongs to $\widetilde{R}$, so $y' > y$. In other words, the vector $f(p) - p$ points vertically upwards. Since this part of the boundary is a part of the graph of the function (8), it is mapped by f to a curve above itself. By symmetry, the part of the boundary of $\widehat{R}$ below the diagonal is mapped by f to a curve to the right of itself. This proves that $f(\widehat{R}) \supset \widehat{R}$.

By Lemma 4.3, all points of $\widehat{R}$ are mapped by f upwards and to the right. Therefore their trajectories either converge to q or leave $\widehat{R}$. If $\alpha \le 2(1+\sqrt{2})$, then by Lemma 4.6, q is a saddle, so its stable manifold is one-dimensional. We know already this manifold, it is the diagonal (without $(0,0)$). Thus the trajectories of the points not on the diagonal cannot converge to q. Hence, they have to leave $\widehat{R}$. If $\alpha = 2(1+\sqrt{2})$, we get to the same conclusion using Lemma 4.9.

If a point $p = (x,y)$ is above the diagonal and not in $\widehat{R}$, then by Lemma 4.3 $x' < x$. Its trajectory cannot enter $\widehat{R}$, because $f(\widehat{R}) \supset \widehat{R}$ and f is one-to-one (by Lemma 4.8). Therefore, this trajectory either leaves the square $[0,1]^2$ or converges to one of the fixed points $(0,0)$ and q. It cannot converge to q by the argument from the preceding paragraph. It cannot converge to $(0,0)$, because when close to $(0,0)$, it would be in $\widetilde{R}$, where f sends points upwards. Thus, it has to leave the square. By the symmetry, the trajectory of every point below the diagonal and not in $\widehat{R}$ has to leave the square. This completes the proof. □

Now we move to the case $\alpha > 2(1+\sqrt{2})$ (see Fig. 5).

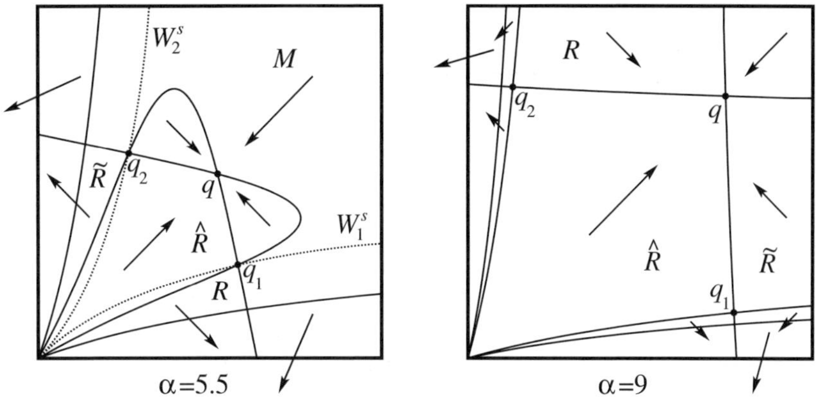

Fig. 5. Map f for $\alpha = 5.5$ and $\alpha = 9$.

Lemma 4.11. *Assume that $\alpha > 2(1+\sqrt{2})$. Then for f the stable manifold W_1^s of q_1 goes from $(0,0)$ (excluding this point) via q_1 to a point on the right side of the square $[0,1]^2$. The stable manifold W_2^s of q_2 goes from $(0,0)$ (excluding this point) via q_2 to a point on the upper side of the square $[0,1]^2$. The trajectories of all points of the region between those manifolds converge to the fixed point q. The trajectories of all points below W_1^s and all points to the left of W_2^s leave the square $[0,1]^2$.*

Proof. The reader is advised to consult Figs. 5 and 6 when reading this proof.

By a similar argument as in the proof of Lemma 4.10 (except that now we have to look at 4 pieces of the boundary of the set $\widehat{R}$, defined as in that lemma), we get $f(\widehat{R}) \supset \widehat{R}$. Moreover, arguments of the same type (looking where the vector $f(p) - p$ points on the pieces of the boundary; see Fig. 6) show that each component A of $R \setminus \widetilde{R}$ and of $\widetilde{R} \setminus R$ is mapped into itself or outside the square (that is, $f(A) \subseteq A \cup (\mathbb{R}^2 \setminus [0,1]^2)$. Here we additionally use the fact that since f is an orientation preserving diffeomorphism, the circular order of the pieces of the boundaries of R and $\widetilde{R}$ coming out from a fixed point remains the same after the application of f. The fact that for large values of α two of the regions have pieces of the boundary lying on the boundary of the square $[0,1]^2$ makes no difference, since as we noted in the proof of Lemma 4.8, on the right side of the square y' is a monotone function of y (and by symmetry, on the upper side of the square x' is a monotone function of x).

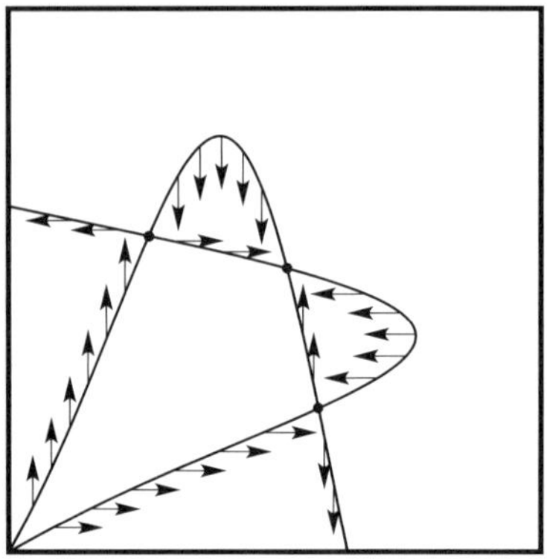

Fig. 6. The directions at which the points on the boundaries of R and $\widetilde{R}$ are mapped by f.

By Lemma 4.6, the fixed points q_1 and q_2 are saddles. They lie at the intersection points of the boundaries of R and $\widetilde{R}$. Taking into account the directions in which f maps the points (see Fig. 5), we see that W_1^s in a small neighborhood of q_1 has to lie in $\widehat{R}$ and $[0,1]^2 \setminus (R \cup \widetilde{R})$. To see what happens to it globally, we have to take the preimages of this short piece of W_1^s. The set $\widehat{R}$ is invariant under f^{-1}, so one component of $W_1^s \setminus \{q_1\}$ lies in $\widehat{R}$. All points of $\widehat{R}$ are mapped by f^{-1} down and to the left, so all trajectories converge to $(0,0)$. Therefore, this component of $W_1^s \setminus \{q_1\}$ ends up at $(0,0)$ (but of course $(0,0)$ does not belong to it). The other component starts in the region where all points are mapped by f^{-1} up and to the right. It cannot enter a component of $\widetilde{R} \setminus R$, since that component is invariant for f. Therefore it ends up at a point of the upper or right side of the square. However, W_1^s cannot cross the diagonal, so it has to end up at a point of the right side of the square. By symmetry, W_2^s stretches from $(0,0)$ via q_2 to a point on the upper side of the square.

Let M be the region of $[0,1]^2$ between W_1^s and W_2^s. The set $f(\widehat{R} \cap M)$ is bounded by a piece of W_1^s, a piece of W_2^s, a curve contained in the component of $R \setminus \widetilde{R}$ that lies above $\widehat{R}$ and a curve contained in the component of $R \setminus \widetilde{R}$ that lies to the right of $\widehat{R}$. The trajectories of points in those components converge to q by the already standard argument about directions in which the points are mapped by f. The same argument shows that the trajectories contained entirely in $\widehat{R} \cap M$ do the same (they cannot converge to q_1 or q_2 because they do not lie on their stable manifolds). The same argument applies to the remaining part of M, that is, to $M \setminus (R \cup \widetilde{R})$. Here we use additionally the observation that such a trajectory cannot leave M because it cannot cross W_1^s or W_2^s (remember that f is an orientation preserving homeomorphism).

The trajectories of all points below W_1^s and all points to the left of W_2^s leave the square $[0,1]^2$ by the already completely standard argument and the observation that such trajectories cannot converge to $(0,0)$ (looking again at the directions). This completes the proof. □

5. Proof of the Main Theorem

Now we are ready to prove Theorem 2.1.

Proof of (a). By Lemma 3.2 (b), n has three fixed points: $(0,0)$, $(1,1)$ and $(1/2,1/2)$. By equation (4), the derivative of n at $(0,0)$ and $(1,1)$ is the identity, so those points preserve orientation. By Lemma 3.2 (c), they are topologically attracting. □

Proof of (b). When $0 < \alpha \leq 1/2$, then n is a homeomorphism onto its image by Lemma 3.7. When $1/2 < \alpha < 2$, by Lemma 3.3 the anti-diagonal is invariant for n, but by Lemma 3.5 n restricted to it is not a homeomorphism. When $\alpha \geq 2$, by Lemma 3.5 n restricted to the anti-diagonal reverses orientation. Since by Lemma 3.3 n preserves the regions above and below the anti-diagonal, it reverses orientation (in two dimensions) at the points of the anti-diagonal. However, by Theorem 2.1 (a) the fixed points $(0,0)$ and $(1,1)$ are orientation preserving. Thus, n cannot be a homeomorphism onto its image. □

Proof of (c). Assume that $0 < \alpha < 2$. By Lemma 3.4 (c), $(1/2,1/2)$ is an orientation preserving saddle fixed point. By Lemmas 3.3 and 3.5 (a), both the anti-diagonal and diagonal are invariant, so by Lemma 3.4 (b) the stable and unstable manifolds of $(1/2,1/2)$ are contained respectively in the anti-diagonal and diagonal. Since there are no other fixed points on the anti-diagonal, the stable manifold is equal to it. Since the only other fixed points on the diagonal are its endpoints, the unstable manifold is equal to the diagonal without endpoints. □

Proof of (d). Assume that $2 < \alpha < 4$. By Lemma 3.4 (c), $(1/2,1/2)$ is an orientation reversing saddle fixed point. By Lemmas 3.3 and 3.5 (a), both the anti-diagonal and diagonal are invariant, so by Lemma 3.4 (b) the stable and unstable manifolds of $(1/2, 1/2)$ are contained respectively in the anti-diagonal and diagonal. By Lemma 4.1, the stable manifold is equal to the anti-diagonal. Since the only other fixed points on the diagonal are its endpoints, the unstable manifold is the diagonal without endpoints. □

Proof of (e)**.** This is Lemma 4.4. □

Proof of (f)**.** Assume that $4 < \alpha < 2(1+\sqrt{2})$. By Lemma 3.4 (c), $(1/2, 1/2)$ is an orientation reversing repelling fixed point. By Lemma 4.6 (a), there is a period 2 orbit on the anti-diagonal (corresponding to the fixed point q of f). By Lemma 4.6 (b), it is a saddle. One point of this orbit lies below $(1/2, 1/2)$. By Lemma 4.10, its stable manifold is the part of the anti-diagonal with $x < 1/2$. By symmetry, the other point lies above $(1/2, 1/2)$ and its stable manifold is the part of the anti-diagonal with $x > 1/2$. By Lemmas 4.10, 4.2, 3.2 (c) and 3.3, all points below the anti-diagonal are attracted to $(0, 0)$ and all points above the anti-diagonal are attracted to $(1, 1)$. □

Note that if $\alpha = 2(1 + \sqrt{2})$, we get the same results, except that the period 2 orbit on the anti-diagonal is not a saddle.

Proof of (g)**.** Assume that $\alpha > 2(1 + \sqrt{2})$. By Lemma 3.4 (c), $(1/2, 1/2)$ is an orientation reversing repelling fixed point. By Lemma 4.6 (a), there is a period 2 orbit P on the anti-diagonal (corresponding to the fixed point q of f) and there are two period 2 orbits that follow P (corresponding to the fixed points q_1 and q_2 of f). By Lemma 4.6 (b) and (c), P is attracting, and the other two period 2 orbits are saddles. By the symmetries, one saddle orbit lies below the anti-diagonal and one above. By Lemmas 4.11, 4.2 and 3.2 (c), every other point is attracted to one of the periodic orbits mentioned (including $(0, 0)$ and $(1, 1)$). □

Acknowledgments

M. Misiurewicz's research was partially supported by NSF grant DMS 0456526.

References

1. J. F. Nash, Non-cooperative games, Ph.D. dissertation, Princeton University (1950).

2. J. F. Nash, Equilibrium points in n-person games, *Proceedings of the National Academy of Sciences* **36**, 48-49 (1950).
3. J. F. Nash, Non-cooperative games, *Annals of Mathematics* **54**, 286-295 (1951).
4. R. A. Becker and S. K. Chakrabarti, Satisficing Behavior, Brouwer's Fixed Point Theorem and the Nash Equilibrium, *Economic Theory* **26**, 63-83 (2005).
5. R. A. Becker, S. K. Chakrabarti, W. Geller, B. Kitchens, and M. Misiurewicz, Dynamics of the Nash Map in the game of Matching Pennies, *Journal of Difference Equations and Applications* **13**, 223-235 (2007).
6. D. Barash, *The Survival Game* (Owl Books, New York, 2003).
7. B. Skyrms, *The Stag Hunt and the Evolution of Social Structure* (Cambridge University Press, Cambridge, UK, 2004).
8. K. Ritzberger and J. Weibull, Evolutionary selection in normal-form games, *Econometrica* **63**, 1371-1399 (1995).
9. W. R. Derrick, A condition under which a map is a homeomorphism, *American Mathematical Monthly* **80**, 554-555 (1973).

Attractive Cycles of an Artificial Neural Network*

Sui Sun Cheng† and Xi-lan Liu‡

†*Department of Mathematics, Tsing Hua University, Hsinchu, Taiwan 30043, R. O. China*

‡*College of Science, Hangzhou Normal University, Hangzhou, Zhejiang 310036, P. R. China*

This paper is concerned with the attractive cycles of an artificial neural network consisting of n neurons situated on the vertices of a regular polygon and connected by its edges. The weight matrix and the bias are dependent on two paramters α and β. By elementary analysis, we are able to give a complete account on the relations between the parameters and the existence and stability of cycles in these networks.

1. Introduction

Most complex systems such as the nervous system are composed of multiple elements interacting with each other as time evolves. Because of the complexity of these systems, accurate mathematical description are usually impossible, and one thus resorts to simplifications. One simplification is that time is discrete so that simulation can be carried out by digital devices. Another is that the behavior of a system at one time depends on the state of the system at a preceding time. Yet another one is that the elements of the system have only a limited number of different values. Despite the rather gross simplifying assumptions, the resulting models can nevertheless be extraordinary complex, yet analysis may lead to useful information on the original system as well as applications in designing digital devices.

In this paper, we are concerned with one such simple artificial neural network model consisting of n neurons situated on the vertices of a regular polygon with n vertices, and are connected by its edges. Furthermore, each neuron unit can take on two values designated by $+1$ and -1. To avoid trivial cases, we will assume that $n \geq 3$.

---

*Partially supported by the NSC of R. O. China

More specifically, let us label the vertices of the polygon in a clockwise manner by $1, 2, ..., n$, and the corresponding state values of the neurons $u_1, ..., u_n$ by $u_1^{(t)}, ..., u_n^{(t)}$. We suppose the i-th neuron is connected only to the two neurons situated on the neighboring vertices and that it interacts with its neighboring neuron units in such a manner that the weight matrix is of the form

$$W = \alpha \begin{pmatrix} -2 & 1 & 0 & ... & 0 & 1 \\ 1 & -2 & 1 & ... & 0 & 0 \\ 0 & 1 & -2 & ... & 0 & 0 \\ ... & ... & ... & ... & ... & ... \\ 1 & 0 & 0 & ... & 1 & -2 \end{pmatrix}_{n\times n},$$

and the bias vector is of the form

$$\omega = \beta(-1, -1, ..., -1)^\dagger,$$

where α, β are real parameters. We will assume that $u_i^{(t)}$ takes on bipolar values -1 or 1, and the update function F is defined by

$$F\left((u_1, ..., u_n)^\dagger\right) = (\text{sgn }(u_1), ..., \text{sgn }(u_n))^\dagger,$$

where $\text{sgn }(u) = +1$ if $u \geq 0$ and $\text{sgn }(u) = -1$ if $u < 0$. Under these assumptions, our neural network now takes the form

$$u^{(t+1)} = F(Wu^{(t)} + \omega),\ t = 0, 1, 2, ..., \tag{1}$$

which depends on the two real parameters α and β. Therefore, for any given $\alpha, \beta \in R$, and an initial state vector $\left(u_1^{(0)}, ..., u_n^{(0)}\right)^\dagger$, (1) defines a vector sequence $\left\{u^{(t)}\right\}_{t=0}^{\infty}$ such that

$$u^{(1)} = F(Wu^{(0)} + \omega),\ u^{(2)} = F(Wu^{(1)} + \omega), ...$$

It is of great interest to study the qualitative behavior of this solution sequence as the parameters α and β vary. Indeed, for different F, W and ω, the corresponding system has different properties and some of them can be incorporated in the design of logic, association, learning or memory networks, Boolean cellular automata (see last section), etc.

It is easy to write a computer program to simulate our neural network. A number of interesting long term behaviors of the solution sequences can then be observed. Here we will report what have been observed and explain the reasons behind these observed behaviors.

To facilitate descriptions of our findings, we first recall some of the terminologies in elementary discrete iteration theory (see e.g. [2]). Let X

be a finite set with cardinality $|X|$ and f a mapping of the set X into X. If $f(u) = v$, then v is called the image of u and u a preimage of v. If $w \in X$ does not have a preimage, then it is called a terminal element of X.

The iterates of f are defined by

$$f^0 = I,\ f^1 = f,\ f^2 = f \circ f,\ f^3 = f \circ f \circ f,\ ...$$

and the iterates of an element u in X under f are defined by

$$u^{(i)} = f^i(u),\ i = 0, 1, 2, ...$$

In particular, $u^{(0)}$ is called the zeroth iterate, $u^{(1)}$ is called the first iterate, etc. of u. If we form a sequence $\{u^{(0)}, u^{(1)}, u^{(2)}, ...\}$ of iterates of u, then since X is finite, it is clear that the sequence $\{u^{(n)}\}_{n=0}^{\infty}$ must repeat itself after a finite number of steps. It is therefore not difficult to see that every sequence $\{u^{(0)}, u^{(1)}, ...\}$ is of the form

$$\left\{u^{(0)}, u^{(1)}, ..., u^{(N-1)}, u^{(N)}, u^{(N+1)}, ..., u^{(N+\omega-1)}, u^{(N+\omega)} = u^{(N)}, u^{(N+1)}, ...\right\},$$

where $N + \omega < |X|$ (the numbers N and ω may depend on $u^{(0)}$), and $u^{(0)}, u^{(1)}, ..., u^{(N+\varpi-1)}$ are pairwise distinct.

It is natural to call the subsequence $\{u^{(N)}, u^{(N+1)}, ..., u^{(N+\omega-1)}\}$ a cycle. More precisely, a point v in X is a periodic point of period ω if its iterates $v^{(0)}, v^{(1)}, ..., v^{(\omega-1)}$ are pairwise distinct but $v^{(0)} = v^{(\omega)}$. If v is a periodic point of period ω, then $\langle v^{(0)}, v^{(1)}, ..., v^{(\omega-1)}\rangle$ is said to form a ω-cycle. Note that any cycle in X has period less than or equal to $|X|$. A point u in X is said to be attracted to a cycle if one of its iterates is equal to some element in the cycle. The set of points in X which are attracted to a cycle is said to be the *basin of attraction* of this cycle. In case every point in X is attracted to a cycle of X, this cycle is said to be globally attractive.

A periodic point of period 1 is also called a fixed point of f. In other words, v is a fixed point of f if $f(v) = v$. A fixed point v is also said to be a global attractor if the cycle $\langle v \rangle$ is globally attractive. Note that when a global attractor exists in X, there will not exist any cycles of period greater than or equal to 2.

Analysis of bipolar state value discrete time neural network is difficult due to its combinatorial nonlinear nature. By focusing on specific artificial network models, we may, however, be able to give a detailed analysis. In the following, *we will be able to provide a complete analysis of the existence of cycles as well as the basin of attraction of these cycles.*

2. Classification

We make the following observations. First, (1) can be written as a system of equations

$$u_1^{(t+1)} = \text{sgn}\ \left(\alpha\left(u_n^{(t)} - 2u_1^{(t)} + u_2^{(t)}\right) - \beta\right),$$
$$u_2^{(t+1)} = \text{sgn}\ \left(\alpha\left(u_1^{(t)} - 2u_2^{(t)} + u_3^{(t)}\right) - \beta\right),$$
$$... = ...$$
$$u_{n-1}^{(t+1)} = \text{sgn}\ \left(\alpha\left(u_{n-2}^{(t)} - 2u_{n-1}^{(t)} + u_n^{(t)}\right) - \beta\right),$$
$$u_n^{(t+1)} = \text{sgn}\ \left(\alpha\left(u_{n-1}^{(t)} - 2u_n^{(t)} + u_1^{(t)}\right) - \beta\right).$$

Therefore, if we define $u_0^{(t)} = u_n^{(t)}$ and $u_{n+1}^{(t)} = u_1^{(t)}$ for $t = 0, 1, 2, ...$, then we can write

$$u_i^{(t+1)} = \text{sgn}\ \left(\alpha\left(u_{i-1}^{(t)} - 2u_i^{(t)} + u_{i+1}^{(t)}\right) - \beta,\right),\ i = 1, 2, ..., n.$$

The values $u_1^{(t)}, u_2^{(t)}$ are naturally said to be *neighboring*, so are $u_2^{(t)}, u_3^{(t)}; ...;$ and $u_{n-1}^{(t)}, u_n^{(t)}$. Furthermore, since the neuron units are on the vertices of a regular polygon, it is also natural to call $u_n^{(t)}, u_1^{(t)}$ neighboring values.

When $u_{i-1}^{(t)} = 1, u_i^{(t)} = 1$ and $u_{i+1}^{(t)} = 1$ for $i \in \{1, ..., n\}$, we have $u_i^{(t+1)} = \text{sgn}\ (-\beta)$. Similarly, the state value $u_i^{(t+1)}$ can take on sgn $(-2\alpha - \beta)$, sgn $(-4\alpha - \beta)$, sgn $(4\alpha - \beta)$ or sgn $(2\alpha - \beta)$. These values can only take on -1 or 1 and therefore it is natural to discuss the behavior of our network in separate cases by regarding (α, β) as a point in the plane. There are 10 mutually exclusive and exhaustive cases:

1. $\beta > 0, 2\alpha - \beta \geq 0$;
2. $2\alpha - \beta < 0, 4\alpha - \beta \geq 0$;
3. $4\alpha - \beta < 0, 4\alpha + \beta > 0$;
4. $4\alpha + \beta \leq 0, 2\alpha + \beta > 0$;
5. $2\alpha + \beta \leq 0, \beta > 0$;
6. $\beta \leq 0, 2\alpha + \beta > 0$;
7. $2\alpha + \beta \leq 0, 4\alpha + \beta > 0$;
8. $4\alpha + \beta \leq 0, 4\alpha - \beta \geq 0$;
9. $4\alpha - \beta < 0, 2\alpha - \beta \geq 0$;
10. $2\alpha - \beta < 0, \beta \leq 0$.

A vector u is said to be bipolar if each of its components is either $+1$ or -1. For the sake of convenience, the column vector $\delta(n) = (\delta_1, ..., \delta_n)^\dagger$ is defined by $\delta_i = (-1)^i$ for $i = 1, 2, ..., n$; the column vector $\overline{1}(n)$ is defined by $(1, 1, ..., 1)^\dagger$; and the column vector $-\overline{1}(n)$ is defined by $(-1, -1, ..., -1)^\dagger$.

If no confusion is caused, $\delta(n)$, $\overline{1}(n)$ and $-\overline{1}(n)$ are also denoted by δ, $\overline{1}$ and $-\overline{1}$ respectively. Let $a = (a_1, ..., a_k)$ and $b = (b_1, ..., b_j)$ be two vectors. The augmented vector $(a_1, ..., a_k, b_1, ..., b_j)$ will be denoted by $a|b$, and the column vector $(a_1, ..., a_k, b_1, ..., b_j)^\dagger$ by $a^\dagger|b^\dagger$. The notations $a|b|c$, etc. will have similar meanings.

Since the vertices of a regular polygon can be relabeled, we need to introduce the following mathematical tools to describe allowable labeling. The cyclic permutation of a bipolar vector $(u_1, u_2, ..., u_{n-1}, u_n)$ is $(u_n, u_1, u_2, ..., u_{n-1})$ and the reflection is $(u_n, u_{n-1}, ..., u_2, u_1)$. Two bipolar vectors u and v are said to be equivalent if v can be obtained by performing a finite number of cyclic permutations or reflections on u. Such a relation is easily shown to be an equivalence relation.

If u is a bipolar vector different from $\overline{1}$ and $-\overline{1}$, then the components of some equivalent vector of u can be broken up into consecutive parts with alternate signs. More precisely, since u is different from $\overline{1}$ and $-\overline{1}$, there will be a pair of consecutive components with different signs. By performing a finite number of cyclic permutations and reflections if necessary, we may suppose without loss of generality that $u_n = -1$ and $u_1 = +1$. Since $u \neq \overline{1}$, we must have $u_1 = ... = u_{i_1} = +1$ and $u_{i_1+1} = -1$ for some $i_1 \in \{1, ..., n-1\}$. Next, if $i_1 \leq n-1$, there must be some $i_2 \in \{i_1+1, ..., n\}$ such that $u_{i_1+1} = ... = u_{i_2} = -1$ and $u_{i_2+1} = +1$. This process can be repeated again and again until we run out of numbers in $\{1, ..., n\}$ to choose from. Then clearly we will end up with a partition $\{S_1, S_2, ..., S_{2m}\}$ of $\{1, ..., n\}$ such that $S_1 = \{1, ..., i_1\}$, $S_2 = \{i_1+1, ..., i_2\}$, ..., $S_{2m} = \{i_{2m-1}+1, ..., n\}$ and $u_j = +1$ for $j \in S_1 \cup S_3 \cup ... \cup S_{2m-1}$ and $u_j = -1$ for $j \in S_2 \cup ... \cup S_{2m}$.

Lemma 1. *Let u be a bipolar n-vector different from $\overline{1}$ and $-\overline{1}$. Then there is an equivalent vector v of u and a partition $\{S_1, S_2, ..., S_{2m}\}$ of $\{1, ..., n\}$ such that $S_1 = \{1, ..., i_1\}$, $S_2 = \{i_1+1, ..., i_2\}$, ..., $S_{2m} = \{i_{2m-1}+1, ..., n\}$ and $v_j = +1$ for $j \in S_1 \cup S_3 \cup ... \cup S_{2m-1}$ and $v_j = -1$ for $j \in S_2 \cup ... \cup S_{2m}$.*

In the following discussions, in cases where Lemma 1 is used, we will, for the sake of convenience, set $S_{2m+1} = S_1$ where S_1 is given in Lemma 1. Also, given a n-vector $u = (u_1, u_2, ..., u_n)^\dagger$, and a subsequence $S = \{i_1, i_2, ..., i_j\}$ of the sequence $\{1, ..., n\}$, we will set

$$u(S) = (u_{i_1}, u_{i_2}, ..., u_{i_j})^\dagger.$$

In particular, the vector v in the above Lemma 1 can be written as

$$v = v(S_1)|v(S_2)|v(S_3)|...|v(S_{2m}).$$

In case $m \geq 1$, the vectors $v(S_1), v(S_3), ..., v(S_{2m-1})$ are naturally called the positive arches of v and the vectors $v(S_2), ..., v(S_{2m})$ the negative arches. More generally, let $u = (u_1, u_2, ..., u_n)^\dagger$ be an arbitrary bipolar vector. A sub-vector of the form $(u_\alpha, u_{\alpha+1}, ..., u_\beta)^\dagger$, $1 \leq \alpha \leq \beta \leq n$, is said to be a positive arch of u if $u_{\alpha-1} = -1, u_\alpha = u_{\alpha+1} = ... = u_\beta = +1$ and $u_{\beta+1} = -1$. A negative arch of u is similarly defined.

As a final notation, we will denote

$$f(u) = F(Wu + \omega),$$

where F, W and ω are defined in the first section.

3. Cases 3 and 8

Among the different possible cases, the one where $4\alpha - \beta < 0$ and $4\alpha + \beta > 0$ is relatively simple.

Indeed, when $u_{i-1}^{(t)} = 1, u_i^{(t)} = 1$ and $u_{i+1}^{(t)} = 1$, we can calculate $u_i^{(t+1)} =$ sgn $(-\beta) = -1$. Similarly, we can construct the following table

$$\begin{pmatrix} u_{i-1}^{(t)} & +1 & +1 & -1 & -1 & +1 & +1 & -1 & -1 \\ u_i^{(t)} & +1 & +1 & +1 & +1 & -1 & -1 & -1 & -1 \\ u_{i+1}^{(t)} & +1 & -1 & +1 & -1 & +1 & -1 & +1 & -1 \\ u_i^{(t+1)} & -1 & -1 & -1 & -1 & -1 & -1 & -1 & -1 \end{pmatrix},$$

which can also be expressed as

$$\begin{pmatrix} u_{i-1}^{(t)} & * \\ u_i^{(t)} & \& \\ u_{i+1}^{(t)} & \# \\ u_i^{(t+1)} & -1 \end{pmatrix},$$

where $*, \&$ and $\#$ can either be -1 or $+1$. Given any initial vector $u^{(0)}$,

$$u^{(1)} = (-1, -1, ..., -1)^\dagger = u^{(2)} = u^{(3)} = ...$$

Therefore, we have the following obvious result.

Theorem 1. *Suppose $4\alpha - \beta < 0$ and $4\alpha + \beta > 0$. The n-vector $-\bar{1}$ is a fixed point and also a global attractor.*

We remark that it will take at most one step for every vector to be attracted to $-\bar{1}$.

Similarly, when $4\alpha + \beta \leq 0$ and $4\alpha - \beta \geq 0$, we may construct the following table

$$\begin{pmatrix} u_{i-1}^{(t)} & * \\ u_i^{(t)} & \$ \\ u_{i+1}^{(t)} & \# \\ u_i^{(t+1)} & +1 \end{pmatrix},$$

where $*, \$$ and $\#$ can either be -1 or $+1$. Given any initial vector $u^{(0)}$,

$$u^{(1)} = (+1, +1, ..., +1)^{\dagger} = u^{(2)} = u^{(3)} = ...$$

As a corollary, the n-vector $\overline{1}$ is a global attractor.

4. Cases 2 and 7

In case 2 where $2\alpha - \beta < 0$ and $4\alpha - \beta \geq 0$, we can construct the following table

$$\begin{pmatrix} u_{i-1}^{(t)} & * & +1 & \$ & -1 \\ u_i^{(t)} & +1 & -1 & -1 & -1 \\ u_{i+1}^{(t)} & \# & +1 & -1 & +1 \\ u_i^{(t+1)} & -1 & +1 & -1 & -1 \end{pmatrix}, \tag{2}$$

where $*, \#$ and $\$$ can either be $+1$ or -1.

There are several immediate consequences:

(i) For each $i \in \{1, ..., n\}$, $u_i^{(t+1)} = +1$ if and only if $u_{i-1}^{(t)} = +1$, $u_i^{(t)} = -1$ and $u_{i+1}^{(t)} = +1$.

(ii) It is clear from (i) that $-\overline{1}$ is a fixed point. We assert that it is the only fixed point. Indeed, if $u^{(0)} = (u_1^{(0)}, ..., u_n^{(0)})$ is a fixed point with $u_k^{(0)} = +1$, then in view of the second column of (2), $-1 = u_k^{(1)} = u_k^{(0)} = +1$, which is a contradiction.

(iii) Let $u^{(0)}$ be a n-vector with two consecutive components equal to -1. Then $u^{(t)} = -\overline{1}$ for all $t \geq n/2$. By performing a finite number of cyclic permutations and reflections if necessary, we may assume without loss of generality that $u_n^{(0)} = u_1^{(0)} = -1$. Then in view of (i),

$$u_{n-1}^{(1)} = u_n^{(1)} = u_1^{(1)} = u_2^{(1)} = -1.$$

By induction, we see that

$$u_{n/2+1}^{(n/2-1)} = ... = u_n^{(n/2-1)} = u_1^{(n/2-1)} = ... = u_{n/2}^{(n/2-1)} = -1$$

if n is even and

$$u_{(n+1)/2}^{((n-1)/2)} = ... = u_n^{((n-1)/2)} = u_1^{((n-1)/2)} = ... = u_{(n+1)/2}^{((n-1)/2)} = -1$$

if n is odd.

(iv) When n is even, $\langle \delta, -\delta \rangle$ is an unique 2-cycle and each vector $u^{(0)} \notin \{\delta, -\delta\}$ is attracted to $-\bar{1}$ in at most $n/2 + 1$ steps. To see this, note that

$$\delta_1^{(0)} = -1,\ \delta_2^{(0)} = +1,\ ...\ ,\ \delta_n^{(0)} = +1.$$

In view of (2),

$$\delta_1^{(1)} = +1,\ \delta_2^{(1)} = -1,\ ...\ ,\ \delta_n^{(1)} = -1,$$

and

$$\delta_1^{(2)} = -1,\ \delta_2^{(2)} = +1,\ ...\ ,\ \delta_n^{(2)} = +1,$$

that is $\delta^{(1)} = -\delta^{(0)} = -\delta$, and $\delta^{(2)} = \delta^{(0)}$ as required. Next, if $u^{(0)}$ is different from δ and $-\delta$, then it must have a pair of neighboring neuron units which have the same values. If $u_i^{(0)} = u_{i+1}^{(0)} = -1$ for some $i \in \{1, ..., n\}$, then in view of (iii), $u^{(t)} = -\bar{1}$ for $t \geq n/2$. If $u_i^{(0)} = u_{i+1}^{(0)} = +1$ for some $i \in \{1, ..., n\}$, then in view of the second column of (2), $u_i^{(1)} = u_{i+1}^{(1)} = -1$. Again, in view of (iii), we see that $u^{(t)} = -\bar{1}$ for $t \geq n/2+1$. Finally, $\{\delta, -\delta\}$ is unique since each vector $u^{(0)} \notin \{\delta, -\delta\}$ is attracted to $-\bar{1}$.

(v) When n is odd, every vector u is attracted to $-\bar{1}$ in at most $(n+1)/2$ steps. That is to say $-\bar{1}$ is a global attractor. Indeed, since $u^{(0)}$ has an odd number of components, it must have two consecutive components with the same sign. The conclusion now follows from (iii).

Theorem 2. *Suppose $2\alpha - \beta < 0$ and $4\alpha - \beta \geq 0$. Then $-\bar{1}$ is the only fixed point. If in addition n is even, then $\langle \delta, -\delta \rangle$ is the unique 2-cycle, and all bipolar n-vectors distinct from δ and $-\delta$ are attracted to $-\bar{1}$ (in at most $n/2 + 1$ steps). If in addition n is odd, then $-\bar{1}$ is a global attractor (and any bipolar n-vector is attracted to it in at most $(n + 1)/2$ steps).*

As a direct consequence, we remark that when n is odd, since $-\bar{1}$ is a global attractor, no ω-cycles with $\omega \geq 2$ can exist. Similarly, when n is even, no ω-cycles with $\omega > 2$ can exist.

To illustrate our result, take $n = 3$. There are 8 bipolar vectors. Let $u_1 = (+1, +1, +1)^\dagger = \bar{1}$ and $u_4 = (-1, -1, -1)^\dagger = -\bar{1}$. Then the remaining vectors belong to two different classes with representing vectors

$u_2 = (+1, -1, +1)^\dagger$ and $u_3 = (+1, -1, -1)^\dagger$. In case $2\alpha - \beta < 0$ and $4\alpha - \beta \geq 0$, we can check directly that all vectors are attracted to $-\overline{1}$.

Similarly, take $n = 4$. Other than $u_1 = (+1, +1, +1, +1)^\dagger = \overline{1}$ and $u_6 = (-1, -1, -1, -1)^\dagger = -\overline{1}$, there are four essentially different bipolar vectors $u_2 = (+1, +1, +1, -1)^\dagger$, $u_3 = (+1, +1, -1, -1)^\dagger$, $u_4 = (+1, -1, -1, -1)^\dagger$ and $u_5 = (+1, -1, +1, -1)^\dagger$. The vectors u_1, u_2, u_3, u_4, u_6 are attracted to $-\overline{1}$, while $\langle u_5, -u_5 \rangle$ is an unique 2-cycle.

In case 7 where $2\alpha + \beta \leq 0$ and $4\alpha + \beta > 0$, we can construct the following table

$$\begin{pmatrix} u_{i-1}^{(t)} & \$ & +1 & -1 & * \\ u_i^{(t)} & +1 & +1 & +1 & -1 \\ u_{i+1}^{(t)} & +1 & -1 & -1 & \# \\ u_i^{(t+1)} & +1 & +1 & -1 & +1 \end{pmatrix},$$

where $*, \#$ and $\$$ can either be $+1$ or -1.

By means of the same arguments discussed above, it is easily seen that $\overline{1}$ is the only fixed point. If n is even, then $\langle \delta, -\delta \rangle$ is the unique 2-cycle, and all vectors distinct from δ and $-\delta$ are attracted to $\overline{1}$. If n is odd, then $\overline{1}$ is a global attractor.

5. Cases 5 and 10

In case 5 where $2\alpha + \beta \leq 0$ and $\beta > 0$, we can construct the following table:

$$\begin{pmatrix} u_{i-1}^{(t)} & +1 & \& & -1 & \$ \\ u_i^{(t)} & +1 & +1 & +1 & -1 \\ u_{i+1}^{(t)} & +1 & -1 & +1 & * \\ u_i^{(t+1)} & -1 & +1 & +1 & -1 \end{pmatrix}, \tag{3}$$

where $\&$, $*$ and $\$$ can either be $+1$ or -1.

There are several immediate consequences:

(i) For each $i \in \{1, ..., n\}$, in view of the second column of (3), $u_i^{(t+1)} = -u_i^{(t)}$ if and only if $u_{i-1}^{(t)} = +1$, $u_i^{(t)} = +1$ and $u_{i+1}^{(t)} = +1$, and in view of the third, fourth and the fifth column of (3), $u_i^{(t+1)} = u_i^{(t)}$ if and only if at least one of $u_{i-1}^{(t)}$, $u_i^{(t)}$ or $u_{i+1}^{(t)}$ is -1.

(ii) If $u^{(0)}$ is a bipolar vector such that for any $i \in \{1, ..., n\}$, at least one of $u_{i-1}^{(0)}$, $u_i^{(0)}$ or $u_{i+1}^{(0)}$ is -1, then by (i), $u^{(1)} = u^{(0)}$. In other words, $u^{(0)}$ is a fixed point. Conversely, if $u^{(0)}$ is a fixed point, then for any $i \in \{1, ..., n\}$, at least one of $u_{i-1}^{(0)}$, $u_i^{(0)}$ or $u_{i+1}^{(0)}$ is -1.

For otherwise $u_{j-1}^{(0)} = u_j^{(0)} = u_{j+1}^{(0)} = +1$ for some $j \in \{1, ..., n\}$ would imply $u_j^{(1)} = -1$ which is contrary to the assumption that $u_j^{(1)} = +1$.

(iii) In view of the fifth column of (3), $-\bar{1}$ is a fixed point. We assert that its basin of attraction is $\{\bar{1}, -\bar{1}\}$. Indeed, If $u^{(0)} = \bar{1}$, then, by (i), $u^{(1)} = -\bar{1}$, so $\bar{1}$ is attracted to the fixed point $-\bar{1}$. If $u^{(0)} \notin \{\bar{1}, -\bar{1}\}$, then $u^{(0)}$ has at least a pair of neighboring components with different state value. By applying a finite number of cyclic permutations and reflections if necessary, we may assume without loss the generality that the first two components of $u^{(0)}$ satisfy

$$\left(u_1^{(0)}, u_2^{(0)}\right) = (+1, -1).$$

In view of (i),

$$\left(u_1^{(1)}, u_2^{(1)}\right) = (+1, -1),$$
$$\left(u_1^{(2)}, u_2^{(2)}\right) = (+1, -1),$$
$$... = ...$$

and in general,

$$\left(u_1^{(t)}, u_2^{(t)}\right) = (+1, -1),\ t = 0, 1, 2, ...$$

This shows that $u^{(t)} \notin \{\bar{1}, -\bar{1}\}$ for any $t \in \{1, 2, ...\}$.

(iv) Let $u^{(0)} \notin \{\bar{1}, -\bar{1}\}$ be a bipolar n-vector such that $u_{i-1}^{(0)} = u_i^{(0)} = u_{i+1}^{(0)} = +1$ for some $i \in \{1, ..., n\}$. By Lemma 1, we may assume without loss of any generality that there is a partition $\{S_1, S_2, ..., S_{2m}\}$ of $\{1, ..., n\}$ such that $S_1 = \{1, ..., i_1\}$, $S_2 = \{i_1 + 1, ..., i_2\}$, ..., $S_{2m} = \{i_{2m-1} + 1, ..., n\}$ and $u_j^{(0)} = +1$ for $j \in S_1 \cup S_3 \cup ... \cup S_{2m-1}$ and $u_j^{(0)} = -1$ for $j \in S_2 \cup ... \cup S_{2m}$. Since $u_{i-1}^{(0)} = u_i^{(0)} = u_{i+1}^{(0)} = +1$ for some $i \in \{1, ..., n\}$, we may assume further that $i_1 \geq 3$. Then by (i),

$$u_0^{(1)} = -1, u_1^{(1)} = +1, u_2^{(1)} = u_3^{(1)} = ... = u_{i_1-1}^{(1)} = -1,$$
$$u_{i_1}^{(1)} = +1, u_{i_1+1}^{(1)} = -1.$$

Clearly, for each $i \in S_1$, at least one of the components $u_{i-1}^{(1)}, u_i^{(1)}$ or $u_{i+1}^{(1)}$ is -1. By (i) again, we see further that

$$u_{i_1}^{(0)} = +1, u_{i_1+1}^{(0)} = u_{i_1+2}^{(0)} = ... = u_{i_2}^{(0)} = -1, u_{i_2+1}^{(0)} = +1$$

imply

$$u_{i_1}^{(1)} = +1, u_{i_1+1}^{(1)} = u_{i_1+2}^{(1)} = ... = u_{i_2}^{(1)} = -1, u_{i_2+1}^{(1)} = +1.$$

Thus, for each $i \in S_2$, at least one of the components $u_{i-1}^{(1)}, u_i^{(1)}$ or $u_{i+1}^{(1)}$ is -1. By similar analysis, we see that for each $i \in \{1, ..., n\}$, at least one of the components $u_{i-1}^{(1)}, u_i^{(1)}$ or $u_{i+1}^{(1)}$ is -1. In view of (ii), $u^{(1)}$ is a fixed point. Finally, since $u_n^{(1)} = u_0^{(1)} = -1$ and $u_1^{(1)} = +1$, we see that $u^{(1)}$ is different from $\bar{1}$ and from $-\bar{1}$.

(v) Let v be a preimage of u under f, that is, let $u = f(v)$. If $u_i = +1$, then $v_i = +1$. This is true in view of the third and the fourth columns of (3).

(vi) Let $u \neq -\bar{1}$ be a fixed point. By Lemma 1, we may assume without loss the generality that there is a partition $\{S_1, S_2, ..., S_{2m}\}$ of $\{1, ..., n\}$ such that $S_1 = \{1, ..., i_1\}$, $S_2 = \{i_1 + 1, ..., i_2\}$, ..., $S_{2m} = \{i_{2m-1} + 1, ..., n\}$ and $u_j = +1$ for $j \in S_1 \cup S_3 \cup ... \cup S_{2m-1}$ and $u_j = -1$ for $j \in S_2 \cup ... \cup S_{2m}$. Let u be written as

$$u = u(S_1)|u(S_2)|u(S_3)|...|u(S_{2m}).$$

Since u is a fixed point, in view of (ii), $1 \leq |u(S_{2k-1})| \leq 2$ for $k \in \{1, ..., m\}$. Let v be a preimage of u under f, that is, let $f(v) = u$. Suppose $m = 1$. We assert that

$$v = u = u(S_1)|u(S_2).$$

To see this, we consider two cases: $|S_1| = 1$ and $|S_1| = 2$. Suppose the former case holds, assume $|S_2| = 2$, then

$$u = (+1, -1, -1)^\dagger.$$

In view of (v), $v_1 = +1$. We assert that $v_2 = -1$. If $v_2 = +1$, then in view of the second and the third column of (3), $v_3 = +1$ for otherwise $u_2 = +1$ which is contrary to our assumption. But if $v_1 = v_2 = v_3 = +1$, then in view of the second column of (3), $u_1 = -1$, which is contrary to our assumption. The case where $|S_1| = 1$ and $|S_2| > 2$ is similarly proved. Next, suppose the latter case $|S_1| = 2$ hold. If $|S_2| = 1$, then

$$u = (+1, +1, -1)^\dagger.$$

By the third and the fourth columns of (3),

$$v = (+1, +1, x)^\dagger.$$

We assert that $x = -1$. Otherwise, by the second column of (3) $u_2 = -1$, which is contrary to our assumption. The case where $|S_1| = 2$ and $|S_2| > 1$ can be proved by induction. Suppose $m > 1$. If $|S_{2k-1}| = 2$ for some $k \in \{1, ..., m\}$, we assert that $v(S_{2k-1}) = u(S_{2k-1})$, $v(S_{2k}) = u(S_{2k})$ and $v(S_{2k-2}) = u(S_{2k-2})$. Indeed, if $|S_{2k}| = 1$, then

$$u(S_{2k-1})|u(S_{2k}) = (+1, +1, -1)^\dagger.$$

By the third and the fourth columns of (3),

$$v(S_{2k-1})|v(S_{2k}) = (+1, +1, x)^\dagger.$$

We assert that $x = -1$. Otherwise, by the second column of (3) the second component of $u(S_{2k-1})|u(S_{2k})$ is -1, which is contrary to our assumption. The case where $|S_{2k-1}| = 2$ and $|S_{2k}| > 1$ can be proved by induction. Similarly, we may show that $v(S_{2k-2})|v(S_{2k-1}) = u(S_{2k-2})|u(S_{2k-1})$. (In particular, if $m = 2$, and if $|S_1| = 2$, or, $|S_3| = 2$, then $v = u$.) If ($m > 1$ and) $|S_{2k-1}| = 1$ for some $k \in \{1, ..., m\}$, we assert that $v(S_{2k}) = u(S_{2k})$ or $v(S_{2k}) = -u(S_{2k})$. Indeed, the first component of $v(S_{2k})$ is $+1$ or -1. If the former case holds, then the second component of $v(S_{2k})$ must be $+1$, otherwise, the second component of $u(S_{2k})$, in view of the third column of (3), is $+1$, which is contrary to our assumption on $u(S_{2k})$. By induction, we may then show that all the components of $v(S_{2k})$ are -1. If the latter case holds, we assert that the second component of $v(S_{2k})$ is $+1$. Otherwise, in view of the third and the fourth columns of (3), the second component of $u(S_{2k})$ must be $+1$, which is again contrary to our assumption on $u(S_{2k})$. By induction, we see that each component of $v(S_{2k})$ is -1.

Example 5.1. Take $n = 12$ and $u = (+1, -1, -1+1, -1, -1, -1, +1, +1, -1, -1)^\dagger$. In view of (ii), it is a fixed point. Furthermore, it is of the form

$$\begin{aligned} u &= u(S_1)|u(S_2)|u(S_3)|u(S_4)|u(S_5)|u(S_6) \\ &= (+1)^\dagger|(-1,-1)^\dagger|(+1)^\dagger|(-1,-1,-1)^\dagger|(+1,+1)^\dagger|(-1,-1)^\dagger. \end{aligned}$$

In view of (v), we see that any vector v that is being mapped to u is of the form

$$v = (+1, *, *, +1, *, *, *, +1, +1, *, *)^\dagger,$$

where each $*$ can be either $+1$ or -1. But since $|S_5| = 2$, in view of (vi),

$$v = (+1, *, *, +1, -1, -1, -1, +1, +1, -1, -1)^{\dagger}.$$

Finally, since $|S_1| = 1$, in view of (vi), v can either be

$$(+1, +1, +1, +1, -1, -1, -1, +1, +1, -1, -1)^{\dagger}$$

or

$$(+1, -1, -1, +1, -1, -1, -1, +1, +1, -1, -1)^{\dagger}.$$

Example 5.2. Take $n = 4$. Let $u = (+1, -1, +1, -1)^{\dagger}$. Then u is a fixed point in view of (ii). Furthermore, its basin of attraction is $\{v, u\}$, where $v = (+1, +1, +1, -1)^{\dagger}$. Let $w = (+1, -1, -1, -1)^{\dagger}$. Then its basin of attraction is $\{w\}$. Let $p = (+1, +1, -1, -1)^{\dagger}$. Then its basin of attraction is $\{p\}$.

Example 5.3. Take $n = 5$. Let $u = (+1, -1, +1, -1, -1)^{\dagger}$. Then u is a fixed point in view of (ii). Furthermore, its basin of attraction is $\{w, v, u\}$, where $w = (+1, +1, +1, -1, -1)^{\dagger}$ and $v = (+1, -1, +1, +1, +1)^{\dagger}$. Let $p = (+1, +1, -1, +1, -1)^{\dagger}$. Then its basin of attraction is $\{p\}$. Let $q = (+1, -1, -1, -1, -1)^{\dagger}$. Then its basin of attraction is $\{q\}$. Let $r = (+1, +1, -1, -1, -1)^{\dagger}$. Then its basin of attraction is $\{r\}$.

Example 5.4. Take $n = 6$. Let $u = (+1, -1, +1, -1, +1, -1)^{\dagger}$. Then u is a fixed point in view of (ii). If $f(v) = u$, then by (v), $v_1 = v_3 = v_5 = +1$. In view of (3), it is easily determined that $v = u$ or $v = (+1, +1, +1, -1, +1, -1)^{\dagger}$. Thus the basin of attraction of u is $\{u, (+1, +1, +1, -1, +1, -1)^{\dagger}\}$. Let $u = (+1, +1, -1, +1, -1, -1)^{\dagger}$. In view of (v) and (vi), we see that any v that satisfies $f(v) = u$ is of the form

$$v = (+1, +1, -1, +1, x, y)^{\dagger}.$$

Again, by (3), it is easily determined that $x = -1$ and $y = -1$. Let $u = (+1, -1, +1, -1, -1, -1)^{\dagger}$. By similar arguments, we may show that its basin of attraction is $\{w, v, u\}$, where $w = (+1, +1, +1, -1, -1, -1)^{\dagger}$ and $v = (+1, -1, +1, +1, +1, +1)^{\dagger}$. Let $u = (+1, -1, -1, +1, -1, -1)^{\dagger}$. Then its basin of attraction is $\{p, u\}$, where $p = (+1, +1, +1, +1, -1, -1)^{\dagger}$. Finally, let $u = (+1, -1, -1, -1, -1, -1)^{\dagger}$. Its basin of attraction is $\{u\}$.

Theorem 3. *Suppose $2\alpha + \beta \leq 0$ and $\beta > 0$.*

(1) *A bipolar vector u is a fixed point if, and only if, for any $i \in \{1, ..., n\}$, at least one of u_{i-1}, u_i or u_{i+1} is -1. Furthermore, any bipolar vector u is attracted, in at most one step, to a fixed point (which can explicitly be given if the sign distribution of u is known).*

(2) *The basin of attraction of the fixed point* $-\overline{1}$ *is* $\{\overline{1}, -\overline{1}\}$. *Let* $u \notin \{\overline{1}, -\overline{1}\}$ *be a fixed point and* $\{S_1, S_2, ..., S_{2m}\}$ *is a partition of* $\{1, ..., n\}$ *such that* $S_1 = \{1, ..., i_1\}$, $S_2 = \{i_1 + 1, ..., i_2\}$, ..., $S_{2m} = \{i_{2m-1} + 1, ..., n\}$ *and* $u_j = +1$ *for* $j \in S_1 \cup S_3 \cup ... \cup S_{2m-1}$ *and* $u_j = -1$ *for* $j \in S_2 \cup ... \cup S_{2m}$. *If* $m = 1$, *then the only preimage of* u *is itself. If* $m > 1$, *then for any preimage* v *of* u,

$$v(S_{2k-1}) = u(S_{2k-1}), \ k \in \{1, ..., m\},$$

$$|S_{2k-1}| = 1 \Rightarrow v(S_{2k}) = u(S_{2k}) \ \textit{or} \ -u(S_{2k}), \ k \in \{1, ..., m\},$$

and

$$|S_{2k-1}| = 2 \Rightarrow v(S_{2k-2})|v(S_{2k-1})|v(S_{2k})$$
$$= u(S_{2k-2})|u(S_{2k-1})|u(S_{2k}), \ k \in \{1, ..., m\}.$$

We remark that once the preimages of a fixed point u is determined, then the basin of attraction of u is just the set of all its preimages which are completely described by Theorem 3.

In case 10 where $2\alpha - \beta < 0$ and $\beta \leq 0$, we can construct the following table

$$\begin{pmatrix} u_{i-1}^{(t)} & * & \# & +1 & -1 \\ u_i^{(t)} & +1 & -1 & -1 & -1 \\ u_{i+1}^{(t)} & \& & +1 & -1 & -1 \\ u_i^{(t+1)} & +1 & -1 & -1 & +1 \end{pmatrix},$$

where $*, \#$ and & can either be $+1$ or -1.

Similar to case 5, a bipolar vector is a fixed point if, and only if, for any $i \in \{1, ..., n\}$, at least one of u_{i-1}, u_i or u_{i+1} is $+1$. $\overline{1}$ is a fixed point and the its basin of attraction contains $\overline{1}$ and $-\overline{1}$ only. Furthermore, any bipolar vector is attracted to a fixed point.

If $u \neq \overline{1}$ is a fixed point and $\{S_1, S_2, ..., S_{2m}\}$ is a partition of $\{1, ..., n\}$ such that $S_1 = \{1, ..., i_1\}$, $S_2 = \{i_1 + 1, ..., i_2\}$, ..., $S_{2m} = \{i_{2m-1} + 1, ..., n\}$ and $u_j = +1$ for $j \in S_1 \cup S_3 \cup ... \cup S_{2m-1}$ and $u_j = -1$ for $j \in S_2 \cup ... \cup S_{2m}$. If $m = 1$, then the basin of attraction of u is $\{u\}$. If $m > 1$, then for any v in the basin of attraction of u,

$$v(S_{2k}) = u(S_{2k}), \ k \in \{1, ..., m\},$$

$$|S_{2k}| = 1 \Rightarrow v(S_{2k+1}) = u(S_{2k+1}) \text{ or } -u(S_{2k+1}), \ k \in \{1, ..., m\},$$

and

$$|S_{2k}|=2 \Rightarrow v(S_{2k-1})|v(S_{2k})|v(S_{2k+1})=v(S_{2k-1})|u(S_{2k})|u(S_{2k+1}), k\in\{1,...,m\},$$

where we have defined $S_{2m+1} = S_1$.

6. Cases 4 and 9

In case 4 where $4\alpha + \beta \leq 0$ and $2\alpha + \beta > 0$, we can construct the following table

$$\begin{pmatrix} u_{i-1}^{(t)} & \$ & +1 & -1 & * \\ u_i^{(t)} & +1 & +1 & +1 & -1 \\ u_{i+1}^{(t)} & +1 & -1 & -1 & \& \\ u_i^{(t+1)} & -1 & -1 & +1 & -1 \end{pmatrix}, \tag{4}$$

where $*$, \$ and & can either be $+1$ or -1.

There are some immediate consequences:

(i) For each $i \in \{1,...,n\}$, $u_i^{(t+1)} = +1$ if and only if $u_{i-1}^{(t)} = -1$, $u_i^{(t)} = +1$ and $u_{i+1}^{(t)} = -1$; and $u_i^{(t+1)} = -1$ if $u_i^{(t)} = -1$. Therefore, if $u_{j-1}^{(0)} = -1$, $u_j^{(0)} = +1$ and $u_{j+1}^{(0)} = -1$ for some $j \in \{1,...,n\}$, then

$$u_{j-1}^{(t)} = -1,\ u_j^{(t)} = +1,\ u_{j+1}^{(t)} = -1$$

for all $t \geq 0$.

(ii) A bipolar vector u is a fixed point if, and only if, for any $i \in \{1,...,n\}$, at least one of u_i or u_{i+1} is -1. Indeed, if $u_i^{(0)} = -1$, then in view of the fifth column of (4), $u_i^{(1)} = -1$; and if $u_i^{(0)} = +1$, then by assumption, $u_{i-1}^{(0)} = u_{i+1}^{(0)} = -1$, so that, in view of (i), $u_i^{(1)} = +1$. Conversely, if u is a fixed point but there is some $j \in \{1,...,n\}$ such that $u_j^{(0)} = u_{j+1}^{(0)} = +1$. Then in view of the second column of (4), $u_j^{(1)} = -1$ which is contrary to our assumption that $u_j^{(0)} = u_j^{(1)}$.

(iii) In view of (ii), $-\bar{1}$ is a fixed point. Furthermore, in view of the second column of (4), $\bar{1}^{(1)} = -\bar{1}$. Let $u^{(0)}$ be a bipolar vector different from $\bar{1}$ and $-\bar{1}$. Then by Lemma 1, we may assume without loss of generality that there is a partition $\{S_1, S_2, ..., S_{2m}\}$ of $\{1,...,n\}$ such that $S_1 = \{1,...,i_1\}$, $S_2 = \{i_1+1,...,i_2\}$, ..., $S_{2m} = \{i_{2m-1}+1,...,n\}$ and $u_j^{(0)} = +1$ for $j \in S_1 \cup S_3 \cup ... \cup S_{2m-1}$ and $u_j^{(0)} = -1$ for $j \in S_2 \cup ... \cup S_{2m}$. If in addition $|S_{2k-1}| \geq 2$ for all $k = 1, 2, ..., m$, then in view of (i) and the second, third and the fifth columns of (4), $u^{(1)} = -\overline{1}$, which means that $u^{(0)}$ is attracted to $-\overline{1}$. If some $|S_{2k-1}| = 1$, then by (i), $u^{(0)}$ is not attracted to $-\overline{1}$.

(iv) Let $u^{(0)}$ be a bipolar vector different from $\overline{1}$ and $-\overline{1}$. Then by Lemma 1, we may assume without loss of generality that there is a partition $\{S_1, S_2, ..., S_{2m}\}$ of $\{1, ..., n\}$ such that $S_1 = \{1, ..., i_1\}$, $S_2 = \{i_1 + 1, ..., i_2\}$, ..., $S_{2m} = \{i_{2m-1} + 1, ..., n\}$ and $u_j^{(0)} = +1$ for $j \in S_1 \cup S_3 \cup ... \cup S_{2m-1}$ and $u_j^{(0)} = -1$ for $j \in S_2 \cup ... \cup S_{2m}$. If $i_1 = 1$, then $u_0^{(0)} = -1$, $u_1^{(0)} = +1$ and $u_2^{(0)} = -1$. In view of (i),

$$u_0^{(t)} = u_0^{(0)} = -1,\ u_1^{(t)} = u_1^{(0)} = +1,\ u_2^{(t)} = u_2^{(0)} = -1$$

for $t \geq 0$. If $i_1 > 1$, then in view of the second and the third columns of (4),

$$u_0^{(t)} = u_1^{(t)} = ... = u_{i_1+1}^{(t)} = -1,\ t \geq 1.$$

In either case, for any $i \in \{1, ..., i_1\}$, at least one of $u_i^{(1)}$ or $u_{i+1}^{(1)}$ is -1. Next, in view of the fifth column of (4),

$$u_{i_1+1}^{(t)} = u_{i_1+2}^{(t)} = ... = u_{i_2}^{(t)} = -1,\ t \geq 1.$$

Again, for any $i \in \{i_1 + 1, ..., i_2\}$, at least one of $u_i^{(1)}$ or $u_{i+1}^{(1)}$ is -1. In general, we see that for any $i \in \{1, ..., n\}$, at least one of $u_i^{(1)}$ or $u_{i+1}^{(1)}$ is -1. Therefore, by (ii), $u^{(1)}$ is a fixed point. Furthermore, if $i_3 - i_2, i_5 - i_4, ..., i_{2m-1} - i_{2m-2} > 1$, then $u^{(1)} = -\overline{1}$, otherwise, for $i_1 = 1$ and each $i_{2k+1} \in \{i_3, i_5, ..., i_{2m-1}\}$ with $i_{2k+1} - i_{2k} = 1$, then $u_{i_{2k+1}}^{(t)} = +1$ for $t \geq 0$.

(v) Let v be a preimage of u under f, that is, $u = f(v)$. If $u_i = +1$, then $v_i = +1$. This is true in view of the fourth column of (4).

(vi) Let $u \neq -\overline{1}$ be a fixed point. Since $\overline{1}$ is not a fixed point, by Lemma 1, we may assume without loss the generality that there is a partition $\{S_1, S_2, ..., S_{2m}\}$ of $\{1, ..., n\}$ such that $S_1 = \{1, ..., i_1\}$, $S_2 = \{i_1 + 1, ..., i_2\}$, ..., $S_{2m} = \{i_{2m-1} + 1, ..., n\}$ and $u_j = +1$ for $j \in S_1 \cup S_3 \cup ... \cup S_{2m-1}$ and $u_j = -1$ for $j \in S_2 \cup ... \cup S_{2m}$. Let u be written as

$$u = u(S_1)|u(S_2)|u(S_3)|...|u(S_{2m}).$$

Since u is a fixed point, in view of (ii), $|S_{2k-1}| = 1$ for any $k \in \{1, ..., m\}$. Let v be a preimage of u under f, that is, $f(v) = u$. Suppose $|S_{2k}| \leq 3$ for some $k \in \{1, ..., m\}$. We assert that

$$v(S_{2k-1})|v(S_{2k})|v(S_{2k+1}) = u(S_{2k-1})|u(S_{2k})|u(S_{2k+1}),$$

where $u(S_{2m+1}) = u(S_1)$ and $v(S_{2m+1}) = v(S_1)$. To see this, we consider two cases: $|S_{2k}| = 3$ and $|S_{2k}| \leq 2$. Suppose the former case holds, then

$$u(S_{2k-1})|u(S_{2k})|u(S_{2k+1}) = (+1, -1, -1, -1, +1)^{\dagger}.$$

In view of (v),

$$v(S_{2k-1})|v(S_{2k})|v(S_{2k+1}) = (+1, x, y, z, +1)^{\dagger}.$$

We assert that $x = -1$. Otherwise, by (i), the first component of $u(S_{2k-1})|u(S_{2k})|u(S_{2k+1})$ is -1, which is contrary to our assumption. Thus the first component of $v(S_{2k})$ must be -1, so that

$$v(S_{2k-1})|v(S_{2k})|v(S_{2k+1}) = (+1, -1, y, z, +1)^{\dagger}.$$

Next, we assert that $y = z = -1$. If $y = +1$, then $z = +1$, for otherwise, by the fourth column of (4), the third component of $u(S_{2k-1})|u(S_{2k})\ |u(S_{2k+1})$ is $+1$, which is a contradiction. But if $y = z = +1$, by the second and the third columns of (4), the last component of $u(S_{2k-1})|u(S_{2k})|u(S_{2k+1})$ is -1, which is contrary to our assumption. That is to say

$$v(S_{2k-1})|v(S_{2k})|v(S_{2k+1}) = u(S_{2k-1})|u(S_{2k})|u(S_{2k+1}).$$

The case where $|S_{2k-1}| = 1$ and $|S_{2k}| \leq 2$ is similarly proved. Suppose $|S_{2k}| \geq 4$ for some $k \in \{1, ..., m\}$. Then by reasons just shown, the first and the last components of $v(S_{2k})$ are -1. Furthermore, by (i), $v(S_{2k})$ cannot have three consecutive components of the form $-1, +1, -1$.

Example 6.1. Take $n = 6$. Let $u = (+1, -1, +1, -1, +1, -1)^{\dagger}$. Then u is a fixed point in view of (ii) and its preimage is itself. Let $u = (+1, -1, -1, -1, -1, -1)^{\dagger}$, which is a fixed point in view of (ii). Then $S_1 = \{1\}$ and $S_2 = \{2, 3, 4, 5, 6\}$,

$$u = (+1)^{\dagger} \,|\, (-1, -1, -1, -1, -1)^{\dagger}.$$

Now $|S_2| = 5$. In view of (vi), the preimages are of the form

$$(+1, -1, *, \$, \#, -1)^{\dagger},$$

and the subvector $(-1, *, \$, \#, -1)^{\dagger}$ cannot have three consecutive components of the form $-1, +1, -1$. Thus the preimages are

$$(+1, -1, -1, -1, -1, -1)^{\dagger}, (+1, -1, +1, +1, -1, -1)^{\dagger},$$
$$(+1, -1, -1, +1, +1, -1)^{\dagger}, (+1, -1, +1, +1, +1, -1)^{\dagger}.$$

Theorem 4. *Suppose $4\alpha+\beta\le 0$ and $2\alpha+\beta>0$.*

(1) A bipolar vector u is a fixed point if, and only if, for any $i\in\{1,...,n\}$, at least one of u_i or u_{i+1} is -1.

(2) Every bipolar vector u is, in at most one step, attracted to a fixed point (which can explicitly be given when the sign distribution of u is known).

(3) Let $u\ne -\bar{1}$ be a fixed point and let $\{S_1,S_2,...,S_{2m}\}$ is a partition of $\{1,...,n\}$ such that $S_1=\{1,...,i_1\}$, $S_2=\{i_1+1,...,i_2\}$, ..., $S_{2m}=\{i_{2m-1}+1,...,n\}$ and $u_j=+1$ for $j\in S_1\cup S_3\cup...\cup S_{2m-1}$ and $u_j=-1$ for $j\in S_2\cup...\cup S_{2m}$. Then $|S_{2k-1}|=1$ for all $k\in\{1,...,m\}$, and for any preimage v of u,

$$v(S_{2k-1})=u(S_{2k-1}),\ k\in\{1,...,m\},$$

and the first as well as the last components of each $v(S_{2k})$ are -1. Furthermore, if $|S_{2k}|\le 3$ for some $k\in\{1,...,m\}$, then for preimage v of u,

$$v(S_{2k-1})|v(S_{2k})|v(S_{2k+1})=u(S_{2k-1})|u(S_{2k})|u(S_{2k+1}),$$

and if $|S_{2k}|\ge 4$, $v(S_{2k})$ cannot have three consecutive components of the form $-1,+1,-1$.

In particular $-\bar{1}$ is a fixed point and its basin of attraction is the union of the set $\{\bar{1},-\bar{1}\}$ and the set of all bipolar vectors whose positive arches have two or more components.

We remark that once the preimages of a fixed point u is determined, then the basin of attraction of u is just the set of all its preimages.

In case 9 where $4\alpha-\beta<0$ and $2\alpha-\beta\ge 0$, we can construct the following table

$$\begin{pmatrix} u_{i-1}^{(t)} & * & +1 & \& & -1 \\ u_i^{(t)} & +1 & -1 & -1 & -1 \\ u_{i+1}^{(t)} & \$ & +1 & -1 & +1 \\ u_i^{(t+1)} & +1 & -1 & +1 & +1 \end{pmatrix},$$

where $*,\&$ and \$ can either be $+1$ or -1.

By means of the same arguments shown in case 4, we may show that a bipolar vector u is a fixed point if, and only if, for each $i\in\{1,...,n\}$, at least one of $u_i^{(0)}$ or $u_{i+1}^{(0)}$ is $+1$. Furthermore, every bipolar vector is attracted to a fixed point in at most one step. If $u^{(t)}\ne\bar{1}$ is a fixed point and there is a partition $\{S_1,S_2,...,S_{2m}\}$ of $\{1,...,n\}$ such that $S_1=\{1,...,i_1\}$, $S_2=\{i_1+1,...,i_2\}$, ..., $S_{2m}=\{i_{2m-1}+1,...,n\}$ and $u_j^{(0)}=+1$ for $j\in$

$S_1 \cup S_3 \cup ... \cup S_{2m-1}$ and $u_j^{(0)} = -1$ for $j \in S_2 \cup ... \cup S_{2m}$, then $|S_{2k}| = 1$ for any $k \in \{1, ..., m\}$ and

$$v(S_{2k}) = u(S_{2k}),\ k \in \{1, ..., m\}$$

and the first and the last components of each $v(S_{2k-1})$ are $+1$. For any preimage v of u, if $|S_{2k-1}| \le 3$ for some $k \in \{1, ..., m\}$, then

$$v(S_{2k-2})|v(S_{2k-1})|v(S_{2k}) = u(S_{2k-2})|u(S_{2k-1})|u(S_{2k}),$$

and if $|S_{2k-1}| \ge 4$, then $v(S_{2k-1})$ cannot have three consecutive components of the form $+1, -1, +1$. In particular $\bar{1}$ is a fixed point and its basin of attraction is the union of the set $\{\bar{1}, -\bar{1}\}$ and the set of bipolar vectors whose negative arches have two or more components.

7. Cases 1 and 6

In case 1 where $\beta > 0$ and $2\alpha - \beta \ge 0$, we can construct the following table

$$\begin{pmatrix} u_{i-1}^{(t)} & * & \$ & +1 & -1 \\ u_i^{(t)} & +1 & -1 & -1 & -1 \\ u_{i+1}^{(t)} & \# & +1 & -1 & -1 \\ u_i^{(t+1)} & -1 & +1 & +1 & -1 \end{pmatrix}, \tag{5}$$

where $*, \#$ and $\$$ can either be $+1$ or -1.

There are several immediate consequences:

(i) For each $i \in \{1, ..., n\}$, $u_i^{(t+1)} = u_i^{(t)}$ if, and only if, $u_{i-1}^{(t)} = -1$, $u_i^{(t)} = -1$ and $u_{i+1}^{(t)} = -1$. Thus if one of $u_{i-1}^{(t)}$, $u_i^{(t)}$ or $u_{i+1}^{(t)}$ is $+1$, then $u_i^{(t+1)} = -u_i^{(t)}$.

(ii) It is clear from the fifth column of (5) that $-\bar{1}$ is a fixed point. We assert that it is the only fixed point and that $\bar{1}^{(1)} = -\bar{1}$. Indeed, any vector $u^{(0)} \ne -\bar{1}$ must have a component $u_i^{(0)} = +1$. In view of the second column of (5), $u_i^{(1)} = -1$, which means that $u^{(0)}$ cannot be a fixed point. The fact that $\bar{1}^{(1)} = -\bar{1}$ follows from the second column of (5).

(iii) If $u^{(0)}$ is a vector such that for all $i \in \{1, ..., n\}$, neither $u_{i-1}^{(0)} = u_i^{(0)} = u_{i+1}^{(0)} = -1$ nor $u_{i-1}^{(0)} = u_i^{(0)} = u_{i+1}^{(0)} = +1$, then $\langle u^{(0)}, -u^{(0)} \rangle$ is a 2-cycle. To see this, we show that for each $i \in \{1, ..., n\}$, $u_i^{(1)} = -u_i^{(0)}$. Indeed, if $u_i^{(0)} = +1$, then by (i), $u_i^{(1)} = -1$; and if $u_i^{(0)} = -1$, then by assumption, either $u_{i-1}^{(0)} = +1$ or $u_{i+1}^{(0)} = +1$, so that

$u_i^{(1)} = +1$ by (i). Conversely, if $\langle w^{(0)}, w^{(1)} \rangle$ is a 2-cycle, we assert that for all $i \in \{1, ..., n\}$, neither $w_{i-1}^{(0)} = w_i^{(0)} = w_{i+1}^{(0)} = -1$ nor $w_{i-1}^{(0)} = w_i^{(0)} = w_{i+1}^{(0)} = +1$. Indeed, we first note, in view of (ii), that $\{w^{(0)}, w^{(1)}\} \cap \{\bar{1}, -\bar{1}\}$ must be empty. Now suppose to the contrary that $w_{j-1}^{(0)} = w_j^{(0)} = w_{j+1}^{(0)} = -1$. By applying a finite number of cyclic permutations and reflections if necessary, we may assume without loss of generality that $j = 2$ so that $w_1^{(0)} = w_2^{(0)} = w_3^{(0)} = -1$. Since $w^{(0)} \neq -\bar{1}$, we see that $w_1^{(0)} = w_2^{(0)} = w_3^{(0)} = ... = w_i^{(0)} = -1$ while $w_{i+1}^{(0)} = +1$ for some $i \in \{3, ..., n-1\}$. But in view of (i), $w_{i-1}^{(1)} = -1$ and $w_i^{(1)} = +1$ and $w_{i-1}^{(2)} = +1$. This is contrary to the assumption that $w_{i-1}^{(0)} = w_{i-1}^{(2)}$. Thus there is no $j \in \{1, ..., n\}$ such that $w_{j-1}^{(0)} = w_j^{(0)} = w_{j+1}^{(0)} = -1$. The other case can similarly be shown. We remark that as a consequence, if $\langle w^{(0)}, w^{(1)} \rangle$ is a 2-cycle, then $w^{(1)} = -w^{(0)}$.

(iv) Let $u^{(0)}$ be a bipolar vector such that for some $i \in \{1, ..., n\}$, $u_i^{(0)}$ and $u_{i+1}^{(0)}$ have different signs, then in view of (i),

$$\left(u_i^{(1)}, u_{i+1}^{(1)}\right)^\dagger = -\left(u_i^{(0)}, u_{i+1}^{(0)}\right)^\dagger,$$

$$\left(u_i^{(2)}, u_{i+1}^{(2)}\right)^\dagger = -\left(u_i^{(1)}, u_{i+1}^{(1)}\right)^\dagger = \left(u_i^{(0)}, u_{i+1}^{(0)}\right)^\dagger,$$

and in general,

$$\left(u_i^{(t)}, u_{i+1}^{(t)}\right)^\dagger = (-1)^t \left(u_i^{(0)}, u_{i+1}^{(0)}\right)^\dagger, \ t = 0, 1, ...$$

(v) The basin of attraction of the fixed point $-\bar{1}$ is the set $\{\bar{1}, -\bar{1}\}$. Indeed, as explained in (ii), $\bar{1}^{(1)} = -\bar{1}$. If $u^{(0)}$ is different from $\bar{1}$ or $-\bar{1}$, then by (iv), for any $t \geq 0$, $u^{(t)}$ must have a consecutive pair of components which are different. Thus $u^{(0)}$ cannot be attracted to $\bar{1}$ nor $-\bar{1}$.

(vi) Let $u^{(0)}$ be a bipolar vector such that $u_i^{(0)} = +1$, $u_{i+1}^{(0)} = u_{i+2}^{(0)} = ... = u_{i+k}^{(0)} = -1$ and $u_{i+k+1}^{(0)} = +1$ for some i and k satisfying $0 \leq i \leq i + k \leq n$. If $k = 1$, then by (iv),

$$\left(u_i^{(t)}, u_{i+1}^{(t)}, u_{i+2}^{(t)}\right)^\dagger = (-1)^t (+1, -1, +1)^\dagger = (-1)^{t+1}\delta(3).$$

If $k = 2$, then by (iv) again,

$$\left(u_i^{(t)}, u_{i+1}^{(t)}, u_{i+2}^{(t)}, u_{i+3}^{(t)}\right)^\dagger = (-1)^t (+1, -1, -1, +1)^\dagger$$
$$= (-1)^{t+1} (\delta(2)| - \delta(2)),\ t \geq 0.$$

If $k = 3$, then in view of (i),

$$\left(u_i^{(1)}, u_{i+1}^{(1)}, u_{i+2}^{(1)}, u_{i+3}^{(1)}, u_{i+4}^{(1)}\right)^\dagger = (-1, +1, -1, +1, -1)^\dagger = \delta(5),$$

so that by (iv),

$$\left(u_i^{(t)}, u_{i+1}^{(t)}, u_{i+2}^{(t)}, u_{i+3}^{(t)}, u_{i+4}^{(t)}\right)^\dagger = (-1)^{t+1}\delta(5),\ t = 1, 2, \ldots .$$

If $k = 4$, then in view of (i),

$$\left(u_i^{(1)}, u_{i+1}^{(1)}, u_{i+2}^{(1)}, u_{i+3}^{(1)}, u_{i+4}^{(1)}, u_{i+5}^{(1)}\right)^\dagger = (-1, +1, -1, -1, +1, -1)^\dagger,$$

$$\left(u_i^{(2)}, u_{i+1}^{(2)}, u_{i+2}^{(2)}, u_{i+3}^{(2)}, u_{i+4}^{(2)}, u_{i+5}^{(2)}\right)^\dagger = (+1, -1, +1, +1, -1, +1)^\dagger,$$

and

$$\left(u_i^{(t)}, u_{i+1}^{(t)}, u_{i+2}^{(t)}, u_{i+3}^{(t)}, u_{i+4}^{(t)}, u_{i+5}^{(t)}\right)^\dagger = (-1)^{t+1} (\delta(3)|\delta(3)),\ t \geq 1.$$

By induction, we may then see that when k is odd,

$$\left(u_i^{\left(\frac{k-1}{2}\right)}, u_{i+1}^{\left(\frac{k-1}{2}\right)}, \ldots, u_{i+k}^{\left(\frac{k-1}{2}\right)}, u_{i+k+1}^{\left(\frac{k-1}{2}\right)}\right)^\dagger = (-1)^{\frac{k+1}{2}}\delta(k+2)$$

and

$$\left(u_i^{\left(\frac{k-1}{2}+t\right)}, u_{i+1}^{\left(\frac{k-1}{2}+t\right)}, \ldots, u_{i+k}^{\left(\frac{k-1}{2}+t\right)}, u_{i+k+1}^{\left(\frac{k-1}{2}+t\right)}\right)^\dagger$$
$$= (-1)^{t+\frac{k+1}{2}}\delta(k+2),\ t \geq 0;$$

when k is even,

$$\left(u_i^{\left(\frac{k-2}{2}\right)}, u_{i+1}^{\left(\frac{k-2}{2}\right)}, \ldots, u_{i+k}^{\left(\frac{k-2}{2}\right)}, u_{i+k+1}^{\left(\frac{k-2}{2}\right)}\right)^\dagger$$
$$= \left((-1)^{\frac{k}{2}}\delta\left(\frac{k+2}{2}\right)|\delta\left(\frac{k+2}{2}\right)\right)$$

and

$$\left(u_i^{\left(\frac{k-2}{2}+t\right)}, u_{i+1}^{\left(\frac{k-2}{2}+t\right)}, ..., u_{i+k}^{\left(\frac{k-2}{2}+t\right)}, u_{i+k+1}^{\left(\frac{k-2}{2}+t\right)}\right)^\dagger$$

$$= (-1)^t \left((-1)^{\frac{k}{2}} \delta\left(\frac{k+2}{2}\right) | \delta\left(\frac{k+2}{2}\right)\right), \quad t \geq 0.$$

(vii) If $u^{(0)} \notin \{\bar{1}, -\bar{1}\}$, then by Lemma 1, we may assume without loss of generality that there is a partition $\{S_1, S_2, ..., S_{2m}\}$ of $\{1, ..., n\}$ such that $S_1 = \{1, ..., i_1\}$, $S_2 = \{i_1 + 1, ..., i_2\}$, ..., $S_{2m} = \{i_{2m-1} + 1, ..., n\}$ and $u_j^{(0)} = +1$ for $j \in S_1 \cup S_3 \cup ... \cup S_{2m-1}$ and $u_j^{(0)} = -1$ for $j \in S_2 \cup ... \cup S_{2m}$. We now consider the iterates of the vector

$$\left(u_n^{(0)}, u_1^{(0)}, ..., u_{i_1}^{(0)}, u_{i_1+1}^{(0)}\right)^\dagger = (-1, +1, ..., +1, -1)^\dagger.$$

In view of (i), we see that

$$\left(u_n^{(1)}, u_1^{(1)}, u_2^{(1)}, ..., u_{i_1-1}^{(1)}, u_{i_1}^{(1)}, u_{i_1+1}^{(1)}\right)^\dagger = (+1, -1, -1, ..., -1, -1, +1)^\dagger.$$

Thus, by (v), if i_1 is odd, then

$$\left(u_n^{(t)}, u_1^{(t)}, u_2^{(t)}, ..., u_{i_1-1}^{(t)}, u_{i_1}^{(t)}, u_{i_1+1}^{(t)}\right)^\dagger = \pm\delta(i_1 + 2) \qquad (6)$$

for $t \geq n$, and if i_1 is even, then

$$\left(u_n^{(t)}, u_1^{(t)}, u_2^{(t)}, ..., u_{i_1-1}^{(t)}, u_{i_1}^{(t)}, u_{i_1+1}^{(t)}\right)^\dagger$$

$$= \pm\left((-1)^{i_1/2}\delta\left(\frac{i_1+2}{2}\right) | \delta\left(\frac{i_1+2}{2}\right)\right) \qquad (7)$$

for $t \geq i_1/2$. In both cases, for each $i \in \{1, ..., i_1\}$, neither $u_{i-1}^{(t)} = u_i^{(t)} = u_{i+1}^{(t)} = -1$ nor $u_{i-1}^{(t)} = u_i^{(t)} = u_{i+1}^{(t)} = +1$. Next, consider the iterates of the vector

$$\left(u_{i_1}^{(0)}, u_{i_1+1}^{(0)}, ..., u_{i_2}^{(0)}, u_{i_2+1}^{(0)}\right)^\dagger = (+1, -1, ..., -1, +1)^\dagger.$$

In view of (v) again, when t is sufficiently large, for each $i \in \{i_1 + 1, ..., i_2\}$, neither $u_{i-1}^{(t)} = u_i^{(t)} = u_{i+1}^{(t)} = -1$ nor $u_{i-1}^{(t)} = u_i^{(t)} = u_{i+1}^{(t)} = +1$. These show that $\langle u^{(t)}, -u^{(t)} \rangle$ is a 2-cycle for all large t. In conclusion, we have shown that any $u^{(0)}$ not in $\{\bar{1}, -\bar{1}\}$ will be attracted to a 2-cycle.

(viii) Let $f(v) = u$, that is, v is a preimage of u. In view of (5), if $u_i = +1$, then it is necessary that $v_i = -1$.

(ix) A bipolar vector u is terminal if for some $i \in \{1, ..., n\}$, $u_{i-1} = u_i = u_{i+1} = +1$. Indeed, suppose to the contrary that $f(v) = u$. Then $v_{i-1} = v_i = v_{i+1} = -1$. By in view of the fifth column of (5), the fact that $v_{i-1} = v_i = v_{i+1} = -1$ would imply $u_i = -1$, which is contrary to our assumption that $u_i = +1$.

(x) Let w be a bipolar vector different from $\overline{1}$ and $-\overline{1}$ and $\{S_1, S_2, ..., S_{2m}\}$ is a partition of $\{1, ..., n\}$ such that $S_1 = \{1, ..., i_1\}$, $S_2 = \{i_1 + 1, ..., i_2\}, ..., S_{2m} = \{i_{2m-1} + 1, ..., n\}$ and $w_j = +1$ for $j \in S_1 \cup S_3 \cup ... \cup S_{2m-1}$ and $w_j = -1$ for $j \in S_2 \cup ... \cup S_{2m}$. Let $f(v) = w$, that is, v is a preimage of w. Suppose there is some $k \in \{1, ..., m\}$ such that $|S_{2k-1}| = 2$, then

$$v(S_{2k-1})|v(S_{2k}) = -\left(w(S_{2k-1})|w(S_{2k})\right).$$

First, suppose $|S_{2k}| = 1$, that is,

$$w(S_{2k-1})|w(S_{2k}) = (+1, +1, -1)^\dagger,$$

we assert that

$$v(S_{2k-1})|v(S_{2k}) = (-1, -1, +1)^\dagger = -\left(w(S_{2k-1})|w(S_{2k})\right).$$

Indeed, in view of (viii),

$$v(S_{2k-1})|v(S_{2k}) = (-1, -1, x)^\dagger.$$

If $x = -1$, then by the fifth column of (5), the second component of $w(S_{2k-1})|w(S_{2k})$ would be -1, which is contrary to our assumption. Thus $x = +1$. Similarly, suppose $|S_{2k}| = 2$, that is,

$$w(S_{2k-1})|w(S_{2k}) = (+1, +1, -1, -1)^\dagger.$$

then

$$v(S_{2k-1})|v(S_{2k}) = (-1, -1, +1, +1)^\dagger = -\left(w(S_{2k-1})|w(S_{2k})\right).$$

Indeed, as above, we may show

$$v(S_{2k-1})|v(S_{2k}) = (-1, -1, +1, y)^\dagger.$$

If $y = -1$, then by the third and the fourth column of (5), the fourth component of $w(S_{2k-1})|w(S_{2k})$ would be $+1$, which is again contrary to our assumption. The rest of our proof can now be completed by induction. By dual arguments, we may easily see that if for some $k \in \{0, 1, ..., m - 1\}$, $|S_{2k+1}| = 2$, then

$$v(S_{2k})|v(S_{2k+1}) = -\left(w(S_{2k})|w(S_{2k+1})\right).$$

If for some $k \in \{1, ..., m\}$, $|S_{2k-1}| \geq 1$ and $|S_{2k+1}| \geq 1$, then

$$v(S_{2k-1})|v(S_{2k})|v(S_{2k+1}) = -\left(w(S_{2k-1})|w(S_{2k})|w(S_{2k+1})\right)$$

or

$$v(S_{2k-1})|v(S_{2k})|v(S_{2k+1}) = -\left(w(S_{2k-1})| - w(S_{2k})|w(S_{2k+1})\right).$$

Indeed, assume without loss of generality that $|S_{2k-1}| = 1 = |S_{2k+1}|$ and that

$$w(S_{2k-1})|w(S_{2k})|w(S_{2k+1}) = (+1, -1, -1, ..., -1, -1, +1)^{\dagger}.$$

Then

$$v(S_{2k-1})|v(S_{2k})|v(S_{2k+1}) = (-1, *, *, ..., *, *, -1)^{\dagger},$$

where each $*$ can be $+1$ or -1. Assume the second component of $v(S_{2k-1})|v(S_{2k})|v(S_{2k+1})$ is -1, then we assert its third component is also -1. Otherwise, by (i) the second component of $w(S_{2k-1})|w(S_{2k})|w(S_{2k+1})$, as the first iterate of $v(S_{2k-1})|v(S_{2k})|v(S_{2k+1})$, must be $+1$, which is contrary to our assumption on w. By induction, it is then easily seen that the fourth component, etc. are all -1. On the other hand, if we assume that the second component of $v(S_{2k-1})|v(S_{2k})|v(S_{2k+1})$ is $+1$, then we assert its third component is also $+1$. Otherwise, by (i), the third component of $w(S_{2k-1})|w(S_{2k})|w(S_{2k+1})$, as the first iterate of $v(S_{2k-1})|v(S_{2k})|v(S_{2k+1})$, must be $+1$, which is again a contradiction. By induction, it is then easily seen that the fourth component, etc. are all $+1$. The proof of our assertion is complete.

Let us now consider some examples. Let $\langle w, -w \rangle$ be a 2-cycle and let $\{S_1, S_2, ..., S_{2m}\}$ is a partition of $\{1, ..., n\}$ such that $S_1 = \{1, ..., i_1\}$, $S_2 = \{i_1 + 1, ..., i_2\}, ..., S_{2m} = \{i_{2m-1} + 1, ..., n\}$ and $w_j = +1$ for $j \in S_1 \cup S_3 \cup ... \cup S_{2m-1}$ and $w_j = -1$ for $j \in S_2 \cup ... \cup S_{2m}$. Since $\langle w, -w \rangle$ is a 2-cycle, in view of (iii), $|S_{2k}| \leq 2$ and $|S_{2k-1}| \leq 2$ for all $k = 1, 2, 3, ..., m$. By means of (viii), (ix) and (x), we may calculate the basin of attraction of w. For instance, suppose $|S_{2k}| = |S_{2k-1}| = 2$ for all $k = 1, 2, 3, ..., m$. In view of (ix), $f^{-1}(w) = -w$ and $f^{-1}(-w) = w$. Its basin of attraction is therefore the set $\{w, -w\}$. As another example, let $n = 12$ and $w = (+1, -1, +1, +1, -1, -1, +1, -1, -1, +1, -1, +1)^{\dagger}$ which is depicted as follows:

$$w = \begin{array}{ccccccc} & & + & - & + & & \\ & & 1 & 2 & 3 & & \\ + & 12 & & & & 4 & + \\ - & 11 & & & & 5 & - \\ + & 10 & & & & 6 & - \\ & & 9 & 8 & 7 & & \\ & & - & - & + & & \end{array}.$$

Let

$$a = \begin{array}{ccccccc} & & - & + & - & & \\ & & 1 & 2 & 3 & & \\ - & 12 & & & & 4 & - \\ + & 11 & & & & 5 & + \\ - & 10 & & & & 6 & + \\ & & 9 & 8 & 7 & & \\ & & + & + & - & & \end{array}, \quad b = \begin{array}{ccccccc} & & - & + & - & & \\ & & 1 & 2 & 3 & & \\ - & 12 & & & & 4 & - \\ + & 11 & & & & 5 & + \\ - & 10 & & & & 6 & + \\ & & 9 & 8 & 7 & & \\ & & - & - & - & & \end{array}$$

$$x = \begin{array}{ccccccc} & & + & - & + & & \\ & & 1 & 2 & 3 & & \\ + & 12 & & & & 4 & + \\ - & 11 & & & & 5 & - \\ + & 10 & & & & 6 & - \\ & & 9 & 8 & 7 & & \\ & & + & + & + & & \end{array}, \quad y = \begin{array}{ccccccc} & & - & - & + & & \\ & & 1 & 2 & 3 & & \\ - & 12 & & & & 4 & + \\ - & 11 & & & & 5 & - \\ + & 10 & & & & 6 & - \\ & & 9 & 8 & 7 & & \\ & & + & + & + & & \end{array}$$

and

$$v = \begin{array}{ccccccc} & & - & - & + & & \\ & & 1 & 2 & 3 & & \\ - & 12 & & & & 4 & + \\ - & 11 & & & & 5 & - \\ + & 10 & & & & 6 & - \\ & & 9 & 8 & 7 & & \\ & & - & - & + & & \end{array}, \quad z = \begin{array}{ccccccc} & & + & + & - & & \\ & & 1 & 2 & 3 & & \\ + & 12 & & & & 4 & - \\ + & 11 & & & & 5 & + \\ - & 10 & & & & 6 & + \\ & & 9 & 8 & 7 & & \\ & & - & - & - & & \end{array}$$

and

$$
u = \begin{array}{ccccccc}
 & & + & + & - & & \\
 & & 1 & 2 & 3 & & \\
+ & 12 & & & & 4 & - \\
+ & 11 & & & & 5 & + \\
- & 10 & & & & 6 & + \\
 & & 9 & 8 & 7 & & \\
 & & + & + & - & &
\end{array}
$$

Then $w = -a$ and $\langle w, a \rangle$ is a cycle.

The preimages of w are a and b, the preimages of b are x and y, the preimages of a are w and v, and the preimages of v are z and u. For instance, since $w_{12} = w_1 = w_3 = w_4 = +1$, we see that a preimage of w must have the form

$$
\begin{array}{ccccccc}
 & & - & + & - & & \\
 & & 1 & 2 & 3 & & \\
- & 12 & & & & 4 & - \\
+ & 11 & & & & 5 & + \\
* & 10 & & & & 6 & + \\
 & & 9 & 8 & 7 & & \\
 & & * & * & * & &
\end{array}.
$$

Next, since $(w_7, w_8, w_9, w_{10}) = (+1, -1, -1, +1)$, we see that a preimage of w must either be of the form

$$
\begin{array}{ccccccc}
 & & * & * & * & & \\
 & & 1 & 2 & 3 & & \\
* & 12 & & & & 4 & * \\
* & 11 & & & & 5 & * \\
- & 10 & & & & 6 & * \\
 & & 9 & 8 & 7 & & \\
 & & + & + & - & &
\end{array}
$$

or of the form

$$
\begin{array}{ccccccc}
 & & * & * & * & & \\
 & & 1 & 2 & 3 & & \\
* & 12 & & & & 4 & * \\
* & 11 & & & & 5 & * \\
- & 10 & & & & 6 & * \\
 & & 9 & 8 & 7 & & \\
 & & - & - & - & &
\end{array}.
$$

Thus the only preimages that satisfy these requirements are a and b.

Note that the bipolar vectors x, y, z and u are terminal since they have three or more consecutive components which are equal to $+1$. Therefore, the basin of attraction of w is exactly $\{w, a, b, x, y, z, u\}$.

Theorem 5. *Suppose $\beta > 0$ and $2\alpha - \beta \geq 0$. Then*

(1) $-\bar{1}$ is the only fixed point and its basin of attraction is $\{\bar{1}, -\bar{1}\}$.

(2) $\{w^{(0)}, w^{(1)}\}$ is a 2-cycle if, and only if, for any $i \in \{1, ..., n\}$, neither $w_{i-1}^{(0)} = w_i^{(0)} = w_{i+1}^{(0)} = -1$ nor $w_{i-1}^{(0)} = w_i^{(0)} = w_{i+1}^{(0)} = +1$. Any 2-cycle $\{w^{(0)}, w^{(1)}\}$ must have property $w^{(1)} = -w^{(0)}$.

(3) Any bipolar vector u not in $\{\bar{1}, -\bar{1}\}$ is attracted to a 2-cycle (which can be explicitly given when the sign distribution of u is known).

(4) A bipolar vector u is terminal if for some $i \in \{1, ..., n\}$, $u_{i-1} = u_i = u_{i+1} = +1$.

(5) Let w be a bipolar vector and suppose there is a partition $\{S_1, S_2, ..., S_{2m}\}$ of $\{1, ..., n\}$ such that $S_1 = \{1, ..., i_1\}$, $S_2 = \{i_1 + 1, ..., i_2\}, ...,$ $S_{2m} = \{i_{2m-1} + 1, ..., n\}$ and $w_j = +1$ for $j \in S_1 \cup S_3 \cup ... \cup S_{2m-1}$ and $w_j = -1$ for $j \in S_2 \cup ... \cup S_{2m}$. Let v be a preimage of w. If there is some $k \in \{1, ..., m\}$ such that $|S_{2k-1}| = 2$, then

$$v(S_{2k-2})|v(S_{2k-1})|v(S_{2k}) = -\left(w(S_{2k-2})|w(S_{2k-1})|w(S_{2k})\right).$$

If for some $k \in \{1, ..., m\}$, $|S_{2k-1}| = 1 = |S_{2k+1}|$, then

$$v(S_{2k-1})|v(S_{2k})|v(S_{2k+1}) = -\left(w(S_{2k-1})|w(S_{2k})|w(S_{2k+1})\right)$$

or

$$v(S_{2k-1})|v(S_{2k})|v(S_{2k+1}) = -\left(w(S_{2k-1})| - w(S_{2k})|w(S_{2k+1})\right).$$

In case 6 where $\beta \leq 0$ and $2\alpha + \beta > 0$, we can construct the following table

$$\begin{pmatrix} u_{i-1}^{(t)} & +1 & \# & -1 & \& \\ u_i^{(t)} & +1 & +1 & +1 & -1 \\ u_{i+1}^{(t)} & +1 & -1 & +1 & * \\ u_i^{(t+1)} & +1 & -1 & -1 & +1 \end{pmatrix},$$

where $*$, $\#$ and $\&$ can either be $+1$ or -1.

By means of the same argument shown in case 1, we may show that $\bar{1}$ is the only fixed point and its basin of attraction is $\{\bar{1}, -\bar{1}\}$. Furthermore, $\langle w^{(0)}, w^{(1)}\rangle$ is a 2-cycle if, and only if, $w^{(1)} = -w^{(0)}$ and for any $i \in \{1, ..., n\}$, neither $w_{i-1}^{(k)} = w_i^{(k)} = w_{i+1}^{(k)} = +1$ nor $w_{i-1}^{(k)} = w_i^{(k)} = w_{i+1}^{(k)} = -1$. Any vector not in $\{\bar{1}, -\bar{1}\}$ is attracted to a 2-cycle. The preimages of a bipolar vector can always be calculated, so is the basin of attraction of a 2-cycle.

8. Remarks

The long time behaviors of the solution space of our artificial neural network (1) depend on both the location of the parameter pair (α, β) in the plane and the initial neuron states. By means of the straight lines

$$\beta = 0; \ \beta = 2\alpha; \ \beta = 4\alpha \text{ and } \beta = -4\alpha,$$

the plane can be divided into ten different regions so that when (α, β) is 'switched' into one of them, the behaviors of the corresponding solution space (in terms of the cycles and their basins of attractions) will also change accordingly.

We may further elaborate on the previous remark in terms of Boolean cellular automata. The so called cellular automata is a concept formulated by scientists to simulate complex systems. In a cellular automaton, there are identical elements, usually located in a regular array of cells. The update rule for each cell depends on that cell and some of its nearest neighbors. In a Boolean cellular automaton, the state of each cell is either 0 or 1, and is updated according to certain Boolean rules. Therefore our neural network fits in this setting. Since there are three inputs, the network takes 2^3 possible input states, and can have 2^6 Boolean functions which determine the output of a cell. For example, the following Table is one of the possible Boolean functions

Input	000	001	010	011	100	101	110	111
Output	0	0	0	0	0	0	0	0

If we identify each 0 with -1 and each 1 with $+1$ in the above table, we see that this is just the same table in Case 3.

Since the 10 different cases in our previous analysis are generated by choosing different combination of α and β, we now have a neural network that may simulate 10 different Boolean cellular automata simply by manipulation of two 'analog' parameters. Furthermore, our results in the previous sections can now be applied for understanding this 'switchable' automata without too much difficulty.

Acknowledgment

The second author is supported by the China Postdoctoral Science Foundation (20070421106) and the Key Project of Chinese Ministry of Education (208154).

References

1. M. T. Hagan, H. B. Demuth and M. H. Beale, Neural Network Design, PWS Publishing Co., 1996.
2. F. Robert, Discrete Iterations: A Metric Study, Springer-Verlag, 1986.

Hyperbolic and Minimal Sets

A. A. Pinto

Escola de Ciências, Universidade do Minho
Braga, Portugal
** E-mail: aapinto@math.uminho.pt*

The goal of this work is to characterize the fine structures of fractals that are in the frontier between order and chaos or whose dynamics are chaotic.

Keywords: Gibbs measures, Hausdorff measures, hyperbolic dynamics, unimodal maps, renormalization.

1. Introduction

The goal of this work is to characterize the fine structures of fractals and its dynamics determined by hyperbolic and chaotic dynamical systems and by dynamical systems that are in the frontier between order and chaos. The fine structures of fractals are determined by the geometric properties of the fractals when studied at infinitesimal scales. This leads us to say that two fractals have the same fine structures and similar dynamics on them if the corresponding dynamical systems are differentiable conjugate. We will use the renormalization operator to understand the fine structures of the dynamical systems in the frontier between order and chaos (see E. de Faria, W. de Melo and A. Pinto (ref. 21)). In hyperbolic dynamics, these fine structures are characterized by affine structures (see A. Pinto and D. Rand (ref. 70), and A. Pinto and D. Sullivan (ref. 75)).

1.1. *Hyperbolic dynamics*

In Section 2, we construct Teichmüller spaces which characterize the smooth conjugacy classes of unidimensional hyperbolic dynamics (see D. Sullivan (ref. 85), A. Pinto and D. Sullivan (ref. 75) and A. Pinto and D. Rand (ref. 64)). These Teichmüller spaces have infinite dimension which implies that there is a great diversity of fine structures. Hence, we say that these dynamical systems are flexible. These dynamical systems also show the following feature that we call explosion of smoothness: if there is a topological conjugacy between two of these dynamical systems with a non-zero derivative at a point then these systems are smooth conjugate (see F. Ferreira and A. Pinto (ref. 26)). Similar results are proved for uniformly asymptotically affine (uaa) unidimensional hyperbolic dynamics whose (uaa) conjugacy classes form the completion of the smooth conjugacy classes (see D. Sullivan (ref. 87) and F. Ferreira (ref. 23)). These (uaa) systems are not necessarily smooth, however their eigenvalues are well-defined (see F. Ferreira and A. Pinto (ref. 25)).

In Section 3, like as in the unidimensional case, we construct Teichmüller spaces which characterize the smooth conjugacy classes of surface diffeomorphisms with hyperbolic basic sets. These Teichmüller spaces have infinite dimension and so we say that these dynamical systems are flexible (see E. Cawley (ref. 12) and A. Pinto and D. Rand (ref. 67)). Since, the holonomies of these dynamical systems are smooth (see A. Pinto and D. Rand (ref. 68)), we get that all the stable leaves have the same fine structures and, similarly, all the unstable leaves also have the same fine structures.

In contrast with the above results showing the flexibility of surface diffeomorphisms with hyperbolic basic sets, we also show the existence of rigid features for these dynamical systems: if the degree of the smoothness of the holonomies is greater than 1 plus the maximal Hausdorff dimension of the hyperbolic basic set along the stable and unstable leaves, then we obtain that the dynamical system is smooth conjugate to a rigid (or affine) model (see E. Ghys (ref. 28) and A. Pinto and D. Rand (ref. 69)). For Anosov diffeomorphisms the rigid models are the Anosov automorphisms in the torus. For Plykin attractors there is no rigid models which implies that the smoothness of the holonomies is not greater than 1 plus the Hausdorff dimension of the basic set along the stable leaves (see F. Ferreira, A. Pinto and D. Rand (ref. 72)).

The construction of the Teichmüller spaces for smooth conjugacy classes of surface diffeomorphisms with hyperbolic basic sets shows that the smooth structures of the stable leaves do not impose any restrictions on the smooth structures along the unstable leaves, and vice-versa. However, using Gibbs Theory and the ratio decomposition of Gibbs measures introduced by A. Pinto and D. Rand in (ref. 66), if we demand that the basic sets have an invariant geometric measure (for instance a measure equivalent to the Hausdorff measure), then we obtain explicit relations between the smooth structures of the stable and unstable leaves (see also E. Cawley (ref. 12) and Ya. Sinai (ref. 81)).

In the future, we hope that these technics can be used with success for studying laminations in surfaces, non-uniformly hyperbolic dynamics and flows in 3-dimensional manifolds.

1.2. *The frontier between order and chaos*

In Section 4, we will study the frontier between order and chaos for unimodal families of quadratic type. We start describing the route from order to chaos and the discovering of the universal constants $\alpha = 0.399\ldots$ and $\delta = 4.669\ldots$ (by M. Feigenbaum in (ref. 22) and by P. Coullet and C. Tresser in (ref. 13)). These universal constants appear in several scientific experiments (see, for instance, (ref. 14), (ref. 32) and (ref. 78)). To explain the existence of these universal constants, M. Feigenbaum, and, independently, P. Coullet and C. Tresser introduced the period doubling operator inspired in the works of K. Wilson (see (ref. 89)). They conjectured the existence of a unique hyperbolic fixed point for the period doubling operator with the property that α and δ are the two eigenvalues with greater modulus at the fixed point. Furthermore, they conjectured that the unimodal maps in the stable manifold of the period doubling operator form the frontier between order and chaos. O. Lanford in (ref. 37) proved the existence of the hyperbolic fixed point in a space of real analytic maps, using interval arithmetics and the support of a computer for doing the numerical computations (see also J.-P. Eckmann and P. Wittwer (ref. 17)). D. Sullivan in (ref. 86) proved that the stable manifold of the period doubling operator consists of all infinitely renormalizable unimodal maps of quadratic type. The existence of the universal constant α is associated to the fine structure of the infinitely renormalizable unimodal maps. D. Rand in (ref. 76) proved that the unimodal maps contained in the stable manifold of the renormalization are conjugate in the closure of their critical orbits by a map with a smooth extension to the real line which implies that they are rigid. A. Pinto and D. Rand in (ref. 63) proved that the degree of smoothness of the extension to the reals of the conjugacy is $2.11\ldots$. The universal constant $2.11\ldots$ is determined by a formula which involves the second and third eigenvalues with greater modulus of the period doubling operator. R. Mackay, A. Pinto and J. Zeijts in (ref. 47) generalized this last result to bimodal maps.

In Section 5, we study with the help of the renormalization operator the geometric properties of the unimodal maps with quadratic type. The conjectures of M. Feigenbaum, P. Coullet and C. Tresser for the period doubling operator were extended to the renormalization operator as we pass to describe: The infinitely renormalizable unimodal maps converge under renormalization to a limit set A with the following properties: (i) the renormalization operator restricted to the limit set A is topologically conjugate to a shift; (ii) the limit set A is hyperbolic with unstable manifolds with dimension 1 and stable manifolds with codimension 1. Restricting the renormalization operator to the set of all C^2 infinitely renormalizable unimodal maps with bounded geometric type, D. Sullivan in (ref. 87) and in (ref. 86) proved the existence of a limit set $A_L \subset A$ for all maps with the following properties: (i) the renormalization operator restricted to A_L is topologically conjugate to a shift of finite type; (ii) the maps in A_L are real analytic unimodal maps with quadratic-like holomorphic extensions; (iii) the stable set of a unimodal map $g \in A_L$ consists of all unimodal maps topological conjugate to g on the closure of their critical orbits. For infinitely renormalizable maps with bounded geometry and quadratic-like holomorphic extensions, C. McMullen in (ref. 52) proved the exponential convergence of the renormalization operator along the stable sets. As a corollary of this result, C. McMullen proved that the topological conjugacy between two unimodal maps in the same stable set and with quadratic-like holomorphic extensions has a smooth extension to the real line. W. de Melo and A. Pinto in (ref. 55) generalized this result for C^2 infinitely renormalizable unimodal maps with bounded geometry and of quadratic type. M. Lyubich in (ref. 44) proved the hyperbolicity of the set A_L in a space of equivalence classes of germs of maps quotient by affine diffeomorphisms. Inspired by the work of A. Davie in (ref. 15), E. de Faria, W. de Melo and A. Pinto in (ref. 21) generalized the result of M. Lyubich showing that A_L has the properties of a hyperbolic set for the renormalization operator in the space U^3 of C^3 unimodal maps with quadratic type. The first difficulty of this generalization is the fact that the renormalization operator is not Frechet differentiable in U^3. However, they proved that the unstable sets are real analytic submanifolds with dimension 1, and that the stable sets are C^1 submanifolds with codimension 1. Furthermore, they proved the following rigidity result in the parameter space: the holonomy between any two transversals to the stable lamination is $C^{1+\alpha}$, with $\alpha > 0$.

The renormalization operator naturally appears in several other families of maps, such as, families of critical circle maps and families of annulus maps (see, for instance, E. de Faria and W. de Melo (ref. 19) and (ref. 20), R. Mackay (ref. 46), M. Martens (ref. 51), W. de Melo (ref. 54), S. Ostlund, D. Rand, J. Sethna and E. Siggia (ref. 61) and M. Yampolsky (ref. 90)). In the future, we hope that similar results can be proven for these families.

2. Hyperbolic unidimensional dynamics

Hyperbolic dynamics is one of the main sources of examples of dynamical systems with chaotic properties. In this section, we are going to concentrate in the study of hyperbolic endomorphisms in 1-dimensional manifolds and, in next section, we will study hyperbolic diffeomorphisms on 2-dimensional manifolds.

2.1. *Examples: cookie-cutters and tent maps*

We are going to study two representative classes of hyperbolic dynamical systems consisting of cookie-cutters and tent maps (which we will define later). The results that we will present for these maps are also satisfied in the larger context of Markov maps on train-tracks. The reason that we have chosen to study the cookie-cutters as examples of Markov maps is that the invariant set of these maps are cantori. This fact demands a more careful study of the meaning of smoothness for the conjugacies between these maps, since the concept of smoothness in Cantor sets is not common. Here, we will use the same definition as the one presented for instance in (ref. 85). The reason that we also have chosen to study the tent maps is that they have an extra difficulty coming from the fact of not being differentiable or even continuous at a point. The study of cookie-cutters and tent maps will also be useful in Sections 4 and 5 to understand the description of the route from order to chaos.

Let us define cookie-cutters and tent maps. Let $F : I_0 \cup I_1 \to I$ be a $C^{1+\alpha}$ map, where $I_0 = [0, a]$, $I_1 = [b, 1]$ and $I = [0, 1]$, with $a \leq b$ and $\alpha > 0$, satisfying the following *hyperbolicity condition*: There exist constants $\lambda < 1$ and $C > 0$ with the property that if $x \in I_0 \cup I_1$, and $n \geq 1$ is such that $F(x), \ldots, F^{n-1}(x) \in I_0 \cup I_1$, then $|(F^n)'(x)| > C\lambda^n$. If $a < b$, we say that F is a *cookie-cutter*. If $a = b$, we say that F is a *tent map* (see Figure 1).

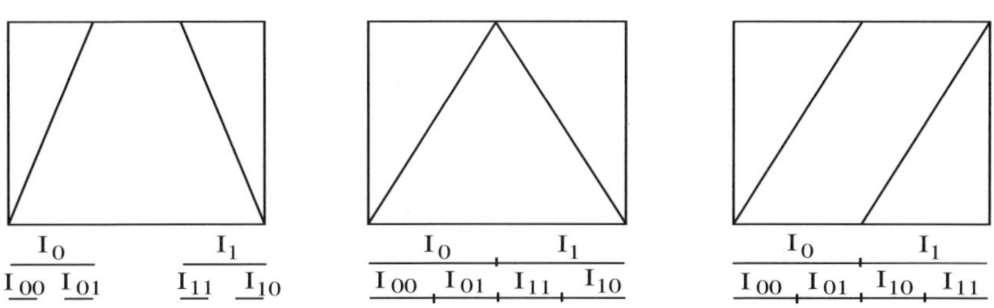

Fig. 1. Cookie-cutters and tent maps.

2.2. *Chaotic dynamics*

We start by giving a set of properties that we will use to characterize the concept of chaotic dynamics. We would like to note that there is no universal definition of chaotic dynamics. The notion of chaotic dynamics that we will use is simple to state and shows the complexity of the dynamics of these systems (see also R. Devaney (ref. 16) and C. Robinson (ref. 77)). Then, we introduce the shift map in a symbolic space that we will use to prove that the cookie-cutters and the tent maps are chaotic.

We say that a map $G : C \to C$ is *chaotic*, with respect to a metric d in C, if it satisfies the following properties:

(i) The closure of the periodic orbits of G is C;
(ii) There exists a point in C whose orbit is dense in C;
(iii) There exists $\delta > 0$ such that for every $\beta > 0$ and every $x \in C$, there exist $y \in C$ and $n > 0$, such that $d(x, y) < \beta$ and $d(G^n(x), G^n(y)) > \delta$.

Let $\sigma : \{0, 1\}^{\mathbb{N}} \to \{0, 1\}^{\mathbb{N}}$ be the *shift* given by

$$\sigma(\varepsilon_1 \varepsilon_2 \ldots) = \alpha_1 \alpha_2 \ldots ,$$

where $\alpha_i = \varepsilon_{i+1}$, for all $i \in \mathbb{N}$. Let $d : \{0, 1\}^{\mathbb{N}} \times \{0, 1\}^{\mathbb{N}} \to \mathbb{R}$ be the metric given by $d(\varepsilon_1 \varepsilon_2 \ldots , \alpha_1 \alpha_2 \ldots) = \sum_{n=1}^{\infty} 2^{-n} |\varepsilon_n - \alpha_n|$. The shift map is chaotic (see a proof for instance in (ref. 16)).

Let $F : I \to I$ be a cookie-cutter or a tent map. We will call *n-cylinders of I* the intervals of the form

$$I_{\varepsilon_1 \ldots \varepsilon_n} = \{x \in I_{\varepsilon_1} : F^j x \in I_{\varepsilon_{j+1}}, 1 \le j < n\}.$$

The *invariant set* of F is given by $C_F = \cap_{n=1}^{\infty} \cup_{\varepsilon \in \{0,1\}^n} I_{\varepsilon}$. Hence, if F is a coockie-cutter the invariant set C_F is a Cantor set, and if F is a tent map the invariant set C_F is the interval I. The map $h : \{0, 1\}^{\mathbb{N}} \to C_F$ which associates to a sequence $\varepsilon_1 \varepsilon_2 \ldots$ the unique point in $\cap_{n \ge 1} I_{\varepsilon_1 \ldots \varepsilon_n}$ is well-defined (see (ref. 16)). Furthermore, the map h is Hölder-continuous, onto and $F \circ h(\varepsilon) = h \circ \sigma(\varepsilon)$, for every $\varepsilon \in \{0, 1\}^{\mathbb{N}}$. Thus, we note that the map h transfers the chaotic properties of the shift map to the map F restricted to the invariant set C_F.

2.3. *Teichmüller spaces*

We are going to characterize the topological conjugacy classes of cookie-cutters and tent maps. Then, we study the fine structures of the invariant sets of these maps which lead us to the smooth classification of the smooth conjugacy classes of these maps. This classification was essentially developed in the works of M. Feigenbaum, D. Sullivan, A. Pinto and D. Rand. Finally, we present a result of F. Ferreira and A. Pinto on explosion of smoothness for conjugacies between unidimensional hyperbolic dynamics.

We say that two maps F and G are *topologically conjugate*, if there is a homeomorphism $h : C_F \to C_G$ such that $h \circ F(x) = G \circ h(x)$, for all $x \in C_F$, and h extends as a homeomorphism to the interval. If h has a $C^{1+\alpha}$ extension to the interval I, with $0 < \alpha \leq 1$, we say that F and G are $C^{1+\alpha}$ *conjugate.* We remind the reader that a diffeomorphism $h : I \to J$ is $C^{1+\alpha}$, if h is smooth and if there is a constant $C > 0$ such that $|h'(x) - h'(y)| \leq C|x - y|^\alpha$, for all $x, y \in I$. Since the cookie-cutters are topologically conjugate to the shift we obtain that two cookie-cutters F e G (or two tent maps) are topologically conjugate, if and only if, for each $j \in \{0,1\}$, $F'(x)$ and $G'(y)$ have the same sign for $x \in I_j^F$ and $y \in I_j^G$.

Now, we are going to introduce the solenoid functions which will allow us to determine when F and G are also $C^{1+\alpha}$ conjugate. A cookie-cutter F determines a *solenoid function* $s_F : \{0,1\}^{\mathbb{N}} \to \mathbb{R}^+$ defined by

$$s_F(\varepsilon_1\varepsilon_2\ldots) = \lim_{n\to\infty} \frac{|J_{\varepsilon_n\ldots\varepsilon_2}|}{|I_{\varepsilon_n\ldots\varepsilon_1}|},$$

where the interval $J_{\varepsilon_n\ldots\varepsilon_2}$ is such that (i) $I_{\varepsilon_n\ldots\varepsilon_2 0} \cup J_{\varepsilon_n\ldots\varepsilon_2} \cup I_{\varepsilon_n\ldots\varepsilon_2 1} = I_{\varepsilon_n\ldots\varepsilon_2}$ and (ii) the interior of $J_{\varepsilon_n\ldots\varepsilon_2}$ is disjoint of the sets $I_{\varepsilon_n\ldots\varepsilon_2 0}$ and $I_{\varepsilon_n\ldots\varepsilon_2 1}$.

Theorem 2.1. *The Hölder functions $s : \{0,1\}^{\mathbb{N}} \to \mathbb{R}^+$ form a Teichmüller space for the smooth conjugacy classes of cookie-cutters, i.e.*

(i) the maps F and G are smooth conjugate if, and only if, $s_F = s_G$;
(ii) given a Hölder function $s : \{0,1\}^{\mathbb{N}} \to \mathbb{R}^+$, there exists a cookie-cutter F such that $s_F = s$.

Let us associate to each $C^{1+\alpha}$ tent map F the corresponding solenoid function. For simplicity of the exposition we will consider just the case where $F'(x) > 0$, for every $x \in I_0 \cup I_1$. We define the *solenoid function* determined by F through the following equality

$$s_F(\varepsilon_1\varepsilon_2\ldots) = \lim_{n\to\infty} \frac{|I_{\alpha_n\ldots\alpha_1}|}{|I_{\varepsilon_n\ldots\varepsilon_1}|},$$

where $I_{\alpha_n\ldots\alpha_1}$ is the cylinder on the right of $I_{\varepsilon_n\ldots\varepsilon_1}$ (we say that a cylinder J is on the right of a cylinder K if K and J have a common endpoint and for every $x \in K$ and $y \in J$ we have that $x \leq y$.). Hence, the solenoid function $s : \{0,1\}^{\mathbb{N}} \to \mathbb{R}^+$ satisfies the following matching condition (see Figure 2):

$$\begin{aligned} s_F(\varepsilon_1\varepsilon_2\ldots\varepsilon_n\ldots) &= \lim_{n\to\infty} \frac{|I_{\alpha_n\ldots\alpha_1 0}| + |I_{\alpha_n\ldots\alpha_1 1}|}{|I_{\varepsilon_n\ldots\varepsilon_1 0}| + |I_{\varepsilon_n\ldots\varepsilon_1 1}|} \\ &= \frac{s_F(0\varepsilon_1\ldots)s_F(1\varepsilon_1\ldots)(1+s_F(0\alpha_1\ldots))}{1+s_F(0\varepsilon_1\ldots)} . \end{aligned} \tag{1}$$

Let Σ_0 be the subset of $\{0,1\}^{\mathbb{N}}$ whose elements have the form $0\varepsilon_1\varepsilon_2\ldots$.

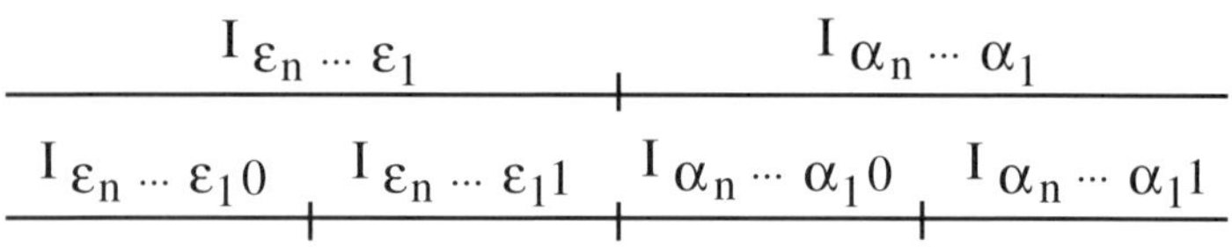

Fig. 2. Matching condition.

Let Σ_n be the subset of $\{0,1\}^{\mathbb{N}}$ whose elements have the form $1(n)0\varepsilon_1\varepsilon_2\ldots$ for every $n \geq 1$, where $1(n)$ is a word with n symbols 1. We note that the subsets of Σ_n form a partition of $\{0,1\}^{\mathbb{N}}$.

Theorem 2.2. *Let $s : \{0,1\}^{\mathbb{N}} \to \mathbb{R}^+$ be a function satisfying the following properties:*

(i) The function $s|\Sigma_n$ is Hölder continuous for every $n \geq 0$;
(ii) The function s satisfies the matching condition introduced in (1).

The set of all these functions form a Teichmüller space for the conjugacy classes of tent maps.

Theorems 2.1 and 2.2, were proved for a more general class of maps consisting of Markov maps on train tracks in A. Pinto and D. Rand (ref. 64). These results also hold in a weaker regularity called uniformly asymptotically affine (uaa). An important feature of the (uaa) conjugacy classes is that they form the completion of the smooth conjugacy classes (see F. Ferreira (ref. 23) and D. Sullivan (ref. 87)). We note that the (uaa) maps are not smooth, however the eigenvalues are well-defined (see F. Ferreira and A. Pinto (ref. 25)).

In the context of $C^{1+\alpha}$ cookie-cutters, M. Feigenbaum and D. Sullivan introduced the notion of scaling functions that they used to classify the smooth conjugacy classes of cookie-cutters. D. Sullivan asked how to generalize this result to expanding circle maps. To answer this question D. Rand and A. Pinto create the solenoid functions which are a simple modification of the scaling function. Together with D. Sullivan they associate to the solenoid functions affine structures on the leaves of the inverse limit of the expanding circle maps which allowed a better understanding of these Teichmüller spaces (see A. Pinto and D. Sullivan in (ref. 75), and A. Pinto and D. Rand in (ref. 64)).

2.4. *Explosion of smoothness*

The results in the previous subsection show the great diversity of fine structures for Markov maps on train-tracks. However, it is interesting to note the following result which we call explosion of smoothness.

Theorem 2.3. *If there exists a topological conjugacy between two Markov maps on train-tracks that has a non-zero derivative at a point, then the Markov maps are smooth conjugate.*

This theorem was first proven for expanding circle maps by D. Sullivan in (ref. 87). E. de Faria in (ref. 18) generalized this result supposing that the conjugacy is just (uaa) at a point. F. Ferreira and A. Pinto, in (ref. 26), generalized these two results to Markov maps on train-tracks. One of the difficulties was to deal with the case where the invariant sets are cantori. This generalization also applies to (uaa) Markov maps showing that the conjugacy is (uaa). J. F. Alves, V. Pinheiro and A. Pinto, in (ref. 2), extended the above theorem for non-uniformly expanding maps.

3. Hyperbolic diffeomorphisms on surfaces

Before proceding to explain the route from order to chaos, we will present the generalizations of the previous results on unidimensional hyperbolic dynamics to diffeomorphisms on surfaces with hyperbolic basic sets.

Some classical examples of diffeomorphisms on surfaces are Anosov diffeomorphisms, Plykin attractors and Smale horseshoes. In the case of Anosov diffeomorphisms the basic set coincides with the manifold. In the case of Plykin attractors the basic set is locally the product of a Cantor set, contained in an interval, cartesian product with an interval. In the case of Smale horseshoes the basic set is locally a Cantor set, contained in an interval, cartesian product with a Cantor set, also contained in an interval.

3.1. *Smoothness of the holonomies*

We start by introducing some basic definitions. A *hyperbolic basic set* Λ of a $C^{1+\gamma}$ diffeomorphism $f : M \to M$, with $\gamma > 0$, is a compact set, invariant under f, with a dense orbit and a product structure (see definitions 4.1 and 8.10 in M. Shub (ref. 80)). For every x contained in a basic set Λ, and every small $\varepsilon > 0$, the *local stable set of* x is given by

$$W^s(x, \varepsilon) = \{y \in M : d(f^n(x), f^n(y)) \leq \varepsilon, \forall n \geq 0\} ,$$

where $d : M \times M \to \mathbb{R}$ is the distance induced by some riemannian metric ρ on M. By the Stable Manifold Theorem, we get that the *global stable set of* x

$$W^s(x) = \cup_{n \geq 0} f^{-n}(W^s(f^n(x), \varepsilon))$$

is the image of a $C^{1+\gamma}$ immersion $i_x : \mathbb{R} \to M$. We say that a set $I \subset W^s(x)$ is a *stable leaf* if I is the image by i_x of an open interval. A *stable segment* K is the intersection of a stable leaf with the basic set Λ. Similarly, we have the notions of *local unstable leaf, global unstable leaf and unstable segment.*

Since an hyperbolic set Λ has a product structure, there are $\varepsilon > 0$ and $\delta > 0$ sufficiently small, such that, for every $x, y \in \Lambda$ with $d(x, y) < \delta$, the set $W^u(x, \varepsilon) \cap W^s(y, \varepsilon)$ is a single point belonging to Λ. We denote this point by $[x, y]$. We call a subset R of Λ a *rectangle* if R is proper and for every $x, y \in R$ the point $[x, y]$ belongs to R.

Given a point $x \in R$, there exists an unique stable segment $\ell^s(x, R)$ and an unique unstable segment $\ell^u(x, R)$ such that $[\ell^s(x, R), \ell^u(x, R)] = R$. Given a rectangle R and two points $x, y \in R$, the *holonomy* $h : \ell^s(x, R) \to \ell^s(y, R)$ is well-defined by $h(z) = [z, y]$ (see Figure 3). D. Rand and A.

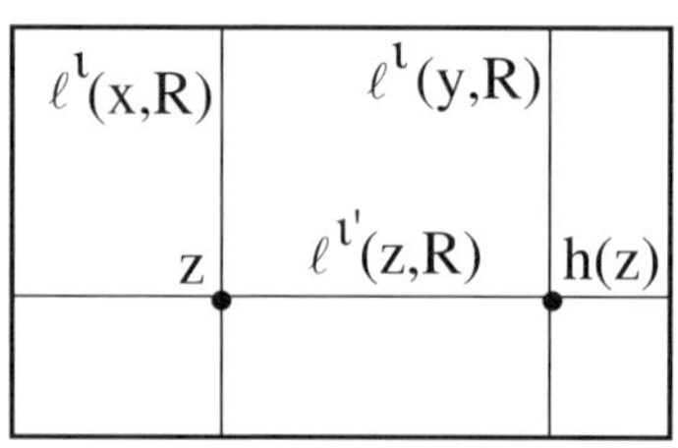

Fig. 3. Holonomy.

Pinto in (ref. 68) have proved the following result on the regularity of the degree of smoothness of the holonomies:

Theorem 3.1. *The holonomy* $h : \ell^s(x,R) \to \ell^s(y,R)$ *has a* $C^{1+\alpha}$ *extension to the stable leaves containing* $\ell^s(x,R)$ *and* $\ell^s(y,R)$, *for some* $\alpha > 0$.

However, there exists an upper limit for the degree of regularity of the smoothness of the holonomies in the case where Λ is a Plykin attractor. We recall that Λ is an *attractor* if there exists an open set $U \subset M$ containing Λ such that $\Lambda = \cap_{n\geq 0} f^n(U)$. We say that f is a *Plykin attractor* if the basic set Λ is an attractor and Λ is locally an interval cartesian product with a Cantor set, contained in an interval. By Theorem 3.1, we obtain that all stable segments have the same Hausdorff dimension HD^s, and that all unstable segments also have the same Hausdorff dimension HD^u. F. Ferreira, A. Pinto and D. Rand in (ref. 72) have proved the following result:

Theorem 3.2. *Let* f *be a Plykin attractor. The stable holonomies are not* $C^{1+\beta}$, *for* $\beta > HD^s$.

J. Harrison in (ref. 29) has conjectured that the degree of smoothness of Denjoy maps is bounded by 1 plus the Hausdorff dimension of the non-wandering set. A related result was proven by A. Norton in (ref. 60) using "box dimension" instead of Hausdorff dimension. The above result on the degree of smoothness of the holonomies of the Plykin attractors lead us to prove that there are no fixed points of renormalization for Denjoy maps such that the degree of smoothness is greater than the sum of 1 with the Hausdorff dimension of the non-wandering set (see F. Ferreira, A. Pinto and D. Rand (ref. 73)).

3.2. *Teichmüller spaces*

Surface diffeomorphisms are chaotic on their hyperbolic basic sets, since they are semi-conjugated to the shift of finite type. The proof of this result uses the existence of Markov rectangles for hyperbolic basic sets (see theorems 10.28, 10.33 and 10.34 in M. Shub (ref. 79), and see also R. Bowen (ref. 10) and Ya. Sinai(ref. 81)). Furthermore, these dynamical systems are structural stable (see theorems 8.3 and 8.22 in M. Shub (ref. 79), and see, also, Anosov (ref. 3), Bowen (ref. 11), M. Hirsch, J. Palis, C. Pugh and M. Shub (ref. 30), M. Hirsch and C. Pugh (ref. 31), M. Peixoto (ref. 62) and S. Smale (ref. 83)). Like as in the unidimensional hyperbolic case, we can ask if there is a Teichmüller space for these diffeomorphisms. Here, we give a positive answer to this question by constructing stable and unstable ratio functions and, also, stable and unstable solenoid functions.

The diffeomorphisms f and g with hyperbolic basic sets Λ_f and Λ_g are *topologically conjugate*, if there is a homeomorphism $h : \Lambda_f \to \Lambda_g$ such that $h \circ f = g \circ h$ and h preserves the order along the stable leaves and unstable leaves. If h has a $C^{1+\alpha}$ diffeomorphic extension to an open set of M containing Λ_f, then we say that f and g are $C^{1+\alpha}$ *diffeomorphic conjugate*. We denote by $\mathcal{T}(f, \Lambda)$ the set of all C^{1+} hyperbolic diffeomorphisms (g, Λ_g) such that (g, Λ_g) and (f, Λ) are topologically conjugated by a homeomorphism $h_{f,g}$.

Let f be a diffeomorphism with a hyperbolic basic set Λ_f. If K is a stable or an unstable segment, we define $|K|_\rho$ to be the length of the smallest leaf containing K. Let T^s be the set of all pairs (I, J), where I and J are stable segments contained in a same leaf. We define the set T^u with respect to the unstable leaves, similarly to the set T^s. Let $g \in \mathcal{T}(f, \Lambda)$ be topologically conjugated to f by $h = h_{f,g}$. By (ref. 67), the *stable ratio function* $r_g^s : T^s \to \mathbb{R}^+$ is given by

$$r_g^s(I : J) = \lim_{n\to\infty} \frac{|f^n(h(I))|_\rho}{|f^n(h(J))|_\rho} ,$$

and the *unstable ratio function* $r_g^u : T^u \to \mathbb{R}^+$ is given by

$$r_g^u(I : J) = \lim_{n\to\infty} \frac{|f^{-n}(h(I))|_\rho}{|f^{-n}(h(J))|_\rho} .$$

Furthermore, for $\iota \in \{s, u\}$, we have that $r^\iota = r_g^\iota$ satisfies the following properties:

(i) the ratios determine an *affine atlas on the segments*, i.e.,

$$r^\iota(I : J) = r^\iota(J : I)^{-1} \quad \text{and} \quad r^\iota(I_1 \cup I_2 : K) = r^\iota(I_1 : K) + r^\iota(I_2 : K),$$

where the segments I_1 and I_2 are contained in the segment K and intersect at most in an unique point;

(ii) the ratios are kept invariant by f, i.e., $r^\iota(I : J) = r^\iota(f(I), f(J))$;

(iii) there exist constants $0 < \alpha < 1$ and $C > 0$ such that, given a holonomy $h : I_h \to J_h$, we have that

$$\left| \log \frac{r^\iota(I : J)}{r^\iota(h(I) : h(J))} \right| \le C((d(I_h, J_h))^\alpha |K|_\rho^\alpha),$$

where the leaves I and J are contained in a same leaf $K \subset I_h$.

A. Pinto and D. Rand in (ref. 67) proved the following result on the classification of the smooth conjugacy classes of diffeomorphisms with hyperbolic basic sets.

Theorem 3.3. *The set of all pairs ($r^s : T^s \to \mathbb{R}^+$, $r^u : T^u \to \mathbb{R}^+$) of functions satisfying the above properties (i), (ii) and (iii) form a Teichmüller space for the smooth conjugacy classes of diffeomorphisms in the same topological conjugacy class of the diffeomorphism f with hyperbolic set Λ.*

We note that the proof of this result uses Theorem 3.1 which says that the holonomies of these dynamical systems are smooth.

Let S^s be the set of all pairs $(I, J) \in T^s$ such that (i) each segment I and J cross an unique Markov rectangle and have endpoints belonging to the unstable boundaries of these rectangles; (ii) I and J intersect only in a common endpoint. Similarly, we define S^u with respect to the unstable segments. The *stable solenoid function* is the restriction of the stable ratio function to the set S^s and the *unstable solenoid function* is the restriction of the unstable ratio function to the set S^u. The invariance under f of the ratio functions, allow us to rebuilt the ratio functions from the solenoid functions. Hence, the pairs of stable and unstable solenoid functions also form a Teichmüller space for the smooth conjugacy classes of diffeomorphisms in the same topological class of f. This implies that the Teichmüller space of Smale horseshoes is the set of all pairs ($s^s : \{0,1\}^{\mathbb{N}} \to \mathbb{R}^+, s^u : \{0,1\}^{\mathbb{N}} \to \mathbb{R}^+$) of Hölder functions.

As we have seen, the stable ratio function determines an affine structure along the stable segments, which is invariant under f and varies Hölder continuous along transversals. The same properties hold for the unstable ratio function with respect to the unstable segments. Since, the stable ratio function does not impose any restriction to the unstable ratio function and vice-versa, we obtain that the affine structure along the stable segments does not make any restriction on the affine structure along the unstable segments, and vice-versa.

However, the affine structure along the stable segments completely determines the Lipschitz structure of the affine structure along the unstable leaves for diffeomorphisms with hyperbolic sets, when we impose the existence of an invariant measure which is absolutely continuous with respect to the Hausdorff measure (see A. Pinto and D. Rand (ref. 70)).

3.3. *Extension of A. N. Livšic and Ja. G. Sinai's eigenvalue formula*

For every $g \in \mathcal{T}(f,\Lambda)$, we denote by $\delta_{g,s}$ (resp. $\delta_{g,u}$) the Hausdorff dimension of the local stable (resp. local unstable) leaves of g intersected with Λ. Let $\lambda_{g,s}(x)$ and $\lambda_{g,u}(x)$ denote the stable and unstable eigenvalues of the periodic orbit of g containing a point x. A. N. Livšic and Ja. G. Sinai (ref. 39) proved that an Anosov diffeomorphism g has an invariant measure that is absolutely continuous with respect to Lesbegue measure if, and only if, $\lambda_{g,s}(x)\lambda_{g,u}(x) = 1$ for every periodic point x. In (ref. 70), we extend the theorem of A. N. Livšic and Ja. G. Sinai to C^{1+} hyperbolic diffeomorphism with hyperbolic sets on surfaces such as Smale horseshoes and codimension one attractors.

Theorem 3.4. *A C^{1+} hyperbolic diffeomorphism $g \in \mathcal{T}(f,\Lambda)$ has a g-invariant probability measure which is absolutely continuous to the Hausdorff measure on Λ_g if and only if for every periodic point x of $g|\Lambda_g$,*

$$\lambda_{g,s}(x)^{\delta_{g,s}}\lambda_{g,u}(x)^{\delta_{g,u}} = 1 \ .$$

Since (f,Λ) is a C^{1+} hyperbolic diffeomorphism it admits a Markov partition $\mathcal{R} = \{R_1,\ldots,R_k\}$. This implies the existence of a two-sided subshift $\tau : \Theta \to \Theta$ of finite type, Θ in the symbol space $\{1,\ldots,k\}^{\mathbb{Z}}$, and an inclusion $i : \Theta \to \Lambda$ such that (a) $f \circ i = i \circ \tau$ and (b) $i(\Theta_j) = R_j$ for every $j = 1,\ldots,k$, where Θ_j is the cylinder containing all words $\ldots\epsilon_{-1}\epsilon_0\epsilon_1\ldots \in \Theta$ with $\epsilon_0 = j$. For every $g \in \mathcal{T}(f,\Lambda)$, the inclusion $i_g = h_{f,g} \circ i : \Theta \to \Lambda_g$ is such that $g \circ i_g = i_g \circ \tau$. We call such a map $i_g : \Theta \to \Lambda_g$ a *marking* of (g,Λ_g).

Definition 3.1. If $g \in \mathcal{T}(f,\Lambda)$ is a C^{1+} hyperbolic diffeomorphism as above and ν is a Gibbs measure on Θ then we say that (g,Λ_g,ν) is a *Hausdorff realisation of* ν if $(i_g)_*\nu$ is absolutely continuous with respect to the Hausdorff measure on Λ_g. If this is the case then we will often just say that ν is a Hausdorff realisation for (g,Λ_g).

We note that if $g \in \mathcal{T}(f,\Lambda)$ the Hausdorff measure on Λ_g exists and is unique. However, a Hausdorff realisation need not exist for (g,Λ_g).

Let $\mathcal{T}_{f,\Lambda}(\delta_s,\delta_u)$ be the set of all C^{1+} hyperbolic diffeomorphisms (g,Λ_g) in $\mathcal{T}(f,\Lambda)$ such that (i) $\delta_{g,s} = \delta_s$ and $\delta_{g,u} = \delta_u$; (ii) there is a g-invariant measure μ_g on Λ_g which is absolutely continuous with respect to the Hausdorff measure on Λ_g. We denote by $[\nu] \subset \mathcal{T}_{f,\Lambda}(\delta_s,\delta_u)$ the subset of all C^{1+}-realisations of a Gibbs measure ν in $\mathcal{T}_{f,\Lambda}(\delta_s,\delta_u)$.

De la Llave, Marco and Moriyon (refs. 40,41,49,50) have shown that the set of stable and unstable eigenvalues of all periodic points is a complete invariant of the C^{1+} conjugacy classes of Anosov diffeomorphisms. In (ref. 70), we extend their result to the sets $[\nu] \subset \mathcal{T}_{f,\Lambda}(\delta_s, \delta_u)$.

Theorem 3.5. *(i) Any two elements of $[\nu] \subset \mathcal{T}_{f,\Lambda}(\delta_s, \delta_u)$ have the same set of stable and unstable eigenvalues and these sets are a complete invariant of $[\nu]$ in the sense that if $g_1, g_2 \in \mathcal{T}_{f,\Lambda}(\delta_s, \delta_u)$ have the same eigenvalues if, and only if, they are in the same subset $[\nu]$.*

(ii) The map $\nu \to [\nu] \subset \mathcal{T}_{f,\Lambda}(\delta_s, \delta_u)$ gives a $1-1$ correspondence between C^{1+}-Hausdorff realisable Gibbs measures ν and Lipschitz conjugacy classes in $\mathcal{T}_{f,\Lambda}(\delta_s, \delta_u)$.

In (ref. 70), we also prove that the set of stable and unstable eigenvalues of all periodic orbits of a C^{1+} hyperbolic diffeomorphism $g \in \mathcal{T}(f, \Lambda)$ is a complete invariant of each Lipschitz conjugacy class. We note that for Anosov diffeomorphisms every Lipschitz conjugacy class is a C^{1+} conjugacy class.

Remark 3.1. We have restricted our discussion to Gibbs measures because it follows from Theorem 3.5 that, if $g \in \mathcal{T}_{f,\Lambda}(\delta_s, \delta_u)$ has a g-invariant measure μ that is absolutely continuous with respect to the Hausdorff measure then μ is a C^{1+}-Hausdorff realisation of a Gibbs measure ν so that $\mu = (i_g)_*\nu$.

E. Cawley (ref. 12) characterised all C^{1+}-Hausdorff realisable Gibbs measures as Anosov diffeomorphisms using cohomology classes on the torus. While it is possible that her cocycles could give enough information to characterise other hyperbolic systems on surfaces up to lippeomorphism, it is clear that they cannot encode enough for C^{1+} conjugacy because, for example, they do not encode enough information about gaps and so do not determine the smooth structure of stable leaves in the case where they are Cantor sets. To deal with all these cases in an integrated way, in (ref. 70), we use measure solenoid functions and gap-cocycle pairs to classify C^{1+}-Hausdorff realisable Gibbs measures of all C^{1+} hyperbolic diffeomorphisms on surfaces.

3.4. *Explosion of smoothness*

As in the unidimensional case, we have the following result that we entitle by explosion of smoothness:

Theorem 3.6. *Let f and g be any two diffeomorphisms with hyperbolic basic sets Λ_f and Λ_g, respectively. If f and g are topologically conjugate and the conjugacy has a derivative with non-zero determinant at a point, then f and g are smooth conjugate.*

We note that Theorem 3.1 is used in the proof of the above theorem (see F. Ferreira and A. Pinto in (ref. 27)).

4. The frontier between order and chaos

The transition between order and chaos, through a cascade of period doubling, was observed in several unimodal families of quadratic type and, in particular, in the quadratic family $f_\lambda(x) = -\lambda x^2 + \lambda - 1$, with $\lambda \in [0, 2]$. One of the most surprising discoveries on this transition from order to chaos was to find the existence of universal constants. To explain this phenomenon we pass to describe some properties of unimodal maps that we use to classify them as p-ordered and p-tent maps.

4.1. *p-ordered and p-tent maps*

We say that $f : [-1, 1] \to [-1, 1]$ is a C^r *unimodal map of quadratic type* if $f(x) = \phi(x^2)$, where $\phi : [0, 1] \to [-1, 1]$ is a C^r map with non-zero derivative, and $f(-1) = -1$. Let U^r be the set of all C^r unimodal maps.

We say that an unimodal map $f \in U^r$ is *p-ordered* if f has a periodic orbit which attracts almost every point in the interval $[-1, 1]$ with respect to Lebesgue measure. We say that f is *p-super stable* if f is p-ordered and the critical point of f belongs to the attracting periodic orbit.

We say that an unimodal map $f \in U^r$ is a *p-cookie-cutter* if there are p intervals $I_1, \ldots, I_p$ such that:

(i) f is a homeomorphism from I_i to I_{i+1}, for $1 \leq i < p$;
(ii) I_p contains the critical point of f;
(iii) there exist two closed and disjoint intervals J_0 and J_1 contained in I_p, not containing the critical point, and containing the endpoints of I_p;
(iv) the images of the maps $f^p|J_0$ and $f^p|J_1$ coincide with I_1.

For p-cookie-cutters, the result of Mañé in (ref. 48) implies the existence of $n \geq p$ and of two intervals $J'_0 \subset J_0$ and $J'_1 \subset J_1$ such that $f^n|J'_0 \cup J'_1$ is a cookie-cutter in the sense of Section 2.1.

We say that an unimodal map $f \in U^r$ is a *p-tent map* if there are p intervals $I_1, \ldots, I_p$ such that:

(i) $f \in U^r$ is a homeomorphism from I_i to I_{i+1}, for $1 \leq i < p$;
(ii) I_p contains a critical point of $f \in U^r$ and the image of the map $f^p|I_1$ coincides with I_1.

The p-tent maps with negative Schwarzian derivative (see definition in pg. 264 of (ref. 56)) and such that the point -1 is a hyperbolic expanding fixed point have the following properties (see chapters III and V in (ref. 56)):

(i) The p-tent maps are chaotic in the sense of Section 2.1 (see Theorem of D. Singer in (ref. 82) and Theorem 6.2 in Chapter III in (ref. 56));
(ii) The p-tent maps have an invariant measure with respect to Lebesgue (see Misiurewicz (ref. 59) and S. van Strien (ref. 84));
(iii) This measure satisfies the Sinai-Bowen-Ruelle property for almost every point x in $[-1, 1]$ with respect to Lebesgue measure (see Theorem 1.5 in Chapter V in (ref. 56));
(iv) The Lyapunov exponent λ_x is well-defined, is positive and its value does not depend on x, for almost every point x in $[-1, 1]$ with respect to Lebesgue measure (see G. Keller (ref. 35));
(v) The topological and metric entropy are both positive (applying Rochlin formula).

Since the p-tent maps of the quadratic family f_λ have negative Schwarzian and the point -1 is an expanding hyperbolic fixed point, we obtain that these unimodal maps satisfy the above (i)-(v) properties.

4.2. *Feigenbaum-Coullet-Tresser universality*

We are going to explain the existence of universal constants linked to the transition from order to chaos for unimodal families of quadratic type using the period doubling operator.

M. Feigenbaum in (ref. 22), and, independently, P. Coullet and C. Tresser in (ref. 13), discovered the existence of sequences of values $\alpha_1 < \alpha_2 < \ldots$ and $\beta_1 > \beta_2 > \ldots$ of the quadratic family f_λ with the following properties:

(i) f_{α_p} is 2^p-super stable;
(ii) f_α is 2^p-ordered, for all $\alpha_p \leq \alpha < \alpha_{p+1}$;
(iii) f_{β_p} is a 2^p-tent;
(iv) f_β is a 2^p-cookie-cutter, for all $\beta_{p+1} > \beta \geq \beta_p$;

(v) the value $\gamma = \lim_{p\to\infty} \alpha_p$ coincide with the value $\lim_{p\to\infty} \beta_p$;

(vi) the following limits exist and determine two universal constants:

$$\lim_{p\to\infty} \frac{\alpha_p - \alpha_{p-1}}{\alpha_{p+1} - \alpha_p} = \lim_{p\to\infty} \frac{\beta_p - \beta_{p-1}}{\beta_{p+1} - \beta_p} = 4.6692016091029\ldots$$

and

$$\lim_{n\to\infty} \frac{|f^{2^{n+1}}(0)|}{|f^{2^n}(0)|} = 0.3995\ldots .$$

Hence, the unimodal maps f_λ are ordered for $0 \leq \lambda < \gamma$ and chaotic for $\gamma < \lambda \leq 2$ which implies that the unimodal map f_λ is in the frontier between order and chaos.

We remind the reader that the topological entropy (see definition in pg.164 of (ref. 56)) is also an important tool to determine the complexity of the dynamical behavior of the unimodal maps. M. Misiurewicz in (ref. 58) proved the following: if the topological entropy $h(f)$ of f is 0, then the period of every periodic orbit of f is a power of 2. Hence, the p-cookie-cutters have positive entropy (however they can also be p'-ordered maps). J. Milnor and W. Thurston (ref. 57), have shown that the topological entropy $h(f_\lambda)$ varies monotonically and continuously with the parameter λ such that $h(f_\lambda) = 0$ for all $0 \leq \lambda \leq \gamma$ and $h(f_\lambda) > 0$ for all $\gamma < \lambda \leq 2$ (see also J.F. Alves and J. Sousa Ramos (ref. 1) and chapter II in (ref. 56)).

Since the properties (i)-(vi) of the quadratic family are also satisfied by other unimodal families of quadratic type, M. Feigenbaum and, independently, P. Coullet and C. Tresser, have introduced the concept of period doubling operator T which we define below, and conjectured the following: (a) there exists an hyperbolic fixed point for the period doubling operator with a unique unstable direction whose eigenvalue is $\delta = 4.669\ldots$; (b) the sets Σ_p^E consisting of all 2^p-super stable maps and the sets Σ_p^T consisting of all 2^p-tent maps are transversal to the local unstable manifold. These conjectures explain why the unimodal families with quadratic type satisfy the above (i)-(vi) properties. O. Lanford in (ref. 37) proved these conjectures in a Banach space of real analytical unimodal maps (see also J.-P. Eckmann and P. Wittwer in (ref. 17)).

Let us define the period doubling operator. We consider from now on that the unimodal maps in U^r are normalized such that $f(0) = 1$ (instead of $f(-1) = 1$). This normalization allow us to construct the period doubling operator and to do computations in a easier way than the previous normalization. An unimodal map $f \in U^r$ is *2-renormalizable* if $\lambda^{-1} f^2(\lambda x) \in U^r$, where $\lambda = f^2(0)$. Let us denote by U^r_T the set of all unimodal maps in U^r which are 2-renormalizable. The period doubling operator $T : U^r_T \to U^r$ is defined by $Tf(x) = \lambda^{-1} f^2(\lambda x)$.

4.3. *Rigidity in the frontier between order and chaos*

We are going to explain the dynamical relevance of the set consisting of the closure of the critical orbit of the unimodal maps contained in the stable set of the period doubling operator. We will explain that these maps are rigid. In fact, the conjugacy between any two of these maps defined in the closure of the critical orbits has a $C^{2.11\dots}$ extension to the reals, where $2.11\dots$ is a universal constant.

The unimodal maps f of quadratic type in the stable manifold of the period doubling operator T and so in the frontier between order and chaos are infinitely 2-renormalizable. The unimodal maps f infinitely 2-renormalizable and with negative Schwarzian derivative have the following properties (see chapters III and V in (ref. 56)):

(i) the closure of the critical orbit C_f of these unimodal maps is a Cantor set;
(ii) the maps f restricted to C_f are topological conjugate to the adding machine in the set of the 2-adic numbers (see also L. Jonker and D. Rand (ref. 34));
(iii) the ω-limit set of a point x is C_f for almost every point in $[-1, 1]$ with respect to the Lebesgue measure (see also A. Blokh and M. Lyubich (ref. 9));
(iv) there exists a probability measure whose support is C_f;
(v) this measure satisfies the Sinai-Bowen-Ruelle property for almost every point in $[-1, 1]$ with respect to the Lebesgue measure.

These properties show the dynamical relevance of the set C_f in the study of the infinitely 2-renormalizable unimodal maps. We note that given any two unimodal maps f and g in the stable manifold of the period doubling operator there exists a homeomorphism $h : C_f \to C_g$ which topologically conjugates f and g. D. Rand in (ref. 76) proved that h has a smooth extension to the reals proving the existence of rigidity for the fine structures of these Cantor sets. D. Rand and A. Pinto in (ref. 63) improved this result showing the following:

Theorem 4.1. *Let f and g be unimodal maps of quadratic type, real analytic, and infinitely 2-renormalizable. The conjugacy $h : C_f \to C_g$ between f and g has a $C^{2.11\dots}$ extension to the interval $[-1, 1]$. The universal constant $2.11\dots$ is given by the formula*

$$2.11\cdots = \frac{\ln 0.13\dots}{2\ln 0.3995\dots} + 1$$

where $0.3995\dots$ and $0.13\dots$ are the second and third largest eigenvalues in module of the period doubling operator, respectively.

This result is proven in (ref. 63) for real analytic unimodal maps contained in the domain V defined by O. Lanford in (ref. 37) to prove the hyperbolicity of the fixed point of the period doubling operator. By the results of G. Levin and S. van Strien in (ref. 38) and of D. Sullivan in (ref. 86) the unimodal maps of quadratic type and infinitely 2-renormalizable after a finite number of iterations under renormalization belong to V which implies the above theorem.

4.4. *The renormalization operator*

We are going to study the dynamical properties of the topological attractors for unimodal maps with quadratic type. An important tool in this study is the renormalization operator which is an extension of the period doubling operator and that we pass to construct.

A unimodal map $f \in U^r$ is *renormalizable* if there exists $p \geq 2$ such that $R_p f(x) = \lambda^{-1} f^p(\lambda x) \in U^r$, where $\lambda = f^p(0)$. If f is renormalizable we choose the smallest value possible of $p(f) \geq 2$ of p such that $R_{p(f)} f \in U^r$. We call $R_{p(f)} f$ the *renormalization* of f. Hence, the intervals $f^j([-|\lambda|, |\lambda|])$, for $j = 0, \dots, p-1$, are disjoint and their embedding in $[-1, 1]$ determine a *unimodal permutation* $\theta : \{0, \dots, p-1\} \to \{0, \dots, p-1\}$. We denote by P the set of all these unimodal permutations and by U^r_θ the set of all renormalizable unimodal maps with permutation θ. We define the *renormalization operador* $R : \cup_{\theta \in P} U^r_\theta \to U^r$ by $Rf = R_{p(f)} f$. Let us fix a finite subset L of P. We say that an infinitely renormalizable unimodal map f has *geometric type bounded by* L if $R^n f$ determines a permutation in L for every $n \geq 1$. We say that two infinitely renormalizable unimodal maps f and g determine the same sequence of permutations, if $R^n f$ and $R^n g$ have the same permutation θ_n, for every $n \geq 1$.

We say that a set A is a *topological attractor* for $f \in U^2$, if (i) the closure of its basin of attraction $B(A)$ contains intervals, and (ii) there are no small subset $A' \subset A$ such that $\overline{B(A)} \setminus \overline{B(A')}$ contains intervals. L. Jonker and

D. Rand in (ref. 34), proved that the topological attractors A for $f \in U^r$ are of the following type:

(i) a periodic orbit;
(ii) a finite union of intervals with the following properties: (a) A contains the critical point of f, (b) f acts transitively in A, (c) f has sensitivity to the initial conditions in A, (d) f restricted to A is topological conjugate to a piecewise affine map,
(iii) a Cantor set C_f in which f acts as an adding machine (see definition in chapter II.5 in (ref. 56)) and, in this case, f is infinitely renormalizable.

The renormalization operator allow us to study the transition from p-ordered unimodal maps to tent unimodal maps through period doubling (where p does not need to be a power of 2 and so the p-ordered unimodal map can simultaneously be a q-cookie-cutter map). G. Świątek in (ref. 88) proved that the set of parameters λ for which the maps f_λ in the quadratic family are p-ordered for some $p \geq 1$ (and the attracting periodic orbit is hyperbolic) form an open and dense set in $[0, 1]$ (see also M. Lyubich (ref. 45) and (ref. 43)). O. Kozlovski in (ref. 36) generalized this result to unimodal maps of quadratic type in C^2. In contrast with these results, M. Jakobson in (ref. 33) proved that there exists a set of parameters λ with positive Lebesgue measure such that the maps f_λ in the quadratic family have positive Lebesgue exponent and satisfy properties (i)-(vi) of the p-tent maps (see also M. Benedicks and L. Carleson in (ref. 8) and chapter V in (ref. 56)). M. Lyubich in (ref. 42) proved that for a set of total measure of the parameter values λ the maps f_λ of the quadratic family are p-ordered or have positive Lyapunov exponent. A. Ávila, M. Lyubich, W. de Melo and C. Moreira in (refs. 4, 5, 6) and (ref. 7) generalized this result of Lyubich to generic families of unimodal maps in U^r, with $r \geq 3$ and negative shwarzian derivative, adding also that these maps are stochastic stable. In (ref. 6), they proved the existence of a set X "with total measure" of real analytic unimodal maps such that if $f, g \in X$, f and g are topologically conjugate and are not p-ordered then they are analytic conjugate.

4.5. *Hyperbolicity of the renormalization operator*

We are going to study the hyperbolicity of the renormalization operator together with the rigidity features of the unimodal maps in the stable sets and also the rigid features appearing in the parameter space of unimodal families.

O. Lanford and others generalized the hyperbolicity conjecture of the period doubling operator to the renormalization operator. D. Sullivan in (ref. 87) and in (ref. 86) proved the following:

(i) there exists a limit set A_L such that all infinitely renormalizable maps in U^2 with geometric type bounded by L converge under iteration by the renormalization operator to A_L;
(ii) the renormalization operator restricted to the set A_L is topologically conjugate to a shift in $\{1, \ldots, \#L\}^{\mathbf{Z}}$; where $\#L$ is the cardinal of the set L;
(iii) the elements of A_L belong to the set U^ω of all real analytic maps with holomorphic quadratic-like extensions.

We remind the reader that a holomorphic map $f : V \to W$ is *quadratic-like* if the the following properties are satisfied: (a) V and W are topological disks; (b) V is compactly contained in W; (c) f is a ramified double covering with a continuous extension to the boundary of V. Using the above results of D. Sullivan, C. McMullen in (ref. 52) and in (ref. 53) proved that if two infinitely renormalizable unimodal maps $f, g \in U^\omega$ with bounded geometric type have the same sequence of unimodal permutations, then the distance $\|R^n f - R^n g\|_{C^0([-1,1])}$ converges exponentially fast to 0, when n tends to infinity. As a corollary of this result, C. McMullen proved that the conjugacy $h : C_f \to C_g$ between f and g has a $C^{1+\alpha}$ extension, for some $\alpha > 0$. This rigidity result was generalized by W. de Melo and A. Pinto in (ref. 55) to unimodal maps in U^2.

Theorem 4.2. *Let $f, g \in U^2$ be infinitely renormalizable unimodal maps with bounded geometric type and with the same sequence of permutations. The topological conjugacy $h : C_f \to C_g$ between f and g has a $C^{1+\alpha}$ extension to the real line, for some $\alpha > 0$.*

The proof of this result uses the work of D. Sullivan and C. McMullen above mentioned, together with the following theorem of M. Lyubich (see ref. 44): The set A_L is embedded in a space of equivalence classes of germs of maps in U^ω quotient by affine maps, where A_L forms a hyperbolic set with respect to the induced renormalization operator. M. Lyubich in (ref. 42), proved that this result also holds in this space for the limit set A of all infinitely renormalizable unimodal maps. E. de Faria, W. de Melo and

A. Pinto in (ref. 21), using the above result of M. Lyubich in (ref. 44) and inspired in the work of A. Davie in (ref. 15), proved that A_L forms a "hyperbolic" set in U^3.

Theorem 4.3. *The renormalization operator $R : U^3_L \to U^3$ satisfies the following hyperbolic properties:*

(i) The local unstable sets of A_L are real analytic submanifolds with dimension 1;
(ii) The local stable sets of A_L are C^1 submanifolds transversal to the unstable submanifolds;
(iii) The global stable sets of A_L in U^4 are C^1 immersed submanifolds;

We note that the renormalization operator is not even smooth in U^3. However, it is a smooth operator from U^3 to U^2 which allow us to construct a formula for its derivative. This formula shows that the image of the maps in U^3 are still in U^3 which allowed the development of estimates in the U^3 norm that are used to prove the above theorem. Several of the main estimates used in the proof had to be done using different methods from A. Davie in (ref. 15). To prove the estimates needed it was used that the maps in A_L are real analytic and satisfy the real and complex bounds as proved by D. Sullivan in (ref. 86). It is also used in the proof the fact that the local unstable sets of A_L have dimension 1 which implies that they are the same as the ones obtained in the real analytic case.

5. Conclusion

In this work, we have characterized the fine structures of fractals determined by hyperbolic and chaotic dynamical systems and by dynamical systems which are in the frontier between order and chaos.

5.1. *Hyperbolic dynamics*

We used solenoid functions in one dimensional hyperbolic dynamics to characterize the fine structures of these systems (see Theorems 2.1 and 2.2). We have stated that if there is a topological conjugacy which has a non-zero derivative at a point then it is smooth everywhere (see Theorem 2.3).

Similarly, we have characterized the fine structures for basic sets of diffeomorphisms on surfaces using ratio functions (see Theorem 3.3). We have stated that if a topological conjugacy is differentiable at a point then is smooth everywhere (see Theorem 3.6). Since the holonomies are smooth, the fine structures along the leaves are similar. Furthermore, if the degree of smoothness of the holonomy is sufficiently larger then the dynamical system is rigid, i.e. it is smooth conjugate to an affine dynamical system with affine holonomies. We also have explained that there are no affine dynamical systems with affine holonomies for Plykin attractors which implies that the degree of smoothness of the stable holonomies is bounded by 1 plus the Hausdorff dimension of the basic set along the stable leaves.

The classification of the diffeomorphisms on surfaces allowed us to understand that the fine structures along the stable lamination does not impose any restriction on the fine structures along the unstable lamination, and vice-versa. However, if we ask that these dynamical systems have an invariant geometric measure (for instance, equivalent to the Hausdorff measure) then the fine structures along the stable lamination determine the fine structures along the unstable direction up to Lipschitz class, and vice-versa.

We hope that in the future there will be similar progresses as the above ones for laminations on surfaces and for flows in manifolds with dimension 3.

5.2. *The frontier between order and chaos*

We have described the existence of ordered and chaotic behaviour for unimodal families of quadratic type and that the maps in the frontier between order and chaos have the property of being infinitely renormalizable. We explained that the infinitely renormalizable maps form the stable manifolds of a limit set with hyperbolic behaviour for the renormalization operator. We stated the rigidity of the fine structures for infinitely renormalizable unimodal maps with bounded geometric type (see Theorems 4.1 and 4.2). We also stated the existence of rigidity on the parameter space and the existence of universal constants (see Theorem 4.3).

The renormalization operator appears in a natural way in several other families of maps, such as, unimodal families with type $\alpha > 1$, multimodal families, families of critical circle maps and of annulus maps (see, for example, E. de Faria and W. de Melo (refs. 19 and 20), R. Mackay (ref. 46), M. Martens (ref. 51), W. de Melo (ref. 54), S. Ostlund, D. Rand, J. Sethna and E. Siggia (ref. 61) and M. Yampolsky (ref. 90)). In the future, we hope that similar results to the ones presented here can be proved for some of these families.

Acknowledgments

I would like to thank the useful discussions with Carlos Rocha, David Rand, Dennis Sullivan, Flávio Ferreira, Henrique Oliveira, João Almeida, João Alves, Luis Silva, Saber Elaydi and Sousa Ramos.

I also thank Calouste Gulbenkian Foundation, PRODYN-ESF, Programs POCTI and POCI by FCT and Ministério da Ciência e da Tecnologia, CIM, Escola de Ciências da Universidade do Minho, Faculdade de Ciências da Universidade do Porto, Centros de Matemática da Universidade do Minho e da Universidade do Porto for their financial support.

References

1. J. F. Alves & J. Sousa Ramos, Kneading Theory: A Functorial Approach. *Commun. Math. Phys.* **204**, 89-114 (1999).
2. J. F. Alves, V. Pinheiro, A. A. Pinto, Explosion for non-uniformly expanding maps. In preparation.
3. D. V. Anosov, Geodesic flows on compact manifolds of negative curvature. *Trudy Mat. Inst. Steklov* **90** (1967); *Proc. Steklov Inst. Math. (transl.)* (1969).
4. A. Ávila, M. Lyubich, & W. de Melo, Regular or stochastic dynamics in real analytic families of unimodal maps. *Inventiones Mathematicae* **154** (2003) 451-550.
5. A. Ávila & C. G. Moreira, Statistical properties of unimodal maps: The quadratic family. *Annals of Math.* **161** (2005) 831-881.
6. A. Ávila & C. G. Moreira, Statistical properties of unimodal maps: Physical measures, periodic orbits and pathological laminations. *Pub. Math. IHES* **101** (2005) 1-67.
7. A. Ávila, *Bifurcations of unimodal maps: the topological and metric picture.* Ph.D. Thesis, IMPA, Brasil (2001).
8. M. Benedicks & L. Carleson, On iterations of $1 - ax^2$ on $(-1, 1)$. *Ann. Math.* **122** (1985) 1-25.
9. A. M. Blokh & M. Yu. Lyubich, Measure and dimension of solenoidal attractors of one-dimensional dynamical systems. *Commun. Math. Phys.* **127** (1990) 573-583.
10. R. Bowen, Markov partitions and minimal sets for axiom A diffeomorphisms. *Amer. J. Mat.* **92** (1970) 907.
11. R. Bowen, *Equilibrium States and the Ergodic Theory of Axiom A diffeomorphisms.* Lecture Notes in Mathematics, N° 470, Springer-Verlag, New York (1975).
12. E. Cawley, The Teichmüller space of an Anosov diffeomorphism of T^2. *Inventiones Mathematicae* **112** (1993) 351-376.
13. P. Coullet & C.Tresser, Itérations d'endomorphismes et groupe de renormalisation. *J. Phys. C* **539** C5-25 (1978).
14. P. Cvitanović, *Universality in Chaos.* Adam Hilger, Bristol (1989).
15. A. M. Davie, Period doubling for $C^{2+\epsilon}$ mappings. *Commun. Math. Phys.* **176** (1996) 262-272.

16. R. Devaney, *Chaotic Dynamical Systems.* Addison-Wesley Publ. CO., New York & Reading, MA (1989).
17. J.-P. Eckmann & P. Wittwer, A complete proof of Feigenbaum conjectures. *J. of Stat. Phys.* **46** (1987) 455-475.
18. E. de Faria, Quasisymmetric distortion and rigidity of expanding endomorphisms of S^1. *Proceedings of the American Mathematical Society* **124** (1996) 1949-1957.
19. E. de Faria & W. de Melo, Rigidity of critical circle mappings I. *J. Eur. Math. Soc.* **1** (1999) 339-392.
20. E. de Faria & W. de Melo, Rigidity of critical circle mappings II. *J. Amer. Math. Soc.* **13** (2000) 343-370.
21. E. de Faria, W. de Melo and A. A. Pinto, Global hyperbolicity of renormalization for C^r unimodal mappings. *Annals of Math.* **164** (2006) 731-824.
22. M. J. Feigenbaum, Qualitative universality for a class of nonlinear transformation. *J. Statist. Phys.* **19** (1978) 25-52.
23. F. Ferreira, *Diferenciabilidade de Aplicações de Markov não Diferenciáveis.* MSc Thesis, University of Porto, Portugal (1994) 1-40.
24. F. Ferreira, *Diferenciabilidade, Universalidade e Renormalização de Sistemas Dinâmicos Hiperbólicos.* PhD Thesis, University of Porto, Portugal (2000).
25. F. Ferreira & A. A. Pinto, Existence of eigenvalues for (uaa) Markov maps. *Portugaliae Mathematica*, vol. 54 **4** (1997) 421-429.
26. F. Ferreira and A. A. Pinto, Explosion of smoothness from a point to everywhere for conjugacies between Markov families. *Dyn. Sys.* **16** 2 (2001) 193-212.
27. F. Ferreira and A.A. Pinto, Explosion of smoothness from a point to everywhere for conjugacies between diffeomorphisms on surfaces. *Ergod. Th. & Dynam. Sys.* **23** (2003) 509-517.
28. E. Ghys, Rigidité différentiable des groupes fuchsiens. *Publ. IHES* **78** (1993) 163-185.
29. J. Harrison, An introduction to fractals. *Proceedings of Symposia in Applied Mathematics, vol. 39*, AMS (1989) 107-126.
30. M. W. Hirsch, J. Palis, C. C. Pugh & M. Shub, Neighborhoods of hyperbolic sets. *Invent. Math.* **9** (1969-70) 121.
31. M. W. Hirsch & C. C. Pugh, Stable manifolds and hyperbolic sets. *Global Analysis* Vol. XIV (Proceedings of Symposia in Pure Mathematics), AMS, Providence, R. I. (1970) 133.
32. E. A. Jackson, *Perspectives of Nonlinear Dynamics*, (Two volumes), Cambridge Univ. Press (1989/1990).
33. M. V. Jakobson, Absolutely continuous invariant measures for one-parameter families of one-dimensional maps. *Commun. Math. Phys.* **81** (1981) 39-88.
34. L. B. Jonker & D. A. Rand, Bifurcation in one-dimension. I. The nonwandering set. *Invent. Math.* **62** (1981) 347-365; II. A versal model for bifurcations. *Invent. Math.* **63** (1981) 1-16.
35. G. Keller, Exponents, attractors, and Hopf decompositions for interval maps. *Ergod. Th. & Dynam. Sys.* **10** (1990) 717-744.
36. O. S. Kozlovski, *Structural Stability in One-dimensional Dynamics.* Ph.D. Thesis, University of Amsterdam, Holanda (1997) 1-82.

37. O. E. Lanford III, A computer assisted-proof of the Feigenbaum conjecture. *Bull. A. M. S.* **6** (1982) 427-434.
38. G. Levin & S. van Strien, Local connectivity of the Julia set of real polynomials. *Annals of Math.* **147** (1998) 471-541.
39. A. N. Livšic and Ja. G. Sinai, Invariant measures that are compatible with smoothness for transitive C-systems. (Russian) *Dokl. Akad. Nauk SSSR* **207** (1972), 1039–1041.
40. R. Llave, J. M. Marco and R. Moriyon, Canonical perturbation theory of Anosov systems and regularity results for the Livšic cohomology equations. *Annals of Math.*, **123** (1986) 537-612.
41. R. Llave, Invariants for Smooth conjugacy of hyperbolic dynamical systems II. *Commun. Math. Phys.*, **109** (1987) 369-378.
42. M. Lyubich, Almost every real quadratic map is either regular or stochastic. *Stony Brook preprint* (1997).
43. M. Lyubich, Dynamics of quadratic polynomials, I-II. *Acta Math.* **178** (1997) 185-297.
44. M. Lyubich, Feigenbaum-Coullet-Tresser universality and Milnor's hairiness conjecture. *Annals of Math.* **149** (1999) 319–420.
45. M. Lyubich, Dynamics of quadratic polynomials, III. Parapuzzle and SRB measure. *"Géometrie complexe et systémes dynamiques"*. Asterique volume in honor of Douady's 60th birthday **261** (2000) 173-200.
46. R. S. Mackay, *Renormalization in area preserving maps.* Thesis, Princeton (Univ. Microfilms Int., Ann Arbor, Michigan) (1982).
47. R. S. Mackay, A. A. Pinto & J. B. J. van Zeijts, Coordinate change eigenvalues for bimodal period doubling renormalization. *Phys. Lett. A* **190** (1994) 412-416.
48. R. Mañé, Hyperbolicity, sinks and measure in one dimensional dynamics. *Commun. Math. Phys.* **100** (1985) 495-524 and Erratum. *Commun. Math. Phys.* **112** (1987) 721-724.
49. J. M. Marco and R. Moriyon, Invariants for Smooth conjugacy of hyperbolic dynamical systems I. *Commun. Math. Phys.*, **109** (1987) 681-689.
50. J. M. Marco and R. Moriyon, Invariants for Smooth conjugacy of hyperbolic dynamical systems III. *Commun. Math. Phys.*, **112** (1987) 317-333.
51. M. Martens, The periodic points of renormalization. *Annals of Math.* (1996).
52. C. McMullen, *Complex Dynamics and Renormalization.* Annals of Math. Studies **135**, Princeton University Press, Princeton, NJ (1994).
53. C. McMullen, *Renormalization and 3-manifolds which Fiber over the Circle.* Annals of Math. Studies **142**, Princeton University Press, Princeton, NJ (1996).
54. W. de Melo, Rigidity and Renormalization in One Dimensional Dynamical Systems. *Proc. Int. Congr. Mathematicians, Berlim, 1998*, Documenta Mathematica, Extra Volume ICM 98, Vol. II 765–778.
55. W. de Melo & A. A. Pinto, Smooth conjugacies between C^2 infinitely renormalizable quadratic maps with the same bounded type. *Commun. Math. Phys.* **208** (1999) 91-105.
56. W. de Melo & S. van Strien, *One-Dimensional Dynamics.* Springer-Verlag, New York (1992).

57. J. Milnor & W. Thurston, On iterated maps of the interval: I, II. *Lect. Notes in Math.* **1342**. Springer, Berlin, New York (1979) 465-563.
58. M. Misiurewicz, Horseshoes for mappings of the interval. *Boll. Acad. Pol. Sci., Sér. Sci. Math.* **27** (1979) 167-169.
59. M. Misiurewicz, Absolutely continuous measures for certain maps of an interval. *Publ. Math. I.H.E.S.* **53** (1981) 17-51.
60. A. Norton, Denjoy's Theorem with exponents. *Proc. Amer. Math. Soc.* vol. 127 **10** (1999) 3111-3118.
61. S. Ostlund, D. A. Rand, J. Sethna, E. Siggia, Universal properties of the transition from quasi-periodicity to chaos in dissipative systems. *Physica D* **8** (1983) 303-342.
62. M. Peixoto, Structural stability on two-dimensional manifolds. *Topology* **1** (1962) 101.
63. A. A. Pinto & D. A. Rand, Global phase space universality, smooth conjugacies and renormalisation: 2. The $C^{k+\alpha}$ case using rapid convergence of Markov families. *Nonlinearity* **5** (1992) 49-81.
64. A. A. Pinto and D. A. Rand, Classifying C^{1+} structures on dynamical fractals: 1 The moduli space of solenoid functions for Markov maps on train-tracks. *Erg. Th. Dyn. Syst.* **15** (1995) 697-734.
65. A. A. Pinto and D. A. Rand, Classifying C^{1+} structures on dynamical fractals: 2 Embedded trees. *Erg. Th. Dyn. Syst.* **15** (1995) 969-992.
66. A. A. Pinto and D. A. Rand, Existence, uniqueness and ratio decomposition for Gibbs states via duality. *Erg. Th. Dyn. Syst.* **21** (2001) 533-544.
67. A. A. Pinto and D. A. Rand, Teichmüller spaces and HR structures for hyperbolic surface dynamics. *Erg. Th. Dyn. Syst.* **22** (2002) 1905-1931.
68. A. A. Pinto and D. A. Rand, Smoothness of holonomies for codimension 1 hyperbolic dynamics. *Bull. London Math. Soc.* **34** (2002) 341-352.
69. A. A. Pinto and D. A. Rand, Rigidity of hyperbolic surface dynamics. *J. Lond. Math. Soc.* **71** (2) (2004) 481-502.
70. A. A. Pinto and D. A. Rand, Eigenvalue formula of A. N. Livšic and Ja. G. Sinai for hyperbolic sets on surfaces. *Stony Brook preprint* 2006.
71. A. A. Pinto and D. A. Rand, Solenoid functions for hyperbolic sets on surfaces. *Dynamics, Ergodic Theory and Geometry: Dedicated to Anatole Katok* (ed. Boris Hasselblatt) MSRI Publications **54** (2007) 145-178.
72. A. A. Pinto, D. A. Rand and F. Ferreira, Hausdorff dimension bounds for smoothness of holonomies for codimension 1 hyperbolic dynamics. *J. Differential Equations.* To be published (2007) 1-15.
73. A. A. Pinto, D. A. Rand and F. Ferreira, Cantor exchange systems and renormalization. *J. Differential Equations.* To be published (2007) 1-27.
74. A. A. Pinto, D. A. Rand and F. Ferreira, Fine structures of hyperbolic diffeomorphisms. *Springer-Verlag Monograph* (in preparation) (2008) 1-311.
75. A. A. Pinto and D. Sullivan, The circle and the solenoid. Dedicated to Anatole Katok On the Occasion of his 60th Birthday. *DCDS*-A **16** 2 (2006) 463-504.
76. D. A. Rand, Global phase space universality, smooth conjugacies and renormalisation: 1. The $C^{1+\alpha}$ case. *Nonlinearity* **1** (1988) 181-202.
77. C. Robinson, *Dynamical Systems: Stability, Symbolic Dynamics, and Chaos.* CRC Press, Inc. (1995).

78. H. G. Schuster, *Deterministic Chaos, an Introduction.* VCH Publishers, Weinheim (1989).
79. M. Shub, Endomorphisms of compact differentiable manifolds. *Amer. J. of Math.* (1969).
80. M. Shub, *Global Stability of Dynamical Systems.* Springer-Verlag (1987).
81. Ja. G. Sinai, Markov Partitions and C-diffeomorphisms. *Anal. and Appl.* **2** (1968) 70-80.
82. D. Singer, Stable orbits and bifurcations of maps of the interval. *SIAM J. Appl. Math.* **35** 260-267.
83. S. Smale, Differentiable dynamical systems. *Bull. Amer. Math. Soc.* **74** (1967) 747.
84. S. van Strien, Hyperbolicity and invariant measures for general C^2 interval maps satisfying the Misiurewicz condition. *Commun. Math. Phys.* **128** (1990) 437-496.
85. D. Sullivan, Differentiable structures on fractal-like sets determined by intrinsic scaling functions on dual Cantor sets. *Proceedings of Symposia in Pure Mathematics* vol. **48** (1988), AMS.
86. D. Sullivan, Bounds, quadratic differentials, and renormalization conjectures. *AMS Centennial Publications.* Volume 2: Mathematics into the Twenty-first Century (1988 Centennial Symposium, August 8-12), American Mathematical Society, Providence, RI., 1991.
87. D. Sullivan, Linking the universalities of Milnor-Thurston, Feigenbaum and Ahlfors-Bers. Proccedings of the Symposium held in honor of John Milnor's 60th birthday, SUNY at Stony Brook, June 14-21 (1991). L. R. Goldberg and A. V. Philips, editors, *Topological Methods in Modern Mathematics*, Publish or Perish, Inc. (1993) 543-563.
88. G. Świątek, Hyperbolicity is dense in the real quadratic family. *Preprint Stony Brook* (1992).
89. K. Wilson, The renormalization group: Critical phenomena and the Kondo problem. *Reviews of Modern Physics* Vol. 47 **4** (1975) 773-839.
90. M. Yampolsky, Complex bounds for renormalization of dritical circle maps. *Ergod. Th. & Dynam. Sys.* **19** (1999) 227-257.

Difference Equations with Continuous Time: Theory and Applications

E. Yu. Romanenko* and A. N. Sharkovsky†

Institute of Mathematics, National Academy of Sciences of Ukraine,
01601, Tereschenkivska, 3, Kyiv, Ukraine
**E-mail: eromanenko@bigmir.net*
†*E-mail: asharkov@imath.kiev.ua*

We built basics of the qualitative theory of the continuous-time difference equations $x(t+1) = f(x(t))$, $t \in \mathbb{R}^+$, with the method of going to the infinite-dimensional dynamical system induced by the equation. For a study of this system we suggest an approach to analyzing the asymptotic dynamics of general nondissipative systems on continuous functions spaces. The use of this approach allows us to derive properties of the solutions from that of the ω-limit sets of trajectories of the corresponding dynamical system. In particular, typical continuous solutions are shown to tend (in Hausdorff metric for graphs) to upper semicontinuous functions whose graphs are, in wide conditions, fractal; there may exist especially nonregular solutions described asymptotically exactly by random processes. We introduce the notion of self-stochasticity in deterministic systems — a situation when the global attractor contains random functions. Substantiated is a scenario for a spatial-temporal chaos in distributed parameters systems with regular dynamics on attractor: The attractor consists of cycles only and the onset of chaos results from the very complicated structure of attractor "points" which are elements of some function space (different from the space of smooth functions). We develop a method to research into boundary value problems for partial differential equations, that bases on their reduction to difference equations.

Keywords: Difference equation with continuous time, infinite-dimensional dynamical system, 1D map, attractor, upper semicontinuous function, fractal, random function, finite-dimensional distributions, deterministic chaos.

1. Introduction

We will deal with the continuous-time difference equation

$$x(t+1) = f(x(t)), \qquad t \in \mathbb{R}^+, \tag{1}$$

with f being a continuous map of a bounded closed interval I into itself.

We would like to begin with feasible applications of Eq.(1): *For which purposes are these equations wanted? Where can they be used? Are there convincing reasons for mathematicians to pay attention to this object?* Mention in passing that back 40 years the same questions arose regarding the "normal" difference equation

$$x(n+1) = f(x(n)), \quad n \in \mathbb{Z}^+, \tag{2}$$

and then the response of "serious" experts was negative: The properties of this "simple" difference equation are of no importance in the view of Fundamental Science.[a] Now it is well known that Eq.(2) has given a vital impetus to the progression of discrete dynamical systems theory — one of the principal mathematical instruments of modern nonlinear dynamics.

What can be said about the applications of continuous-time difference equations now?

The first "point of application" is DIFFERENTIAL-DIFFERENCE EQUATIONS THEORY, that is actively developed since the 40s of the last century. This theory should contain, at least formally, the theory of continuous-time difference equations and, certainly, use that, especially to differential-difference equations of neutral type.

The second field of application is PARTIAL DIFFERENTIAL EQUATIONS THEORY. There are ample classes of boundary value problems for partial differential equations, that can be reduced directly to continuous-time difference equations or to equations close these, for instance, to differential-difference equations.

Lastly, continuous-time difference equations find many applications in MODELLING SPATIAL-TEMPORAL CHAOS (TURBULENCE). These equations are surprisingly well suitable for simulating such phenomena as 'chaos', 'cascade process of emergence of coherent structures', 'fractal structures', 'intermixing', and etc.

Why do we summarize investigations of Eq.(1) just today? In our (with Yu.L.Maistrenko) book,[3] published in Russian as early as 1986, one of

[a] A prominent example of this is the following. On the 4th International Conference on Nonlinear Oscillations (Prague 1967), one of the authors of this paper reported his results,[1] that concern the coexistence of periodic solutions of Eq.(2) in accord with the ordering $3 \succ 5 \succ 7 \succ \ldots \succ 4 \succ 2 \succ 1$; in particular, the statement: The existence of a solution with period 3 implies the existence of periodic solutions with any period $m \in \mathbb{Z}^+$, was presented. These results, which now have been rendered as fundamental in nonlinear dynamics, had been outlined only in the conference abstracts[2] but the report text represented by the author had not been included into the conference proceedings.

the four chapters was named 'Difference Equations with continuous time'. Nevertheless, we were exploring difference equations further at all the time, and a number of fundamental steps have been done quite recently. In particular, a proof of the fact that the attractor of a deterministic system is liable to contain random functions was published only in the last year.[4] Now there are strong grounds for saying about *the completion of a certain stage of our research into continuous-time difference equations.* One more reason is that the achievements in continuous-time difference equations theory, as our experience implies, are not sufficiently widely known. Therefore, it will be useful to briefly describe our approach to the study of such equations and present some relevant results, both those obtained earlier and recent ones.

Certainly, the development of the theory of the difference equations (1) should lean upon the theory of difference equations (2). At the same time, both the theories have essential distinctions, if for the fact that: Eq.(2) induces a one-dimensional dynamical system $x \mapsto f(x)$, whose phase space consists of points $x \in I$, and Eq.(1) induces an infinite-dimensional dynamical system $\varphi \mapsto f \circ \varphi$, whose phase space consists of functions $\varphi : [0,1] \to I$. For this reason, *of importance for Eq.(1) is the behavior of the trajectories not of points but of the neighborhoods of points under the map f.* Typically, the trajectories of neighborhoods of points are found asymptotically periodic. For instance,[5] the trajectory of any neighborhood is asymptotically periodic in the following cases:

- f is a C^2-smooth map that is non-flat at critical points,
- f is a piecewise linear map such that $f(x) \not\equiv \text{const}$ on any interval.

As a result, *the generic solutions of Eq.(1) are likewise asymptotically periodic.*[b] An explanation of why this is the case can be found in Ref. 6.

At present the Qualitative Theory of Continuous-Time Difference Equations includes the following basic topics:

(1) Two Main Types of Solutions: Uniformly Continuous and Asymptotically Discontinuous Solutions.
(2) Limit Properties of the Semigroup Generated by a Continuous Interval Map.

[b] In this regard Eq.(1) is simpler than Eq.(2), among generic solutions of which exist, as is well known, solutions that are not asymptotically periodic or asymptotically almost periodic.

(3) Dynamical Systems associated with Difference Equations:
Noncompactness of Trajectories. Completion with Upper Semicontinuous Functions. ω-Limit Sets of Generic Trajectories. Attractor. Topological Entropy. Bifurcations and Stability of Trajectories. Completion with Random Functions. Self-Stochasticity Phenomenon.

(4) Asymptotically Discontinuous Solutions:
Generator of Jumps and Classification of Solutions. Nonstandard Properties of Solutions: Self-Similarity, Fractality and Self-Stochasticity.

(5) Difference Equations with Unimodal Nonlinearity:
Limit Semigroup, Separator and Spectrum of Jumps. Solutions of Relaxation and Turbulent Types. Stability and Bifurcations of Solutions.

Here we put forth only the central ideas and results of this theory, therewith theorems are formulated in shortened and simplified form. For a deeper discussion of this theory and its applications we refer the reader to our works 3–28 and the references given there.

2. Idea of Investigation, Problems arising in so doing, and Main Results

Eq.(1) is proposed to investigate *with the method of going to its associated infinite-dimensional dynamical system*

$$S: \varphi \mapsto f \circ \varphi, \qquad \varphi \in C([0,1], I). \tag{3}$$

This method is used extensively in evolutionary problems. But here the employment of the method is faced with obstacles, caused by the *noncompactness* of the phase space $C([0,1], I)$. As a consequence, some trajectories may turn out to be noncompact, in which case their ω-limit sets are noncompact or even empty in $C([0,1], I)$. Such trajectories are "trying" to come out from the phase space: The behavior of functions $f^n(\varphi(t))$, that constitute the trajectory, becomes increasingly complicated when n grows, in particular, the gradient of $f^n(\varphi(t))$ increases infinitely as $n \to \infty$ and *gradient catastrophe* happens.

This gives no way of representing the limit behavior of $f^n(\varphi(t))$ with continuous functions. *The dynamical system needs to be extended on a wider function space and the notion of global attractor should be modified.* This calls for the finding of a metric (different from the *sup*-metrics) such that after completing the phase space $C([0,1], I)$ via this new metric:

(1) We obtain *a continuous extension* of the dynamical system (3) on the completed (extended) phase space;

(2) *Almost all trajectories* of the dynamical system (3) *are compact trajectories of the extended dynamical system.*

Here the significance of the term 'almost all' is as follows: By the words 'almost all trajectories have a property $\mathcal{A}$' we mean that the set of the functions $\varphi \in C([0,1], I)$ whose trajectories have the property $\mathcal{A}$ is massive in one sense or another, as a task in hand requires (for instance, the term 'massive' set can be used in reference to a set of full or positive measure, a dense set, a set of second Baire category, and etc).

If such a metric function space has been found, then we can modify the notion of global attractor, fitting Milnor's ideas[29] for noncompact phase spaces.

Definition 2.1. By the *global attractor* of a dynamical system on noncompact phase space we mean the smallest invariant closed set in the phase space of the extended dynamical system, that contains ω-limit sets of almost all trajectories of the original system.

In the course of solving these problems — "the egress" from the phase space in some wider function space and the construction of attractor — the following results have been established.

- We have suggested *a general approach to the study of dynamical systems on (noncompact) spaces of continuous or smooth functions.* This approach, in particular, proposes two special metrics which allow one to complete the phase space respectively with upper semicontinuous and random functions.
- We have introduced, based on the possibility to complete the phase space with random functions, *the notion of self-stochasticity in deterministic systems: The global attractor of a deterministic system contains random functions.* This notion is meaningful — self-stochasticity occurs in systems of the form (3), and 'physically realized' — self-stochasticity takes place over a positive measure set of parameter values, once f is parameter-depending.[c]
- We have substantiated *a scenario of spatial-temporal chaos in infinite-dimensional systems with regular dynamics on attractor:* The attractor consists of fixed points and cycles, and the system chaotization is due to

[c] How much would this model of changing from a deterministic behavior to a quasi-random one be in conformity with the reality? Of course, this model is usable for an adequate description of real processes only within limits and should most likely be adjusted beginning with some spatial-temporal scales.

a very complicated inherent structure of the attractor points, which are elements of some function space.

3. Metrics and Extended Phase Spaces

To realize the above-advanced general "strategy" in conformity to the dynamical system (3), we suggest two extended phase spaces endowed with special metrics.

The former space, denoted by C^{Δ}, is the completion of the phase space $C([0,1], I)$ with upper semicontinuous functions $\xi : [0,1] \to 2^I$ via the metric

$$\rho_{\mathrm{v}}^{\Delta}(\xi_1, \xi_2) = \sup_{\varepsilon>0} \min\left\{\varepsilon, \sup_{t\in[0,1]} \mathrm{dist}_H\left(V_{\xi_1}^{\varepsilon}(t), V_{\xi_2}^{\varepsilon}(t)\right)\right\}, \tag{4}$$

where $\mathrm{dist}_H(\cdot,\cdot)$ is Hausdorff distance between sets, $V_{\xi}^{\varepsilon}(t) = \xi(V_{\varepsilon}(t))$, $V_{\varepsilon}(\cdot)$ is the ε-neighborhood of a point.

The metric $\rho_{\mathrm{v}}^{\Delta}$ is equivalent to Hausdorff metric for graphs of functions

$$\rho^{\Delta}(\xi_1, \xi_2) = \mathrm{dist}_H(\mathrm{gr}\,\xi_1, \mathrm{gr}\,\xi_2), \qquad \mathrm{gr}\,\xi \text{ is the graph of } \xi,$$

which is of frequent use in research into continuous and upper semicontinuous functions, and is, in many cases, handier than the metric $\rho_{\mathrm{v}}^{\Delta}$. This fact allows one to understand the meaning of the convergence in the space C^{Δ}: The convergence of a function sequence ξ_i to a function ξ is equivalent to the following:

$$\mathrm{Lt}_{\,i\to\infty}\, \mathrm{gr}\,\xi_i = \mathrm{gr}\,\xi, \tag{5}$$

where Lt is for the topological limit of a sequence of sets.

The latter space, denoted by $C^{\#}$, is the completion of $C([0,1], I)$ with deterministic (measured) and random functions $\zeta : [0,1] \to I$, given by their finite-dimensional distributions $F_{\zeta}^{r}(z,t)$, $z \in I^{r}$, $t \in [0,1]^{r}$, $r = 1, 2, \ldots,$ via the metric[d]

$$\rho^{\#}(\zeta_1, \zeta_2) = \sup_{\varepsilon>0} \min\left\{\varepsilon, \sum_{r=1}^{\infty} \frac{1}{2^{r}}\, \mathrm{dist}_R\left(F_{\zeta_1}^{r,\varepsilon}(z,t), F_{\zeta_2}^{r,\varepsilon}(z,t)\right)\right\}, \tag{6}$$

where

$$\mathrm{dist}_R\left(F_{\zeta_1}^{r,\varepsilon}, F_{\zeta_2}^{r,\varepsilon}\right) = \sup_{(z,t)\in I^{r}\times[0,1]^{r}} \left|F_{\zeta_1}^{r,\varepsilon}(z,t) - F_{\zeta_2}^{r,\varepsilon}(z,t)\right|, \tag{7}$$

[d] Originally we used a somewhat different metric as $\rho^{\#}$. As a recent close look revealed, that metric can be replaced with the stronger metric (6), (7).

$F^{r,\varepsilon}_{\zeta}(z,t) = F^{r,\varepsilon}_{\zeta}(z_1,\dots,z_r;\, t_1,\dots,t_r)$ is the average of the r-dimensional distribution $F_\zeta(z_1,\dots,z_r;\,\cdot,\dots,\cdot)$ over the ε-neighborhood of the point $t = (t_1,\dots,t_r)$. Random functions with the same finite-dimensional distributions are thought to be identical. It is worth noting that by a random function (process) we actually mean the distributions of this function (in other words, the measure on the space of selectors). We exploit the term 'random function' in the above sense for convenience sake (which is not commonly accepted).

The meaning of the convergence in the space $C^\#$ is as follows:
If a function sequence ζ_i converges to a function ζ, then

$$\nu_i(\cdot,t) \implies \nu(\cdot,t), \tag{8}$$

where $\nu_i(\cdot,t)$ and $\nu(\cdot,t)$ are the probability measures generated respectively by the distributions $F^{r,\varepsilon}_{\zeta_i}(z,t)$ and $F^{r,\varepsilon}_{\zeta}(z,t)$ (for any given $r \in \mathbb{N}^+$ and $\varepsilon > 0$) and $\implies$ is for the weak convergence of probability measures.

The space C^Δ always "works": Completing the phase space $C([0,1],I)$ via the metric ρ^Δ_{v} results in that *all the trajectories of the dynamical system (3) become compact in the extended phase space* C^Δ (with respect to ρ^Δ_{v}). This is a consequence of the fact that the space $2^{[0,1]\times I}$ is compact with respect to Hausdorff metric. As to the space $C^\#$, its "capacity for work" depends on the properties of the map f — the action of the dynamical system. Below we present the conditions on f under which the space $C^\#$ "works". The principal condition is the existence of an ergodic smooth invariant measure for f.

4. Attractor in the Space $\mathbf{C}^\Delta$

As noted above, the global attractor for the dynamical system (3) in the space C^Δ always exists. Moreover, it consists, in sufficiently wide conditions, of periodic and almost periodic trajectories of the C^Δ-extended dynamical system. "Points" of the attractor — upper semicontinuous functions — possess an intricate inherent structure. In particular, it may occur that *the values of these functions are (nondegenerated) intervals on a Cantor-like set or on an interval*.

Let $\omega_\Delta[\varphi]$ be the ω-limit set (nonempty and compact) of the trajectory of the C^Δ-extended dynamical system, which starts from the "point" φ, and $Q_f(z)$ be the domain of influence[e] of the point z, i.e.,

$$Q_f(z) = \cap_{\varepsilon>0} \cap_{j\geq 0} \overline{\cup_{n\geq j} f^n\big(V_\varepsilon(z)\big)}, \quad V_\varepsilon(z) = (z-\varepsilon,\, z+\varepsilon) \cap I.$$

In Sec. 4, we will use the term 'almost all' in the topological sense. In particular, in Definition 2.1 'almost all trajectories' now means that the initial functions generated these trajectories form a set of second Baire category in $C([0,1], I)$.

Theorem 4.1. *For almost all* $f \in C(I,I)$ *and* $\varphi \in C([0,1], I)$, *the set* $\omega_\Delta[\varphi]$ *is a periodic trajectory of the* C^Δ *- extended dynamical system. More precisely, there exists an integer* $N > 0$ *such that*

$$\omega_\Delta[\varphi] \;=\; \{f^\Delta \circ \varphi, \quad f \circ f^\Delta \circ \varphi, \quad \ldots, \quad f^{N-1} \circ f^\Delta \circ \varphi\}, \tag{9}$$

where $f^\Delta : I \to 2^I$ *is the upper semicontinuous function*

$$f^\Delta(z) \;=\; Q_{f^p}(z), \qquad z \in I. \tag{10}$$

Now we need the following sets. Let

- $Q_f^+(z)$ and $Q_f^-(z)$ be respectively the right and the left domains of influence of z, i.e.,

$$Q_f^+(z) = \cap_{\varepsilon>0} \cap_{j\geq 0} \overline{\cup_{n\geq j} f^n\big(V_\varepsilon^+(z)\big)}, \quad V_\varepsilon^+(z) = [z,\, z+\varepsilon) \cap I,$$
$$Q_f^-(z) = \cap_{\varepsilon>0} \cap_{j\geq 0} \overline{\cup_{n\geq j} f^n\big(V_\varepsilon^-(z)\big)}, \quad V_\varepsilon^-(z) = (z-\varepsilon,\, z] \cap I;$$

- $\mathcal{D}(f)$ be the separator of the map f, i.e.,

$$\mathcal{D}(f) = \{z \in I : \text{ the trajectory of } z \text{ under } f \text{ is Lyapunov unstable}\};$$

- $\mathcal{D}^*(f) = \{z \in \mathcal{D}(f) : \; Q_f^+(z) \neq Q_f^-(z)\}$.
- $\Phi(f)$ be the set of initial functions $\varphi \in C([0,1], I)$ such that:

$$\varphi(t_*) \notin \mathcal{D}(f) \text{ if } \varphi(t) \text{ is a constant in the vicinity of } t = t_*;$$
$$\varphi(t_*) \notin \mathcal{D}^*(f) \text{ if } \varphi(t) \text{ attains extremum at } t = t_*. \tag{11}$$

The principal significance of the conditions (11) is to ensure the fulfilment of the relationship

$$\cap_{\varepsilon>0} \cap_{j\geq 0} \overline{\cup_{n\geq j} f^n\big(\varphi(V_\varepsilon(t))\big)} = \cap_{\varepsilon>0} \cap_{j\geq 0} \overline{\cup_{n\geq j} f^n\big(V_\varepsilon(\varphi(t))\big)}.$$

[e] The notion 'domain of influence' made its appearance in,[3] wherein it was introduced as a development of the well-known notion 'prolongation'. Later the notion of domain of influence was treated by another authors, but without resorting to this name.

Theorem 4.2. *If $\mathcal{D}^*(f)$ is a set of first Baire category,*[f] *then the global attractor of the dynamical system (3) in the space C^Δ consists of periodic and almost periodic trajectories of the C^Δ-extended dynamical system, namely:*

$$\mathcal{A}^\Delta = \overline{\cup_{\varphi \in \Phi(f)} \cup_{i \geq 0} \; f^i \circ f^\Delta \circ \varphi}. \tag{12}$$

5. Attractor in the Space $\mathbf{C}^\#$

The employment of the metric $\rho^\#$ allows us to substantiate the *self-stochasticity phenomenon*: In sufficiently general conditions, the global attractor of the dynamical system (3) in the space $C^\#$ exists and consists of cycles of the $C^\#$-extended dynamical system, therewith the "points" of the attractor are random functions.

Let $MC_{\mu,\,p}$ be the set of piecewise monotone maps $f \in C(I, I)$ such that:

- the map f is *nonsingular*, i.e.,
$$\text{if } \operatorname{mes} B = 0, \quad \text{then } \operatorname{mes} f^{-1}(B) = 0;$$
- the map f has a probabilistic *smooth invariant measure* μ, i.e.,
$$\mu(I){=}1, \; \mu(\varnothing){=}0, \; \mu(f^{-1}(B)){=}\mu(B); \quad \text{if } \operatorname{mes} B{=}0, \; \text{ then } \mu(B){=}0;$$
- the support $\operatorname{supp} \mu$ of the measure μ is the union of closed intervals $E_1, \ldots, E_p$ that form a period-p (transitive) cycle of intervals;[g]
- the measure μ is *equivalent to Lebesgue measure on its support*, i.e.,
$$\mu(B) = 0 \quad \text{if and only if} \quad \operatorname{mes} B = 0, \quad B \subset \operatorname{supp} \mu;$$
- the map f^p is *intermixing* on each interval E_i, $i = 1, \ldots, p$, i.e.,
$$\lim_{j\to\infty} \mu(B_1 \cap f^{-pj}(B_2)) = p \cdot \mu(B_1)\,\mu(B_2), \qquad B_1, B_2 \subset E_i;$$
- $\operatorname{mes} E_* = 0$, where E_* is the boundary of the basin of the measure μ, i.e., E_* is the boundary of the set
$$\mathcal{E}_f(\mu) = \cup_{i=0}^{p-1} \cup_{j\geq 0} \operatorname{int} f^{-j}(E_i).$$

---

[f] This condition is sufficiently general; in particular,[3] $\mathcal{D}_*(f)$ is nowhere dense for almost all unimodal C^3-smooth f.

[g] Closed intervals $E_1, E_2, \ldots, E_p$ are referred to as a period-p cycle of intervals for a map f if these intervals are moved cyclicly by the map f and are disjoint by interior in pairs; if, in addition, the union $\cup_{s=1}^{p} E_s$ contains an everywhere dense trajectory of the map f, then the cycle of intervals is referred to as transitive.

For ample classes of continuous interval maps dependent on parameter, the set of parameter values marked by the fulfilment of these conditions is of positive Lebesgue measure (see for instance Ref. 30). In particular, the conditions are realized when f is a unimodal C^3-smooth map with negative Schwartz derivative, that satisfies the Collet–Eckmann relationship:

$$\lim_{k\to\infty} \inf \frac{1}{k} \log \left| \frac{d}{dt} f^k(c) \right| > 0, \qquad c \text{ is the critical point of } f.$$

Let $\omega_{\#}[\varphi]$ be the ω-limit set (which may turn out to be empty) of the trajectory of the $C^{\#}$-extended dynamical system, which starts from the "point" φ;

We restrict our consideration to the simplest case:

(i) $f \in MC_{\mu,1}$, that is, $\operatorname{supp}\mu$ consists of a single interval;

(ii) $I = \overline{\mathcal{E}_f(\mu)}$, that is, $\operatorname{supp}\mu$ attracts under f all the points from I outside of a zero Lebesgue measure set.

In Definition 2.1, we now take 'almost all trajectory' to mean 'the trajectories generated by nonsingular initial functions'.

Theorem 5.1. *In conditions* (i) *and* (ii), *for any nonsingular function* $\varphi \in C([0,1], I)$, *the set* $\omega_{\#}[\varphi]$ *consists of a single random function, which is a fixed point of the* $C^{\#}$*-extended dynamical system. More precisely,*

$$\omega_{\#}[\varphi] = \{\varphi^{\#}\}, \tag{13}$$

where $\varphi^{\#} : [0,1] \to I$ *is the random process given by the distributions*

$$F_{\varphi^{\#}}(z_1, \ldots, z_r; t_1, \ldots, t_r) = \prod_{i=1}^{s} \mu\Big((\infty, z(i)] \cap \operatorname{supp}\mu \Big), \quad z(i) = \min_{k\in M_i} \{z_k\},$$

where $M_1, \ldots, M_s$, $s \le r$, *are the equivalence classes into which the set* $\{1, \ldots, r\}$ *is subdivided by the equivalence relation defined as follows:* $i \sim j$ *if* $\varphi(t_i) = \varphi(t_j)$, $i, j = 1, \ldots, r$.

Theorem 5.2. *In conditions* (i) *and* (ii), *the global attractor* $\mathcal{A}^{\#}$ *of the dynamical system (3) in the space* $C^{\#}$ *consists of the only point* $\{\varphi^{\#}\}$, *which is a fixed point of the* $C^{\#}$*-extended dynamical system.*

If $f \in MC_{\mu,p}$, where $p > 1$, but (ii) remains valid, then every nonsingular function φ is matched by its own ω-limit set; more precisely, every set $\omega_{\#}[\varphi]$ is a period-p trajectory of the $C^{\#}$-extended dynamical system and consists of random functions, whose distributions are given by the invariant measure μ and depend on φ. The attractor $\mathcal{A}^{\#}$ consists just of these period-p trajectories.

Theorems 4.2 and 5.2 show that *the hypothetic scenario of spatial-temporal chaos in parameter distributed systems with regular dynamics on attractor do can occur in dynamical systems of the form (3).* Moreover, Theorem 5.1 and its generalization on the case $p > 1$ provide *a means of describing deterministic chaos in the probabilistic terms.*

6. Long-Term Properties of Solutions

Based on the results obtained for the dynamical system (3), we can describe the long-term behavior of solutions for the nonlinear continuous-time difference equations (1)

$$x(t+1) \;=\; f(x(t)), \qquad t \in \mathbb{R}^{+}.$$

The relation between the solution $x_{\varphi}(t)$, generated by the initial function $\varphi : [0,1] \to I$, and the trajectory $S^{n}[\varphi] = \{\varphi,\ f \circ \varphi,\ \ldots\}$, originating at the "point" φ, is as follows:

$$x_{\varphi}(t) \;=\; S^{\langle t \rangle}[\varphi](\{t\}), \quad t \in \mathbb{R}^{+}, \tag{14}$$

with $\langle \cdot \rangle$ and $\{\cdot\}$ being respectively for the integral and fractional parts of a number. Hence the long-term properties of the solutions can be characterized with help of the ω-limit sets of the trajectories.

Definition 6.1. We say that *a continuous function* $u : \mathbb{R}^{+} \to I$ *approaches to a periodic (or almost periodic) upper semicontinuous function* $\mathcal{P} : \mathbb{R}^{+} \to 2^{I}$, if

$$\rho^{\Delta}\left(u(t+T),\ \mathcal{P}(t+T)\right) \to 0 \quad \text{as} \quad T \to \infty \qquad \text{for } t \in [0,1]. \tag{15}$$

If (15) holds, we refer to $\mathcal{P}(t)$ as the *limit function* of $u(t)$.

Formula (14) and the above theorems concerning the dynamical system (3) allow us to the characterize long-term properties of the solutions of Eq. (1). Here we present only some results that steam from the asymptotic dynamics of the system (3) in the space C^{Δ}.

Theorem 6.1. *For any* φ *meeting (11), the solution* $x_{\varphi}(t)$ *of Eq. (1) is asymptotically periodic or asymptotically almost periodic. In particular, if* $\omega_{\Delta}[\varphi]$ *is a period-*N *trajectory of the* C^{Δ}*-extended dynamical system,* *then* $x_\varphi(t)$ *approaches to the period-*N *upper semicontinuous function*

$$\mathcal{P}_\varphi(t) = \left(f^{\langle t\rangle \bmod N} \circ f^\Delta \circ \varphi\right)(\{t\}), \qquad t \in \mathbb{R}^+. \tag{16}$$

Thus, the limit function $\mathcal{P}_\varphi(t)$ of the solution $x_\varphi(t)$ is obtained by "sewing tail to head" the elements of the corresponding ω-limit set $\omega_\Delta[\varphi]$. Each limit function can be treated as a generalized solution for Eq. (1).

Theorem 6.1 and the properties of the upper semicontinuous function f^Δ imply the following peculiar features of the long-term behavior of the solutions.

• *Representative of Eq.(1) are asymptotically discontinuous solutions* — continuous bounded functions that are not uniformly continuous on $\mathbb{R}^+$. In typical situations, the number of undamped oscillations and the gradient of such a solution on $[T, T+1]$ increase indefinitely as $T \to \infty$. Solutions of this kind are named *turbulent.* They can be classified by the cardinality of the *generator of jumps* — the set of points $t \in [0,1]$ at which the values of the solution limit function are intervals. The amplitudes of undamped oscillations for a turbulent solution are characterized by the *spectrum of jumps* — the collection of intervals that are the values of the solution limit function on the generator of jumps. The conditions for existence of one or other type of turbulent solutions is best shown by Eq. (1) with f being a unimodal map. In this case, a description of long-term properties of solutions are derived from the spectral decomposition of the nonwandering points set of the map f.

• Turbulent solutions have a number of nonstandard properties, which testify that the solutions behave very irregular. *The graphs of limit functions for turbulent solutions are locally self-similar, and, in wide conditions, fractal.* These properties take place, in particular, where the generator of jumps contains intervals. Then the graph of a solution becomes with time similar to a space-filling curve, as a result of which *the solution comes out the horizon of predictability*: The solution values at t large enough cannot be calculated with assurance.

• The results on the asymptotic dynamics of the system (3) in the space $C^\#$ make, in many cases, it possible to give a probabilistic representation of the above-mentioned "unpredictable" solutions. When $f \in MC_{\mu,p}$, *the long-term behavior of an "unpredictable" solution of Eq.(1) can be described with a certain random process*, defined in terms of the invariant measure of f. More specifically, the averaged finite-dimensional distributions of the solution are close to the corresponding finite-dimensional distributions of the random process when time is large enough.[28]

7. Application to boundary value problems

In Introduction we have mentioned three major areas of application of the continuous-time difference equations. Now we dwell slightly on the second of these — the applications to the theory of boundary value problems (BVPs) for partial differential equations (PDEs) — because it is quite promising. Further reason is that the advances in the field of boundary value problems provide a highly efficient mathematical means for the third area of application — modelling spatial-temporal chaos (indeed, the most frequently used type of mathematical models for nonlinear wave processes are just BVPs for PDEs).

Examples of BVPs reducible to difference, functional, differential-functional and other relevant equations have long been known[h], but their effective study has been made possible only in the last years due to the progress of the theory of continuous-time difference equations. Hyperbolic PDEs are a rich source of reducible BVPs. Here we will not detail this question (which is fairly fully considered in[3,21,24]) and only present the simplest example:

$$\frac{\partial u}{\partial t} = \frac{\partial u}{\partial y}, \quad y \in [0,1], \quad t \in \mathbb{R}^+, \tag{17}$$

$$u\,|_{y=1} = f(u)\,|_{y=0}, \tag{18}$$

where $f \in C^1(I, I)$ is a nonlinear map, I is a closed bounded interval. By substituting the general solution of Eq. (17)

$$u(y,t) = w(y+t) \quad \text{with } w \text{ bing an arbitrary } C^1\text{-smooth function,}$$

into (18), the BVP is reduced to the difference equation

$$w(\tau + 1) = f(w(\tau)), \quad \tau \in \mathbb{R}^+, \tag{19}$$

Every initial condition $u(y,0) = \varphi(y)$ for Eq. (17, (18) induces the initial condition $w(\tau) = \varphi(\tau)$, $\tau \in [0,1]$, for Eq. (19). Thus, if u_φ and w_φ stand for the solutions of the corresponding equations, which are generated by the initial function φ, then

$$u_\varphi(y,t) = w_\varphi(y+t). \tag{20}$$

---

[h] Such a reduction is due to the fact that the general solution for the PDE entering into a BVP can be written as an analytic formulae involving only elementary operations on arbitrary functions.

Formulae (20) allows one to extend without trouble the results obtained for Eq. (1) to the BVP (17), (18). In particular, a reformulation of Theorem 6.1 is as follows.

Theorem 7.1. *For any* φ *meeting (11), the solution* $u_\varphi(y,t)$ *of the BVP (17), (18) is asymptotically periodic or asymptotically almost periodic in* t. *In particular, if* $\omega_\Delta[\varphi]$ *is a period-*N *trajectory of the* C^Δ*-extended dynamical system, then* $u_\varphi(y,t)$ *approaches to the upper semicontinuous function*

$$\mathcal{P}_\varphi(y,t) = \left(f^{\langle y+t\rangle \bmod N} \circ f^\Delta \circ \varphi \right)(\{y+t\}), \quad y \in [0,1],\ t \in \mathbb{R}^+, \tag{21}$$

in the sense that

$$\rho^\Delta\left(u_\varphi(y,T),\ \mathcal{P}_\varphi(y,T) \right) \to 0 \quad as \quad T \to \infty.$$

The problem (17), (18) is the simplest conceivable nonlinear BVP. As a model problem, we have taken just it for the ease of explanations. There are, of course, many other one- and many-dimensional BVPs reducible to continuous-time difference equations. The reduction of such BVPs and the subsequent study, on this basis, of their solutions calls, generally speaking, for much more efforts.

8. Conclusion

Continuous-time difference equations are in our view respectively simple and very profitable instrument for the study of various evolutionary problems modeling wave processes. This paper have hopefully given a rough idea of continuous-time difference equations and shown that even the simplest nonlinear equations of the form (1) have very complicated behavioral solutions up to quasirandom ones whose asymptotic properties are described in terms of random processes. All this not only leads to the no less complicated dynamics of the original evolutionary problems but also provides comparatively simple scenarios for intricate nonlinear phenomena such as the cascade process of structures emergence, producing fractal sets, chaotic mixing, and self-stochasticity.

Further investigation should be aimed at the following lines:

(1) Systems of continuous-time difference equations;
(2) Perturbed and nonautonomous difference equations;
(3) Differential-difference equations;

(4) Boundary value problems for partial difference equations, which are reducible to equations close the difference ones.

Something has already been done in these lines: There have been made the study of the so-called completely integrable differential-difference equations[3] and some classes of perturbed difference equations.[3,31,32]

References

1. A. N. Sharkovsky, *Ukr. Mat. Zurn.* **16** (1964) [in Russian]; *Int. J. Bifurcation and Chaos* **5**, 5 (1995).
2. A. N. Sharkovsky, On the periodic solutions of a class of finite differences equations, in *Proc. Fourth Conference on Nonlinear Oscillations* (Prague, Czechoslovak, 1967), (Academia, CSAV, Prague, 1968).
3. A. N. Sharkovsky, Yu. L. Maistrenko and E. Yu. Romanenko, *Difference Equations and Their Applications*, Ser. Mathematics and Its Applications **250** (Kluwer Academic Publishers, Dordrecht, 1993).
4. E. Yu. Romanenko, *Ukrainian Math. J.* **58**, 7 (2006).
5. V. V. Fedorenko, E. Yu. Romanenko and A. N. Sharkovsky, *J. Difference Equations and Appl.* **13**, 8–9 (2007).
6. E. Yu. Romanenko, *Int. J. Difference Equations and Appl.* **9**, 3–4 (2003).
7. A. N. Sharkovsky and E. Yu. Romanenko, *Int. J. Bifurcation and Chaos* **2**, 1 (1992).
8. A. N. Sharkovsky, *IEEE Transactions on Circuits and Systems.* **40**, 10 (1993).
9. A. N. Sharkovsky, Yu. L. Maistrenko, Ph. Deregel and L. O. Chua, *Systems and Computers* **3**, 2 (1993).
10. A. N. Sharkovsky, *Int. J. Bifurcation and Chaos* **4**, 2 (1994).
11. A. N. Sharkovsky and A. G. Sivak, *Nonlinear Math. Physics* **1**, 2 (1994).
12. E. Yu. Romanenko, A. N. Sharkovsky and M. B. Vereikina, *Int. J. Bifurcation and Chaos* **5**, 5 (1995).
13. A. N. Sharkovsky, *Int. J. Bifurcation and Chaos* **5**, 5 (1995).
14. A. N. Sharkovsky, From one-dimensional to infinite-dimensional dynamical systems, in *Proc. Int. Workshop on Nonlinear Dynamics of Electronic Systems (NDES'95)*, (Dublin, Ireland, 1995).
15. A. N. Sharkovsky, Ph. Deregel and L. O. Chua, *Int. J. Bifurcation and Chaos* **5**, 5 (1995).
16. E. Yu. Romanenko and A. N. Sharkovsky, *Ukrainian Math. J.* **48**,12 (1996).
17. E. Yu. Romanenko, A. N. Sharkovsky and M. B. Vereikina, Structural turbulence in boundary value problems, in *Proc. Conf. on Control of Oscillations and Chaos*, (St.Petersburg, Russia, 1997).
18. A. D. Fedorenko, V. V. Fedorenko, A. F. Ivanov and A. N. Sharkovsky, *Bull. Australian Math. Soc.* **57** (1998).
19. E. Yu. Romanenko, *Computers and Mathematics with Appl.* **36**, 10–12 (1998).
20. E. Yu. Romanenko and A. N. Sharkovsky, *Ukrainian Math. J.* **51**, 6 (1999).
21. E. Yu. Romanenko and A. N. Sharkovsky, *Int. J. Bifurcation and Chaos* **9**, 7 (1999).

22. E. Yu. Romanenko, A. N. Sharkovsky and M. B. Vereikina, *Nonlinear Boundary Value Problems* (Donetsk, Ukraine) **9** (1999).
23. A. N. Sharkovsky, *Ann. Math. Silesianae* **13** (1999).
24. E. Yu. Romanenko and A. N. Sharkovsky, *Proc. Steklov Inst. Math.* **244** (2004).
25. A. N. Sharkovsky and E.Yu. Romanenko, *Turbulence: Ideal*, in *Encyclopedia of Nonlinear Science*, ed. Alwyn Scott, (Routledge, New York–London, 2005).
26. A. N. Sharkovsky, *Nonlinear Dynamics* **44** (2006).
27. A. N. Sharkovsky, E.Yu. Romanenko and V. V. Fedorenko, *Regular and Chaotic Dynamics* **11**, 2 (2006).
28. E. Yu. Romanenko, *Int. J. Difference Equations and Appl.* (2009), to appear.
29. J. Milnor, *Commun. Math. Phys.* **99** (1985).
30. A. Avila, M. Lyubich and W. de Melo, *Inventiones Mathematicae* **154** (2003).
31. A. N. Sharkovsky, Yu. L. Maistrenko and E.Yu. Romanenko, *Differential equations and their applications* (Lit.SSR, Vilnus) **29** (1981) [in Russian].
32. R. P. Agarwal and E.Yu. Romanenko, *Appl. Math. Lett.* **11**, 4 (1998).

Lyapunov-Schmidt: A Discrete Revision

André Vanderbauwhede
Department of Pure Mathematics, Ghent University, Belgium
E-mail: avdb@cage.ugent.be
http://cage.ugent.be/~avdb

We present an adapted version of the classical Lyapunov-Schmidt reduction method to study, for some given integer $q \geq 1$, the bifurcation of q-periodic orbits from fixed points in discrete autonomous systems. The approach puts some particular emphasis on the $\mathbb{Z}_q$-equivariance of the reduced problem. We also discuss the relation with normal form theory, consider special cases such as equivariant, reversible or symplectic mappings, and obtain some results on the stability of the bifurcating periodic orbits. We conclude with an application of the approach to the generic bifurcation of q-periodic orbits for $q \geq 3$, and showing how for $q \geq 5$ Arnol'd tongues appear as an immediate consequence of the $\mathbb{Z}_q$-equivariance.

Keywords: Bifurcation; periodic orbits; Lyapunov-Schmidt reduction; discrete systems; normal forms; stability.

1. Introduction

In this review paper we describe how the Lyapunov-Schmidt reduction method can be used to study the bifurcation of periodic orbits from fixed points in discrete autonomous systems. The approach which we will present here is not so well known, as the method was mainly developped to study the bifurcation of periodic orbits in continuous systems. We will emphasize the advantages of the method, in particular the symmetry induced by it, and show how the method can be combined with normal form reductions to give additional information (such as stability properties) on the bifurcating periodic orbits.

Our starting point is a parametrized family of local diffeomorphisms with a fixed point at the origin; more precisely we consider smooth mappings $f : \mathbb{R}^n \times \mathbb{R}^m \to \mathbb{R}^n$ with $f(0, \lambda) = 0$ and $A_\lambda := D_x f(0, \lambda) \in \mathcal{L}(\mathbb{R}^n)$ invertible for all $\lambda \in \mathbb{R}^m$. Fix some integer $q \geq 1$ and some critical parameter value $\lambda_0 \in \mathbb{R}^m$; for convenience of notation we will take $\lambda_0 = 0$. The problem we want to discuss is then the following:

($\mathbf{P}_q$) Determine, for all values of λ near the critical value $\lambda_0 = 0$, all small q-periodic orbits of $f_\lambda := f(\cdot\,,\lambda)$.

The classical approach to this problem is to look for (small) fixed points of the q-th iterate $f_\lambda^q := f_\lambda \circ f_\lambda \circ \cdots \circ f_\lambda$ (q times) of f_λ. Such approach overlooks some of the hidden symmetries of the problem and hence can not take advantage of these symmetries. The approach which we will describe here reformulates the problem in such a way that the symmetry becomes explicit from the outset.

Our formulation is based on the notion of the *orbit space* for the given problem. Every q-periodic orbit of f_λ can be seen as a bi-infinite sequence $z = (x_j)_{j\in\mathbb{Z}}$ of points in $\mathbb{R}^n$ which is q-periodic: $x_{j+q} = x_j$ for all $j \in \mathbb{Z}$. Therefore we introduce the orbit space

$$\mathcal{O}_q := \{z = (x_j)_{j\in\mathbb{Z}} \mid x_j \in \mathbb{R}^n \text{ and } x_{j+q} = x_j,\ \forall j \in \mathbb{Z}\}. \tag{1}$$

Observe that this space is finite-dimensional, isomorphic to $(\mathbb{R}^n)^q$. The fact that an element z of $\mathcal{O}_q$ forms an orbit under the mapping f_λ is expressed by the relation $x_{j+1} = f_\lambda(x_j)$, valid for all $j \in \mathbb{Z}$. To write this as an equation in the orbit space we define for each mapping $g : \mathbb{R}^n \to \mathbb{R}^n$ the *lift* of g to $\mathcal{O}_q$ as the mapping $\hat{g} : \mathcal{O}_q \to \mathcal{O}_q$ given by

$$\hat{g}(z) := (g(x_j))_{j\in\mathbb{Z}}, \quad \forall z = (x_j)_{j\in\mathbb{Z}} \in \mathcal{O}_q.$$

We also define the linear *shift operator* $\sigma \in \mathcal{L}(\mathcal{O}_q)$ by

$$(\sigma \cdot z)_j := x_{j+1}, \quad \forall j \in \mathbb{Z},\ \forall z = (x_j)_{j\in\mathbb{Z}} \in \mathcal{O}_q.$$

The relation $x_{j+1} = f_\lambda(x_j)$ (for all $j \in \mathbb{Z}$) then takes the form

$$\hat{f}(z,\lambda) = \sigma \cdot z, \tag{2}$$

where we have written $\hat{f}(z,\lambda)$ for $\hat{f}_\lambda(z)$. The problem ($\mathbf{P}_q$) is equivalent to finding all solutions $(z,\lambda) \in \mathcal{O}_q \times \mathbb{R}^m$ of (2) near $(0,0)$.

An important property of (2) is its $\mathbb{Z}_q$-equivariance, as follows. Clearly $\sigma^q = \mathrm{Id}_{\mathcal{O}_q}$, and hence σ generates a $\mathbb{Z}_q$-action on the orbit space $\mathcal{O}_q$ (here $\mathbb{Z}_q \cong \mathbb{Z}/q\mathbb{Z}$ is the cyclic group with q elements). Moreover it is trivial to verify that for each mapping $g : \mathbb{R}^n \to \mathbb{R}^n$ we have that $\hat{g} \circ \sigma = \sigma \circ \hat{g}$; in particular

$$\hat{f}(\sigma \cdot z,\lambda) = \sigma \cdot \hat{f}(z,\lambda), \quad \forall (z,\lambda) \in \mathcal{O}_q \times \mathbb{R}^m.$$

Hence both sides of (2) commute with σ, and if $(z,\lambda) \in \mathcal{O}_q \times \mathbb{R}^m$ is a solution of (2), then so is $(\sigma^j \cdot z,\lambda)$ for all $j \in \mathbb{Z}$. In the next section we will apply a

Lyapunov-Schmidt reduction to (2), and when doing so it is important to make sure that the reduced problem retains this $\mathbb{Z}_q$-equivariance.

Not only the existence part of the problem $(\mathbf{P}_q)$ can be reformulated as solving an equation in the orbit space $\mathcal{O}_q$, but also the stability properties of the bifurcating q-periodic orbits can be obtained from this equation. It is well known that the stability properties of a q-periodic orbit $z = (x_j)_{j\in\mathbb{Z}}$ of f_λ are determined by the eigenvalues of

$$Df_\lambda^q(x_0) = Df_\lambda(x_{q-1})\circ\cdots\circ Df_\lambda(x_1)\circ Df_\lambda(x_0) \in \mathcal{L}(\mathbb{R}^n);$$

if all eigenvalues are strictly inside the unit circle the the periodic orbit is *asymptotically stable*, if some of the eigenvalues are strictly outside the unit circle then the periodic orbit is *unstable*. (In certain particular cases, such as for symplectic or reversible maps, one will rather distinguish between *elliptic* and *hyperbolic* periodic orbits, depending on whether all eigenvalues are on the unit circle or some are off the unit circle). Some easy manipulation of the eigenvalue-eigenvector equation shows the following.

Theorem 1.1. *For each smooth mapping $g : \mathbb{R}^n \to \mathbb{R}^n$, each $z = (x_j)_{j\in\mathbb{Z}} \in \mathcal{O}_q$ and each $\mu \in \mathbb{C}$ $(\mu \neq 0)$ we have the following: μ^q is an eigenvalue of $Dg(x_{q-1})\circ\cdots\circ Dg(x_1)\circ Dg(x_0) \in \mathcal{L}(\mathbb{R}^n)$ if and only if μ is an eigenvalue of $\sigma^{-1}\circ D\hat{g}(z) \in \mathcal{L}(\mathcal{O}_q)$, that is, if and only if the nullspace of*

$$D\hat{g}(z) - \mu\sigma \in \mathcal{L}(\mathcal{O}_q)$$

is nontrivial.

In relation to this result (and for counting multiplicities) one should observe that if $\mu \in \mathbb{C}$ is an eigenvalue of $\sigma^{-1}\circ D\hat{g}(z)$, say with eigenvector $\tilde{z} = (\tilde{x}_j)_{j\in\mathbb{Z}}$, then so is $\mu\exp(2\pi ip/q)$ for each $p \in \mathbb{Z}$; the corresponding eigenvector is $(\exp(2\pi ijp/q)\tilde{x}_j)_{j\in\mathbb{Z}}$. It follows that if $z \in \mathcal{O}_q$ is close to zero then the eigenvalues of $\sigma^{-1}\circ D\hat{g}(z)$ will be close to the set

$$\{\mu\exp(2\pi ip/q) \mid p \in \mathbb{Z},\ \mu = \text{eigenvalue of } Dg(0)\}.$$

In the next sections we will apply a Lyapunov-Schmidt reduction to the equation (2), taking into account the $\mathbb{Z}_q$-equivariance of that equation, and use Theorem 1.1 to obtain some information on the stability of bifurcating periodic orbits. We will also explore the relation between the reduced equation and the normal form of the mappings f_λ. More details on certain parts of the exposition can be found in our earlier paper.[5]

2. Lyapunov-Schmidt reduction

The first step in studying the solution set of (2) near $(0,0)$ is to linearize (in the z-variable) at that point. This gives the linear equation

$$\hat{A}_0 z = \sigma \cdot z, \qquad \text{with } A_0 := D_x f(0,0) \in \mathcal{L}(\mathbb{R}^n). \tag{3}$$

If this equation has only the zero solution $z = 0$ then, by the implicit function theorem, the branch of trivial solutions $\{(0,\lambda) \mid \lambda \in \mathbb{R}^m\}$ of (2) is isolated for small λ, and there is no bifurcation of non-trivial q-periodic orbits near $\lambda = 0$. So we will further on implicitly assume that (3) has some non-trivial solutions. If $z = (x_j)_{j\in\mathbb{Z}} \in \mathcal{O}_q$ is a solution of (3) then $x_0 \in \ker(A_0^q - \mathrm{Id}_{\mathbb{R}^n})$ and $x_{j+1} = A_0 x_j$ for all $j \in \mathbb{Z}$. Conversely, if $x_0 \in \ker(A_0^q - \mathrm{Id}_{\mathbb{R}^n})$ then $z := (A_0^j x_0)_{j\in\mathbb{Z}}$ belongs to $\mathcal{O}_q$ and is a solution of (3). Hence there exists an isomorphism between the subspace $\ker(A_0^q - \mathrm{Id}_{\mathbb{R}^n})$ of $\mathbb{R}^n$ and the solution space of (3). Assuming this space is nontrivial an application of the standard Lyapunov-Schmidt method then reduces the problem to solving an equation on $\ker(A_0^q - \mathrm{Id}_{\mathbb{R}^n})$ (see e.g. Vanderbauwhede[7] for a survey on the Lyapunov-Schmidt method). However, the precise details of this reduction will depend on whether the eigenvalue 1 of A_0^q is semi-simple or not. Since we do not want to make any hypotheses on the spectrum of A_0 (except for the fact that A_0 should be invertible) we choose a slightly modified approach in which we do not reduce to an equation on $\ker(A_0^q - \mathrm{Id}_{\mathbb{R}^n})$ but to an equation on the *generalized nullspace* of $A_0^q - \mathrm{Id}_{\mathbb{R}^n}$.

Each linear operator $\tilde{A} \in \mathcal{L}(\mathbb{R}^n)$ has a unique *Jordan-Chevalley decomposition* $\tilde{A} = \tilde{S} + \tilde{N}$ such that $\tilde{S}$ is semi-simple (i.e. complex diagonalizable), $\tilde{N}$ is nilpotent, and $\tilde{S}\tilde{N} = \tilde{N}\tilde{S}$. There exist a polynomial $P(s)$ with $P(0) = 0$ such that $\tilde{S} = P(\tilde{A})$ and $\tilde{N} = \tilde{A} - P(\tilde{A})$; as a consequence $\ker(\tilde{A}) = \ker(\tilde{S}) \cap \ker(\tilde{N})$. Also $\mathbb{R}^n = \ker(\tilde{S}) \oplus \mathrm{im}(\tilde{S})$, and this decomposition is invariant under $\tilde{S}$, $\tilde{N}$ and $\tilde{A}$.

Denote the Jordan-Chevalley decomposition of $A_0 = D_x f(0,0)$ as $A_0 = S_0 + N_0$; then the Jordan-Chevalley decomposition of $(\hat{A}_0 - \sigma) \in \mathcal{L}(\mathcal{O}_q)$ is given by $(\hat{A}_0 - \sigma) = (\hat{S}_0 - \sigma) + \hat{N}_0$, and

$$\mathcal{O}_q = \ker(\hat{S}_0 - \sigma) \oplus \mathrm{im}(\hat{S}_0 - \sigma). \tag{4}$$

This splitting is invariant under σ, $\hat{S}_0$, $\hat{N}_0$ and $\hat{A}_0$. Observe that such splitting will in general no longer be valid if we replace $\hat{S}_0$ by $\hat{A}_0$; this is the reason for introducing the Jordan-Chevalley decomposition of A_0. If $z = (x_j)_{j\in\mathbb{Z}} \in \mathcal{O}_q$ belongs to $\ker(\hat{S}_0 - \sigma)$ then $x_0 \in \ker(S_0^q - \mathrm{Id}_{\mathbb{R}^n})$ and $x_{j+1} = S_0 x_j$ for all $j \in \mathbb{Z}$. Conversely, for each $x_0 \in \ker(S_0^q - \mathrm{Id}_{\mathbb{R}^n})$ we have that $z := (S_0^j x_0)_{j\in\mathbb{Z}}$ belongs to $\mathcal{O}_q$ and is actually an element of $\ker(\hat{S}_0 - \sigma)$

(observe that S_0 is invertible since it was assumed that A_0 is invertible). This allows us to introduce the subspace

$$U := \ker\left(S_0^q - \mathrm{Id}_{\mathbb{R}^n}\right) \tag{5}$$

of $\mathbb{R}^n$ which will play a basic role in the further reduction; for reasons that will become clear later we will call U the *reduced phase space*. This space is invariant under S_0, N_0 and A_0; we denote the restrictions of S_0, N_0 and A_0 to U by respectively S, N and A. One can easily show that $\ker(A_0^q - \mathrm{Id}_{\mathbb{R}^n}) = \{u \in U \mid Nu = 0\} \subset U$. The argument which led to the definition (5) shows that the linear mapping

$$\zeta : U \longrightarrow \ker(\hat{S}_0 - \sigma), \quad u \longmapsto \zeta(u) := (S^j u)_{j \in \mathbb{Z}}$$

forms an isomorphism between U and $\ker(\hat{S}_0 - \sigma)$, such that we can rewrite (4) as

$$\mathcal{O}_q = \zeta(U) \oplus \mathrm{im}(\hat{S}_0 - \sigma). \tag{6}$$

The splitting (6) will form the basis for our reduction, but before working this out we should pay some attention to the $\mathbb{Z}_q$-symmetry. We have already observed the splitting (4) is invariant under σ, and therefore also under the $\mathbb{Z}_q$-action generated by σ on $\mathcal{O}_q$. Moreover, S generates a $\mathbb{Z}_q$-action on U (since $S^q = \mathrm{Id}_U$ by definition of U and S), and $\zeta \circ S = \sigma \circ \zeta$, i.e. the isomorphism ζ commutes with the $\mathbb{Z}_q$-actions on U and $\ker(\hat{S}_0 - \sigma)$. This will allow us to preserve the $\mathbb{Z}_q$-equivariance of (2) under the Lyapunov-Schmidt reduction.

It follows from (6) that we can write each $z \in \mathcal{O}_q$ as $z = \zeta(u) + v$ for some (unique) $u \in U$ and $v \in V := \mathrm{im}(\hat{S}_0 - \sigma)$. Similarly there exist smooth mappings $g : U \times V \times \mathbb{R}^m \to U$ and $h : U \times V \times \mathbb{R}^m \to V$ such that

$$\hat{f}(\zeta(u) + v, \lambda) = \zeta(g(u, v, \lambda)) + h(u, v, \lambda), \quad \forall (u, v, \lambda) \in U \times V \times \mathbb{R}^m.$$

Bringing this in (2) shows that this equation can be rewritten as a system of two equations

$$\begin{cases} \text{(a)} \quad g(u, v, \lambda) = Su, \\ \text{(b)} \quad h(u, v, \lambda) = \sigma \cdot v, \end{cases} \tag{7}$$

which has to be solved for (u, v, λ) near $(0, 0, 0)$ in $U \times V \times \mathbb{R}^m$. It is easy to obtain the following properties of the mappings g and h appearing in the equations (7):

(i) $g(0, 0, \lambda) = 0$ and $h(0, 0, \lambda) = 0$ for all $\lambda \in \mathbb{R}^m$;

(ii) $D_u g(0, 0, 0) = A = S + N$ and $D_v g(0, 0, 0) = 0$;

(iii) $D_u h(0,0,0) = 0$ and $D_v h(0,0,0) = \hat{A}_0\big|_V$;

(iv) $g(Su, \sigma \cdot v, \lambda) = Sg(u,v,\lambda)$ and $h(Su, \sigma \cdot v, \lambda) = \sigma \cdot h(u,v,\lambda)$ for all $(u,v,\lambda) \in U \times V \times \mathbb{R}^m$.

This last property expresses the $\mathbb{Z}_q$-equivariance of the equations (7).

Next we consider the equation (7)-(b). Since $V = \mathrm{im}(\hat{S}_0 - \sigma)$ is invariant under $(\hat{A}_0 - \sigma)$ and $\ker(\hat{A}_0 - \sigma) \subset \ker(\hat{S}_0 - \sigma)$ it follows that $\ker(\hat{A}_0 - \sigma) \cap V \subset \ker(\hat{S}_0 - \sigma) \cap V = \{0\}$ and that $(\hat{A}_0 - \sigma)$ is an isomorphism on V. Hence we can apply the implicit function theorem, starting at the solution $(0,0,0)$, to solve (7)-(b) for $v = v^*(u,\lambda)$. The mapping $v^* : U \times \mathbb{R}^m \to V$ is smooth, with $v^*(0,\lambda) = 0$ for all λ, $D_u v^*(0,0) = 0$, and $v^*(Su,\lambda) = \sigma \cdot v^*(u,\lambda)$ for all (u,λ) ($\mathbb{Z}_q$-equivariance). Bringing the solution $v = v^*(u,\lambda)$ of (7)-(b) into (7)-(a) gives us the *determining equation*

$$f_{\mathrm{red}}(u,\lambda) = Su, \tag{8}$$

to be solved for (u,λ) near $(0,0)$ in $U \times \mathbb{R}^m$, and with

$$f_{\mathrm{red}} : U \times \mathbb{R}^m \longrightarrow U, \quad (u,\lambda) \longmapsto f_{\mathrm{red}}(u,\lambda) := g(u, v^*(u,\lambda), \lambda). \tag{9}$$

The *reduced mapping* f_{red} is smooth, with $f_{\mathrm{red}}(0,\lambda) = 0$ for all λ, and $D_u f_{\mathrm{red}}(0,0) = A = S + N$; moreover, f_{red} is $\mathbb{Z}_q$-equivariant:

$$f_{\mathrm{red}}(Su,\lambda) = S f_{\mathrm{red}}(u,\lambda), \quad \forall (u,\lambda) \in U \times \mathbb{R}^m. \tag{10}$$

The foregoing means that we have reduced our problem to solving the $\mathbb{Z}_q$-equivariant determining equation (8) on the reduced phase space U. To each (sufficiently small) solution (u,λ) of (8) there corresponds a solution of (2), given by $(z^*(u,\lambda),\lambda)$ with $z^*(u,\lambda) := \zeta(u) + v^*(u,\lambda)$. Conversely, if $(z,\lambda) \in \mathcal{O}_q \times \mathbb{R}^m$ is a sufficiently small solution of (2), with $z = \zeta(u) + v$, then $v = v^*(u,\lambda)$ and (u,λ) is a solution of (8). Clearly the solutions of (8) come in $\mathbb{Z}_q$-orbits, and the same holds for the corresponding solutions of (2). Returning to the original problem $(\mathbf{P}_q)$ we can conclude that for all sufficiently small $\lambda \in \mathbb{R}^m$ all small q-periodic points of f_λ have the form $x = x^*(u,\lambda)$, where $(u,\lambda) \in U \times \mathbb{R}^m$ is a sufficiently small solution of (8), and where $x^* : U \times \mathbb{R}^m \to \mathbb{R}^n$ is given by $x^*(u,\lambda) = u + v_0^*(u,\lambda)$ (here $v_0^*(u,\lambda)$ is the 0-component of $v^*(u,\lambda) \in \mathcal{O}_q$).

There is, however, a different way to formulate the reduction, as we explain next. Consider the following problem:

$(\mathbf{P}_q^{\mathrm{red}})$ Determine, for all values of λ near the critical value $\lambda_0 = 0$, all small q-periodic orbits of $f_{\mathrm{red},\lambda} := f_{\mathrm{red}}(\cdot\,,\lambda)$, where $f_{\mathrm{red}} : U \times \mathbb{R}^m \to U$ is the reduced mapping obtained from the foregoing reduction for the problem $(\mathbf{P}_q)$.

This problem is analogous to the problem $(\mathbf{P}_q)$, and hence we can apply the reduction procedure we just explained. When we work out this reduction we find that the new reduced phase space U_{new} is just U itself (so no further reduction in dimension), $S_{\text{new}} = S$, $v^*_{\text{new}}(u,\lambda) = 0$, and the new determining equation is just the old one, i.e it coincides with (8). Moreover, we have for each $(u,\lambda) \in U \times \mathbb{R}^M$ that (u,λ) is a solution of (8) if and only if the $f_{\text{red},\lambda}$-orbit $\{f^j_{\text{red},\lambda}(u) \mid j \in \mathbb{Z}\}$ coincides with the $\mathbb{Z}_q$-orbit $\{S^j u \mid j \in \mathbb{Z}\}$. Combining these observations with our earlier conclusion of the Lyapunov-Schmidt reduction for the problem $(\mathbf{P}_q)$ we obtain the following result.

Theorem 2.1 (Main reduction theorem). *Let $f : \mathbb{R}^n \times \mathbb{R}^m \to \mathbb{R}^n$ be a smooth mapping such that $f(0,\lambda) = 0$ and $D_x f(0,\lambda) \in \mathcal{L}(\mathbb{R}^n)$ is invertible for each $\lambda \in \mathbb{R}^m$. Fix some $q \geq 1$. Let $A_0 = S_0 + N_0$ be the Jordan-Chevalley decomposition of $A_0 := D_x f(0,0)$, let $U := \ker(S_0^q - Id_{\mathbb{R}^n})$, and let A, S and N be the restrictions of respectively A_0, S_0 and N_0 to U. Then there exist smooth mappings $f_{red} : U \times \mathbb{R}^m \to U$ and $x^* : U \times \mathbb{R}^m \to \mathbb{R}^n$ such that the following holds:*

(i) *$f_{red}(0,\lambda) = 0$ for all $\lambda \in \mathbb{R}^m$, and $D_u f_{red}(0,0) = A$;*

(ii) *$f_{red}(Su,\lambda) = S f_{red}(u,\lambda)$ for all $(u,\lambda) \in U \times \mathbb{R}^m$, i.e. f_{red} is equivariant with respect to the $\mathbb{Z}_q$-action generated by S on U;*

(iii) *$x^*(0,\lambda) = 0$ for all $\lambda \in \mathbb{R}^m$, and $D_u x^*(0,0) \cdot u = u$ for all $u \in U$;*

(iv) *for each sufficiently small $(x,\lambda) \in \mathbb{R}^n \times \mathbb{R}^m$ we have that x is a q-periodic point of $f_\lambda := f(\cdot,\lambda)$ if and only $x = x^*(u,\lambda)$, where u is a q-periodic point of $f_{red,\lambda} := f_{red}(\cdot,\lambda)$;*

(v) *for each sufficiently small $\lambda \in \mathbb{R}^m$ all sufficiently small q-periodic orbits of $f_{red,\lambda}$ are also orbits under the $\mathbb{Z}_q$-action on U, i.e. they are generated by the solutions of (8), and for such solutions we have that $f(x^*(u,\lambda),\lambda) = x^*(Su,\lambda)$.*

The advantage of the foregoing formulation of the Lyapunov-Schmidt method (in contrast to more classical formulations) is twofold: first, there is no explicit reference to the splitting (6) behind the reduction, but second, there is an explicit reference to the $\mathbb{Z}_q$-equivariance of the reduced problem.

In the next sections we will further elaborate on the reduced problem, and more in particular on the reduced mapping $f_{\text{red}}(u,\lambda)$. Indeed, there remain a number of questions one can ask:

(1) what is the relation of the foregoing formulation, and in particular the determining equation (8), to the bifurcation equation obtained from a classical Lyapunov-Schmidt reduction?

(2) how does the reduction described by theorem 2.1 deal with additional structures of the mapping f, such as equivariance, reversibility, symplecticity, and so on?

(3) how can one calculate or approximate the reduced mapping $f_{\mathrm{red}}(u, \lambda)$?

(4) is it possible to obtain a relationship between the stability properties of a bifurcating periodic orbit of f_λ and the stability properties of the corresponding periodic orbit of $f_{\mathrm{red},\lambda}$?

In the next sections (some of them quite short) we will try to give some answers to these questions; in particular the answer to question (3) will give us a relationship between the Lyapunov-Schmidt reduction and normal form theory.

3. The bifurcation equation

When $N \neq 0$ then $\ker(A_0^q - \mathrm{Id}_{\mathbb{R}^n})$ is a strict subspace of $U = \ker(S_0^q - \mathrm{Id}_{\mathbb{R}^n})$, and $\ker(\hat{A}_0 - \sigma)$ is a strict subspace of $\ker(\hat{S}_0 - \sigma)$. As a consequence the determining equation (8) has a higher dimension than the bifurcation equation obtained from a "strict" Lyapunov-Schmidt reduction. This is the reason for calling (8) the "determining equation" and not the "bifurcation equation".

However, it is rather easy to further reduce (8) and obtain a "real" bifurcation equation. Indeed, (8) has for $\lambda = 0$ the form

$$Nu + \mathrm{O}(\|u\|^2) = 0.$$

One has then to consider two different splittings of U, namely

$$U = \ker(N) \oplus U_{\mathrm{aux}} \qquad \text{and} \qquad U = \mathrm{im}(N) \oplus \bar{U},$$

where the subspaces U_{aux} and $\bar{U}$ should be chosen to be invariant under S, such that both splittings are invariant under the $\mathbb{Z}_q$-action on U generated by S; since $\ker(N) \cap \mathrm{im}(N)$ is nontrivial it is not possible to take $U_{\mathrm{aux}} = \mathrm{im}(N)$ and $\bar{U} = \ker(N)$. Using the first splitting one writes the unknown $u \in U$ in (8) as $u = (\tilde{u}, u_{\mathrm{aux}})$, with $\tilde{u} \in \tilde{U} := \ker(N)$ and $u_{\mathrm{aux}} \in U_{\mathrm{aux}}$. The second splitting is used to rewrite the equation (8) itself as a system of two equations. The $\mathrm{im}(N)$-part of this system can then be solved by the implicit function theorem for $u_{\mathrm{aux}} = u^*_{\mathrm{aux}}(\tilde{u}, \lambda)$. Bringing this solution in the remaining $\bar{U}$-part of the system gives the bifurcation equation

$$b(\tilde{u}, \lambda) = 0, \tag{11}$$

with $b : \tilde{U} \times \mathbb{R}^m \to \bar{U}$ a smooth mapping such that $b(0, \lambda) = 0$ and $D_{\tilde{u}} b(0, 0) = 0$. The mapping b is also $\mathbb{Z}_q$-equivariant, in the sense that $b(\tilde{S}\tilde{u}, \lambda) = \bar{S} b(\tilde{u}, \lambda)$ for all $(\tilde{u}, \lambda)$, where $\tilde{S}$ and $\bar{S}$ denote the restrictions of S to respectively $\tilde{U}$ and $\bar{U}$. So the remaining $\mathbb{Z}_q$-equivariance involves in principle two different $\mathbb{Z}_q$-actions, namely one on $\tilde{U}$ and one on $\bar{U}$, but it is possible to show that these actions are equivalent, i.e. they coincide after a proper identification of $\tilde{U}$ and $\bar{U}$.

The foregoing shows that in case $N \neq 0$ the classical Lyapunov-Schmidt reduction to a problem on the nullspace of the linearized problem is less straightforward, requiring appropriate choices of complements and identifications. This is one of the reasons why we think a reduction to the generalized nullspace is more natural; a second reason has to do with the stability properties of bifurcating q-periodic orbits, see Section 6.

4. Additional structures

In a very loose way one can say that when the original family of local diffeomorphisms f_λ has some additional structure, then this structure is inherited by the reduced family $f_{\mathrm{red},\lambda}$, with the $\mathbb{Z}_q$-equivariance as a surplus. In this section we make this more precise for three particular structures: equivariance, reversibility and symplecticity.

We begin with the easiest one, equivariance. Actually, one can treat equivariance and reversibility in one common framework by adopting the following definition.

Definition 4.1. We say that the family of local diffeomorphisms f_λ is *equivariant-reversible* (or more precisely (Γ, χ)-*equivariant-reversible*) if there exist a compact subgroup Γ of $\mathrm{GL}(n, \mathbb{R})$ and a group character (i.e. group homomorphism) $\chi : \Gamma \to \{+1, -1\}$ such that

$$\gamma^{-1} \circ f_\lambda \circ \gamma = f_\lambda^{\chi(\gamma)}, \qquad \forall \gamma \in \Gamma,\ \forall \lambda \in \mathbb{R}^m. \tag{12}$$

There are two important particular cases: the pure equivariant case when $\chi(\gamma) = +1$ for all $\gamma \in \Gamma$, and the pure reversible case when $\Gamma = \{\mathrm{Id}_{\mathbb{R}^n}, R_0\}$ and $\chi(R_0) = -1$, with R_0 a linear involution on $\mathbb{R}^n$, i.e. $R_0^2 = \mathrm{Id}_{\mathbb{R}^n}$. In order not to make the notation too heavy we will discuss here these two particular cases separately; it is then easy to formulate and prove the corresponding results for the general equivariant-reversible case as described in definition 4.1.

We will need two conclusions which can be drawn from the reduction in section 2 and which we reformulate here for convenience:

(A) if $(z, \lambda) = \zeta(u) + v, \lambda) \in \mathcal{O}_q \times \mathbb{R}^m$ is a sufficiently small solution of (2) then $v = v^*(u, \lambda)$ and $f_{\text{red}}(u, \lambda) = Su$;

(B) for each sufficiently small $(u, \tilde{u}, v, \lambda) \in U \times U \times V \times \mathbb{R}^m$ we have $\hat{f}(\zeta(u) + v, \lambda) = \zeta(\tilde{u}) + \sigma \cdot v$ if and only if $v = v^*(u, \lambda)$ and $\tilde{u} = f_{\text{red}}(u, \lambda)$.

4.1. *Equivariance*

In the equivariant case we assume that

$$\gamma^{-1} \circ f_\lambda \circ \gamma = f_\lambda, \qquad \forall \gamma \in \Gamma,\ \forall \lambda \in \mathbb{R}^m, \tag{13}$$

where Γ is a compact subgroup of $\text{GL}(n, \mathbb{R})$. This clearly implies that if $x \in \mathbb{R}^n$ is a q-periodic point of f_λ, then so is γx for each $\gamma \in \Gamma$; this property should of course also appear in the reduced problem. It follows from (13) that A_0, S_0 and N_0 commute with every $\gamma \in \Gamma$, and that the reduced phase space U is invariant under the Γ-action on $\mathbb{R}^n$; we will denote the restriction of $\gamma \in \Gamma$ to U again by γ, hoping that the precise status of γ is clear from the context. Also $\hat{f}_\lambda \circ \hat{\gamma} = \hat{\gamma} \circ \hat{f}_\lambda$, $\sigma \circ \hat{\gamma} = \hat{\gamma} \circ \sigma$ and $\zeta(\gamma u) = \hat{\gamma} \cdot \zeta(u)$ for all $\gamma \in \Gamma$. Applying $\hat{\gamma}$ to the identity

$$\hat{f}(\zeta(u) + v^*(u, \lambda), \lambda) = \zeta(f_{\text{red}}(u, \lambda)) + \sigma \cdot v^*(u, \lambda) \tag{14}$$

(see property (B) above) and using the commutation relations just mentionned gives

$$\hat{f}(\zeta(\gamma u) + \hat{\gamma} \cdot v^*(u, \lambda), \lambda) = \zeta(\gamma f_{\text{red}}(u, \lambda)) + \sigma \cdot \hat{\gamma} \cdot v^*(u, \lambda).$$

Again using (B) we conclude that

$$v^*(\gamma u, \lambda) = \hat{\gamma} \cdot v^*(u, \lambda) \ \text{ and } \ f_{\text{red}}(\gamma u, \lambda) = \gamma f_{\text{red}}(u, \lambda), \quad \forall \gamma \in \Gamma. \tag{15}$$

This also implies that $x^*(\gamma u, \lambda) = \gamma x^*(u, \lambda)$ for all γ, and we can conclude that both the reduced mapping $f_{\text{red}}(\cdot, \lambda)$ and the mapping x^* relating the reduced problem to the original one inherit the Γ-equivariance of the original mapping.

The reduced mapping is not only Γ-equivariant, but also $\mathbb{Z}_q$-equivariant (where the $\mathbb{Z}_q$-action on U is generated by S). We have then to distinguish two different cases, depending on whether or not there exists some $\gamma_0 \in \Gamma$ such that $\gamma_0 u = Su$ for all $u \in U$. In case such γ_0 exists the $\mathbb{Z}_q$ action on U is already contained in the Γ-action, and f_{red} is just Γ-equivariant. For each solution of the determining equation $f_{\text{red}}(u, \lambda) = Su$ we have then that $f(x^*(u, \lambda), \lambda) = x^*(Su, \lambda) = x^*(\gamma_0 u, \lambda) = \gamma_0 x^*(u, \lambda)$, which shows that each q-periodic orbit of f_λ is contained in a Γ-orbit: all (sufficiently small) q-periodic orbits are *relative fixed points* (i.e. fixed points modulo the Γ-action). In the other case (no such γ_0 exists) the reduced mapping is $\Gamma \times \mathbb{Z}_q$-equivariant (the Γ-action commutes with the $\mathbb{Z}_q$-action since S is Γ-equivariant). When using this symmetry and interpreting results one should keep in mind that the Γ-action corresponds to space symmetries (symmetries on the phase space), while the $\mathbb{Z}_q$-action encodes temporal behaviour, as follows from $f(x^*(u,\lambda),\lambda) = x^*(Su,\lambda)$.

4.2. *Reversibility*

Next we turn to the case of reversible mappings. In addition to our standing hypotheses we assume that there exists some $R_0 \in \mathcal{L}(\mathbb{R}^n)$ such that $R_0^2 = \mathrm{Id}_{\mathbb{R}^n}$ and

$$R_0 \circ f_\lambda \circ R_0 = f_\lambda^{-1}, \qquad \forall \lambda \in \mathbb{R}^m. \tag{16}$$

We say that the family f_λ is R_0-*reversible*, or that R_0 is a *reversor* for the family f_λ. It follows from (16) that also A_0, S_0 and N_0 are R_0-reversible:

$$R_0 \circ A_0 \circ R_0 = A_0^{-1}, \quad R_0 \circ S_0 \circ R_0 = S_0^{-1} \quad \text{and} \quad R_0 \circ N_0 \circ R_0 = N_0^{-1},$$

and that the reduced phase space U is invariant under R_0. We denote the restriction of R_0 to U by R. Then $SR = RS^{-1}$, which together with $S^q = \mathrm{Id}_U$ and $R^2 = \mathrm{Id}_U$ shows that S and R generate on U a $\mathbb{D}_q$-action ($\mathbb{D}_q$ is the symmetry group of a regular q-gone and contains $2q$ elements). When we define $\rho \in \mathcal{L}(\mathcal{O}_q)$ by

$$\rho \cdot z := (R_0 x_{-j})_{j\in\mathbb{Z}} \qquad \forall z = (x_j)_{j\in\mathbb{Z}} \in \mathcal{O}_q, \tag{17}$$

then one can easily verify that $\rho^2 = \mathrm{Id}_{\mathcal{O}_q}$ and $\sigma \circ \rho = \rho \circ \sigma^{-1}$, i.e. σ and ρ generate on $\mathcal{O}_q$ a $\mathbb{D}_q$-action. Moreover

$$\rho \circ \hat{f}_\lambda \circ \rho = \hat{f}_\lambda^{-1} \quad \text{and} \quad \zeta \circ R = \rho \circ \zeta.$$

Applying ρ to the identity (14) and using the foregoing commutation relations gives us

$$\hat{f}_\lambda^{-1}(\zeta(Ru) + \rho \cdot v^*(u,\lambda)) = \zeta(Rf_{\mathrm{red}}(u,\lambda)) + \sigma^{-1} \cdot \rho \cdot v^*(u,\lambda)$$

and hence

$$\zeta(Ru) + \sigma \cdot \left(\sigma^{-1} \cdot \rho \cdot v^*(u,\lambda)\right) = \hat{f}_\lambda(\zeta(Rf_{\mathrm{red}}(u,\lambda)) + \sigma^{-1} \cdot \rho \cdot v^*(u,\lambda)).$$

An application of conclusion (B) above then shows that

$$v^*(Rf_{\mathrm{red}}(u,\lambda),\lambda) = \sigma^{-1} \cdot \rho \cdot v^*(u,\lambda) \tag{18}$$

and

$$f_{\text{red}}(Rf_{\text{red}}(u,\lambda),\lambda) = Ru \quad \Longrightarrow \quad R\circ f_{\text{red},\lambda}\circ R = f_{\text{red},\lambda}^{-1}. \tag{19}$$

This last relation shows that the reduced mapping f_{red} is R-reversible, or, in combination with the $\mathbb{Z}_q$-equivariance, $(\mathbb{D}_q,\chi)$-equivariant-reversible, where the character $\chi:\mathbb{D}_q\to\{+1,-1\}$ is defined by $\chi(S)=+1$ and $\chi(R)=-1$.

The relation (18) is much more complicated and it is hard to give it an interpretation. However, when we assume that (u,λ) satisfies the determining equation $f_{\text{red}}(u,\lambda)=Su$ then (18) shows that

$$v^*(RSu,\lambda)=\sigma^{-1}\cdot\rho\cdot v^*(u,\lambda),$$

and since also $(S^{-1}u,\lambda)$ satisfies the same determining equation we get

$$v^*(Ru,\lambda)=\sigma^{-1}\cdot\rho\cdot v^*(S^{-1}u,\lambda)=\sigma^{-1}\cdot\rho\cdot\sigma^{-1}\cdot v^*(u,\lambda)=\rho\cdot v^*(u,\lambda)$$

and

$$x^*(Ru,\lambda)=R_0x^*(u,\lambda). \tag{20}$$

The same relations can also be obtained in a more direct way. If (u,λ) satisfies (8) then $z^*(u,\lambda)=\zeta(u)+v^*(u,\lambda)$ is a solution of (2), and so is $\rho\cdot z^*(u,\lambda)=\zeta(Ru)+\rho\cdot v^*(u,\lambda)$ (just use the reversibility). Property (A) above implies then that $v^*(Ru,\lambda)=\rho\cdot v^*(u,\lambda)$, which in turn gives (20).

The fact that (20) is only valid for the solutions of the determining equation is somewhat unpleasant, but there is a way out. When we define $\tilde{x}^*:U\times\mathbb{R}^m\to\mathbb{R}^n$ by

$$\tilde{x}^*(u,\lambda):=\frac{1}{2}\left(x^*(u,\lambda)+R_0x^*(Ru,\lambda)\right)$$

then the foregoing shows that $\tilde{x}^*(u,\lambda)=x^*(u,\lambda)$ when (u,λ) satisfies (8). Therefore the statement of the main reduction theorem 2.1 remains valid if we replace x^* by $\tilde{x}^*$, and we have the additional relations

$$R\circ f_{\text{red},\lambda}\circ R=f_{\text{red},\lambda}^{-1} \quad \text{and} \quad \tilde{x}^*_\lambda\circ R=R_0\circ\tilde{x}^*_\lambda, \tag{21}$$

meaning that the reversibilty is fully preserved by the reduction.

There is one further remark which is important enough to make here, namely that it is possible (in the reversible situation discussed in this subsection) to replace the determining equation (8) by an equivalent equation which is fully $\mathbb{D}_q$-equivariant in an appropriate sense. Since both $f_{\text{red},\lambda}$ and S are $\mathbb{Z}_q$-equivariant and R-reversible it follows that if (u,λ) is a solution of the determining equation (8) then so are (S^ju,λ) $(j\in\mathbb{Z})$, by the $\mathbb{Z}_q$-equivariance, and (Ru,λ), by the R-reversibility. However, this last property is not an immediate consequence of some kind of explicit equivariance of (8) (it involves taking inverses of the mappins itself); since explicit equivariances imply restrictions on the possible form of the equation it would be nice if we could translate the reversibility of (8) into such explicit equivariance. This can be done by showing that (8) is equivalent to the equation

$$F(u,\lambda) := S^{-1} f_{\mathrm{red},\lambda}(u) - S f_{\mathrm{red},\lambda}^{-1}(u) = 0. \tag{22}$$

It is easy to verify that $F(Su,\lambda) = SF(u,\lambda)$ and $F(Ru,\lambda) = -RF(u,\lambda)$, i.e. the mapping $F : U \times \mathbb{R}^m \to U$ satisfies the equivariance condition

$$F(\gamma u,\lambda) = \chi(\gamma)\gamma F(u,\lambda), \qquad \forall \gamma \in \mathbb{D}_q, \tag{23}$$

where the $\mathbb{D}_q$-action on U and $\chi : \mathbb{D}_q \to \{+1,-1\}$ are as described before. To prove the equivalence of (8) and (22) we notice that by the $\mathbb{Z}_q$-equivariance of $f_{\mathrm{red},\lambda}$ it is immediate that (8) implies (22). Conversely, (22) implies (by the same $\mathbb{Z}_q$-equivariance) that $f^2_{\mathrm{red},\lambda}(u) = S^2 u$ and hence (since $S^{2q} = \mathrm{Id}_U$) $f^{2q}_{\mathrm{red},\lambda}(u) = u$, i.e. the solutions of (22) are $2q$-periodic points of $f_{\mathrm{red},\lambda}$. Applying our foregoing Lyapunov-Schmidt reduction to the problem of finding (small) $2q$-periodic orbits of the family $f_{\mathrm{red},\lambda}$ shows that this problem reduces to solving the equation (8), which means that all $2q$-periodic orbits of $f_{\mathrm{red},\lambda}$ are automatically q-periodic. Hence (22) implies (8), as we wanted to show.

For further information on the reversible case we refer to the paper of Ciocci et al.[2] which also contains an application which illustrates the advantage of (22) over (8).

4.3. *Symplectic mappings*

The third special class of mappings we consider here are symplectic diffeomorphisms. To define such diffeomorphisms we have to give the phase space $\mathbb{R}^n$ the structure of a *symplectic vectorspace*, i.e. we must have an anti-symmetric and non-degenerate bilinear form $\omega_0 : \mathbb{R}^n \times \mathbb{R}^n \to \mathbb{R}$; the existence of such *symplectic form* requires n to be even. Given any scalar product $\langle\cdot,\cdot\rangle$ on $\mathbb{R}^n$ the symplectic form ω_0 can be written as $\omega_0(x_1,x_2) = \langle x_1, Jx_2\rangle$, where $J \in \mathcal{L}(\mathbb{R}^n)$ is anti-symmetric ($J^T = -J$) and non-singular. (Actually, one can show that there exists a scalar product such that $J^2 = -\mathrm{Id}_{\mathbb{R}^n}$, and hence $J^{-1} = J^T$; for the proof, see e.g. Vanderbauwhede[6]).

Definition 4.2. A smooth mapping $\Phi : \mathbb{R}^n \to \mathbb{R}^n$ is *symplectic* if

$$\omega_0(D\Phi(x)\cdot x_1, D\Phi(x)\cdot x_2) = \omega_0(x_1,x_2), \qquad \forall x, x_1, x_2 \in \mathbb{R}^n. \tag{24}$$

This condition is equivalent to

$$D\Phi(x)^T J D\Phi(x) = J, \qquad \forall x \in \mathbb{R}^n,$$

and can only be satisfied if $D\Phi(x) \in \mathcal{L}(\mathbb{R}^n)$ is invertible for all $x \in \mathbb{R}^n$.

Let us assume now that the family f_λ is symplectic:

$$Df_\lambda(x)^T J Df_\lambda(x) = J, \qquad \forall (x,\lambda) \in \mathbb{R}^n \times \mathbb{R}^m. \tag{25}$$

This implies immediately that also $A_0 = Df_{\lambda=0}(0)$ is symplectic, i.e. $A_0^T J A_0 = J$, or equivalently, $J^{-1} A_0^T J = A_0^{-1}$. Consider the Jordan-Chevalley decomposition $A_0 = S_0 + N_0$ of A_0; the corresponding decompositions of A_0^T and A_0^{-1} are then given by respectively $A_0^T = S_0^T + N_0^T$ and $A_0^{-1} = S_0^{-1} + S_0^{-1}[(\mathrm{Id}_{\mathbb{R}^n} + S_0^{-1}N_0)^{-1} - \mathrm{Id}_{\mathbb{R}^n}]$. Bringing this into $J^{-1}A_0^T J = A_0^{-1}$ and using the uniqueness of the Jordan-Chevalley decomposition one finds that $J^{-1} S_0^T J = S_0^{-1}$, meaning that also S_0 is symplectic. This in turn implies that the reduced phase space $U = \ker(S_0^q - \mathrm{Id}_{\mathbb{R}^n})$ forms a *symplectic subspace*, in the sense that the restriction ω of the symplectic form ω_0 to $U \times U$ is still non-degenerate. To prove this fix some $u_0 \in U$ and assume that $\omega(u_0, u) = 0$ for all $u \in U$. Since $\mathbb{R}^n = \ker(S_0^q - \mathrm{Id}_{\mathbb{R}^n}) \oplus \mathrm{im}(S_0^q - \mathrm{Id}_{\mathbb{R}^n})$ we can write an arbitrary $x \in \mathbb{R}^n$ in the form $x = u + (S_0^q - \mathrm{Id}_{\mathbb{R}^n})w$, with $u \in U$ and $w \in \mathbb{R}^n$; it follows then that

$$\omega_0(u_0, x) = \omega(u_0, u) + \omega_0(u_0, (S_0^q - \mathrm{Id}_{\mathbb{R}^n})w) = \omega_0((S_0^{-q} - \mathrm{Id}_{\mathbb{R}^n})u_0, w) = 0.$$

Since this holds for all $x \in \mathbb{R}^n$ and ω_0 is non-degenerate we conclude that $u_0 = 0$, and hence (U, ω) forms itself a symplectic vectorspace. Since $\omega(Su_1, Su_2) = \omega(u_1, u_2$ the $\mathbb{Z}_q$-action on U is symplectic. Next we will show that the reduced family $f_{\mathrm{red},\lambda}$ is symplectic with respect to the symplectic form ω on U.

The symplectic form ω_0 on $\mathbb{R}^n$ can be lifted to a symplectic form $\hat{\omega}_0 : \mathcal{O}_q \times \mathcal{O}_q \to \mathbb{R}$ on the orbit space $\mathcal{O}_q$ by setting

$$\hat{\omega}_0(z, \tilde{z}) := \frac{1}{q} \sum_{j=0}^{q-1} \omega_0(x_j, \tilde{x}_j), \qquad \forall (z, \tilde{z}) = ((x_j)_{j\in\mathbb{Z}}, (\tilde{x}_j)_{j\in\mathbb{Z}}) \in \mathcal{O}_q \times \mathcal{O}_q.$$

It is then immediate to verify that the family $\hat{f}_\lambda$ is symplectic:

$$\hat{\omega}_0 \left(D\hat{f}_\lambda(z) \cdot z_1, D\hat{f}_\lambda(z) \cdot z_2 \right) = \hat{\omega}_0(z_1, z_2), \;\; \forall z, z_1, z_2 \in \mathcal{O}_q, \forall \lambda \in \mathbb{R}^m. \tag{26}$$

Also $\hat{S}_0$ and σ are symplectic:

$$\hat{\omega}_0(\hat{S}_0 z_1, \hat{S}_0 z_2) = \hat{\omega}_0(z_1, z_2) \quad \text{and} \quad \hat{\omega}_0(\sigma \cdot z_1, \sigma \cdot z_2) = \hat{\omega}_0(z_1, z_2)$$

for all $(z_1, z_2) \in \mathcal{O}_q \times \mathcal{O}_q$; moreover

$$\hat{\omega}_0(\zeta(u_1), \zeta(u_2)) = \omega(u_1, u_2), \qquad \forall (u_1, u_2) \in U \times U.$$

Consider an arbitrary $(u, v) \in U \times V$; writing v as $v = (\hat{S}_0 - \sigma) \cdot z$ for some $z \in \mathcal{O}_q$ we find that

$$\hat{\omega}_0(\zeta(u), v) = \hat{\omega}_0\left(\zeta(u), (\hat{S}_0 - \sigma) \cdot z\right) = \hat{\omega}_0\left((\hat{S}_0^{-1} - \sigma^{-1}) \cdot \zeta(u), z\right) = 0;$$

it follows that for all $u_1, u_2 \in U$ and all $v_1, v_2 \in V$ we have

$$\hat{\omega}_0(\zeta(u_1) + v_1, \zeta(u_2) + v_2) = \omega(u_1, u_2) + \hat{\omega}_0(v_1, v_2). \tag{27}$$

Now fix some $u, u_1, u_2 \in U$ and some $\lambda \in \mathbb{R}^m$, and set

$$z = \zeta(u) + v^*(u, \lambda) \ \text{ and } \ z_i = \zeta(u_i) + D_u v^*(u, \lambda) \cdot u_i \ \ (i = 1, 2)$$

in (26). Differentiating (14) shows that

$$D\hat{f}_\lambda(z) \cdot z_i = \zeta(Df_{\text{red},\lambda}(u) \cdot u_i) + \sigma \cdot D_u v^*(u, \lambda) \cdot u_i, \quad (i = 1, 2).$$

Bringing all the foregoing relations together we finally get

$$\omega(Df_{\text{red},\lambda}(u) \cdot u_1, Df_{\text{red},\lambda}(u) \cdot u_2) = \omega(u_1, u_2), \tag{28}$$

which proves that the reduced family $f_{\text{red},\lambda}$ is indeed symplectic. This simplecticity is complemented with the equivariance with repect to the symplectic $\mathbb{Z}_q$-action generated on U by S.

For more information on this symplectic case and for an application of the reduction results we refer to Ciocci et al.[1]

5. Approximation of the reduced mapping

Since the Lyapunov-Schmidt reduction just uses the Implicit Finction Theorem and appropriate substitutions the obvious choice for approximating the reduced mapping f_{red} is to use Taylor expansions. In principle it is possible to obtain the Taylor expansion of $f_{\text{red}}(u, \lambda)$ around the origin up to any order, but the expressions become quickly unwieldy, so that this approach is impractical for orders higher than two or three. Here we describe a different approach based on normal form reduction; although the main result (see Theorem 5.2) is rather theoretical and admittedly hard to implement on concrete examples we will describe at the end of this section a way in which the result could be used for practical calculations.

We start by formulating a basic result on the normal form reduction of parametrized families of diffeomorphisms such as considered in this paper. Even giving some hints on the proof would bring us too far away from the main topic of this paper; therefore we refer to the appendix of Vanderbauwhede[5] for a detailed proof.

The mappings f_λ studied in the foregoing sections belong to the group $\mathrm{Diff}_0(\mathbb{R}^n)$ of (local) diffeomorphisms on $\mathbb{R}^n$ with a fixed point at the origin (the group operation is just composition of mappings). There is a natural action of the group $\mathrm{Diff}_0(\mathbb{R}^n)$ on itself, given by

$$(\Psi, \Phi) \in \mathrm{Diff}_0(\mathbb{R}^n) \times \mathrm{Diff}_0(\mathbb{R}^n) \mapsto \mathrm{Ad}(\Psi) \cdot \Phi := \Psi \circ \Phi \circ \Psi^{-1} \in \mathrm{Diff}_0(\mathbb{R}^n). \quad (29)$$

This action corresponds to a change of coordinates on the phase space $\mathbb{R}^n$. The general aim of normal form theory is to find, for a given $\Phi \in \mathrm{Diff}_0(\mathbb{R}^n)$, a "sufficiently simple" element $\tilde{\Phi}$ of the group orbit

$$\mathcal{O}(\Phi) := \{\mathrm{Ad}(\Psi) \cdot \Phi \mid \Psi \in \mathrm{Diff}_0(\mathbb{R}^n)\};$$

the precise meaning of "sufficiently simple" depends on a detailed analysis of the elements of $\mathcal{O}(\Phi)$ but typically amounts to imposing certain restrictions on the truncated Taylor expansion of $\tilde{\Phi}$. In the context of this paper the situation is further complicated by the fact that we do not work with a single diffeomorphism but with parametrized families of such diffeomorphisms; in such context one wants the normal form transformation to depend smoothly on the parameter and to be valid in at least some neighborhood of some critical value of the parameter ($\lambda = 0$ in our case). Under the standing hypotheses and notations of this paper one can prove the following.

Theorem 5.1 (Normal Form). *For each $k \geq 1$ there exist a neighborhood Ω_k of the origin in $\mathbb{R}^m$ and a smooth mapping $\Psi_k : \mathbb{R}^n \times \mathbb{R}^m \to \mathbb{R}^n$, with $\Psi_k(0, \lambda) = 0$ for all λ, $D_x\Psi_k(0,0) = 0$, and such that for all $\lambda \in \Omega_k$ the transformed local diffeomorphism $\mathrm{Ad}(\Psi_{k,\lambda}) \cdot f_\lambda$ commutes with S_0 up to terms of order $k+1$, i.e.*

$$(\mathrm{Ad}(\Psi_{k,\lambda}) \cdot f_\lambda)(x) = f_{\mathrm{NF}}(x, \lambda) + R_k(x, \lambda), \quad (30)$$

with

$$f_{\mathrm{NF}}(S_0 x, \lambda) = S_0 f_{\mathrm{NF}}(x, \lambda), \qquad \forall (x, \lambda) \in \mathbb{R}^n \times \Omega_k, \quad (31)$$

and $R_k(x, \lambda) = \mathrm{O}(\|x\|^{k+1})$, uniformly for $\lambda \in \Omega_k$.

When $N_0 \neq 0$ one can impose next to (31) some further conditions on $f_{\mathrm{NF}}(x, \lambda)$ (see Vanderbauwhede[5]); however, these additional conditions will play no role in our main result of this section (Theorem 5.2), but they may simplify the study of the determining equation for particular cases. Also, there exist versions of the foregoing normal form theorem for the special cases when the family f_λ has any of the additional structures discussed in

Section 4; for such cases the normal form reduction can be done in such a way that the additional structure is fully preserved by the transformation. The neighborhood Ω_k appearing in Theorem 5.1 will typically shrink to $\{0\}$ as $k \to \infty$.

Definition 5.1. When the family f_λ is such that

$$f(x, \lambda) = f_{\mathrm{NF}}(x, \lambda) + R_k(x, \lambda), \tag{32}$$

with f_{NF} satisfying (31) and $R_k(x, \lambda) = \mathrm{O}(\|x\|^{k+1})$ uniformly for all λ in some neighborhood Ω_k of $\lambda = 0$ in $\mathbb{R}^m$ then we say that the family f_λ is *in normal form up to order* k, and we call f_{NF} the *normal form part* of f.

The normal form theorem tells us that the family f_λ can be brought into normal form up to any finite order $k \geq 1$ by an appropriate parameter-dependent near-identity transformation.

Assume now that we *first* bring the family f_λ into normal form up to some order $k \geq 1$, and *then* apply the Liapunov-Schmidt reduction as descibed in section 2. What would be the outcome? The normal form transformation does not change A_0, S_0 and N_0, hence the reduced phase space U will remain unchanged. Using (31) it is very easy to check that if the remainder term $R_k(x, \lambda)$ would be missing in (32) then the resulting determining equation would take the form

$$f_{\mathrm{NF}}(u, \lambda) = Su,$$

while $x^*(u, \lambda)$ would be equal to u. Stated differently: if the family f_λ is *fully* in normal form (i.e commutes with S_0) then all sufficiently small q-periodic orbits are contained in the reduced phase space U and are also orbits under the $\mathbb{Z}_q$-action on U. Taking the remainder $R_k(x, \lambda)$ into account will only add terms of order $k + 1$ to the different expressions; this leads to the following result.

Theorem 5.2. *Under the assumptions of the main reduction theorem, assume also that the family f_λ is in normal form up to some order $k \geq 1$, with normal form part $f_{\mathrm{NF}}(x, \lambda)$. Then*

$$x^*(u, \lambda) = u + \mathrm{O}(\|u\|^{k+1}) \tag{33}$$

and

$$f_{red,\lambda} = f_{\mathrm{NF},\lambda}\big|_U + r_{k,\lambda}, \qquad \text{with } r_{k,\lambda}(u) = \mathrm{O}(\|u\|^{k+1}), \tag{34}$$

both uniformly for sufficiently small values of λ.

In order to use theorem 5.2 directly one has to bring the family f_λ into normal form up to some appropriate order k, something which is not an easy task for larger values of n and k. A more practical approach is to obtain an approximation (to a convenient order) of the center manifold and the restriction of the family f_λ to this center manifold, then bring this restricted family in normal form, and then apply the Lyapunov-Schmidt reduction. The reduction to the center manifold results in a new family of mappings living on the tangent space to the center manifold at the origin. The reduced phase space U is contained in this tangent space, in particular cases (namely when all eigenvalues of A_0 on the unit circle are q-th roots of unity) U will coincide with this tangent space. Let us assume for simplicity that we are in this last situation. Then the starting point for the normal form reduction is a family of mappings $\tilde{f} : U \times \mathbb{R}^m \to U$, with $\tilde{f}(0, \lambda) = 0$ and $D_u\tilde{f}(0,0) = A = S + N$, and only known explicitly up to a certain order k in $\|u\|$. We know from center manifold theory that all (sufficiently small) periodic orbits of the original family f_λ will be contained in the center manifold, and will hence be periodic orbits for the family $\tilde{f}_\lambda$. Applying a normal form reduction to the family $\tilde{f}_\lambda$ is more manageable because typically $\dim U$ will be relatively low. Assume that this normal form reduction brings the family $\tilde{f}_\lambda$ in the form

$$\tilde{f}(u, \lambda) = \tilde{f}_{\text{NF}}(u, \lambda) + \tilde{R}_k(u, \lambda),$$

with $\tilde{f}_{\text{NF}}(Su, \lambda) = S\tilde{f}_{\text{NF}}(u, \lambda)$ and $\tilde{R}_k(u, \lambda) = \text{O}(\|u\|^{k+1})$. Finally, by theorem 5.2, an application of the Lyapunov-Schmidt reduction to the family $\tilde{f}_\lambda$ gives the reduced mapping

$$f_{\text{red}}(u, \lambda) = \tilde{f}_{\text{NF}}(u, \lambda) + r_k(u, \lambda),$$

with $r_k(u, \lambda) = \text{O}(\|u\|^{k+1})$.

One can ask about the advantage of adding a Lyapunov-Schmidt reduction on top of the center manifold plus normal form reduction. Without the Lyapunov-Schmidt reduction one has to find the q-periodic points of $\tilde{f}(u, \lambda)$, after the reduction we have to solve the same problem for the mapping $f_{\text{red}}(u, \lambda)$. Since both $\tilde{f}$ and f_{red} are approximated (up to the same order) by the normal form part $\tilde{f}_{\text{NF}}$ the first step will in both cases be to study the q-periodic orbits of $\tilde{f}_{\text{NF}}(u, \lambda)$, which amounts (because of the $\mathbb{Z}_q$-equivariance of $\tilde{f}_{\text{NF}}$) to solving the approximate determining equation

$$\tilde{f}_{\text{NF}}(u, \lambda) = Su. \tag{35}$$

But proving the persistence of the approximate bifurcation picture obtained from (35) under the perturbation by the higher order terms is quite different for the two cases. For the reduced mapping $f_{\rm red}(u,\lambda)$ the perturbation $r_k(u,\lambda)$ is also $\mathbb{Z}_q$-equivariant, and we just have to add it at the left hand side of (35) and prove persistence. This no longer works for the mapping $\tilde{f}(u,\lambda)$ since the perturbation term $\tilde{R}_k(u,\lambda)$ has in general no symmetry; one is forced to use other means to prove the persistence. There is of course a further advantage when the dimension of the center manifold is strictly bigger than $\dim U$, because then the Lyapunov-Schmidt reduction not only gives an added $\mathbb{Z}_q$-symmetry, but also results in a problem with a lower dimension. And clearly the $\mathbb{Z}_q$-equivariance given by the Lyapunov-Schmidt reduction will always be a bonus in theoretical discussions.

6. Stability of bifurcating periodic orbits

In this section we describe some recently obtained results on the stability of bifurcating q-periodic orbits; detailed proofs will be given in a forthcoming paper. According to theorem 1.1 the stability of the q-periodic orbit $z^*(u,\lambda) = \zeta(u) + v^*(u,\lambda)$ of f_λ generated by a solution (u,λ) of the determining equation (8) is determined by the eigenvalues of the linear operator $\sigma^{-1}\circ\hat{f}_\lambda(z^*(u,\lambda)) \in \mathcal{L}(\mathcal{O}_q)$; by the remark after theorem 1.1 these eigenvalues will (for sufficiently small (u,λ)) be close to the set

$$\{\mu \exp(2\pi i p/q) \mid p \in \mathbb{Z},\ \mu = \text{eigenvalue of } A_0\}.$$

If this set contains some elements strictly outside of the unit circle then all sufficiently small bifurcating q-periodic orbits will be unstable. In the other case one has to study the eigenvalues of $\sigma^{-1}\circ\hat{f}_\lambda(z^*(u,\lambda))$ which are on or near the unit circle. In particular, our hypotheses imply that there will be some eigenvalues near $+1$, with total (algebraic) multiplicity equal to $\dim U$; we will call these the *critical eigenvalues*, and the aim of this section is to describe some results on these critical eigenvalues. Since we will not use the fact that (u,λ) should be a solution of (8) these results will be valid for all sufficiently small values of (u,λ); to draw stability conclusions one should restrict to solutions of (8).

We start by writing $L(u,\lambda) := \sigma^{-1}\circ D\hat{f}_\lambda(z^*(u,\lambda)) \in \mathcal{L}(\mathcal{O}_q)$ in block form with respect to the splitting (4) of $\mathcal{O}_q$:

$$L(u,\lambda) = \begin{pmatrix} L_{00}(u,\lambda) & L_{01}(u,\lambda) \\ L_{10}(u,\lambda) & L_{11}(u,\lambda) \end{pmatrix}. \tag{36}$$

Setting $\hat{U} := \ker(\hat{S}_0 - \sigma) = \zeta(U)$ and $V = \operatorname{im}(\hat{S}_0 - \sigma)$ we have $L_{00}(u,\lambda) \in \mathcal{L}(\hat{U})$, $L_{01}(u,\lambda) \in \mathcal{L}(V,\hat{U})$, $L_{10}(u,\lambda) \in \mathcal{L}(\hat{U},V)$ and $L_{11}(u,\lambda) \in \mathcal{L}(V)$. For $(u,\lambda) = (0,0)$ we have that $L_{00}(0,0) - \mathrm{Id}_{\hat{U}} = \hat{S}_0^{-1}\hat{N}_0|_{\hat{U}}$ is nilpotent,

$L_{01}(0,0) = 0$, $L_{10}(0,0) = 0$, and $L_{11}(0,0) - \mathrm{Id}_V = \sigma^{-1}\circ(\hat{A}_0 - \sigma)|_V$ is invertible. By means of the implicit function theorem and a similarity transformation using linear operators of the form

$$\Psi(K) = \begin{pmatrix} \mathrm{Id}_{\hat{U}} & 0 \\ K & \mathrm{Id}_V \end{pmatrix}, \qquad K \in \mathcal{L}(\hat{U}, V),$$

one can show that there exists a smooth mapping $K^* : U \times \mathbb{R}^m \to \mathcal{L}(\hat{U}, V)$ with $K^*(0,0) = 0$ and such that

$$\Psi(K^*(u,\lambda))L(u,\lambda)\Psi(K^*(u,\lambda))^{-1} = \begin{pmatrix} \tilde{L}_0(u,\lambda) & L_{01}(u,\lambda) \\ 0 & \tilde{L}_1(u,\lambda) \end{pmatrix}, \quad \forall(u,\lambda).$$

In this expression

$$\tilde{L}_0(u,\lambda) = L_{00}(u,\lambda) - L_{01}(u,\lambda)K^*(u,\lambda)$$

and

$$\tilde{L}_1(u,\lambda) = L_{11}(u,\lambda) + K^*(u,\lambda)L_{01}(u,\lambda).$$

From this one can easily deduce that the critical eigenvalues of $L(u,\lambda)$ are precisely the eigenvalues of $\tilde{L}_0(u,\lambda)$; defining $\Phi : U \times \mathbb{R}^m \to \mathcal{L}(U)$ by $\Phi(u,\lambda) := S\zeta^{-1}\tilde{L}_0(u,\lambda)\zeta$ we obtain the following result.

Theorem 6.1. *Under the hypotheses of theorem 2.1 there exists a smooth mapping $\Phi : U \times \mathbb{R}^m \to \mathcal{L}(U)$ with $\Phi(0,0) = S + N$ and such that for all sufficiently small $(u,\lambda) \in U \times \mathbb{R}^m$ the critical eigenvalues of $\sigma^{-1}\circ D\hat{f}_\lambda(z^*(u,\lambda))$ coincide with the eigenvalues of $S^{-1}\Phi(u,\lambda)$.*

Next consider the q-periodic orbits of the $\mathbb{Z}_q$-equivariant reduced mapping $f_{\mathrm{red}}(u,\lambda)$, given by the solutions of (8). The stability of such periodic orbit is determined by the eigenvalues of

$$Df^q_{\mathrm{red},\lambda}(u) = Df_{\mathrm{red},\lambda}(S^{q-1}u)\circ\cdots\circ Df_{\mathrm{red},\lambda}(Su)\circ Df_{\mathrm{red},\lambda}(u);$$

differentiating the equivariance relation $f_{\mathrm{red},\lambda}(Su) = Sf_{\mathrm{red},\lambda}(u)$ shows that $D_u f_{\mathrm{red}}(Su,\lambda) = SD_u f_{\mathrm{red}}(u,\lambda)S^{-1}$, from which we get

$$Df^q_{\mathrm{red},\lambda}(u) = \left(S^{-1}D_u f_{\mathrm{red}}(u,\lambda)\right)^q.$$

We conclude that the stability of a q-periodic orbit $\{S^j u \mid j \in \mathbb{Z}\}$ of $f_{\mathrm{red},\lambda}$ is determined by the eigenvalues of $S^{-1}D_u f_{\mathrm{red}}(u,\lambda)$. Comparing with theorem 6.1 one can ask whether there is any relation between the eigenvalues of $S^{-1}\Phi(u,\lambda)$ and the eigenvalues of $S^{-1}D_u f_{\mathrm{red}}(u,\lambda)$. The answer can be obtained by a careful comparison between the expressions for $\Phi(u,\lambda)$ and

$D_u f_{\mathrm{red}}(u,\lambda)$, taking into account the equation satisfied by $K^*(u,\lambda)$; this leads to the following result.

Theorem 6.2. *Under the hypotheses of theorem 2.1 we have for all sufficiently small* $(u,\lambda) \in U \times \mathbb{R}^m$ *that*

$$\Phi(u,\lambda) = (Id_U + \eta(u,\lambda))D_u f_{red}(u,\lambda), \tag{37}$$

where $\eta : U \times \mathbb{R}^m \to \mathcal{L}(U)$ *is a smooth mapping such that*

$$\eta(u,\lambda) = \mathrm{O}((\|u\| + \|\lambda\|)^2).$$

When the original mapping $f(u,\lambda)$ *is in normal form up to order* $k \geq 1$ *(see theorem 5.2) then*

$$\eta(u,\lambda) = \mathrm{O}(\|u\|^{2k}).$$

In combination with theorem 5.2 this suggests to study the eigenvalues of $S^{-1}D_x f_{\mathrm{NF}}(u,\lambda)\big|_U$ as a first approximation for the critical eigenvalues of $\sigma^{-1} \circ D\hat{f}_\lambda(z^*(u,\lambda))$. For an application of this approach in case of reversible diffeomorphisms see Ciocci et al.[2]

7. A basic example

In this last section we describe a simple application of the foregoing Lyapunov-Schmidt reduction which shows how a generic bifurcation of q-periodic orbits with $q \geq 3$ leads to so-called Arnol'd tongues in parameter space; in fact, as we will see the particular shapes of these tongues are just a consequence of the $\mathbb{Z}_q$-equivariance of the reduced problem.

We start by fixing some $q \geq 3$ and some primary q-th root of unity χ_q, i.e.

$$\chi_q = \exp(2\pi i p/q), \quad \text{with } 0 < p < q \text{ and } \gcd(p,q) = 1.$$

We make the following spectral hypothesis about the family f_λ:

- (**S**) The linear operator $A_0 := D_x f(0,0)$ has χ_q and $\bar{\chi}_q$ as simple eigenvalues, and these are the only eigenvalues of A_0 which are q-th roots of unity.

This implies that $A_\lambda := D_x f(0,\lambda)$ has for sufficiently small $\lambda \in \mathbb{R}^m$ a pair of complex conjugate simple eigenvalues $\{\mu(\lambda), \bar{\mu}(\lambda)\}$, with $\mu : \mathbb{R}^m \to \mathbb{C}$ smooth and such that $\mu(0) = \chi_q$. We also assume the following transversality condition:

- (**T**) The mapping $\mu : \mathbb{R}^m \to \mathbb{C}$ is at $\lambda = 0$ transversal to $\{\chi_q\}$.

This transversality condition requires $m \geq 2$ and will be made more explicit further on. Another consequence of (**S**) is that $\dim U = 2$ and that U can be identified with $\mathbb{C}$ (which should be considered as a 2-dimensional real vectorspace). To make this more explicit, let $x_q + iy_q \in \mathbb{C}^n$ be an eigenvector of A_0 corresponding to the eigenvalue χ_q, and define $\varphi : \mathbb{C} \to \mathbb{R}^n$ by $\varphi(u) := \Re(u(x_q + iy_q))$ for all $u \in \mathbb{C}$; then φ is an isomorphism between $\mathbb{C}$ and $U = \varphi(\mathbb{C})$. As we will see the identification of U with $\mathbb{C}$ (via φ) is quite convenient for the notations; in particular, the restriction A of A_0 to U takes the form $Au = \chi_q u$, and therefore $Su = Au = \chi_q u$ and $Nu = 0$. The reduced mapping f_{red} takes the form of a smooth mapping $f_{\text{red}} : \mathbb{C} \times \mathbb{R}^m \to \mathbb{C}$, with $f_{\text{red}}(0, \lambda) = 0$ and

$$f_{\text{red}}(\chi_q u, \lambda) = \chi_q f_{\text{red}}(u, \lambda). \tag{38}$$

This last relation expresses the $\mathbb{Z}_q$-equivariance of f_{red} and imposes some strong restrictions on the form of f_{red}. Indeed, using singularity theory one can show the following.

Theorem 7.1. *The $\mathbb{Z}_q$-equivariance* (38) *implies that the smooth mapping $f_{red} : \mathbb{C} \times \mathbb{R}^m \to \mathbb{C}$ has the form*

$$f_{red}(u, \lambda) = \phi(v, w, \lambda)u + \psi(v, w, \lambda)\bar{u}^{q-1}, \tag{39}$$

where

$$v = v(u) := |u|^2 \qquad \textit{and} \qquad w = w(u) := \Re(u^q),$$

and where $\phi : \mathbb{C} \times \mathbb{C} \times \mathbb{R}^m \to \mathbb{C}$ and $\psi : \mathbb{C} \times \mathbb{C} \times \mathbb{R}^m \to \mathbb{C}$ are smooth functions, with $\phi(0, 0, 0) = \chi_q$.

A proof is given in Golubitsky et al.[3]; a somewhat weaker result but with a classical proof can be found in Vanderbauwhede.[5] It follows from (39) that the linear part of $f_{\text{red}}(u, \lambda)$ is given by $D_u f_{\text{red}}(0, \lambda) \cdot u = \phi(0, 0, \lambda)u$; using the approximation results of sections 5 and 6 we can (by bringing the original mapping $f(x, \lambda)$ in normal form up to any order $k \geq 1$) without loss of generality assume that

$$\phi(0, 0, \lambda) = \mu(\lambda), \qquad \forall \lambda \in \mathbb{R}^m. \tag{40}$$

Suppose (for simplicity) that $m = 2$; the transversality condition (**T**) then takes the form

$$\frac{\partial(\Re\phi, \Im\phi)}{\partial(\lambda_1, \lambda_2)}(0, 0, 0) \neq 0. \tag{41}$$

Now consider (8), with $u \in \mathbb{C}$, f_{red} of the form (39), and $Su = \chi_q u$. Setting $u = \rho e^{i\theta}$ (with $\rho \geq 0$ and $\theta \in S^1 = \mathbb{R}/2\pi\mathbb{R}$), multiplying by

$\bar{z} = \rho e^{-i\theta}$ and dividing by ρ^2 we see that the non-trivial solutions of (8) are obtained by solving the equation

$$\phi(\rho^2, \rho^q \cos(q\theta), \lambda) - \chi_q + \rho^{q-2} e^{-iq\theta} \psi(\rho^2, \rho^q \cos(q\theta), \lambda) = 0. \tag{42}$$

This equation is satisfied for $(\rho, \lambda) = (0, 0)$; using (41) and splitting into real and imaginary parts it can be solved by the implicit function theorem for $\lambda = \lambda^*(\rho, \theta)$. It follows from (42) that $\lambda^*(\rho, \theta)$ must have the form

$$\lambda^*(\rho, \theta) = \tilde{\lambda}(\rho^2) + \mathrm{O}(\rho^{q-2}), \tag{43}$$

with $\tilde{\lambda} : \mathbb{R} \to \mathbb{R}^2$ smooth and such that $\tilde{\lambda}(0) = 0$. The set

$$T_q := \{\lambda^*(\rho, \theta) \mid 0 < \rho < \rho_0,\ \theta \in S^1\} \tag{44}$$

(with $\rho_0 > 0$ sufficiently small) contains those parameter values λ for which the mapping f_λ has a non-trivial q-periodic orbit near its trivial fixed point $x = 0$; this fixed point has then a pair of eigenvalues near χ_q given by $\{\mu^*(\rho, \theta), \bar{\mu}^*(\rho, \theta)\}$, with $\mu^*(\rho, \theta) := \phi(0, 0, \lambda^*(\rho, \theta))$.

For describing the set T_q we will restrict to the case $q \geq 5$ (the cases $q = 3$ and $q = 4$ have to be studied separately) and assume that $D_v\phi(0, 0, 0) \neq 0$ and $\psi(0, 0, 0) \neq 0$ in (39); this implies in particular that $\tilde{\lambda}'(0) \neq 0$ in (43). The set $T_q \cup \{0\}$ has then a typical sharp conelike form originating from the origin and known as a *resonance horn* or *Arnol'd tongue*. The following proposition describes the main feature of these Arnol'd tongues and is easy to prove using (43).

Proposition 7.2. *Let $q \geq 5$ and assume that $\tilde{\lambda}'(0) \neq 0$. Then*

$$\operatorname{diam}(\{\lambda \in T_q \mid \|\lambda\| = d\}) = \mathrm{O}(d^{(q-2)/2}) \quad \text{as } d \to 0\ (d > 0). \tag{45}$$

A more detailed analysis shows that for each λ in the interior of the Arnol'd tongue T_q the diffeomorphism f_λ has two different q-periodic orbits, corresponding to two different solutions (ρ, θ) of the equation $\lambda^*(\rho, \theta) = \lambda$; as one crosses the boundary of T_q these two q-periodic orbits collide and disappear in a saddle-node bifurcation (of periodic orbits). Moreover, the assumptions we have made so far together with a stronger version of the hypothesis (**S**), namely that $\{\chi_q, \bar{\chi}_q\}$ are the only eigenvalues of A_0 on the unit circle, imply that the family f_λ undergoes a *Neimark-Sacker bifurcation* as the pair of eigenvalues $\{\mu(\lambda), \bar{\mu}(\lambda)\}$ crosses the unit circle, i.e there is a bifurcation of an *invariant curve* from the fixed point at the origin. The set T_q is contained in the set of parameter values for which such invariant curve exists, and for each $\lambda \in T_q$ the q-periodic orbits of f_λ are contained in the invariant curve. For each λ in the interior of T_q one of the two q-periodic orbits will be stable *on the invariant curve*, and the other one unstable: the flow on the invariant curve is *phase-locked*. We refer to Kuznetsov[4] for a more detailed discussion of the Neimark-Sacker bifurcation and phase-locking.

References

1. M.-C. Ciocci et al. *Journal of Difference Eqns. and Appl.* **3**, 485 (1998).
2. M.-C. Ciocci et al. *New Progress in Difference Equations.* B. Aulbach et al. (Eds.) Chapman & Hall, CRC Press (2004) pp. 75-93.
3. M. Golubitsky et al. *Singularities and Groups in Bifurcation Theory* Vol. II. Springer-Verlag (1988).
4. Y. Kuznetsov. *Elements of Applied Bifurcation Theory.* Springer-Verlag (1995).
5. A. Vanderbauwhede. *Proceedings of the Mathematics Conference.* S. Elaydi et al. (Eds.) World Scientific (2000), pp. 254–276.
6. A. Vanderbauwhede. *A short tutorial on Hamiltonian systems and their reduction near periodic orbits.* Report Ghent University, 2001. Download: `http://cage.ugent.be/~avdb/`.
7. A. Vanderbauwhede. Subsection in the *Encyclopedia of Complexity and System Science*, Springer Verlag. To appear.

Part 2
Invited Papers

Pinto's Golden Tilings

J. P. Almeida

Departamento de Matemática,
Escola Superior de Tecnologia e de Gestão,
Instituto Politécnico de Bragança,
Bragança, 5301-857, Portugal
E-mail: jpa@ipb.pt

We present an one-to-one correspondence between (i) Pinto's golden tilings; (ii) smooth conjugacy classes of golden diffeomorphism of the circle that are fixed points of renormalization; (iii) smooth conjugacy classes of Anosov difeomorphisms, with an invariant measure absolutely continuous with respect to the Lebesgue measure, that are topologically conjugated to the Anosov automorphism $G(x, y) = (x + y, x)$; (iv) solenoid functions. As we will explain, the solenoid functions give a parametrization of the infinite dimensional space consisting of the mathematical objects described in the above equivalences.

Keywords: Anosov diffeomorphisms; arc exchange systems; renormalization; tilings; train tracks.

1. Introduction

A. Pinto and D. Sullivan (Ref. 5) proved a one-to-one correspondence between: (i) C^{1+} conjugacy classes of expanding circle maps; (ii) solenoid functions and (iii) Pinto-Sullivan's dyadic tilings on the real line. Here, instead of expanding dynamics, we consider dynamics with minimal sets determined by golden diffeomorphisms. In this case, the expanding dynamics are hidden in the renormalization operator that acts on the minimal set.

2. Pinto's golden tilings

Let $(a_i)_{i \geqslant 2}$ be a sequence of positive real numbers. We are going to establish certain conditions that will give rise to the definition of golden tilings.

2.1. *Fibonacci decomposition*

The Fibonacci numbers, F_1, F_2, $F_3, \ldots$, are inductively given by the well-known relation $F_{n+2} = F_{n+1}+F_n$, $n \geqslant 1$, where F_1 and F_2 are both equal to 1. We say that a finite sequence $F_{n_0}, \ldots, F_{n_p}$ is a *Fibonacci decomposition* of a natural number $i \in \mathbb{N}$ if the following properties are satisfied: (i) $i = F_{n_p} + \cdots + F_{n_0}$; (ii) F_{n_p} is the greatest Fibonacci number less or equal to i and, for each $0 \leqslant k < p$, F_{n_k} is the greatest Fibonacci number less or equal to $i - \left(F_{n_p} + \cdots + F_{n_{k+1}}\right)$; (iii) if $n_0 = 1$ then n_1 is even, and if $n_0 = 2$ then n_1 is odd.

Like this, every natural number $i \in \mathbb{N}$ has a unique Fibonacci decomposition.

We define the *Fibonacci shift* $\sigma_F : \mathbb{N} \to \mathbb{N}$ as follows: For every $i \in \mathbb{N}$ let $F_{n_0}, F_{n_1}, \ldots, F_{n_p}$ be the Fibonacci decomposition associated to i. We define $\sigma_F(i) = F_{n_p+1} + \cdots + F_{n_0+1}$. Hence, letting $F_{n_0}, F_{n_1}, \ldots, F_{n_p}$ be the Fibonacci decomposition associated to $i \in \mathbb{N}$, if $n_0 \neq 1$ then $\sigma_F^{-1}(i) = F_{n_p-1} + \cdots + F_{n_0-1}$, and if $n_0 = 1$ then $\sigma_F^{-1}(i) = \emptyset$. For simplicity of notation we will denote $\sigma_F(i)$ by $\sigma(i)$.

2.2. *Matching condition*

We say that a sequence $(a_i)_{i \geqslant 2}$ satisfies the *matching condition* if, for every $i = F_{n_p+1} + \cdots + F_{n_0+1}$, the following conditions hold:

(i) If $n_0 = 1$ or, $n_0 = 3$ and n_1 odd, then $a_{\sigma(i)} = a_i \left(a_{\sigma(i)+1} + 1\right)^{-1}$ (see Fig. 2).

(ii) If $n_0 = 2$ or, $n_0 > 3$ and even, then $a_{\sigma(i)} = a_i \left(a^{-1}_{\sigma(i)-1} + 1\right)$

(iii) If $n_0 = 3$ and n_1 even or $n_0 > 3$ and odd, then

$$a_{\sigma(j)} = \frac{a_i \left(1 + a_{\sigma(i)-1}\right)}{a_{\sigma(i)-1} \left(1 + a_{\sigma(i)+1}\right)}.$$

Let $\mathbb{L} = \{i \in \mathbb{N} : i \geqslant 2\}$. Therefore, every sequence $(b_i)_{i \in \mathbb{L} \setminus \sigma(\mathbb{L})}$ determines, uniquely, a sequence $(a_i)_{i \in \mathbb{L}}$ as follows: for every $i \in \mathbb{L} \setminus \sigma(\mathbb{L})$, we define $a_i = b_i$ and, for every $i \in \mathbb{L}$, we define $a_{\sigma(i)}$ using the matching condition with respect to the elements of the sequence $(a_j)_{2 \leqslant j \leqslant \sigma(i)}$ already determined.

2.3. *Boundary condition*

A sequence $(a_i)_{i \in \mathbb{L}}$ satisfies the *boundary condition*, if the following limits are well-defined and satisfy the inequalities:

(i) $\lim_{n\to+\infty} a_{F_n+2}^{-1}\left(1+a_{F_n+1}^{-1}\right)\neq 0$

(ii) $\lim_{n\to+\infty} a_{F_n}\left(1+a_{F_n+1}\right)\neq 0$

2.4. *Exponentially fast Fibonacci repetitive*

A sequence $(a_i)_{i\in\mathbb{L}}$ is said to be *exponentially fast Fibonacci repetitive*, if there exist constants $C\geqslant 0$ and $0<\mu<1$ such that $|a_{i+F_n}-a_i|\leqslant C\mu^n$ for every $n\geqslant 3$ and $2\leqslant i<F_{n+1}$.

2.5. *Golden tilings*

A *tiling* $\mathcal{T}=\{I_\beta\subset\mathbb{R}:\beta\geqslant 2\}$ of the positive real line is a collection of tiling intervals I_β, with the following properties: (i) the tiling intervals are closed intervals; (ii) the union $\cup_{\beta\geqslant 2} I_\beta$ is equal to the positive real line; (iii) any two distinct intervals have disjoint interiors; (iv) for every $\beta\geqslant 2$ the intersection of the tiling intervals I_β and $I_{\beta+1}$ is only a point, that is an endpoint, simultaneously, of both intervals.

The tilings $\mathcal{T}_1=\left\{I_\beta\subset\mathbb{R}_0^+:\beta\geqslant 2\right\}$ and $\mathcal{T}_2=\left\{J_\beta\subset\mathbb{R}_0^+:\beta\geqslant 2\right\}$ of the positive real line are in the same affine class, if there exists an affine map $h:\mathbb{R}_0^+\to\mathbb{R}_0^+$ such that $h(I_\beta)=J_\beta$, for every $\beta\geqslant 2$. Thus, every positive sequence $(a_i)_{i\in\mathbb{L}}$ determines an affine class of tilings $\mathcal{T}=\left\{I_m\subset\mathbb{R}_0^+:m\geqslant 2\right\}$ such that $a_m=|I_{m+1}|/|I_m|$, and vice-versa.

Definition 1 *A* golden sequence $(a_i)_{i\geqslant 2}$ *is an exponentially fast Fibonacci repetitive sequence that satisfies the matching and the boundary conditions. A tiling $\mathcal{T}=\{I_m\subset\mathbb{R}:m\geqslant 2\}$ of the positive real line is* golden, *if the corresponding sequence $(a_i=|I_{i+1}|/|I_i|)_{i\geqslant 2}$ is a golden sequence.*

Hence, there is a one-to-one correspondence between golden tilings and golden sequences.

3. Golden diffeomorphisms

We will denote by $\mathbb{S}$ a *clockwise oriented circle* homeomorphic to the circle $\mathbb{S}^1=\mathbb{R}/(1+\gamma)\mathbb{Z}$, where γ is the inverse of the golden number $\left(1+\sqrt{5}\right)/2$. We say that $g:\mathbb{S}\to\mathbb{S}$ is a *golden homeomorphism* if g is a quasi-symmetric homeomorphism conjugated to the rigid rotation $r_\gamma:\mathbb{S}^1\to\mathbb{S}^1$, with rotation number equal to γ.

3.1. *Golden train tracks*

Let us mark a point in $\mathbb{S}$ that we will denote by $0 \in \mathbb{S}$ from now on. Let $\mathrm{A} = \left[g(0), g^2(0)\right]$ be the oriented closed arc in $\mathbb{S}$, with endpoints $g(0)$ and $g^2(0)$, containing the point 0, and let $\mathrm{B} = \left[g^2(0), g(0)\right]$ be the oriented closed arc in $\mathbb{S}$, with endpoints $g(0)$ and $g^2(0)$, not containing the point 0. We introduce an *equivalence relation* $\sim$ in $\mathbb{S}$ by identifying the points $g(0)$ and $g^2(0)$. We call the oriented topological space $T(\mathbb{S}, g) = \mathbb{S}/\sim$ by *train track*. We consider $T = T(\mathbb{S}, g)$ equipped with the quotient topology. Let $\pi_g : \mathbb{S} \to T$ be the natural projection. We call the point $\pi_g(g(0)) = \pi_g(g^2(0)) \in T$ the junction of the train track T and w denote it by ξ. Let $\mathrm{A}_T = \mathrm{A}_T(\mathbb{S}, g) \subset T$ be the projection by π_g of the closed arc A, and let $\mathrm{B}_T = \mathrm{B}_T(\mathbb{S}, g) \subset T$ be the projection by π_g of the closed arc B.

3.2. *Golden arc exchange systems*

The C^{1+} golden diffeomorphism $g : \mathbb{S} \to \mathbb{S}$ determines three maximal diffeomorphisms, $g_{(A,A)}$, $g_{(A,B)}$ and $g_{(B,A}$, on the train track, with the property

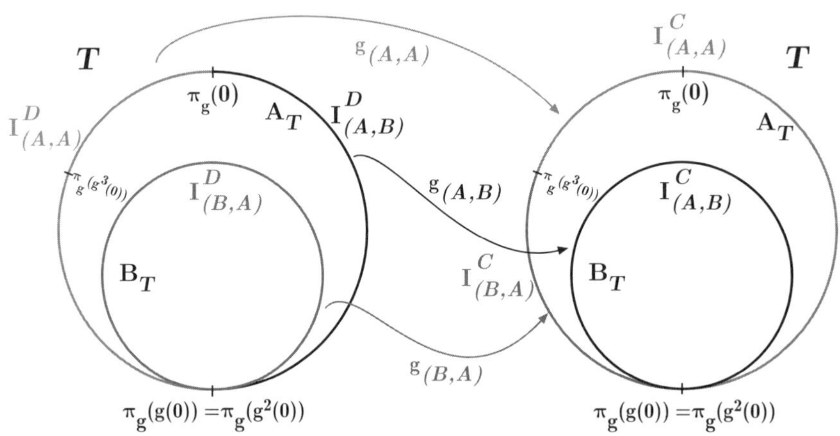

Fig. 1. The arc exchange maps for the train track $T = T(g)$.

that the domain and the counterdomain of each diffeomorphism are either contained in A or in B, in the way that we pass to describe. Let $I^D_{(A,A)}$ be the arc $\pi_g([0, g^2(0)])$, let $I^D_{(A,B)}$ be the arc $\pi_g([g(0), 0])$, and let $I^D_{(B,A)}$ be the arc $\pi_g([g^2(0), g(0)])$. Let $I^C_{(A,A)}$ be the arc $\pi_g([g(0), g^3(0)])$, let $I^C_{(A,B)}$ be the arc $\pi_g([g^2(0), g(0)])$, and let $I^C_{(B,A)}$ be the arc $\pi_g([g^3(0), g^2(0)])$. Let

$g_{(A,A)}: I^D_{(A,A)} \to I^C_{(A,A)}$ be the homeomorphism given by $g_{(A,A)} \circ \pi_g = \pi_g \circ g$, let $g_{(A,B)}: I^D_{(A,B)} \to I^C_{(A,B)}$ be the homeomorphism given by $g_{(A,B)} \circ \pi_g = \pi_g \circ g$, and let $g_{(B,A)}: I^D_{(B,A)} \to I^C_{(B,A)}$ be the homeomorphism given by $g_{(B,A)} \circ \pi_g = \pi_g \circ g$ (see Fig. 1). We call these maps and their inverses by *arc exchange maps*. The *arc exchange system*

$$E = E(g) = \left\{ g_{(A,A)}, g^{-1}_{(A,A)}, g_{(A,B)}, g^{-1}_{(A,B)}, g_{(B,A)}, g^{-1}_{(B,A)} \right\}$$

is the union of all arc exchange maps defined with respect to the train track $T(\mathbb{S}, g)$.

Lemma 1 *(1) If g is a C^{1+} golden diffeomorphism with respect to a C^{1+} atlas $\mathcal{A}$, then the arc exchange system $E(g)$ is C^{1+} with respect to the extended pushforward $(\pi_g)_* \mathcal{A}$ of the C^{1+} atlas $\mathcal{A}$.*

(2) If E is a C^{1+} arc exchange system with respect to a C^{1+} atlas $\mathcal{A}$, then the golden homeomorphism $g(E)$ is C^{1+} with respect to the pullback $(\pi_g)_* \mathcal{A}$ of the C^{1+} atlas $\mathcal{A}$.

See proof in (Ref. 1).

3.3. *Golden renormalization*

The *renormalization of the triple* $(g, \mathbb{S}, \mathcal{A})$ is the triple $(Rg, \mathrm{A}, \mathcal{B}|_\mathrm{A})$ where $\mathrm{A} = [g(0), g^2(0)]/\sim$ is a topological circle, $\mathcal{B}|_\mathrm{A}$ is the restriction of the atlas $\mathcal{B}$ to A, and $Rg : \mathrm{A} \to \mathrm{A}$ is the map given by

$$Rg(x) = \begin{cases} g_{(A,A)}(x) & \text{if } x \in I^D_{(A,A)} \\ g_{(B,A)} \circ g_{(A,B)}(x) & \text{if } x \in I^D_{(A,B)} \end{cases} \tag{1}$$

For simplicity, we will refer to the renormalization of the triple $(g, \mathbb{S}, \mathcal{A})$ by renormalization of g, and we will denote it by Rg.

Let us consider the rigid rotation $r_\gamma : \mathbb{S}^1 \to \mathbb{S}^1$ with the atlas $\mathcal{A}_{iso}$ given by the local isometries with respect to the natural metric in $\mathbb{S}^1$ induced by the Euclidean metric in $\mathbb{R}$. Then there is an isometry $h : \mathbb{S}^1 \to \mathrm{A}$ with respect to the atlas $\mathcal{B}|_\mathrm{A}$ in A and with respect to the atlas $\mathcal{A}$ in $\mathbb{S}^1$, uniquely determined by $h(0) = \pi_{r_\gamma}(0) \in \mathrm{A}$. Furthermore, the map h is a topological (isometric) conjugacy between $(r_\gamma, \mathbb{S}^1, \mathcal{A})$ and $(Rr_\gamma, \mathrm{A}, \mathcal{B}|_\mathrm{A})$.

Let $\mathcal{G}^{1+}$ be the set of all C^{1+} golden diffeomorphisms $(g, \mathbb{S}, \mathcal{A})$.

Lemma 2 *The renormalization R_g of a C^{1+} golden diffeomorphism g is a C^{1+} golden diffeomorphism, i.e. there is a well defined map $R : \mathcal{G}^{1+} \to \mathcal{G}^{1+}$ given by $R(g) = Rg$.*

See the proof in (Ref. 1).

The marked point $0 \in \mathbb{S}$ determines a marked point $\pi_g(0)$ in the circle A. Since Rg is homeomorphic to a golden rigid rotation, there exists $h : \mathbb{S} \to A$, with $h(0) = \pi_g(0)$, such that h conjugates g and Rg. When h is C^{1+} we say that g is a *fixed point of renormalization*. We will denote by $R\mathcal{G}^{1+}$ the set of all C^{1+} fixed points of renormalization.

4. Anosov diffeomorphisms

The *(golden) Anosov automorphism* $G_A : \mathbb{T} \to \mathbb{T}$ is given by $G_A(x, y) = (x+y, x)$, where $\mathbb{T}$ is equal to $\mathbb{R}^2/(v\mathbb{Z}\times w\mathbb{Z})$ with $v = (\gamma, 1)$ and $w = (-1, \gamma)$. Let $\pi : \mathbb{R}^2 \to \mathbb{T}$ be the natural projection. Let $\mathbf{A}$ and $\mathbf{B}$ be the rectangles $[0, 1] \times [0, 1]$ and $[-\gamma, 0] \times [0, \gamma]$ respectively. A Markov partition of G_A is given by $\pi(\mathbf{A})$ and $\pi(\mathbf{B})$. The unstable manifolds of G_A are the projection by π of the vertical lines of the plane, and the stable manifolds of G_A are the projection by π of the horizontal lines of the plane. A C^{1+} *golden Anosov diffeomorphism* $G : \mathbb{T} \to \mathbb{T}$ is a C^{1+} diffeomorphism such that (i) G is topologically conjugated to G_A; (ii) the tangent bundle has a $C^{1+\alpha}$ hyperbolic splitting into a stable direction and an unstable direction. We denote by $\mathcal{C}_G$ the C^{1+} structure on $\mathbb{T}$ in which G is a diffeomorphism.

Theorem 1 *There is a map $G \to g\,(E_G)$ that determines a one-to-one correspondence between C^{1+} conjugacy classes of Anosov diffeomorphisms, with an invariant measure absolutely continuous with respect to the Lebesgue measure, and C^{1+} golden diffeomorphisms that are fixed points of renormalization.*

See the proof in (Ref. 2).

In (Ref. 1), it is proved that the local holonomies determine C^{1+} golden diffeomorphisms.

4.1. *Golden tilings*

Using the mean value theorem and the fact that G is $C^{1+\alpha}$, for some $\alpha > 0$, for all short leaf segments $h(K)$ and all leaf segments $h(I)$ and $h(J)$ contained in it, the unstable realised ratio function r_G given by

$$r_G(I : J) = \lim_{n\to\infty} \frac{|G^{-n}(h(I))|}{|G^{-n}(h(J))|}$$

is well-defined. Let **sol** denote the set of all ordered pairs (I, J) of unstable leaf segments spanning the Markov rectangles such that the intersection of

I and J consists of a single endpoint. The ratio functions r_G restricted to the solenoid set **sol**, determine the solenoid functions $r_G|_{\mathbf{sol}}$. By (Ref. 4), we get the following equivalence:

Theorem 2 *The map $G \to r_G|\mathbf{sol}$ determines a one-to-one correspondence between C^{1+} conjugacy classes of Anosov diffeomorphisms with an invariant measure that is absolutely continuous with respect to the Lebesgue measure and stable ratio functions.*

See the proof in (Ref. 4).

Let W be the unstable leaf with only one endpoint x, that it is the fixed point x of G and passes through all the unstable holonomies of the Markov rectangles $\pi(\mathbf{A})$ and $\pi(\mathbf{B})$. Let I_2, $I_3, \ldots \in W$ be the unstable leaves with the following properties (see Fig. 2): (i) The boundaries ∂I_n of I_n are contained in the stable boundaries of the Markov rectangles, and the interiors $int(I_n)$ of I_n do not intersect the stable boundaries of the Markov rectangles, for every $n \geqslant 2$; (ii) $I_n \cap I_{n+1} = \{x_n\}$ is a common boundary point of both I_n and I_{n+1} for every $n \geqslant 1$. Let $I_1 \in W$ be the union of all the unstable boundaries of the Markov rectangles. By construction, the set

$$\mathcal{L} = \{(I_n, I_{n+1}), n \geqslant 2\}$$

is contained in **sol** and it is dense in **sol**. For every golden sequence $A = (a_n)_{n \geqslant 2}$ let $s_A : \mathcal{L} \to \mathbb{R}^+$ be defined by $s_A((I_n, I_{n+1})) = a_n$.

Theorem 3 *The map $A \to s_A$ gives a one-to-one correspondence between golden sequences and solenoid functions.*

See the proof in (Ref. 1).

5. Conclusion

Putting together Theorem 1, Theorem 2 and Theorem 3 and since, by Definition 1, there is a one-to-one correspondence between golden tilings and golden sequences, we get that there is a one-to-one correspondence between:

(i) golden tilings,
(ii) smooth conjugacy classes of golden diffeomorphism of the circle that are fixed point of renormalization,

(iii) smooth conjugacy classes of Anosov diffeomorphisms, with an invariant measure absolutely continuous with respect to the Lebesgue measure, that are topologically conjugated to the Anosov automorphism $G(x, y) = (x + y, x)$, and
(iv) solenoid functions.

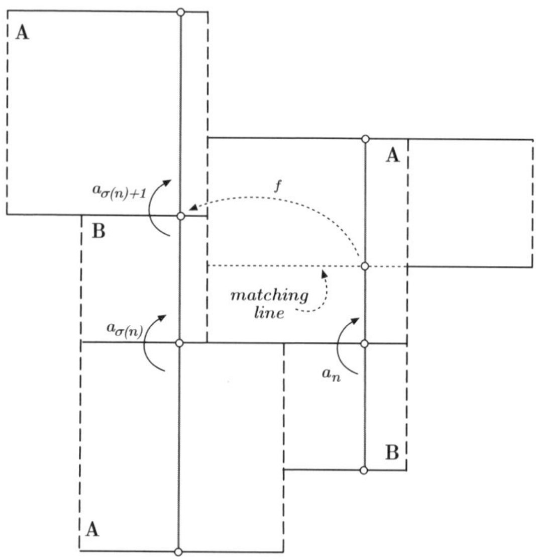

Fig. 2. The matching condition for the case (i).

References

1. J. P. Almeida, A. A. Pinto, A. Portela, D. Rand, *Golden tilings and circle diffeomorphisms, submitted.*
2. A. A. Pinto and D. Rand, *Renormalization gives all surface Anosov diffeomorphisms with a smooth invariant measure, submitted.*
3. A. A. Pinto & D. A. Rand, *Geometric measures for hyperbolic sets on surfaces*, Stony Brook IMS preprint, **3**, pp. 1-69, (2006).
4. A. A. Pinto and D. A. Rand, *Teichmüller spaces and HR structures for hyperbolic surface dynamics*, Ergodic Theory and Dynamical Systems, **22**, pp. 1905-1931, (2002).
5. A. A. Pinto and D. Sullivan, *The Circle And The Solenoid*, Dedicated to Anatole Katok On the Occasion of his 60th Birthday. DCDS-A, **16**, 2, pp. 463-504, (2006).

Non-Exponential Stability and Decay Rates in Nonlinear Stochastic Homogeneous Difference Equations

J. A. D. Appleby

School of Mathematical Sciences, Dublin City University, Dublin 9, Ireland
Email:john.appleby@dcu.ie
http://webpages.dcu.ie/~applebyj

G. Berkolaiko

Department of Mathematics, Texas A&M University, College Station, TX 77843, USA
Email:gregory.berkolaiko@math.tamu.edu
http://www.math.tam.edu/~berko

A. Rodkina

Department of Maths&CSci, The University of the West Indies, Kingston-7, Jamaica
Email:alexandra.rodkina@uwimona.edu.jm
http://www.mona.uwi.edu/dmcs/staff/arodkina/alya.htm

We consider stochastic difference equation

$$x_{n+1} = x_n\Big(1 - hf(x_n) + \sqrt{h}g(x_n)\xi_{n+1}\Big), \quad n = 0, 1, \ldots, \quad x_0 \in \mathbf{R},$$

where functions f and g are nonlinear and bounded, random variables ξ_i are independent and $h > 0$ is a non-random parameter.

We establish results on asymptotic stability and instability of the trivial solution $x_n \equiv 0$. We also show, that when f and g have polynomial behavior in zero neighborhood, the rate of decay of x_n is approximately polynomial: we find $\alpha > 0$ such that x_n decays faster than $n^{-\alpha+\varepsilon}$ but slower than $n^{-\alpha-\varepsilon}$ for any $\varepsilon > 0$.

Keywords: Nonlinear stochastic difference equations, almost sure stability, decay rates, martingale convergence theorem.

1. Introduction

We investigate global stability and the rate of decay of solutions of the difference equation

$$x_{n+1} = x_n\left(1 - hf(x_n) + \sqrt{h}g(x_n)\xi_{n+1}\right), \quad n = 0, 1, \ldots, \quad x_0 \in \mathbf{R}, \quad (1)$$

where ξ_{n+1} are independent random variables. The functions f and g are nonlinear and are assumed to be bounded. The small parameter $h > 0$ usually arises as the step size in numerical schemes. Equation (1) may be viewed as a stochastically perturbed version of a deterministic autonomous difference equation, where the random perturbation is state-dependent. In general, it does not have linear leading order spatial dependence close to the equilibrium. As a consequence of the non-hyperbolicity of the equilibrium, the convergence of solutions of (1) to its equilibrium zero cannot be expected to take place at an exponentially fast rate.

In this note we analyze sufficient and necessary conditions for solutions x_n to converge to zero as $n \to \infty$ ("stability") and the rate at which such convergence happens for different types of the nonlinearities f and g. Our results should be compared to an earlier work [5] (see also [4]) in which similar differential equations had been analyzed. We also mention here papers [2] and [6], where a.s. stability (and also instability) was proved for nonhomogeneous stochastic difference equations.

One of the technical difficulties arising in the study of stability of stochastic difference equations is dealing with unbounded noise. Many results have been only available for the case of bounded noises (e.g. [6]). Yet, one of the most applicable scenarios, discretization of the white noise, involves normally distributed (and thus unbounded) random variables. To overcome this difficulty we apply a special discrete variant of the Itō formula. The corresponding theorem (Theorem 3.1) is presented in Section 3. In particular, it is instrumental in proving the instability result (Theorem 4.2 in Section 4) in this paper and can also be used to prove instability in several related models (e.g. in [7,8]).

Armed with Theorem 3.1, in Section 4 we present criteria for almost sure asymptotic stability and instability of solutions to equation (1). In Section 5 we concentrate on the decay rate of the solutions (assuming they converge to 0). The principal result here is the comparison theorem which provides implicit information on asymptotic behavior of solution x_n via the limit

$$\lim_{n\to\infty} \frac{\ln|x_n|}{\sum_{i=1}^{n} S(x_i)},$$

where $S(x)$ stands for either $g^2(x)$ or $|f(x)|$. In the special (but typical) cases of polynomially decaying f and g, we extract explicit information (see Corollary 5.1) on the decay rate of x_n in the form of the limit

$$\lim_{n\to\infty} \frac{\ln|x_n|}{\ln n} = -\lambda < 0.$$

The above limit allows one to conclude that the decay rate of x_n is of polynomial type. More precisely, for any $\epsilon > 0$, the following bound is valid *eventually* as $n \to \infty$,

$$n^{-\lambda-\epsilon} \le |x_n| \le n^{-\lambda+\epsilon}. \tag{2}$$

More detailed analysis of the rate of decay of x_n is given in [1]. In particular it is shown that when $f(x)$ is dominant, as x tends to 0, the convergence of x_n happens at an exact power-law rate, $n^{-\lambda}$. If the noise term $g(x)\xi$ is significant, an exact rate result is *impossible.*

2. Auxiliary Definitions

In this section we give a number of necessary definitions and a lemma we use to prove our results. A detailed exposition of the definitions and facts of the theory of random processes can be found in, for example, [9].

Let $(\Omega, \mathcal{F}, \{\mathcal{F}_n\}_{n\in\mathbf{N}}, \mathbb{P})$ be a complete filtered probability space. Let $\{\xi_n\}_{n\in\mathbf{N}}$ be a sequence of independent random variables with $\mathbf{E}\xi_n = 0$. We assume that the filtration $\{\mathcal{F}_n\}_{n\in\mathbf{N}}$ is naturally generated: $\mathcal{F}_{n+1} = \sigma\{\xi_{i+1} : i = 0, 1, ..., n\}$.

Among all the sequences $\{X_n\}_{n\in\mathbf{N}}$ of the random variables we distinguish those for which X_n are $\mathcal{F}_n$-measurable $\forall n \in \mathbf{N}$.

A stochastic sequence $\{X_n\}_{n\in\mathbf{N}}$ is said to be an $\mathcal{F}_n$-*martingale*, if $\mathbf{E}|X_n| < \infty$ and $\mathbf{E}\big[X_n\big|\mathcal{F}_{n-1}\big] = X_{n-1}$ for all $n \in \mathbf{N}$ *a.s.*

A stochastic sequence $\{\xi_n\}_{n\in N}$ is said to be an $\mathcal{F}_n$-*martingale-difference*, if $\mathbf{E}|\xi_n| < \infty$ and $\mathbf{E}\big(\xi_n\big|\mathcal{F}_{n-1}\big) = 0$ *a.s.* for all $n \in \mathbf{N}$.

We use the standard abbreviation "*a.s.*" for the wordings "almost sure" or "almost surely" throughout the text.

If $\{X_n\}_{n\in\mathbf{N}}$ is a martingale, in the form $X_n = \sum_{i=1}^{n} \rho_i$, then the *quadratic variation* of X is the process $\langle X\rangle$ defined by

$$\langle X_n\rangle = \sum_{i=1}^{n} \mathbf{E}[\rho_i^2|\mathcal{F}_{i-1}].$$

The following lemma is a variant of martingale convergence theorems (see e.g. [9]) and is proved in [3].

Lemma 2.1. *Let $\{Z_n\}_{n\in\mathbf{N}}$ be a non-negative $\mathcal{F}_n$-measurable process, $\mathbf{E}|Z_n| < \infty$ $\forall n \in \mathbf{N}$, and*

$$Z_{n+1} \le Z_n + u_n - v_n + \nu_{n+1}, \quad n = 0, 1, 2, \dots,$$

where $\{\nu_n\}_{n\in\mathbf{N}}$ is an $\mathcal{F}_n$-martingale-difference, $\{u_n\}_{n\in\mathbf{N}}$, $\{v_n\}_{n\in\mathbf{N}}$ are nonnegative $\mathcal{F}_n$ -measurable processes and $\mathbf{E}|u_n|, \mathbf{E}|v_n| < \infty$ $\forall n \in \mathbf{N}$.

Then

$$\left\{\omega : \sum_{n=1}^{\infty} u_n < \infty\right\} \subseteq \left\{\omega : \sum_{n=1}^{\infty} v_n < \infty\right\} \bigcap \{Z \to\}.$$

Here by $\{X_n \to\}$ we denote the set of all $\omega \in \Omega$ for which $\lim_{n\to\infty} X_n(\omega)$ exists and is finite.

3. Discretized It¯ formula

We will make use of the notation $o(\cdot)$:

$$\alpha(r) = o(\beta(r)) \quad \text{as} \quad r \to r_0 \qquad \Leftrightarrow \qquad \lim_{r\to r_0} \frac{\alpha(r)}{\beta(r)} = 0.$$

here r_0 can be a real number or $\pm\infty$ and the argument r can be both continuous and discrete.

Assumption 3.1. *We will make the following assumptions about the noise ξ_n:*

(1) ξ_n are independent random variables satisfying

$$\mathbf{E}\xi_n = 0, \qquad \mathbf{E}\xi_n^2 = 1, \qquad \mathbf{E}|\xi_n|^3 \text{ are uniformly bounded,}$$

(2) the probability density functions $p_n(\xi)$ exist and satisfy

$$x^3 p_n(x) \to 0 \quad \text{as } |x| \to \infty \quad \text{uniformly in } n.$$

The following theorem can be thought of as a discretized relative of the Itō formula.

Theorem 3.1. *Consider $\varphi : \mathbf{R} \to \mathbf{R}$ such that there exists $\delta > 0$ and $\tilde{\varphi} : \mathbf{R} \to \mathbf{R}$ satisfying*

(1) $\tilde{\varphi} \equiv \varphi$ on $U_\delta = [1 - \delta, 1 + \delta]$,
(2) $\tilde{\varphi} \in C^3(\mathbf{R})$ and $|\tilde{\varphi}'''(x)| \le M$ for some M and all $x \in \mathbf{R}$,

(3) $\int_{\mathbf{R}} |\varphi - \tilde{\varphi}| dx < \infty$.

Let f and g be $\mathcal{F}$-measurable bounded random variables; ξ be an $\mathcal{F}$-independent random variable satisfying Assumption 3.1. Then

$$\mathbf{E}\left[\varphi\left(1 + fh + g\sqrt{h}\xi\right)\Big|\,\mathcal{F}\right] = \varphi(1)+\varphi'(1)fh+\frac{\varphi''(1)}{2}g^2h+hfo(1)+hg^2o(1), \tag{3}$$

where the error terms $o(1)$ satisfy

(1) if $|f|, |g| < K$ then $o(1) \to 0$ as $h \to 0$, uniformly in f and g,
(2) if $h < H$ then $o(1) \to 0$ as $f \to 0$ and $g \to 0$ uniformly in h.

To prove Theorem 3.1 we derive formula (3) for $\mathbf{E}[\tilde{\varphi}]$ and show that $\mathbf{E}[\varphi - \tilde{\varphi}] = hg^2o(1)$.

4. Stability and instability

We consider equation

$$x_{n+1} = x_n\left(1 - hf(x_n) + \sqrt{h}g(x_n)\xi_{n+1}\right), \quad n = 0, 1, \ldots, \tag{4}$$

with nonrandom initial value $x_0 \in \mathbf{R}$, and independent random variables ξ_n satisfying $\mathbf{E}\xi_n = 0$, $\mathbf{E}\xi_n^2 = 1$ for all $n \in \mathbf{N}$. The functions $g, f : \mathbf{R} \to \mathbf{R}$ are nonrandom, continuous and bounded:

$$|g(u)|, |f(u)| \le 1 \quad \forall u \in \mathbf{R}. \tag{5}$$

We formulate two theorems about stability and instability of solution to equation (4). Their proofs are based on Theorem 3.1 and Lemma 2.1.

Theorem 4.1. *Let functions f and g be bounded and ξ_n satisfy Assumption 3.1. Let also*

$$\sup_{u\in\mathbf{R}\setminus\emptyset}\left\{\frac{2f(u)}{g^2(u)}\right\} = \beta < 1. \tag{6}$$

If h is small enough then $\lim_{n\to\infty} x_n(\omega) = 0$ a.s. where x_n is a solution to equation (4).

Remark 4.1. If $g(u) = 0$ for some $u \neq 0$, we consider (6) fulfilled iff $f(u) < 0$. Thus we impose no restrictions on $g(u)$ when $f(u) < 0$ for all nonzero u.

To prove Theorem 4.1 we fix some $\alpha > 0$ and define $\phi_\alpha(y) = |y|^\alpha$, $\Phi_n = \mathbf{E}\left[\phi_\alpha\left(1 + hf(x_n) + \sqrt{h}g(x_n)\xi_{n+1}\right)\right] - 1$. Using Theorem 3.1 we show that $\Phi_n \le \frac{1}{2}\alpha hg^2(x_n)\left(\beta + \alpha - 1 + o(1)\right)$ and then apply Lemma 2.1.

It turns out that condition (6) is close to being necessary for stability. For the same equation we now ask the opposite question: under what conditions on f and g solutions of (4) do *not* tend to zero.

Theorem 4.2. *Let f and g be bounded and ξ_n satisfy Assumption 3.1. Let also*

$$f(u) > 0 \quad \textit{and} \quad g(u) \neq 0 \quad \textit{when} \quad u \neq 0$$

and

$$\liminf_{u \to 0} \left\{ \frac{2f(u)}{g^2(u)} \right\} > 1. \tag{7}$$

If x_n is a solution to equation (4) with an initial value $x_0 \in \mathbf{R}$ and h is small enough then $\mathbb{P}\{\lim_{n\to\infty} x_n(\omega) = 0\} = 0$.

To prove Theorem 4.2 we fix $\alpha < 1$ and define $\Omega_1 = \{\omega : \lim x_n(\omega) = 0\}$,

$$\Phi_i = \mathbf{E}\left[\left| 1 + hf(x_i) + \sqrt{h} g(x_i)\xi_{i+1} \right|^{-\alpha} \Big| \mathcal{F}_i \right].$$

We show that $\Phi_n < 1$ on Ω_1 for big enough n which leads to a contradiction: $x_n > 0$ on Ω_1. Thus $\mathbb{P}\Omega_1 = 0$.

5. Decay Rate

5.1. *A comparison theorem*

Theorem 5.1. *Suppose that f and g are bounded with $f(0) = g(0) = 0$ and the random variables ξ_n satisfy Assumption 3.1. Assume, further, that $x_n \to 0$ a.s., where x_n is a solution of (4).*

a) If

$$\lim_{u\to 0} \frac{2f(u)}{g^2(u)} = L, \quad L \in \mathbf{R}, \tag{8}$$

then, a.s.,

$$\lim_{n\to\infty} \frac{2 \ln |x_n|}{\sum_{i=1}^n g^2(x_i)} = h(L-1). \tag{9}$$

b) If $f(u) \le 0$ in a neighborhood of $u = 0$ and

$$\lim_{u\to 0} \frac{|f(u)|}{g(u)^2} = \infty, \tag{10}$$

then, a.s.,

$$\lim_{n\to\infty} \frac{2 \ln |x_n|}{\sum_{i=1}^n |f(x_i)|} = -h.$$

Remark 5.1. If the initial value x_0 is non-zero and the distributions of ξ_n are non-atomic, then x_n is a.s. non-zero for any n.

Remark 5.2. Theorem 4.2 imposes restrictions on possible values of L. To have convergent x_n we must have $L \leq 1/2$. Then, in equation (9), $-h+2Lh$ is non-positive which one would expect with $x_n \to 0$.

5.2. *Rate of decay of* $\ln|x_n|$

Theorem 5.1 provides some information on the decay of solutions x_n to zero, but does it in a rather implicit way. We will now show how one can extract an explicit estimate on the decay of $\ln|x_n|$ as a function of n.

Consider the following example. Let it be given that $x_n \to 0$ a.s. as $n \to \infty$ (see Theorem 4.1 for a set of sufficient conditions) and let condition (8) be satisfied. Assume that the function $g(u)$ behaves like a power of u around zero,

$$\lim_{u\to 0} \frac{g^2(u)}{u^{\mu_g}} = \text{const.}$$

Then, using the following lemma we can conclude that

$$\lim_{n\to\infty} \frac{\ln|x_n|}{\ln\left(n^{-1/\mu_g}\right)} = 1.$$

Lemma 5.1. *Let $\lambda > 0$ and x_n be a positive sequence satisfying*

$$\lim_{n\to\infty} x_n = 0 \quad \textit{and} \quad \lim_{n\to\infty} \frac{\ln x_n}{\sum_{i=1}^n x_i^\lambda} = -b < 0. \tag{11}$$

Then

$$\lim_{n\to\infty} \frac{\ln x_n}{\ln n} = -1/\lambda.$$

Next we formulate a corollary which extends and formalizes the discussion at the start of the present section.

Corollary 5.1. *Suppose that f and g are bounded with $f(0) = g(0) = 0$ and the random variables ξ_n satisfy Assumption 3.1. Assume, further, that $x_n \to 0$ a.s., where x_n is a solution of (4). If one of the following conditions is fulfilled,*

a)

$$\lim_{u\to 0} \frac{f(u)}{|u|^\lambda} = c < 0, \qquad \lim_{u\to 0} \frac{f(u)}{g(u)^2} = -\infty, \tag{12}$$

or

b)

$$\lim_{u\to 0} \frac{g^2(u)}{|u|^\lambda} = c > 0, \qquad \lim_{u\to 0} \frac{f(u)}{g(u)^2} = L < \frac{1}{2},$$

then, a.s.,

$$\lim_{n\to\infty} \frac{\ln |x_n|}{\ln n} = -\frac{1}{\lambda}.$$

We would also like to mention that Lemma 5.1 can be extended to include other forms of the functions $f(u)$ and $g^2(u)$ around the origin.

Acknowledgement

The authors would like to thank the referee for valuable feedback which helped them to improve this work.

References

1. J. A. D. Appleby, G. Berkolaiko, and A. Rodkina, *Non-exponential stability and decay rates in nonlinear stochastic difference equation with unbounded noises*, Accepted for publication in "Stochastics: An International Journal of Probability and Stochastic Processes".
2. Appleby, J. A. D., Mackey D. and Rodkina, A. (2006). Almost sure polynomial asymptotic stability of stochastic difference equations. *Contemporary Mathematics. Fundamental Research.* **17**, 110-128.
3. Appleby, J. A. D., Mao, X. and Rodkina, A. (2006) On stochastic stabilization of difference equations, *Dynamics of Continuous and Discrete System.* (3)**15**, 843–857.
4. Appleby, J.A.D. and Rodkina, A. (2005) Rates of decay and growth of solutions to linear stochastic differential equations with state-independent perturbations. *Stochastics: An International Journal of Probability and Stochastic Processes.* Vol 77, **3**, 269-290.
5. Appleby, J.A.D., Rodkina, A. and Schurz, H. (2006) Pathwise non-exponential decay rates of solutions of scalar nonlinear stochastic differential equations. *Discrete Continuous Dynam. Systems Ser. B.* **6**. 667-696.
6. Berkolaiko G. and Rodkina A., Almost Sure Convergence of Solutions to Non-Homogeneous Stochastic Difference Equation, *Difference Equations and Applications*, 12:6 (2006), 535-553.
7. Higham, D.J., Mao, X., and Stuart, A.M. (2002) Strong convergence of numerical methods for nonlinear stochastic differential equations, *SIAM J. Num. Anal.* **40**, 1041–1063.
8. Higham, D.J., Mao, X., and Yuan, Y. (2006) Almost sure and moment exponential stability in the numerical simulation of stochastic differential equations, to appear in *SIAM J. Num. Anal.*
9. Shiryaev, A.N. (1996): *Probability* (2nd edition), Springer, Berlin.

Dynamical Consistency of Solutions of Continuous and Discrete Stochastic Equations with a Finite Time Explosion

John A. D. Appleby[†], Cónall Kelly[*] and Alexandra Rodkina[*]

[†]*School of Mathematical Sciences, Dublin City University, Dublin 9, Ireland*
[*]*Department of Mathematics and Computer Science, University of the West Indies, Mona, Kingston 7, Jamaica, W.I.*

We construct a discrete-time stochastic process that mimics the incidence of finite-time explosion in the solutions of a scalar nonlinear stochastic differential equation.

Keywords: Stochastic differential equation; Finite-time explosion; Stochastic difference equation; Discrete-time model.

1. Introduction

The goal of this paper is to construct discrete–time models of solutions of the stochastic differential equation

$$dX(t) = f(X(t))\,dt + g(X(t))\,dB(t), \quad t > 0, \tag{1a}$$

$$X(0) = \psi \in \mathbb{R}^+. \tag{1b}$$

that reliably reflect the occurrence of a finite-time explosion. We require that $f, g \in C(\mathbb{R};\mathbb{R})$ be locally Lipschitz continuous functions. Protter[1] then guarantees the existence of a unique strong solution up to an explosion time τ_e^ψ, where the solution grows without bound when approaching τ_e^ψ. If τ_e^ψ is finite with probability one, then we say that an explosion has occurred a.s. (almost surely).

The simplest possible discretisation of (1) is an Euler-Maruyama approximation (see, for example Kloeden & Platen[2]) over a uniform mesh of step length $h > 0$:

$$X_{n+1} = X_n + hf(X_n) + \sqrt{h}\,g(X_n)\xi_{n+1}, \quad t > 0, \quad X_0 = \psi. \tag{2}$$

Here, $\{\xi_n\}_{n\geq 0}$ is a sequence of independent $N(0,1)$ random variables. However it is easy to see that that solutions of (2) cannot achieve an infinite value in a finite number of time steps. An alternative discretisation must be applied in order to yield a discrete process that can mimic a finite-time explosion.

This question is addressed in Dávila et al.[3] There, the authors apply Euler-Maruyama over a state-dependent, and therefore random, mesh. It is seen that, in circumstances where solutions of (1) explode a.s., the state-dependent meshpoints converge to a random finite value. Therefore, even though solutions of the discrete process will remain finite after any finite number of time steps, the convergence of the mesh can be interpreted as an a.s. explosion in discrete time. The authors also examine various notions of convergence of the numerical method as the overall density of mesh points increases. Additionally they show that, under certain circumstances, the growth rate of solutions of (1) in the neighbourhood of the explosion time can be reproduced by the discrete approximation.

In this paper, we attempt to develop this analysis in two directions. First, our analysis should yield a test for explosion. This requires that, not only should the discrete process mimic an a.s. explosion in the solutions of (1), but it should also mimic the absence of such explosions. Second, our analysis should apply over a wider range of coefficients f and g. However, we do not restrict ourselves to Euler-type numerical discretisations - we seek to construct a discrete model for explosion, rather than developing a numerical method.

In Section 2, we outline the problem in further detail. In Section 3 we describe the method of discretisation used in the paper, and define the general form of a state-dependent mesh. In Section 4 we present our main results. Finally, in Section 5 we draw conclusions and outline the future development of the work.

2. Statement of the Problem

Let $(\Omega, \mathcal{F}, (\mathcal{F}(t))_{t\geq 0}, \mathbb{P})$ be a complete probability space with a filtration $(\mathcal{F}(t))_{t\geq 0}$ satisfying the usual conditions (increasing and right continuous while $\mathcal{F}(0)$ contains all $\mathbb{P}$-null sets). Let $B = (B(t))_{t\geq 0}$ be a scalar Brownian motion defined on the probability space. Since we will consider equations with deterministic initial conditions, we set $\mathcal{F}(t) = \mathcal{F}^B(t)$, where $\mathcal{F}^B(t) = \sigma(B(s) : 0 \leq s \leq t)$.

Let $f, g \in C([0,\infty);\mathbb{R})$ be locally Lipschitz continuous on $[0,\infty)$ and satisfy

$$f(x) > 0, \quad x > 0; \quad f(0) = 0; \quad g^2(x) > 0, \quad x > 0; \quad g(0) = 0. \qquad (3)$$

This guarantees the existence of a unique continuous positive $\mathcal{F}$–adapted process X satisfying

$$X(t\wedge\tau_k,\psi)=\psi+\int_0^{t\wedge\tau_k} f(X(s))\,ds+\int_0^{t\wedge\tau_k} g(X(s))\,dB(s),\quad t\geq 0,\quad \text{a.s.},$$

for each $k\in\mathbb{N}$, where $\tau_k=\inf\{t>0 : X(t,\psi)=k\}$. The explosion time is defined to be

$$\tau_e^\psi=\lim_{k\to\infty}\tau_k=\inf\{t>0\ :\ \lim_{s\uparrow t}X(s,\psi)=\infty\}. \tag{4}$$

We generally choose to suppress the dependence of the solution of (1) on the initial function by writing the solution at time t as $X(t)$.

Definition 2.1. Solutions of (1) are said to undergo an a.s. finite time explosion if $\tau_e^\psi<\infty$ a.s.

In the following theorem, we classify the solutions of (1) according to Definition 2.1 across a parameter range which describes the relative asymptotic properties of f and g.

Theorem 2.1. *Suppose that (3) holds, and that there exists $L\in(0,\infty]$ such that*

$$L:=\lim_{x\to\infty}\frac{xf(x)}{g^2(x)}. \tag{5}$$

Then solutions of (1) can be classified in L as follows:

(i) If $L\in(0,1/2)$ then $\tau_e^\psi=\infty$ a.s.
(ii) If $L\in(1/2,\infty]$ then $\lim_{t\uparrow\tau_e^\psi}X(t,\psi)=\infty$, a.s., and additionally,
  (a) if

$$\int_\psi^\infty\frac{1}{f(x)}\,dx<\infty, \tag{6}$$

  then $\tau_e^\psi<\infty$ a.s.,
  (b) if

$$\int_\psi^\infty\frac{1}{f(x)}\,dx=\infty, \tag{7}$$

  then $\tau_e^\psi=\infty$ a.s.

Proof. The first two statements follow from an analysis of a scale function of X by use of e.g., Proposition 5.5.22 in Karatzas and Shreve.[4] The results in parts (a) and (b) follow from Feller's test (see e.g., Theorem 5.5.29 and Proposition 5.5.32 in Karatzas & Shreve.[4] □

Our goal in this paper is to construct a discrete process that captures this behaviour in the parameter L.

3. Method of Discretisation

We start out by defining the general form of a discretisation on a random mesh. Suppose that $\{\xi_n\}_{n\geq 0}$ is a sequence of i.i.d random variables, with zero mean and unit variance. Since we are now operating in discrete time, we define a discrete-time filtration $\{\mathcal{F}_n\}_{n\geq 0}$ associated with the sequence $\{\xi_n\}_{n\geq 0}$, where $\mathcal{F}_n = \sigma\{\xi_j : 0 \leq j \leq n\}$.

3.1. *Defining the mesh*

Define a sequence of $\mathcal{F}_n$-measurable random variables $\{h(n)\}_{n\geq 0}$. We define the general form of a state-dependent, and therefore random, mesh to be $\mathcal{M} = \{t_n : n \in \mathbb{N}\}$, where $t_0 = 0$ and

$$t_n = \sum_{j=0}^{n-1} h(j), \quad n \geq 1. \tag{8}$$

Thus each t_n is $\mathcal{F}_{n-1}$–measurable, and $h(n-1) = t_n - t_{n-1}$ is the *step length* of the n-th interval $[t_{n-1}, t_n]$. The precise form of the mesh is specified by defining the sequence $\{h(n)\}_{n\geq 0}$.

3.2. *Defining the discrete-time model*

For an appropriate choice of $\{\xi_n\}_{n\geq 0}$, the equation

$$X_{n+1} = X_n + h(n)f(X_n) + \sqrt{h(n)}g(X_n)\xi_{n+1}, \quad n \geq 0, \tag{9a}$$

$$X_0 = \psi > 0, \tag{9b}$$

can be considered a discretisation of (1) over $\mathcal{M}$. For example, setting $t_0 = 0$ and $t_{n+1} = t_n + h$ for $n \geq 1$, and choosing $\{\xi_n\}$ to be a sequence of i.i.d. $N(0,1)$ random variables yields the usual Euler-Maruyama discretisation of (1). However, we do not restrict ourselves to the Euler class of discretisations. For any mesh $\mathcal{M}$ and random sequence $\{\xi_n\}$, we view each X_n as an approximation, to a greater or lesser extent, of $X(t_n)$, the solution of (1) at time t_n. We judge the quality of the approximation according to whether or not the incidence of explosion and non-explosion in solutions of (1) is captured in the solutions of (9).

3.3. *Discrete-time explosions*

Definition 3.1. We say that (9) exhibits *an almost sure explosion* on the mesh $\mathcal{M}$ if

$$\lim_{n\to\infty} X_n = \infty, \quad a.s., \quad \text{and} \quad \lim_{n\to\infty} t_n < \infty, \quad a.s. \tag{10}$$

The random variable $\mathcal{T} := \lim_{n\to\infty} t_n$ is called the positive *a.s. discrete explosion time*, and satisfies

$$\mathcal{T} = \sum_{n=0}^{\infty} h(n). \tag{11}$$

3.4. *Prior analysis in the literature*

The following result can be found in Dávila et al.[3] The authors apply an Euler-Maruyama discretisation to solutions of (1) over the mesh $\mathcal{M}$ defined in Subsection 3.1, where they specify, for any $h > 0$,

$$h(n) = \frac{h}{f(X_n)}. \tag{12}$$

Theorem 3.1. *Let f be a positive, nondecreasing function. Assume that there exists $k_1, k_2 > 0$ such that*

$$k_1 \le g(x)^2 \le k_2 f(x), \text{ for all } x > 0, \tag{13}$$

and

$$\int_0^\infty \frac{1}{f(s)} ds < \infty. \tag{14}$$

Then the discretisation (9), *where $\{\xi_n\}_{n\ge0}$ is a sequence of independent $N(0,1)$ random variables, exhibits an a.s. discrete finite-time explosion on the mesh $\mathcal{M}$.*

Remark 3.1. Note that (13) implies that $L = \infty$. Therefore by (14), all solutions of (1) undergo a finite-time explosion. However, Dávila et al[3] do not address the behaviour of (9) when

$$\int_0^\infty \frac{1}{f(s)} ds = \infty,$$

a condition which guarantees that the solutions of (1) do not explode, nor is the parameter range $0 < L < \infty$ considered. Finally, we remark that condition $L = \infty$ allows the diffusion term to grow more rapidly as $x \to \infty$ for a given f than (13).

By addressing some of these questions, we seek to develop a more general theory that excludes the possibility of false positives and which applies over a wider range of values of L.

4. Main Results

Throughout what follows we will assume that

$$f : x \mapsto f(x) \text{ and } \tilde{f} : x \mapsto \frac{f(x)}{x} \text{ are non–decreasing functions on } (0, \infty), \tag{15}$$

and

$$\text{there exists } \kappa > 0 \text{ such that } \xi_n > -\kappa \text{ for all } n \in \mathbb{N}, \text{ a.s.} \tag{16}$$

We present two main theorems, but prove only the second, as the proofs are quite similar. In the first, we address the case where $L = \infty$. This represents an improvement over Theorem 3.1 in that the discrete explosion time of the discrete process will be finite if and only if the explosion time of the solutions of (1) are finite. However, we do not apply an Euler-Maruyama discretisation, since we must impose the condition (16) on the sequence $\{\xi_n\}_{n\geq 0}$.

We specify our first new mesh $\mathcal{M}$ following the template described in Subsection 3.1 by defining for a deterministic $C > 0$ the $\mathcal{F}_n$–adapted sequence

$$h(n) = C\frac{X_n}{f(X_n)}. \tag{17}$$

It transpires that for appropriate C, $X_n > 0$ for all $n \in \mathbb{N}$ a.s., so $h(n) > 0$ for all $n \in \mathbb{N}$, a.s.

Theorem 4.1. *Suppose f, g are locally Lipschitz continuous on $[0, \infty)$, obey (3), and that (5) holds with $L = \infty$. Suppose that f obeys (15), and that the i.i.d. sequence $\{\xi(n)\}_{n\geq 0}$ obeys (16).*

(i) If f obeys (6), then there is a deterministic $C > 0$ such that the solution of (9), evolving on the mesh $\mathcal{M}$ given by (17), exhibit an a.s. positive explosion at the a.s. discrete explosion time $\mathcal{T}$ defined by (11).

(ii) If f obeys (7), then there is a deterministic $C > 0$ such that the solution of (9), evolving on the mesh $\mathcal{M}$ given by (17) obeys $\lim_{n\to\infty} X_n = \infty$, a.s., and

$$\mathcal{T} = \lim_{n\to\infty} t_n = \infty, \quad a.s.$$

In the second of our main theorems, we address the case where $L < \infty$.

We specify our second new mesh $\mathcal{M}$ following the template described in Subsection 3.1 by defining, for an appropriate deterministic $C > 0$

$$h(n) = C\left(\frac{g(X_n)}{f(X_n)}\right)^2. \tag{18}$$

Theorem 4.2. *Suppose f, g are locally Lipschitz continuous on $[0,\infty)$, obey (3), and that there exists $L \in (0,\infty)$ such that (5) holds. Suppose that f obeys (15), and that the i.i.d. sequence $\{\xi_n\}_{n\geq 0}$ obeys (16).*

(i) If f obeys (6), then there is a deterministic $C > 0$ such that the solution of (9), evolving on the mesh $\mathcal{M}$ given by (18), exhibits an a.s. positive explosion at the a.s. discrete explosion time $\mathcal{T}$ defined by (11).
(ii) If f obeys (7), then there is a deterministic $C > 0$ such that the solution of (9), evolving on the mesh $\mathcal{M}$ given by (18) obeys $\lim_{n\to\infty} X_n = \infty$, a.s., and

$$\mathcal{T} = \lim_{n\to\infty} t_n = \infty, \quad a.s.$$

Remark 4.1. Theorem 4.2 incorrectly predicts an explosion in the solution of (1) in the case when $L \in (0, 1/2)$.

The proof of Theorem 4.2 requires the following lemma.

Lemma 4.1. *Let $\{x_n\}_{n\geq 0}$ be a positive sequence. Suppose f satisfies the conditions of (3) and (15), and that there exist $\alpha_1, \alpha_2 \in (1,\infty)$ and $N > 0$ such that $\alpha_1^n \leq x_n \leq \alpha_2^n$, $n \geq N$.*

(i) If f obeys (6), then $\sum_{j=1}^{\infty} x_j/f(x_j) < +\infty$.
(ii) If f obeys (7), then $\sum_{j=1}^{\infty} x_j/f(x_j) = \infty$.

Proof. If $\alpha_1^n \leq x_n \leq \alpha_2^n$, $n \geq N$ by then (15) for any $M > N$ we have

$$\sum_{j=N}^{M} \frac{\alpha_2^j}{f(\alpha_2^j)} \leq \sum_{j=N}^{M} \frac{x_n}{f(x_n)} \leq \sum_{j=N}^{M} \frac{\alpha_1^j}{f(\alpha_1^j)}. \tag{19}$$

As f is non–decreasing, $1/f(x) \geq 1/f(\alpha_1^{j+1})$ for $x \in [\alpha_1^j, \alpha_1^{j+1}]$ and so

$$\int_{\alpha_1^j}^{\alpha_1^{j+1}} \frac{1}{f(x)}\,dx \geq (\alpha_1^{j+1} - \alpha_1^j)\frac{1}{f(\alpha_1^{j+1})} = (1 - 1/\alpha_1)\frac{\alpha_1^{j+1}}{f(\alpha_1^{j+1})}.$$

In the case when f obeys (6), we have

$$(1 - 1/\alpha_1)\sum_{j=0}^{\infty} \frac{\alpha_1^{j+1}}{f(\alpha_1^{j+1})} \leq \sum_{j=0}^{\infty} \int_{\alpha_1^j}^{\alpha_1^{j+1}} \frac{1}{f(x)}\,dx = \int_1^{\infty} \frac{1}{f(x)}\,dx < \infty.$$

Therefore by (19) we have

$$\sum_{j=0}^{M} \frac{x_n}{f(x_n)} = \sum_{j=0}^{N-1} \frac{x_n}{f(x_n)} + \sum_{j=N}^{M} \frac{x_n}{f(x_n)} \leq \sum_{j=0}^{N-1} \frac{x_n}{f(x_n)} + \sum_{j=0}^{\infty} \frac{\alpha_1^j}{f(\alpha_1^j)},$$

which is finite and independent of M, proving part (i). The proof of part (ii) is similar, and hence omitted. □

Proof of Theorem 4.2. Let $C > \kappa^2$ in (16). First, we show that $\lim_{n\to\infty} X_n = \infty$ a.s. By assumption, X_0 is $\mathcal{F}_0$-measurable and positive. Suppose X_n is $\mathcal{F}_n$-measurable and positive. Then $h(n)$ defined by (18) is also $\mathcal{F}_n$-measurable and the relation

$$X_{n+1} = X_n(1 + \sqrt{C}\lambda(X_n)[\sqrt{C} + \xi_{n+1}]) \tag{20}$$

holds, where the function λ defined by

$$\lambda(x) = \frac{g^2(x)}{xf(x)}, \quad x \in \mathbb{R}^+, \tag{21}$$

is in $C(\mathbb{R}^+;\mathbb{R}^+)$, by the continuity of f and g and (3). Since $C > \kappa^2$, by (16), $\sqrt{C} + \xi_{n+1} > 0$, and therefore $X_{n+1} > 0$ and is $\mathcal{F}_{n+1}$-measurable. By induction, this holds for all $n \in \mathbb{N}$. Additionally we see that $X_{n+1} \geq X_n$, $n \in \mathbb{N}$. Therefore, $\lim_{n\to\infty} X_n$ exists, and is a positive, possibly infinite, random variable. Assume that the event $A = \{\omega : \lim_{n\to\infty} X_n(\omega) < \infty\}$ has nonzero probability. By (20), for almost all $\omega \in A$,

$$\lim_{n\to\infty} \xi_{n+1} = \lim_{n\to\infty} \frac{1}{\sqrt{C}\lambda(X_n)} \left(\frac{X_{n+1}}{X_n} - 1 \right) - \sqrt{C} = -\sqrt{C}.$$

This implies that $C \leq \kappa^2$, which contradicts (16). Therefore we must conclude that

$$\lim_{n\to\infty} X_n = \infty, \quad \text{a.s.} \tag{22}$$

Next, we look at the behaviour of the discrete explosion time $\mathcal{T}$, and determine the circumstances under which it is finite or infinite. Note that (22), (21) and (5) imply that $\lim_{n\to\infty} \lambda(X_n) = 1/L > 0$, a.s. Therefore, for every $\omega \in \Omega$, there is $N_1(\omega)$ such that $\lambda(X_n) \geq (2L)^{-1}$. By (16), (20), and since each X_n is positive, we have

$$X_{n+1} \geq X_n \left(1 + \frac{C}{2L} + \frac{\sqrt{C}}{2L}\xi_{n+1} \right), \quad n \geq N_1(\omega).$$

Next define each term in a sequence of independent random variables $\{\nu_n\}_{n\geq 0}$ by

$$\nu_n := \log\left(1 + \frac{\sqrt{C}}{2L}(\sqrt{C} + \xi_n)\right) \geq \log\left(1 + \frac{\sqrt{C}}{2L}(\sqrt{C} - \kappa)\right) =: \nu_+,$$

and $\nu_+ > 0$. For $n \geq N_1(\omega)$ we have

$$\log X_{n+1} \geq \log X_{N_1(\omega)} - \sum_{j=0}^{N_1(\omega)-1} \nu_{j+1} + \sum_{j=0}^{n} \nu_{j+1}.$$

By the Strong Law of Large Numbers it follows that $\lim_{n\to\infty} \frac{1}{n}\sum_{j=0}^{n-1} \nu_j = \nu^*$ a.s., where $\nu^* = \mathbb{E}[\log\left(1 + \sqrt{C}(2L)^{-1}[\sqrt{C} + \zeta]\right)]$, and ζ is a random variable with zero mean and unit variance satisfying (16). Thus $\nu^* \geq \nu_+ > 0$ and

$$\liminf_{n\to\infty} \frac{1}{n} \log X_{n+1} \geq \nu^* > 0, \quad \text{a.s.}$$

A similar argument shows that there exists $\mu^* \geq \nu^* > 0$ such that

$$\limsup_{n\to\infty} \frac{1}{n} \log X_{n+1} \leq \mu^*, \quad \text{a.s.}$$

So, for any $0 < \nu_1 < \nu^*$ and $\mu_1 > \mu^*$, there is an a.s. finite random variable N such that

$$e^{\nu_1 n} \leq X_n \leq e^{\mu_1 n}, \quad n \geq N, \quad \text{a.s.} \tag{23}$$

Now, applying Lemma 4.1, and using the facts that $h(n) = C\lambda(X_n)\cdot\frac{X_n}{f(X_n)}$, and $\lim_{n\to\infty}\lambda(X_n) = 1/L \in (0,\infty)$ a.s., the remainder of the proof follows. □

5. Conclusions and Further Work

We have seen that over the range $L \in (1/2, \infty]$, it is possible to construct discrete processes that mimic the phenomena of explosion and non-explosion in the solutions of (1), by choosing an appropriate state-dependent mesh. This process is not an Euler-Maruyama discretisation as we must truncate the left tail of the distribution of the perturbing random variables, as described in (16).

It seems likely that this restriction results in the occurrence of spurious discrete finite-time explosions solutions of (9) in the range $L \in (0, 1/2)$, as this skews the stochastic perturbation in such a way that growth is encouraged. In the future we hope to develop techniques that do not require the restriction on the distribution of the perturbing random variable. The application of a logarithmic transformation to the solutions of (1) prior to discretisation looks promising. This may allow us to develop models that correctly reflect not only the occurrence of explosion and nonexplosion in the parameter region $L \in (0, 1/2)$, but also mimic the associated asymptotic growth rates of solutions in both explosive and non–explosive cases.

Acknowledgements

The first author was partially supported by an Albert College Fellowship, awarded by Dublin City University's Research Advisory Panel. The third author was supported by the Mona Research Fellowship Programme, awarded by the University of the West Indies, Jamaica, and by the International Society of Difference Equations.

References

1. P. E. Protter. On the existence, uniqueness, convergence and explosion of solutions of systems of stochastic integral equations. *Ann. Probab.*, 5:243–261, 1977.
2. P. E. Kloeden and E. Platen. Numerical Solution of Stochastic Differential Equations. Volume 23 of Applications of Mathematics, Springer–Verlag. Berlin, 1992.
3. J. Dávila, J. F. Bonder, J. D. Rossi, P. Groisman and M. Sued. Numercial Analysis of Stochastic Differential Equations with Explosions. *Stoch. Anal. Appl.*, **23**(4): 809–825, 2005.
4. I. Karatzas and S. E. Shreve. Brownian Motion and Stochastic Calculus. *Second Edition, Springer–Verlag.* New York, 1991.

Interval Maps and Cellular Automata*

L. Bandeira, M. J. Martinho and C. Correia Ramos
CIMA-UE, Departamento de Matemática, Universidade de Évora, Portugal
E-mail: lmzb@uevora.pt, mjfbmartinho@sapo.pt, ccr@uevora.pt

We associate to cellular automata elementary rules a class of interval maps defined in [0,1]. The considered configuration space is the space of the one-sided sequences in the set {0,1} and with the appropriate choice of the update procedure the interval maps do not depend on the boundary conditions. We study the rule 184 and obtain a family of transition matrices that characterizes the dynamics of the cellular automaton. We show that these matrices can be obtained recursively by an algorithm that depends on the local rule.

1. Introduction and preliminaries

In our work we deal with cellular automata, interval maps and subshifts of finite type. We associate to an elementary cellular automata rule an interval map, as in Wolfram,[7] and we show that this interval map is characterized by a functional equation. The considered configuration space is the space of the one-sided sequences in the set $\{0,1\}$ and with the appropriate choice of the update procedure the interval maps do not depend on the boundary conditions, obtaining different results from those in a previous work, Bandeira *et. al.*[1] Next, we define a family of partitions of the interval $[0,1]$ and in each partition the effect of the interval map is codified by a transition matrix. Naturally, the structure of the transition matrices is directly related with the elementary cellular automaton rule. Each transition matrix determine a subshift of finite type which approximately describe the behavior of the interval map. In this work we study the interval map that arises in this way from the cellular automata elementary rule 184. The rule 184, is a particular case of the Fukui-Ishibashi model which consist in a family of cellular automata traffic rules. It has been studied in several contexts, see for example Boccara *et. al.*[2] and Fuks.[3]

*This work is supported by FCT/POCTI/FEDER and CIMA.

Let φ be defined by $\varphi : \{0,1\}^{\mathbb{N}} \rightarrow [0,1]$, $\varphi(\xi) := \sum_{k=1}^{\infty} \xi_k 2^{-k}$, $\xi \in \{0,1\}^{\mathbb{N}}$. Every point x in the unit interval $[0,1]$ is an image under φ, that is, $x = \varphi(\xi) = \sum_{k=1}^{\infty} \xi_k 2^{-k}$, where $\xi = (\xi_k)_{k\in\mathbb{N}} \in \{0,1\}^{\mathbb{N}}$ corresponds to the binary expansion of x. Note that this is expansion may not be unique. In fact, the sequences of the type $\xi_1\xi_1...\xi_k 01^{\infty}$ and $\xi_1\xi_1...\xi_k 10^{\infty}$ give origin under φ to the same real number. Since we are interested in dealing with all binary sequences we consider the lateral limits of every point $x \in [0,1]$, denoted by x^- and x^+. Therefore, $\varphi(\xi_1\xi_2...\xi_k 01^{\infty}) = x^-$ and $\varphi(\xi_1\xi_2...\xi_k 10^{\infty}) = x^+$ with $x = \sum_{k=1}^{\infty} \xi_k 2^{-k}$. This expansion is related to the piecewise linear interval map

$$b(x) = \begin{cases} 2x & \text{if } x \in [0,1/2] \\ 2x-1 & \text{if } x \in [1/2,1], \end{cases}$$

since $\xi_k = 0$ if $b^{k-1}(x) \in [0,1/2]$ and $\xi_k = 1$ if $b^{k-1}(x) \in [1/2,1]$, where $b^k(x) = b(b^{k-1}(x))$. Particularly, the map b, acting in $[0,1]$, induces the usually called shift map on $\{0,1\}^{\mathbb{N}}$ given by $\sigma(\xi_k)_{k\in\mathbb{N}} := (\xi_{k+1})_{k\in\mathbb{N}}$. For convenience we assume that b is multivalued at $x = 1/2$. Nevertheless no problem arises since we have to consider the lateral limits of $1/2$ in order to deal with the two different expansions associated with $1/2$. For these lateral limits, the image under b is well defined, in fact, $b(1/2^-) = 2(1/2^-) = 1^- = 1$ and $b(1/2^+) = 2(1/2^+) = 0^+ = 0$. This, in turn, is in agreement with our assumption on φ applied to the boundary of $I = [0,1]$, $\varphi(1^{\infty}) = 1^- = 1 \in I$ and $\varphi(0^{\infty}) = 0^+ = 0 \in I$.

Let I_0 and I_1 be the intervals $[0,1/2]$ and $[1/2,1]$ respectively. Denote the inverse branches of b by $b_0^{-1} : I \rightarrow I_0$ with $b_0^{-1}(x) = x/2$ and $b_1^{-1} : I \rightarrow I_1$ with $b_1^{-1}(x) = x/2 + 1/2$. Observe that $b \circ b_0^{-1}(x) = id(x)$ and $b \circ b_1^{-1}(x) = id(x)$. We denote by I_{ξ} the set of points with binary expansion having the block ξ as a prefix, *i.e.*, $I_{\xi} = \{x \in [0,1] : x = \xi_1 2^{-1} + ... + \xi_k 2^{-k} + ...\}$. The set I_{ξ} is an interval for every $\xi \in \mathcal{E}^k$, for some $k \in \mathbb{N}$, and is given by $b_{\xi_1}^{-1} \circ b_{\xi_2}^{-1} ... \circ b_{\xi_k}^{-1}([0,1])$.

More generally, consider the pair $(\mathcal{E}^{\mathbb{N}}, \sigma)$, where $\mathcal{E} = \{1,...,n\}$. $\mathcal{E}^{\mathbb{N}}$ is the space of one-sided sequences in $\mathcal{E}$, and σ is the usual shift map in $\mathcal{E}^{\mathbb{N}}$, $\sigma : \mathcal{E}^{\mathbb{N}} \rightarrow \mathcal{E}^{\mathbb{N}}$, $\sigma(\xi_1\xi_2...) = (\xi_2\xi_3...)$. The set $\mathcal{E}^{\mathbb{N}}$ is a compact totally disconnected topological space and the shift map is a continuous, onto n-to-one map, see Kitchens.[4] A *block* is a finite sequence of symbols from $\mathcal{E}^{\mathbb{N}}$. The *length* of a block u is the number of symbols it contains and is denoted by $|u|$. We denote the block, in $\xi = (\xi_i)_{i\in\mathbb{N}}$, indexed by the integer interval $[i,j]$ by $\xi_{[i,j]} := \xi_i\xi_{i+1}...\xi_j$, with $i \leq j$

Let $A = (a_{ij})$ be a $n \times n$ $\{0,1\}$-matrix, usually called a transition matrix. A *subshift of finite type* Λ_A is a subset of $\mathcal{E}^{\mathbb{N}}$, invariant for the shift map, defined by $\Lambda_A = \{(\xi_i)_{i\in\mathbb{N}} \in \mathcal{E}^{\mathbb{N}} : a_{\xi_i\xi_{i+1}} = 1,\ i \in \mathbb{N}\}$. The set Λ_A is a compact, totally disconnected topological space and the shift map is a continuous, onto map, see Kitchens.[4] Denoted by Λ_A^n is the collection of blocks of size n in Λ_A. Consider a map $\phi : \Lambda_A^n \to \mathcal{E}$ which we call a *local block map.* Then a *block map of type* (n_L, n_R), with $n_L < n_R$, and $n_L, n_R \in \mathbb{Z}$, associated to the local map ϕ is a map verifying

$$\Phi : \Lambda_A \to \Lambda_A$$

$$\Phi\left(\xi\right)_i := \phi\left(\xi_{[i+n_L, i+n_R]}\right)$$

When $n_L < 0$, in the context of one-sided subshifts, it is necessary to specify boundary conditions in order to the map Φ be well defined. More precisely to compute $\Phi\left(\xi\right)_1$, $\Phi\left(\xi\right)_2$,..., $\Phi\left(\xi\right)_{|n_L|}$ it is necessary to specify ξ_{n_L-1}, ..., ξ_{-1}, ξ_0.

A *one-dimensional cellular automaton* can be defined as a continuous, shift commuting, map from a subshift of finite type to itself. It can be proven that every map satisfying such conditions is a block map, see for example Kitchens[4] or Marcus *et. al.*[6]

Consider the subshift of finite type Λ_A, characterized by a transition matrix A. Suppose that we choose an interval map τ, in $[0, 1]$, that realizes the subshift Λ_A, *i.e.*, there is a map $\varphi : \Lambda_A \to [0, 1]$ so that $\tau \circ \varphi = \varphi \circ \sigma$. Naturally, the properties of τ will depend on the properties of φ. Namely, if φ is a homeomorphism then the pair $(\varphi(\Lambda_A), \tau)$ will be topologically conjugated to (Λ_A, σ), see Kitchens[4] or Marcus *et. al.*[6]

Suppose that for the subshift Λ_A, and for the map $\varphi : \Lambda_A \to [0, 1]$, is given a block map Φ of type (n_L, n_R), with local block map ϕ, and fixed boundary conditions. Then we can find an interval map g depending on the block map and on the boundary conditions chosen, such that the associated cellular automaton is realized, that is $g \circ \varphi = \varphi \circ \Phi$. Similarly to the shift map, σ, and its realization on the interval, τ, some properties of the interval map g depend on the properties of φ.

2. Functional relations for g arising from rule 184

Let $\mathcal{E} = \{0, 1\}$. The rule 184, of type $(0, 2)$, is given by the following local block map

$$f(000) = 0, \quad f(001) = 0, \quad f(010) = 0, \quad f(011) = 1$$
$$f(100) = 1, \quad f(101) = 1, \quad f(110) = 0, \quad f(111) = 1$$

Since we are considering the rule of type $(0,2)$ we have a unique global block map associated to this automata rule $F : \mathcal{E}^{\mathbb{N}} \to \mathcal{E}^{\mathbb{N}}$, with $F(\xi)_i = f\left(\xi_{[i,i+2]}\right)$ and $F(\xi)_0 = f(\xi_0\xi_1\xi_2)$. As it was explained in the introduction the interval map g associated to the global block map F will be defined in a way that $g \circ \varphi = \varphi \circ F$, where φ was introduce in the previous section. In Figure 1 we show the graph of the interval map g.

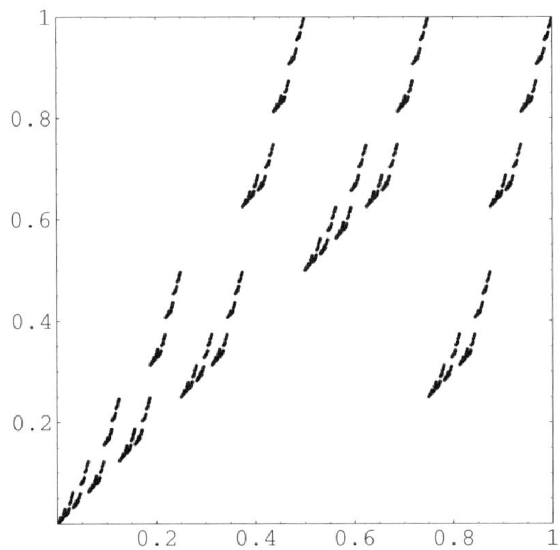

Fig. 1. Graph of the map $g(x)$.

Note that g can be given explicitly by

$$g(x) = \sum_{k=1}^{\infty} f(\xi_k\xi_{k+1}\xi_{k+2})\, 2^{-k}.$$

Lemma 2.1. *The interval map g satisfy the following general functional equation*

$$2g(x) = g(b(x)) + f(\xi_1\xi_2\xi_3)\,, \text{ with } x \in I_{\xi_1\xi_2\xi_3}.$$

Proof. Let $x = \sum\limits_{k=1}^{\infty} \xi_k 2^{-k}$. Then $g(x) = \sum\limits_{k=1}^{\infty} f(\xi_k\xi_{k+1}\xi_{k+2})\, 2^{-k}$. On the other hand $b(x) = \sum\limits_{k=1}^{\infty} \xi_{k+1} 2^{-k}$. Then $g(b(x)) = \sum\limits_{k=2}^{\infty} f(\xi_k\xi_{k+1}\xi_{k+2})\, 2^{-k+1}$.

The expression $\sum_{k=3}^{\infty} f(\xi_k \xi_{k+1} \xi_{k+2}) 2^{-k+1}$ will be equal to $g(b(x)) - f(\xi_2\xi_3\xi_4) 2^{-1}$, and equal to $2\left(g(x) - f(\xi_1\xi_2\xi_3) 2^{-1} - f(\xi_2\xi_3\xi_4) 2^{-2}\right)$. Then we get

$$g(b(x)) - f(\xi_2\xi_3\xi_4) 2^{-1} = 2\left(g(x) - f(\xi_1\xi_2\xi_3) 2^{-1} - f(\xi_2\xi_3\xi_4) 2^{-2}\right)$$

which is equivalent to $2g(x) = g(b(x)) + f(\xi_1\xi_2\xi_3)$. □

Proposition 2.1. *The interval map g satisfies the following functional relation:*

$$2g(x) = \begin{cases} g(2x) & \text{if } x \in I_{00},\ I_{010} \\ g(2x) + 1 & \text{if } x \in I_{011} \\ g(2x-1) & \text{if } x \in I_{110} \\ g(2x-1) + 1 & \text{if } x \in I_{10},\ I_{111} \end{cases}.$$

Proof. Let $x = \sum_{k=1}^{\infty} \xi_k 2^{-k}$. Suppose $x \in I_{00}$, then $\xi_1\xi_2 = 00$. Following the result in the Lemma 2.1, and recalling that $b(x) = 2x$ for $x \in I_{00}$, we get

$$2g(x) = g(2x) + f(00\xi_3).$$

As $f(00\xi_3) = 0$, we get $2g(x) = g(2x)$ for $x \in I_{00}$. The other cases $x \in I_{010}$, $x \in I_{011}$, $x \in I_{110}$, $x \in I_{10}$ and $x \in I_{111}$ are analyzed in a similar way, noting that $b(x) = 2x - 1$ for $x \in I_1$. □

3. Transition matrices

When we consider a piecewise monotone interval map $\tau : I \to I$ such that the orbits of the singular points (critical or discontinuity points) under τ are finite then it is possible to give a Markov partition $P = \{I_1, ..., I_n\}$ of I and a transition matrix A associated to τ in the following way

$$A_{ij} = \begin{cases} 1 & \text{if } \tau(I_i) \supset I_j \\ 0 & \text{otherwise.} \end{cases}$$

The Markov partition is determined by the union of the orbits of the singular points. This procedure is given, for example, in the case of the bimodal interval maps in Lampreia *et. al.*[5] Thus, there is a subshift of finite type (Λ_A, σ) associated to (I, τ) which is codified by the matrix A, and some properties of the dynamical system (I, τ) can be studied with the help of (Λ_A, σ). For example the topological entropy of (I, τ) is given by $\log \lambda_A$ where λ_A is the spectral radius of A.

In a similar way, we will associate to the map g certain transition matrices. Since g has an infinite number of discontinuity points we cannot build a unique finite transition matrix. However, we will show that it is possible to define a family of partitions of the interval $[0, 1]$, and in each partition the map g induces a transition matrix. Each matrix is a coarse-grained approximation to the infinite matrix which codifies the transition of an infinite configuration $\xi \in \mathcal{E}^{\mathbb{N}}$ to its image $F(\xi) \in \mathcal{E}^{\mathbb{N}}$, under the global block map F.

Let us define the family of partitions of the interval $[0, 1]$, given by $P_1 = \{I_0, I_1\}$, and $P_k = \{I_\xi : \xi \in \mathcal{E}^k\}$, where I_ξ are the intervals introduced in the section 2. Note that each P_{k+1} is a refinement of P_k. We will also use the following notation for the intervals on the partition P_k. The interval denoted by $I_{(n)}$, where $n = 0, ..., 2^k - 1$, is equal to the interval I_ξ with the block ξ being the binary expansion of n.

The map g acting on the partition P_k induce a transition matrix denoted by A_k, with $k \in \mathbb{N}$, and defined as follows

$$(A_k)_{ij} = \begin{cases} 1 & \text{if } g\left(I_{(i)}\right) \cap I_{(j)} \neq \emptyset \\ 0 & \text{otherwise,} \end{cases} \quad \text{with} \quad i, j = 0, 1, ..., 2^k - 1.$$

Recall that the interval I_ξ is the set of all numbers in $[0, 1]$ with the binary expansion having the block $\xi = \xi_1 ... \xi_m$ as prefix.

We will show that the matrices A_k have a natural block structure and that this block can be obtained from a recursive algorithm. For $k = 2$ we have

$$A_2 = \begin{pmatrix} 1\,1\,0\,0 \\ 0\,1\,1\,1 \\ 0\,0\,1\,1 \\ 0\,1\,1\,1 \end{pmatrix}.$$

In fact, for I_{00} we have

$$g\left(I_{00}\right) \subset \bigcup_{\xi_3, \xi_4 = 0,1} I_{f(00\xi_3) f(0\xi_3\xi_4)}.$$

By the rule 184, $f(00\xi_3) = 0$. If $\xi_3 = 0$ then $f(0\xi_3\xi_4) = f(00\xi_4) = 0$. If $\xi_3 = 1$ and $\xi_4 = 0$ then $f(0\xi_3\xi_4) = f(010) = 0$ or if $\xi_3 = 1$ and $\xi_4 = 1$ then $f(0\xi_3\xi_4) = f(011) = 1$. We obtain

$$g\left(I_{00}\right) \subset I_{00} \cup I_{01},$$

thus there are no transitions from I_{00} to I_{10} and I_{11}.

For I_{01}, I_{10}, and I_{11} we obtain with the same reasoning

$$g\left(I_{01}\right) \subset I_{01} \cup I_{10} \cup I_{11}, \quad g\left(I_{10}\right) \subset I_{10} \cup I_{11}, \quad g\left(I_{11}\right) \subset I_{01} \cup I_{10} \cup I_{11}.$$

Consider now the intervals of the partition P_k, $I_{ij\xi}$, with $|\xi| = k-2$ and $i,j \in \{0,1\}$. Let us define the blocks $B_{ij,i'j'}(k)$, $i,j,i',j' \in \{0,1\}$, as the submatrices of A_k which codifies the transition, under g, from intervals of type $I_{ij\xi}$ to intervals of type $I_{i'j'\eta}$, for blocks ξ,η of size $k-2$. In the same way the blocks $B_{ij,i'j'}(k+1)$ are the submatrices of A_{k+1} which codifies the transition, under g, from intervals of type $I_{ij\xi}$ to intervals of type $I_{i'j'\eta}$, for blocks ξ,η of size $k-1$. The key point is to determine, for k fixed, the decomposition of the blocks $B_{ij,i'j'}(k+1)$ into the blocks $B_{ij,i'j'}(k)$, $i,j,i',j' \in \{0,1\}$. Consider the intervals $I_{ij\xi}, I_{i'j'\eta} \in P_{k+1}$, with $\xi = \xi_1...\xi_{k-1}$ and $\eta = \eta_1...\eta_{k-1}$. We have

$$B_{ij,i'j'}(k+1) = \begin{pmatrix} \gamma(ij0,i'j'0)\,B_{j0,j'0}(k) & \gamma(ij0,i'j'1)\,B_{j0,j'1}(k) \\ \gamma(ij1,i'j'0)\,B_{j1,j'0}(k) & \gamma(ij1,i'j'1)\,B_{j1,j'1}(k) \end{pmatrix}$$

where $\gamma(ij\xi_1,i'j'\eta_1) = 0,1$ depends on if there is a transition, under g, between the interval $I_{ij\xi_1...\xi_{k-1}}$ to the interval $I_{i'j'\eta_1...\eta_{k-1}}$. A transition from $I_{ij\xi_1...\xi_{k-1}}$ to $I_{i'j'\eta_1...\eta_{k-1}}$, under g, can occur if

$$f(ij\xi_1)\,f(j\xi_1\xi_2)\,f(\xi_1\xi_2\xi_3) = i'j'\eta_1.$$

Then we have

$$\gamma(ij\xi_1,i'j'\eta_1) = \begin{cases} 1 \text{ if } f(ij\xi_1)\,f(j\xi_1 s_1)\,f(\xi_1 s_1 s_2) = i'j'\eta_1,\ s_1,s_2 \in \{0,1\} \\ 0 \text{ otherwise.} \end{cases}$$

It is straightforward to determine every $\gamma(ij\xi_1,i'j'\eta_1)$. Here we show those which are equal to 1.

$$\begin{array}{llll} \gamma(000,000)=1 & \gamma(000,001)=1 & \gamma(001,001)=1 & \gamma(001,010)=1 \\ \gamma(001,011)=1 & \gamma(010,010)=1 & \gamma(010,011)=1 & \gamma(011,101)=1 \\ \gamma(011,110)=1 & \gamma(011,111)=1 & \gamma(100,100)=1 & \gamma(100,101)=1 \\ \gamma(101,101)=1 & \gamma(101,110)=1 & \gamma(101,111)=1 & \gamma(110,010)=1 \\ \gamma(110,011)=1 & \gamma(111,101)=1 & \gamma(111,110)=1 & \gamma(111,111)=1 \end{array}$$

Therefore, we obtain the decomposition of the blocks $B_{ij}(k+1)$ expressed in the following result

Theorem 3.1. *The transition matrix A_k, $k \geq 2$, has the block structure*

$$A_k = \begin{pmatrix} B_{00,00}(k) & B_{00,01}(k) & B_{00,10}(k) & B_{00,11}(k) \\ B_{01,00}(k) & B_{01,01}(k) & B_{01,10}(k) & B_{01,11}(k) \\ B_{10,00}(k) & B_{10,01}(k) & B_{10,10}(k) & B_{10,11}(k) \\ B_{11,00}(k) & B_{11,01}(k) & B_{11,10}(k) & B_{11,11}(k) \end{pmatrix}$$

and the transition matrix A_{k+1} is given by

$$\begin{pmatrix} B_{00,00}(k) & B_{00,01}(k) & 0 & 0 & 0 & 0 & 0 & 0 \\ 0 & B_{01,01}(k) & B_{01,10}(k) & B_{01,11}(k) & 0 & 0 & 0 & 0 \\ 0 & 0 & B_{10,10}(k) & B_{10,11}(k) & 0 & 0 & 0 & 0 \\ 0 & 0 & 0 & 0 & 0 & B_{11,01}(k) & B_{11,10}(k) & B_{11,11}(k) \\ 0 & 0 & 0 & 0 & B_{00,00}(k) & B_{00,01}(k) & 0 & 0 \\ 0 & 0 & 0 & 0 & 0 & B_{01,01}(k) & B_{01,10}(k) & B_{01,11}(k) \\ 0 & 0 & B_{10,10}(k) & B_{10,11}(k) & 0 & 0 & 0 & 0 \\ 0 & 0 & 0 & 0 & 0 & B_{11,01}(k) & B_{11,10}(k) & B_{11,11}(k) \end{pmatrix}$$

Example 3.1. Consider the case $k = 3$. Then the matrix A_3 is given by

$$A_3 = \begin{pmatrix} 1 & 1 & 0 & 0 & 0 & 0 & 0 & 0 \\ 0 & 1 & 1 & 1 & 0 & 0 & 0 & 0 \\ 0 & 0 & 1 & 1 & 0 & 0 & 0 & 0 \\ 0 & 0 & 0 & 0 & 0 & 1 & 1 & 1 \\ 0 & 0 & 0 & 0 & 1 & 1 & 0 & 0 \\ 0 & 0 & 0 & 0 & 0 & 1 & 1 & 1 \\ 0 & 0 & 1 & 1 & 0 & 0 & 0 & 0 \\ 0 & 0 & 0 & 0 & 0 & 1 & 1 & 1 \end{pmatrix}$$

References

1. L. Bandeira, M. J. Martinho, C. Correia Ramos, *Interval maps associated to cellular automaton rule 184*, submitted (2007).
2. N. Boccara, H. Fuks. *Convergence to equilibrium in a class of interacting particle systems evolving in discrete time*, (2001).
3. H. Fuks. *A class of cellular automata equivalent to deterministic particle systems.* Hydrodynamic limits and related topics (Toronto, ON, 1998), 57–70, Fields Inst. Commun., 27, Amer. Math. Soc., Providence, RI, (2000).
4. B. Kitchens. *Symbolic dynamics. One-sided, two-sided and countable state Markov shifts.* Universitext. Springer-Verlag, Berlin, (1998).
5. J.P. Lampreia, J. Sousa Ramos. *Symbolic dynamics of bimodal maps.* Portugal. Math. 54 (1997), no. 1.
6. D. Lind, B. Marcus. *An introduction to symbolic dynamics and coding.* Cambridge University Press, Cambridge, (1995).
7. S. Wolfram *A new kind of science.* Wolfram Media, Inc., Champaign, IL, (2002).

Difference Equations on Matrix Algebras

A. Nascimento Baptista*

Department of Mathematics, ESTG, Polytechnic Institute of Leiria,
Leiria, Campus 2, Morro do Lena - Alto do Vieiro 2411-901, Portugal
**E-mail: acfsn@estg.ipleiria.pt*
www.estg.ipleiria.pt

C. Correia Ramos

University of Évora, CIMA, Department of Mathematics
Rua Romão Ramalho, 59, 7000-671, Évora, Portugal
E-mail: ccr@uevora.pt

Nuno Martins

Centro de Análise Matemática, Geometria e Sistemas Dinâmicos,
Department of Mathematics, Instituto Superior Técnico
Av. Rovisco Pais 1, 1049-001 Lisboa, Portugal
E-mail: nmartins@math.ist.utl.pt

We study non-linear difference equations $X_{n+1} = F(X_n)$, where X_n belongs to a certain matrix algebra A, and F is a polynomial map defined on A. We are interested in analyzing the type of periodic orbits and their stability. We also will discuss the dependence of the dynamical behavior on parameters in different situations, since we can consider the parameters to be in the algebra A or in some sub-algebra of A. We study the concrete cases when F is a quadratic map and A is $M_2(\mathbb{R})$, or some sub-algebra of $M_2(\mathbb{R})$.

Keywords: Matrix dynamics, critical orbits, non-critical attracting cycles.

1. Introduction

We study non-linear difference equations $X_{n+1} = F(X_n)$, where X_n belongs to a certain matrix algebra $\mathcal{A}$, and F is a polynomial map defined on $\mathcal{A}$. In a similar way as in the real or complex situation, we discuss the dependence of the dynamical behavior on parameters in different situations, since we can consider the parameters to be in the algebra $\mathcal{A}$ or in some sub-algebra of $\mathcal{A}$. We are interested in analyzing the type of periodic or aperiodic orbits, bounded or unbounded, and their stability.

The motivation of this work came from some questions: How can the algebraic structure of the algebra of matrices be useful in the study of the dynamics of various applications? Will its structure be strong enough to allow reach general conclusions about the dynamics of these applications on matrices? Can the understanding of these properties and relations allow us to understand some facts about the dynamics in lower dimensions?

The aim of this work is to focus on the dynamical behavior of subalgebras $\mathcal{A}$ of $M_2(\mathbb{R})$ with the quadratic map

$$\begin{gathered} F_C : M_2(\mathbb{R}) \longrightarrow M_2(\mathbb{R}) \\ X \mapsto X^2 + C \end{gathered}$$

associated to the discrete dynamical system $(M_2(\mathbb{R}), F_C)$, where C is the parameter matrix.

Note that, from the beginning, this study in $M_2(\mathbb{R})$ is trickier than the case in $\mathbb{R}$ or $\mathbb{C}$, because of the non-commutative nature of matrices multiplication.

Along the text, we will refer to the one-dimensional case of the quadratic map in $\mathbb{C}$ as f_c, that is, $f_c(z) = z^2 + c$, where c is the complex parameter.

2. Preliminaries

2.1. *The Jordan canonical forms*

There are some powerful results in the algebraic structure of matrices that allow us to obtain important conclusions, for example, the Jordan canonical form classification.

Let $X \in M_2(\mathbb{R})$ be not a multiple of the identity and let $J(X)$ denote the Jordan canonical form of X. In $M_2(\mathbb{R})$, the Jordan classification gives us one of the three following Jordan canonical types:

Type I : $\begin{pmatrix} x & 0 \\ 0 & y \end{pmatrix}$ - matrix with distinct real eigenvalues x and y

Type II : $\begin{pmatrix} x & 1 \\ 0 & x \end{pmatrix}$ - matrix with identical real eigenvalues x

Type III : $\begin{pmatrix} x & y \\ -y & x \end{pmatrix}$ - matrix with complex eigenvalues $x \pm yi$

Now we can consider the inner automorphism of $M_2(\mathbb{R})$ defined by

$$\begin{gathered} \alpha_P : M_2(\mathbb{R}) \longrightarrow M_2(\mathbb{R}) \\ \alpha_P(X) = PXP^{-1} \end{gathered} .$$

The discrete dynamical system $(M_2(\mathbb{R}), F_C)$ is equivalent to $(M_2(\mathbb{R}), F_{PCP^{-1}})$. Therefore, we can assume that the parameter matrix C is in the Jordan canonical form and, along the study, we must consider the three types distinctly.

The commutator $[X_0, C] := X_0C - CX_0$ is decisive for the resulting dynamics. If $[X_0, C] = 0$ then there is a basis of $\mathbb{R}^2$ in which X_0 and C are simultaneously in the Jordan form. That is, there is $P \in GL_2(\mathbb{R})$, so that $PX_0P^{-1} = J(X_0)$ and $PCP^{-1} = J(C)$.

If X_0 and C commute, then we work in three distinct Jordan canonical subspaces, and have the following situations:

- If X_0 and C are of type I, we work in the subalgebra of the diagonal matrices with distinct real eigenvalues. If
$$X_0 = \begin{bmatrix} x_0 & 0 \\ 0 & y_0 \end{bmatrix} \qquad \text{and} \qquad C = \begin{bmatrix} a & 0 \\ 0 & b \end{bmatrix}$$
the dynamical system $X_0 \mapsto F_C(X_0)$ is equivalent to the product space of the dynamics of f_a and f_b,
$$\begin{cases} x_{k+1} = f_a(x_k) \\ y_{k+1} = f_b(y_k) \end{cases},$$
with diagonal entries independent of one another.
- If X_0 and C are of type II, we work in the subalgebra of the triangular matrices with identical real eigenvalues. If
$$X_0 = \begin{bmatrix} x_0 & 1 \\ 0 & x_0 \end{bmatrix} \qquad \text{and} \qquad C = \begin{bmatrix} a & 1 \\ 0 & a \end{bmatrix}$$
the dynamical system $X_0 \mapsto F_C(X_0)$ is equivalent to
$$\begin{cases} x_{k+1} = f_a(x_k) \\ y_{k+1} = f'_a(x_k)y_k + 1 \end{cases}.$$
In this case, the diagonal entries are independent, however the other non zero entry depends on the behavior of the diagonal entry.
- If X_0 and C are of type III, we work in the subalgebra of the real matrices with complex eigenvalues. If
$$X_0 = \begin{bmatrix} x_0 & y_0 \\ -y_0 & x_0 \end{bmatrix} \qquad \text{and} \qquad C = \begin{bmatrix} a & b \\ -b & a \end{bmatrix}$$
the dynamical system $X_0 \mapsto F_C(X_0)$ is equivalent to $z_{k+1} = f_c(z_k)$ with $c = a + bi$ the complex eigenvalue of C and $z_0 = x_0 + y_0 i$ the complex eigenvalue of X_0, following the canonical isomorphism between matrices X and the points $\phi(X)$ in the complex plane, given by $\phi\begin{pmatrix} x & y \\ -y & x \end{pmatrix} = x + iy$.

2.2. *Invariant 3-dimensional spaces*

The Cayley-Hamilton theorem is another strong result in the algebraic structure of matrices. For X in $M_2(\mathbb{R})$, it gives $X^2 = \tau X - \delta I$ where $\tau = \mathrm{tr}X$ and $\delta = \det X$, and has important consequences, because

$X_1 = F(X_0) = X_0^2 + C = \tau_0 X_0 - \delta_0 I + C \in \langle I, X_0, C\rangle$ and

$X_2 = F(X_1) = X_1^2 + C = \tau_1\tau_0 X_0 - (\tau_1\delta_0 + \delta_1)I + (\tau_1 + 1)C \in \langle I, X_0, C\rangle$

where $\tau_i = \mathrm{tr}X_i$, $\delta_i = \det X_i$ and $\langle I, X_0, C\rangle$ the subspace spanned by the matrices I, C and X_0.

In fact, A. Serenevy[1] referred the following result:

Proposition 2.1. *The dynamics of F_C is restricted to the 3-dimensional subspace $\langle I, X_0, C\rangle$ of $M_2(\mathbb{R})$.*

3. The orbit of the critical point 0 in $M_2(\mathbb{R})$

Consider the matrix $\mathbf{0}$ in $M_2(\mathbb{R})$, a critical point of F_C, with F_C seen as a map in $\mathbb{R}^4$. The orbit of this critical point is

$$\mathbf{0} \to C \to C^2 + C \to C^4 + 2C^3 + C^2 + C \to ...,$$

and it remains on the invariant subspace spanned by the identity matrix I and the parameter matrix C.

Note that the algebraic expression is exactly the same as the orbit of the critical point 0 in the one-dimensional case of $f_c(z) = z^2 + c$, with c the complex parameter.

The dynamical behavior of the critical point of F_C is given by the following result:

Theorem 3.1. *Let C be of type I or III. Then the orbit of the zero matrix $\mathbf{0}$ is bounded (periodic) under F_C if and only if:*

- *For C of type I with distinct eigenvalues a, b the orbits of 0 under f_a and f_b are bounded (periodic).*
- *For C of type III with complex eigenvalue c the orbit of 0 under f_c is bounded (periodic).*

Proof. Since the zero matrix commutes trivially with any matrix, we have from previous argument that the iterates of the zero matrix reduces to two different situations, according to the Jordan form of C. If C is of type I, with distinct real eigenvalues a, b the orbit of the zero matrix is reduced to the orbit of 0 under f_a and the orbit of 0 under

f_b. Therefore, the orbit of the zero matrix is bounded (periodic) if and only if the orbits of 0 under f_a and f_b are both bounded (periodic). If C is of type III the dynamics is reduced to the iteration of 0 under f_c, where $c = a + bi$. Therefore, the orbit of 0 is bounded (periodic) if and only if the orbit of 0 under f_c is bounded (periodic). □

Theorem 3.2. *Let C be of type II with eigenvalue a. Then $F_C(\mathbf{0})$ is periodic if and only if the orbit of 0 under f_a is periodic. Moreover, if the orbit of the zero matrix is bounded then the orbit of 0 under f_a is bounded.*

Proof. First note that C can be considered in the form $C = \begin{pmatrix} a & 1 \\ 0 & a \end{pmatrix}$ and the kth-iterate of the zero matrix is denoted by $F_C^k(\mathbf{0}) = \begin{pmatrix} x_k & y_k \\ 0 & x_k \end{pmatrix}$, with $x_{k+1} = f_a(x_k)$ and $y_{k+1} = 2x_k y_k + 1$. For the first part, if $F_C(\mathbf{0})$ is periodic then necessarily 0 is periodic under f_a since the only possible pre-image of a is 0. On the other hand, if 0 is periodic this means there is a natural number k so that $x_k = f_a^k(0) = 0$. Since $y_{k+1} = 2x_k y_k + 1 = 1$ then $F_C(\mathbf{0})$ is periodic. The second part is direct. □

4. Non-critical attractive cycles

4.1. *The existence of non-critical attractive cycles*

In the real and complex case, for polynomial maps of degree at least 2, every attractive cycle must attract the orbit of a critical point. Is this true when we work in matrices, with the map F_C? No, in fact, as pointed by A. Serenevy[1] , the existence of attractive cycles that do not attract the orbit of the critical point $\mathbf{0}$ is possible. We call them **non-critical attractive cycles**.

Recall that the orbit of the critical point $\mathbf{0}$ remains on the invariant subspace spanned by the identity matrix I and the parameter matrix C, so, if it exists an attractive cycle outside the subspace $\langle I, C \rangle$, we prove the existence of attractive cycles which do not attract the orbit of $\mathbf{0} \in M_2(\mathbb{R})$.

It will be interesting to observe the behavior, under the action of F_C, of the orbits with X_0 and C being of distinct types.

Example 4.1. Let's consider the initial matrix X_0 of type I, that is, in the Jordan canonical form with distinct real eigenvalues, and C of type III, in the Jordan canonical form with complex eigenvalues:

$$X_0 = \begin{bmatrix} x & 0 \\ 0 & y \end{bmatrix} \quad \text{and} \quad C = \begin{bmatrix} -1 & 0.22 \\ -0.22 & -1 \end{bmatrix}$$

We verify that, under the action of F_C :

- The orbit of the critical point $X_0 = 0$ is attracted to an attracting cycle with period 2, remaining, as expected, in the subspace $\langle I, C\rangle$:

$$\begin{bmatrix} 0 & 0 \\ 0 & 0 \end{bmatrix} \overset{F_C}{\to} \cdots \overset{F_C}{\to} \begin{bmatrix} 0.039 & -0.204 \\ 0.204 & 0.039 \end{bmatrix} \overset{F_C}{\to} \begin{bmatrix} -1.039 & 0.204 \\ -0.204 & -1.039 \end{bmatrix} \overset{F_C}{\to} \begin{bmatrix} 0.039 & -0.204 \\ 0.204 & 0.039 \end{bmatrix} \overset{F_C}{\to} \cdots$$

- The orbit of some points is attracted to a non-critical attractive cycle with period 2, which doesn't remain on the subspace $\langle I, C\rangle$:

$$\cdots \overset{F_C}{\to} \begin{bmatrix} 0.012 & 0.11 \\ -0.11 & -1.012 \end{bmatrix} \overset{F_C}{\to} \begin{bmatrix} -1.012 & 0.11 \\ -0.11 & 0.012 \end{bmatrix} \overset{F_C}{\to} \begin{bmatrix} 0.012 & 0.11 \\ -0.11 & -1.012 \end{bmatrix} \overset{F_C}{\to} \cdots$$

- The Julia set in Figure 1, the set of starting matrix X_0 for which the iterations under F_C remain bounded, associated with this parameter matrix C, has a checkerboard pattern, pointed by A. Serenevy[1] , that describes the behavior of the orbits: the orbit of the points in the subsets with the circle mark is attracted to the same attracting orbit of the critical point; the orbit of the points in the subsets with the square mark is attracted to the non-critical attractive cycle.

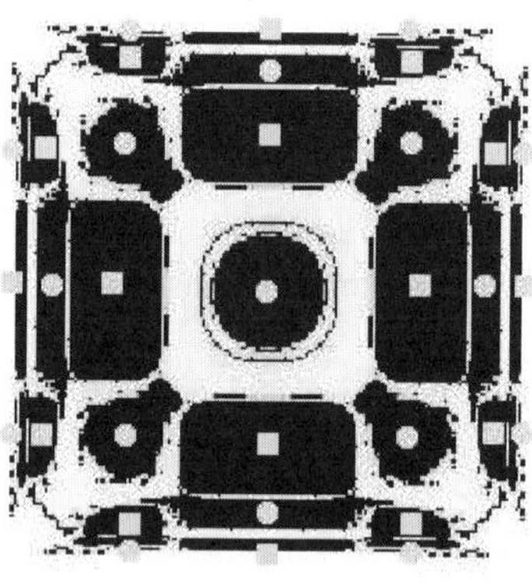

Fig. 1. The geometric description of the orbits

In fact, for some values of the parameter matrix C with complex eigenvalues, the orbit of the critical point $\mathbf{0} \in M_2(\mathbb{R})$ is not bounded, therefore, it is not attracted to any attracting cycle, however an attracting cycle exists.

Example 4.2. Let's consider the initial matrix X_0 of type I, that is, in the Jordan canonical form with distinct real eigenvalues, and C of type III, in the Jordan canonical form with complex eigenvalues:

$$X_0 = \begin{bmatrix} x & 0 \\ 0 & y \end{bmatrix} \quad \text{and} \quad C = \begin{bmatrix} -1.28 & 0.2 \\ -0.2 & -1.28 \end{bmatrix},$$

Consider the Julia set in Figure 2 associated with the parameter matrix C :

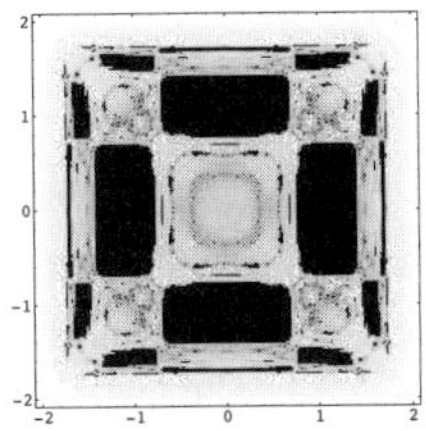

Fig. 2. The Julia set associated with the parameter matrix C of Example 4.2

In this example, we see that the orbit of the critical point is not bounded, however an attracting cycle exists. In this case, the non-critical attractive cycle has period 4.

4.2. *Non-critical attracting cycles of period 2 when C is of type III*

If we consider the one-dimensional complex case, when the quadratic map is $f_c(z) = z^2 + c$, the complex Mandelbrot set gives us information about the period of the attracting cycles in the complex case. The matricial Mandelbrot set, composed by the parameter matrices $C = \begin{bmatrix} a & b \\ -b & a \end{bmatrix}$ for which the orbit of the critical point $\mathbf{0} \in M_2(\mathbb{R})$ is bounded under the action of F_C, is similar to the one on the one-dimensional complex case. In fact, it is possible to get some information about the period of the attracting cycles by observing the geometry of this set and recalling the canonical isomorphism between matrices X and the points $\phi(X)$ in the complex plane, given by $\phi\begin{pmatrix} x & y \\ -y & x \end{pmatrix} = x + iy$. The main cardioid is the region of parameter matrices C for which F_C has an attracting fixed matrix. To the left of the main cardioid, we can see a circular-shaped bulb. This bulb consists of those parameters for which the critical point has an attracting cycle of period 2 under the action of F_C. This attracting cycle, when it exists, attracts the orbit of the critical point and has the form

$$\dots \rightarrow \begin{bmatrix} \mathrm{Re}(z_1) & -\mathrm{Im}(z_2) \\ \mathrm{Im}(z_2) & \mathrm{Re}(z_1) \end{bmatrix} \rightarrow \begin{bmatrix} \mathrm{Re}(z_2) & -\mathrm{Im}(z_1) \\ \mathrm{Im}(z_1) & \mathrm{Re}(z_2) \end{bmatrix} \rightarrow \begin{bmatrix} \mathrm{Re}(z_1) & -\mathrm{Im}(z_2) \\ \mathrm{Im}(z_2) & \mathrm{Re}(z_1) \end{bmatrix} \rightarrow \dots$$

where $z_1 = -\frac{1}{2} - \frac{1}{2}\sqrt{-3-4(a+bi)}$ and $z_2 = -\frac{1}{2} + \frac{1}{2}\sqrt{-3-4(a+bi)}$.

Note that z_1 and z_2 are such that $f_c(z_1) = z_2$ and $f_c(z_2) = z_1$, that is, we have an attracting cycle with period 2 with respect to f_c. Suppose that we have x and y such that the iterates, under f_c, converge to this attractive cycle with period 2: $\{x, f_c(x), ..., z_1, z_2, z_1, ...\}$ and $\{y, f_c(y), ..., z_1, z_2, z_1, ...\}$. If X has distinct real eigenvalues x and y, two situations may occur. In the first one, the orbits are such that they ocurr "synchronized", that is, there is a $n \in \mathbb{N}$ such that $f_c^n(x) = z_1$ and $f_c^n(y) = z_1$. In the second, they don't ocurr "synchronized", that is, from some $n \in \mathbb{N}$, we have $f_c^n(x) = z_1$ and $f_c^{n+1}(y) = z_1$. When X has identical real eigenvalues x, we have similar situations: "synchronized" orbits or "non synchronized" orbits.

Transposing this two different situations to the matricial case, the first situation leads to the case where X converges to the orbit of the critical point, under the action of F_C; the second one leads to the case where X converges to the non-critical attracting cycle. Taking this into account, we reach to the following result:

Theorem 4.1. *If, in the complex case, $c = a+bi$ is a complex parameter for which the orbit of the attracting cycle has period 2, then, for $C = \begin{bmatrix} a & b \\ -b & a \end{bmatrix}$ with complex eigenvalues, for X_0 in type I, $X_0 = \begin{bmatrix} x & 0 \\ 0 & y \end{bmatrix}$, or X_0 in type II, $X_0 = \begin{bmatrix} x & 1 \\ 0 & x \end{bmatrix}$, when the iterates converge to a non-critical attracting cycle of period 2 under the action of F_C, the 2−cycle has the form*

$$\dots \rightarrow \begin{bmatrix} x_k & y_k \\ -y_k & -x_k-1 \end{bmatrix} \rightarrow \begin{bmatrix} -x_k-1 & y_k \\ -y_k & x_k \end{bmatrix} \rightarrow \begin{bmatrix} x_k & y_k \\ -y_k & -x_k-1 \end{bmatrix} \rightarrow \dots \quad \textit{with}$$

$$x_k = \left(-\frac{1}{2} - \frac{1}{2}\sqrt{-3-4a+b^2}\right) \quad \textit{and} \quad y_k = \frac{b}{2}.$$

References

1. A. K. Serenevy. *Dynamics of polynomial maps of 2×2 real matrices*, Topology proceedings, Vol. 26, 763–778, 2001.

The Beverton–Holt Difference Equation

Martin Bohner*

Missouri University of Science and Technology, Department of Economics and Finance
103 Harris Hall, Rolla, MO 65409-1250, USA
E-mail: bohner@mst.edu, *URL* http://web.mst.edu/~bohner

Stevo Stević

Mathematical Institute of the Serbian Academy of Science
Knez Mihailova 36/III, 11000 Beograd, Serbia
E-mail: sstevic@ptt.rs

Howard Warth

Missouri University of Science and Technology
Department of Mathematics and Statistics
314 Rolla Building, Rolla, MO 65409-0020, USA
E-mail: hlw6c2@mst.edu

The Beverton–Holt equation, usually treated as a rational difference equation, is shown in fact to be a logistic difference equation. Based on a crucial transformation connected to logistic equations, an elementary proof of the Cushing–Henson conjectures is given.

Keywords: Beverton–Holt equation; logistic equation; Cushing–Henson conjecture; strictly convex; Subject Classifications 39A10, 39A12, 37N25, 92D25.

1. Introduction

The difference equation

$$x_{n+1} = \frac{\nu K_n x_n}{K_n + (\nu - 1)x_n}, \quad n \in \mathbb{N}_0, \tag{1}$$

where $\nu > 1$, $K_n > 0$, and $x_0 > 0$ is known as the Beverton–Holt equation and has wide applications in population dynamics. The positive sequence $\{K_n\}$ is the *carrying capacity* and ν is the *inherent growth rate.* A periodically forced Beverton–Holt equation is obtained by letting the carrying

*Supported by NSF Grant #0624127

capacity be a periodic positive sequence $\{K_n\}$ with period $p \in \mathbb{N}$ so that $K_{n+p} = K_n$ for all $n \in \mathbb{N}_0$.

The Beverton–Holt equation has been treated in the literature as a rational difference equation (see [1–4]). In this paper we show that the Beverton–Holt equation is in fact a logistic difference equation. A simple substitution transforms this equation into a linear difference equation, which can be solved employing the usual variation of parameters formula. This, via an application of the definition of a strictly convex function, also provides an elementary proof of the so-called Cushing–Henson conjecture, which says that *any periodic variation in the carrying capacity of the environment is deleterious to the average population size.* The results in this paper essentially make use of the key ideas in our previous articles [5,6].

2. Conversion to a Linear Difference Equation

We begin by introducing

$$\alpha := \frac{\nu - 1}{\nu} \quad \text{so that} \quad \nu = \frac{1}{1-\alpha}.$$

Suppose $\{x_n\}$ solves (1). Then we have

$$x_{n+1} = \frac{K_n x_n}{(1-\alpha)K_n + \alpha x_n}.$$

Now the substitution $u_n = 1/x_n$ yields the linear difference equation

$$\Delta u_n = -\alpha u_n + \frac{\alpha}{K_n}. \tag{2}$$

We solve the problem (1) with the p-periodic boundary condition $\overline{x}_0 = \overline{x}_p$ by solving the equivalent problem (2) with the p-periodic boundary condition $\overline{u}_0 = \overline{u}_p$, where we assume

$$K : \mathbb{Z} \to \mathbb{R}^+ \quad \text{is } p\text{-periodic,} \qquad \text{and} \quad 0 < \alpha < 1, \quad \text{i.e.,} \quad \nu > 1. \tag{3}$$

Using the variation of parameters formula [7, Theorem 3.1 on page 45], the general solution of (2) is

$$u_n = (1-\alpha)^n u_0 + \sum_{i=0}^{n-1} (1-\alpha)^{n-(i+1)} \frac{\alpha}{K_i}. \tag{4}$$

(Note here that (4) implies that $u_n > 0$ provided $u_0 > 0$, i.e., $x_n > 0$ provided $x_0 > 0$.) Then, using $\overline{u}_0 = \overline{u}_p$ in (4), yields the initial value

$$\overline{u}_0 = \frac{1}{1-(1-\alpha)^p} \sum_{i=0}^{p-1} (1-\alpha)^{p-(i+1)} \frac{\alpha}{K_i}. \tag{5}$$

Plugging the initial value (5) into the general solution (4), we obtain the particular solution

$$\overline{u}_n = \frac{(1-\alpha)^p}{1-(1-\alpha)^p}\sum_{i=0}^{p-1}(1-\alpha)^{n-(i+1)}\frac{\alpha}{K_i} + \sum_{i=0}^{n-1}(1-\alpha)^{n-(i+1)}\frac{\alpha}{K_i}. \qquad (6)$$

Theorem 2.1. *Assume* (3). *Then* $\overline{x}_n = 1/\overline{u}_n$ *is the only* p*-periodic solution to* (1), *where* $\overline{u}_n$ *is given in* (6).

Proof. We already showed uniqueness. For $n \in \mathbb{N}$, we use again variation of paramenters to find another form of the general solution of (2) as

$$u_{n+p} = (1-\alpha)^n u_p + \sum_{i=p}^{n+p-1}(1-\alpha)^{n+p-(i+1)}\frac{\alpha}{K_i},$$

which together with (4) and the p-periodicity of $\{K_n\}$ yields

$$\overline{u}_{n+p} - \overline{u}_n = (1-\alpha)^n(\overline{u}_p - \overline{u}_0) = 0.$$

Thus the solution $\overline{u}_n$ of (2) is p-periodic. □

Remark 2.1. Note also that, if we assume that $K_n = K_{n+p}$ for all $n \in \mathbb{Z}$, then (2) can be used to define $\overline{u}_n$ for all $n \in \mathbb{Z}$ (since $\alpha \neq 1$). Then

$$\overline{u}_{-1} = \frac{\overline{u}_0 - \frac{\alpha}{K_{-1}}}{1-\alpha} = \frac{\overline{u}_p - \frac{\alpha}{K_{p-1}}}{1-\alpha} = \overline{u}_{p-1}$$

and it is easy to see that $\overline{u}_n$ obtained as such satisfies $\overline{u}_{n+p} = \overline{u}_n$ for all $n \in \mathbb{Z}$. Hence there is a unique both-side periodic solution to (2) and consequently to (1).

Theorem 2.2. *Assume* (3). *The solution* $\{\overline{x}_n\}$ *of* (1) *is globally attractive.*

Proof. Let $\{x_n\}$ be an arbitrary solution of (1) with $x_0 > 0$ and consider $u_n = 1/x_n$. Define

$$\gamma = \gamma(u) := \min\left\{u_0, \min_{0\le i\le p-1}\frac{1}{K_i}\right\} > 0.$$

By (4) we have

$$u_n \geq (1-\alpha)^n\gamma + \sum_{i=0}^{n-1}(1-\alpha)^{n-(i+1)}\alpha\gamma = \gamma$$

so that, again by (4),

$$|x_n - \overline{x}_n| = \frac{(1-\alpha)^n|u_0 - \overline{u}_0|}{u_n\overline{u}_n} \leq \frac{(1-\alpha)^n|u_0-\overline{u}_0|}{\gamma(u)\gamma(\overline{u})},$$

which tends to zero as n tends to infinity. □

Remark 2.2. Although it was not necessary here to use the fact that the Beverton–Holt equation is a logistic difference equation, this observation was the idea for the key substitution $u_n = 1/x_n$. In fact, by letting in turn $x_n = 1/u_n$ in (2), one can easily see that the resulting equation is

$$\Delta x_n = \alpha x_{n+1}\left(1 - \frac{x_n}{K_n}\right),$$

which is the natural discrete analogue of the logistic equation, and which therefore should be called the *logistic difference equation.*

3. The Cushing–Henson Conjecture

Let $\chi_{\{m<n\}}$ be 1 if $m < n$ and 0 otherwise. Define

$$h_{nm} := \alpha(1-\alpha)^{n-(m+1)}\left(\frac{(1-\alpha)^p}{1-(1-\alpha)^p} + \chi_{\{m<n\}}\right),$$

where $h_{nm} > 0$ since $0 < \alpha < 1$. Then from (6) we obtain for $n < p$

$$\begin{aligned}\overline{u}_n &= \frac{(1-\alpha)^p}{1-(1-\alpha)^p}\sum_{i=0}^{p-1}(1-\alpha)^{n-(i+1)}\frac{\alpha}{K_i} + \sum_{i=0}^{p-1}(1-\alpha)^{n-(i+1)}\frac{\alpha}{K_i}\chi_{\{i<n\}} \\ &= \sum_{i=0}^{p-1}(1-\alpha)^{n-(i+1)}\left(\frac{(1-\alpha)^p}{1-(1-\alpha)^p} + \chi_{\{i<n\}}\right)\frac{\alpha}{K_i} = \sum_{i=0}^{p-1}\frac{h_{ni}}{K_i}.\end{aligned}$$

Lemma 3.1. *For all* $0 \le j \le p-1$, *we have* $\sum_{i=0}^{p-1} h_{ij} = \sum_{i=0}^{p-1} h_{ji} = 1.$

Proof. We have

$$\sum_{i=0}^{p-1} h_{ji} = \frac{\alpha(1-\alpha)^j}{1-(1-\alpha)^p}\sum_{i=0}^{p-1}(1-\alpha)^{p-1-i} + \alpha\sum_{i=0}^{j-1}(1-\alpha)^{j-1-i} = 1$$

and

$$\sum_{i=0}^{p-1} h_{ij} = \frac{\alpha(1-\alpha)^{p-(j+1)}}{1-(1-\alpha)^p}\sum_{i=0}^{p-1}(1-\alpha)^i + \alpha\sum_{i=j+1}^{p-1}(1-\alpha)^{i-(j+1)} = 1,$$

which concludes the proof. □

Remark 3.1. In the one-line proof of the Cushing–Henson conjecture below we use the (consequence of the) definition of a strictly convex function $f : (0, \infty) \to \mathbb{R}$ which says that for any $m \in \mathbb{N}$,

$$\begin{cases} f\left(\sum_{k=1}^{m} \lambda_k x_k\right) < \sum_{k=1}^{m} \lambda_k f(x_k) \quad \text{for all} \quad \lambda_k, x_k > 0,\ 1 \le k \le m, \\ \text{where} \quad \sum_{k=1}^{m} \lambda_k = 1 \quad \text{and there exist } 1 \le i < j \le m \text{ with} \quad x_i \ne x_j. \end{cases}$$

Theorem 3.1. *Assume* (3). *If* $\{K_n\}$ *is not constant, then*

$$\frac{1}{p}\sum_{i=0}^{p-1} \overline{x}_i < \frac{1}{p}\sum_{i=0}^{p-1} K_i.$$

Proof. Using Remark 3.1 for $f(x) = 1/x$, Lemma 3.1, and $h_{ij} > 0$ gives

$$\sum_{i=0}^{p-1} \overline{x}_i = \sum_{i=0}^{p-1} \frac{1}{\overline{u}_i} = \sum_{i=0}^{p-1} \frac{1}{\sum_{j=0}^{p-1} \frac{h_{ij}}{K_j}} < \sum_{i=0}^{p-1}\sum_{j=0}^{p-1} h_{ij} K_j = \sum_{j=0}^{p-1} K_j.$$

Dividing by p yields the final result. □

References

1. S. Elaydi and R. J. Sacker. Nonautonomous Beverton–Holt equations and the Cushing–Henson conjectures. *J. Difference Equ. Appl.*, 11(4-5):337–346, 2005.
2. V. Lj. Kocic. A note on the nonautonomous Beverton–Holt model. *J. Difference Equ. Appl.*, 11(4-5):415–422, 2005.
3. J. M. Cushing and S. M. Henson. A periodically forced Beverton–Holt equation. *J. Difference Equ. Appl.*, 8(12):1119–1120, 2002.
4. M. R. S. Kulenović and G. Ladas. *Dynamics of Second Order Rational Difference Equations.* Chapman & Hall/CRC, Boca Raton, FL, 2002. With open problems and conjectures.
5. M. Bohner and H. Warth. The Beverton–Holt dynamic equation. *Applicable Anal.*, 86(8):1007–1015, 2007.
6. S. Stević. A short proof of the Cushing–Henson conjecture. *Discrete Dyn. Nat. Soc.*, 2006:Art. ID 37264, 5 pages, 2006.
7. W. G. Kelley and A. C. Peterson. *Difference Equations: An Introduction with Applications.* Academic Press, San Diego, second edition, 2001.

Global Attractivity of a Multiplicative Delay Population Dynamics Model

E. Braverman

Department of Mathematics and Statistics, University of Calgary, Calgary, Alberta, Canada T2N 1N4
E-mail: maelena@math.ucalgary.ca

S. H. Saker

Department of Mathematics Skills, PYD, King Saud University, Riyadh 11451, Saudi Arabia

We establish some sufficient conditions for the global attractivity of the multiplicative difference equation with variable nonnegative ω-periodic positive coefficients

$$y(n+1) = y(n)\exp\left(-\lambda(n)y(n) + p(n) - \frac{Q(n)y^m(n-\omega)}{R + y^m(n-\omega)}\right), \quad n \geq 0, \; R > 0.$$

Keywords: Periodic solutions; Nonlinear delay difference equations; Attractivity; Population dynamics; Iterative method.

1. Introduction

The differential equation

$$x'(t) + px(t) - \frac{qx(t)}{r + x^m(t-h)} = 0, \quad t \geq 0, \tag{1}$$

was introduced by Nazarenko [8] to model the growth of a single cell population.

However, it is natural to assume that environment parameters are subject to periodic (daily, seasonal or subject to any other rhythms) changes. This corresponds to periodic coefficients of relevant mathematical models. For equations with variable parameters a periodic solution is expected (positive, for ecological models) rather than a constant equilibrium. The existence of such periodic regime, as well is its global attractivity, are the key problems of model study.

Assuming variable environment and a possible perturbation with a nonnegative effect, (1) can be modified to the form

$$x'(t) + p(t)x(t) - \frac{q(t)x(t)}{r + x^m(t-h)} = \lambda(t), \tag{2}$$

which after the substitution $y(t) = \dfrac{1}{x(t)}$ can be rewritten as

$$y'(t) = -\lambda(t)y^2(t) + p(t)y(t) - \frac{q(t)y(t)y^m(t-h)}{ry^m(t-h)+1},$$

or $\dfrac{y'(t)}{y(t)} = -\lambda(t)y(t) + p(t) - \dfrac{Q(t)y^m(t-h)}{R + y^m(t-h)}$, where $Q(t) = \dfrac{q(t)}{r}$, $R = \dfrac{1}{r}$.

Let us introduce a semidiscretization assuming the right hand side is piecewise constant. Without loss of generality we suppose $h = \omega$ is an integer (otherwise, we discretize with step h/ω) :

$$\frac{y'(t)}{y(t)} = -\lambda([t])y([t]) + p([t]) - \frac{Q([t])y^m([t-\omega])}{R + y^m([t-\omega])},$$

where $[t]$ is the integer part of $t \in [0, \infty)$. Integrating over intervals $[n, n+1]$, we obtain the difference equation

$$y(n+1) = y(n)\exp\left(-\lambda(n)y(n) + p(n) - \frac{Q(n)y^m(n-\omega)}{R + y^m(n-\omega)}\right), \quad n \geq 0, \tag{3}$$

which can be treated as a discrete analogy of the modified Nazarenko's equation (2). Our paper will focus on the attractivity of a positive periodic solution of (3).

2. Solution Estimates

We assume the coefficients in (3) are positive and ω-periodic, while the perturbation λ is nonnegative

$$m \in \mathbb{N}, \omega \in \mathbb{N}, R > 0, \ \ p(n) > 0, \ Q(n) > 0, \ \lambda(n) \geq 0, \ n = 0, 1, 2, \dots , \tag{4}$$

$$p(n+\omega) = p(n), \ \ Q(n+\omega) = Q(n), \ \ \lambda(n+\omega) = \lambda(n), \ n = 0, 1, 2, \dots . \tag{5}$$

By a solution of (3), we mean a sequence $\{y(n)\}$ which is defined for $n \geq -\omega$ and satisfies (3) for $n \geq 0$. According to the biological interpretation, (3) has nonnegative initial values

$$y(-\omega), \ y(-\omega+1), \ y(-\omega+2), ..., y(-1) \in [0, \infty) \text{ and } y(0) > 0. \tag{6}$$

Define for any ω-periodic sequence $\{f(n)\}$

$$I_\omega = \{0, 1, 2, ..., \omega - 1\}, \quad f^* = \max_{n \in I_\omega} f(n), \quad f_* = \min_{n \in I_\omega} f(n). \tag{7}$$

The exponential form of (3) yields that the solution $\{y(n)\}$ for any initial conditions (6) remains positive. We assume that (3) has a positive periodic solution. Sufficient conditions for the existence of a positive periodic solution of (3) can be found in Elabbasy, Saker [4].

Definition. The positive periodic solution $\overline{y}(n)$ of (3) is said to be *globally attractive* if $\lim_{n\to\infty} [y(n) - \overline{y}(n)] = 0$ for any positive solution $y(n)$ of (3).

Our aim is to establish sufficient global attractivity conditions for the periodic solution $\overline{y}(n)$ of (3). To this end, let us first obtain lower and upper solution estimates, this method was widely used in Györi, Ladas [6]. Recently lower and upper solutions in stability study were applied, for example, in Cabada *et al* [3].

Consider the equation

$$x(n+1) = x(n) \exp\left[f_n(x(n), x(n-1), \ldots, x(n-\omega))\right]. \tag{8}$$

Theorem 2.1. *Assume that (8) has a positive periodic solution, f_n in (8) are nonincreasing functions in all arguments, $f_{n+\omega}$ coincides with f_n for any $n = 0, 1, \ldots$ and there exist continuous functions $h_1(x)$, $h_2(x)$, satisfying*

$$h_i(0) > 0, \quad \lim_{t\to\infty} h_i(t) < 0, \quad h_i(x) \ \text{are decreasing in } x \text{ for } x \geq 0, \quad i = 1, 2, \tag{9}$$

$$\text{and} \quad h_1(x) \leq f_n(x, x, \ldots, x) \leq h_2(x), \quad n \in I_\omega.$$

Let x_1 and x_2 be positive roots of $h_1(x) = 0$ and $h_2(x) = 0$, respectively (according to (9), such roots exist and are unique). Then for any positive solution $x(n)$ of (8) there exists $n_1 \in I\!N$ such that for $n \geq n_1$ this solution satisfies the estimate

$$\mu_1 \leq x(n) \leq M_1, \quad M_1 = x_2 \exp((\omega+1)h_2(0)), \quad \mu_1 = x_1 \exp((\omega+1)h_1(M_1)). \tag{10}$$

Proof. Let us remark that $x(n+1) \leq x(n)$ as far as all $x(i)$ exceed x_2, $i = n-\omega, n-\omega+1, \ldots, n$. Really, let $x(i) > x_2$, $i = n-\omega, n-\omega+1, \ldots, n$. Since f_n is nonincreasing in all arguments, then $f_n(x(n), x(n-1), \ldots, x(n-\omega)) \leq f_n(x_2, x_2, \ldots, x_2) \leq h_2(x_2) = 0$, thus $x(n+1) = x(n)\exp(f_n(x(n), x(n-1), \ldots, x(n-\omega)) \leq x(n)$.

There may be the following cases: a solution eventually does not exceed x_2, is not less than x_2 and is oscillating about x_2. The first assumption immediately implies the upper bound in (10). In the second case $\{x(n)\}$ is eventually nonincreasing. Assuming $x(n) \geq M_1 > x_2$, $n \geq n_0$, we obtain $x(n+1) \leq x(n)\exp(h_2(M_1))$, $n \geq n_0 + \omega + 1$, so $x(n+k) \to 0$ as $k \to \infty$, since $h_2(M_1) < h_2(x_2) = 0$. The contradiction implies the upper bound in (10).

Next, let $x(n_0) < x_2$, $x(n_0+1) > x_2$. Since there are at most $\omega + 1$ increasing successive points $x(n_0+1), x(n_0+2), \dots, x(n_0+\omega+1)$ exceeding x_2 and

$$\frac{x(n+1)}{x(n)} \leq \exp\left(f_n(0, 0, \dots, 0)\right) \leq \exp(h_2(0)),$$

then after $\omega + 1$ steps we obtain the upper bound in (10) for $n \geq n_0$.

Let us prove the lower bound in (10). Similarly, we consider the last two cases among the following: $x(n)$ is eventually not less than x_1, does not exceed x_1 and is oscillating about x_1. The first case is treated as for the upper bound. Since the upper bound is valid for $n \geq n_0$, then for $n > n_1 = n_0 + \omega + 1$ the ratio

$$\frac{x(n+1)}{x(n)} \geq \exp\left(f_n(M_1, M_1, \dots, M_1)\right) \geq \exp\left(h_1(M_1)\right),$$

which after $\omega + 1$ steps gives the lower bound in (10). □

The proof of Theorem 2.1 emphasizes that there cannot be more than $\omega+1$ increasing (decreasing) successive $x(n)$ which are greater than x_2 (less than x_1).

Let us apply this result to equation (3). To this end, define $h_1(y)$ and $h_2(y)$ as

$$h_1(y) := -\lambda^* y + p_* - \frac{Q^* y^m}{R + y^m}, \quad h_2(y) := -\lambda_* y + p^* - \frac{Q_* y^m}{R + y^m}, \qquad (11)$$

where $\lambda^*, Q^*, p^*, \lambda_*, Q_*, p_*$ were denoted in (7). Let y_1 and y_2 be the roots of $h_1(y) = 0$ and $h_2(y) = 0$, respectively. These positive roots exist and are unique if

$$\text{either} \quad p^* < Q_* \quad \text{or} \quad \lambda_* > 0. \qquad (12)$$

It is possible to prove that (3) has a positive periodic solution as far as (12) holds, which was demonstrated in Elabbasy, Saker [4] for the first condition in (12).

Corollary 2.1. *Assume that (4),(5),(6) and (12) hold and $y(n)$ is a positive solution of (3). Then, there exists $n_1 > 0$ such that for all $n \geq n_0$*

$$\mu_1 := y_1 \exp((\omega+1)h_1(M_1)) \leq y(n) \leq y_2 \exp((\omega+1)h_2(0)) := M_1, \quad (13)$$

where y_1 and y_2 are positive roots of $h_1(y) = 0$ and $h_2(y) = 0$, respectively, h_1, h_2 are defined in (11).

Corollary 2.2. *Assume that (4),(5),(6) and (12) hold and $y(n)$ is a positive solution of (3). Denote*

$$M_2 = y_2 \exp((\omega+1)h_2(\mu_1)), \mu_2 = y_1 \exp((\omega+1)h_1(M_2)), \ldots, \quad (14)$$

$$M_{k+1} = y_2 \exp((\omega+1)h_2(\mu_k)), \mu_{k+1} = y_1 \exp((\omega+1)h_1(M_k)), \ k \in I\!N, \quad (15)$$

where h_1 and h_2 are defined in (11), μ_1 and M_1 are denoted in (13). Then for any k there exists n_k such that $\mu_k \leq y(n) \leq M_k$ for any $n \geq n_k$.

Proof. Let us remark that by the proof of Theorem 2.1 we should refer to oscillatory solutions only. By Corollary 2.1 there exists $n_1 > 0$ such that $\mu_1 \leq y(n) \leq M_1$, $n \geq n_1$. Then for $n > n_2 = n_1 + \omega + 1$ we have (similar to the proof of Theorem 2.1)

$$\frac{x(n+1)}{x(n)} \leq M_2 := \exp\left(h_2(\mu_1)\right), \quad \text{and} \quad \frac{x(n+1)}{x(n)} \geq \mu_2 := \exp\left(h_1(M_2)\right).$$

There are not more than $\omega + 1$ points exceeding y_2 or less than y_1, which implies lower and upper bounds in (14) for $n \geq n_2$. We proceed by induction and obtain that there exists n_k such that $\mu_k \leq y(n) \leq M_k$, $n > n_k$, for any $n \in \mathbb{N}$. □

Remark 1. Since $\mu_1 > 0$, then $h_2(\mu_1) < h_2(0)$, so $h_1(M_2) > h_1(M_1)$ and $\mu_2 > \mu_1$. By induction we can prove that $\{M_k\}$ is a nonincreasing, while $\{\mu_k\}$ is a nondecreasing sequence. Denote

$$Y_1 = \sup \mu_k, \quad Y_2 = \inf M_k. \quad (16)$$

Thus $\mu_k \leq y(n) \leq M_k$, $n \geq n_k$, implies

$$Y_1 \leq \liminf_{n\to\infty} y(n), \quad Y_2 \geq \limsup_{n\to\infty} y(n). \quad (17)$$

3. Global Attractivity

Let us proceed to attractivity conditions.

Theorem 3.1. *Assume that (4),(5),(6) and (12) hold and there exists* $\varepsilon > 0$ *such that the zero solution of the linear equation*

$$x(n+1) - x(n) + \mathcal{F}_1(n)x(n) + \mathcal{F}_2(n)x(n-\omega) = 0 \tag{18}$$

is globally asymptotically stable for any

$$\lambda(n)(Y_1 - \varepsilon) \leq \mathcal{F}_1(n) \leq \lambda(n)(Y_2 + \varepsilon), \tag{19}$$

$$\frac{mRQ(n)(Y_1-\varepsilon)^m}{(R+(Y_2+\varepsilon)^m)^2} \leq \mathcal{F}_2(n) \leq \frac{mRQ(n)(Y_2+\varepsilon)^m}{(R+(Y_1-\varepsilon)^m)^2}, \tag{20}$$

where Y_1, Y_2 *are defined in (13)-(16). If* $y(n)$ *is a positive solution of (3) then*

$$\lim_{n\to\infty} [y(n) - \overline{y}(n)] = 0, \tag{21}$$

where $\overline{y}(n)$ *is a positive periodic solution of (3).*

Proof. First, we prove that every positive solution $y(n)$ which does not oscillate about $\overline{y}(n)$ satisfies (21). Assume that $y(n) > \overline{y}(n)$ for n sufficiently large (the proof when $y(n) < \overline{y}(n)$ is similar and will be omitted). Set

$$y(n) = \overline{y}(n)\exp\{x(n)\}. \tag{22}$$

From (3) and (22) we see that $x(n) > 0$ and satisfies the equation

$$x(n+1) - x(n) + \lambda(n)\overline{y}(n)(e^{x(n)} - 1) + \frac{Q(n)R\overline{y}^m(n)}{(R+\overline{y}^m(n))}\frac{(e^{mx(n-\omega)}-1)}{(R+\overline{y}^m(n)e^{mx(n-\omega)})} = 0. \tag{23}$$

So (21) holds if and only if

$$\lim_{n\to\infty} x(n) = 0. \tag{24}$$

Since $0 < (e^{x(n)} - 1)$, then (23) implies

$$x(n+1) - x(n) + \frac{Q(n)R\overline{y}^m(n)}{(R+\overline{y}^m(n))}\frac{(e^{mx(n-\omega)}-1)}{(R+\overline{y}^m(n)e^{mx(n-\omega)})} < 0, \tag{25}$$

and hence $\{x(n)\}$ is a positive decreasing sequence and therefore it has a limit $\lim_{n\to\infty} x(n) = \alpha \in [0,\infty)$. If $\alpha > 0$, then there exist $\delta > 0$ and $n_\delta > 0$

such that for $n \geq n_\delta$, $0 < \alpha - \delta < x(n-\omega) < \alpha + \delta$. However using (25), we find $x(n+1) - x(n) < -P(n)$, where

$$P(n) = \left[\frac{Q(n) R \overline{y}^m(n)}{(R + \overline{y}^m(n))} \frac{(e^{m(\alpha-\delta)} - 1)}{(R + \overline{y}^m(n) e^{m(\alpha+\delta)})} \right] > 0.$$

Now, since $Q(n)$ and $\overline{y}(n)$ are positive periodic functions of period ω, we see that $P(n)$ is also an ω-periodic positive function and $0 < P_* \leq P(n) \leq P^*$. Thus $x(n+1) - x(n) \leq -P_*$ for $n \geq n_\delta$. The summation of the latter inequality from n_δ to n immediately implies $x(n) \leq x(n_\delta) - P_*(n - n_\delta) \to -\infty$ as $n \to \infty$, which is a contradiction. Hence, $\alpha = 0$ and therefore $x(n)$ tends to zero as $n \to \infty$ and (21) holds for any positive solution which does not oscillate about $\overline{y}(n)$.

To complete the global attractivity results we will prove that every oscillatory about $\overline{y}(n)$ solution satisfies (21). After denoting

$$G_1(n,v) = \lambda(n)\overline{y}(n)(e^v - 1) \quad \text{and} \quad G_2(n,v) = \frac{Q(n)R\overline{y}^m(n)}{R + \overline{y}^m(n)} \frac{(e^{mv} - 1)}{R + \overline{y}^m(n) e^{mv}}$$

(here $G_1(n,0) = G_2(n,0) = 0$ for any $n \in \mathbb{N}$), equation (23) can be rewritten as

$$x(n+1) - x(n) + G_1(n, x(n)) - G_1(n,0) + G_2(n, x(n-\omega)) - G_2(n,0) = 0. \quad (26)$$

By the Mean Value Theorem (26) has the form (18), where due to (22)

$$\mathcal{F}_1(n) = \left. \frac{\partial G_1(n,v)}{\partial v} \right|_{v=\zeta_1(n)} = \lambda(n)\overline{y}(n)e^v \big|_{v=\zeta_1(n)} = \lambda(n)\eta_1(n),$$

$$\mathcal{F}_2(n) = \left. \frac{\partial G_2(n,v)}{\partial v} \right|_{v=\zeta_2(n)} = \left. \frac{mRQ(n)\overline{y}^m(n)e^{mv}}{(R + \overline{y}^m(n)e^{mv})^2} \right|_{v=\zeta_2(n)} = \frac{mRQ(n)\eta_2^m(n)}{(R + \eta_2^m(n))^2}.$$

Here ζ_1 is between zero and $x(n)$, ζ_2 is between zero and $x(n-\omega)$, $\eta_1(n)$ lies between $\overline{y}(n)$ and $y(n)$, and $\eta_2(n)$ lies between $\overline{y}(n)$ and $y(n-\omega)$. By (17), for n large enough we have $Y_1 - \varepsilon \leq \eta_1(n) \leq Y_2 + \varepsilon$, $Y_1 - \varepsilon \leq \eta_2(n) \leq Y_2 + \varepsilon$, so $\mathcal{F}_1$, $\mathcal{F}_2$ satisfy (19), (20), respectively, which completes the proof. □

Let us recall some results on the asymptotic stability of the linear difference equations with a nondelay term

$$x(n+1) - x(n) + a(n)x(n) + b(n)x(n-\omega) = 0, \quad a(n) \geq 0, \quad b(n) \geq 0, \quad (27)$$

$$x(n+1) - x(n) + b(n)x(n-\omega) = 0, \quad b(n) \geq 0. \quad (28)$$

The result of Györi,Pituk [7] states that the following conditions

$$\limsup_{n\to\infty} \sum_{k=n-\omega}^{n-1} b(k) < 1, \quad \sum_{k=1}^{\infty} b(k) = \infty, \tag{29}$$

are sufficient for the asymptotic stability of the zero solution of (28). According to the results by Erbe *et al* [5]

$$\limsup_{n\to\infty} \sum_{i=n-\omega}^{n} b(n) < \frac{3}{2} + \frac{1}{2(\omega+1)} \tag{30}$$

is sufficient for the asymptotic stability of (28).

According to Corollary 6 in Berezansky, Braverman [1] and Remark on p. 787 in Berezansky *et al* [2] each of the following conditions is sufficient for the asymptotic stability of (27):

$$0 < a_0 < a(n) < \alpha < 1, \quad \limsup_{n\to\infty} \frac{b(n)}{a(n)} < 1; \tag{31}$$

$$0 < a(n) < 2, \quad \limsup_{n\to\infty} [|1 - a(n)| + |b(n)|] < 1. \tag{32}$$

Applying the above conditions and Theorem 3.1 we obtain the following result.

Theorem 3.2. *Assume that (4),(5),(6) and (12) hold,* $\lambda^* Y_2 < 1$ *and at least one of the following conditions is satisfied:*

$$\lambda(n) Y_1 > \frac{mRQ(n)Y_2^m}{(R+Y_1^m)^2}, \quad n = 0, \ldots, \omega - 1; \tag{33}$$

$$\sum_{k=n-\omega}^{n-1} \frac{mRQ(k)Y_2^m}{(R+Y_1^m)^2} \prod_{i=k-\omega}^{k} [1 - \lambda(i)Y_2]^{-1} < 1, \quad n = 0, \ldots, \omega - 1; \tag{34}$$

$$\sum_{k=n-\omega}^{n} \frac{mRQ(k)Y_2^m}{(R+Y_1^m)^2} \prod_{i=k-\omega}^{k} [1 - \lambda(i)Y_2]^{-1} < \frac{3}{2} + \frac{1}{2(\omega+1)}, \quad n = 0, \ldots, \omega - 1. \tag{35}$$

If $y(n)$ *is a positive solution of (3) then (21) holds.*

Proof. Condition (33) is an immediate corollary of either (31) or (32), if $a(n) = \mathcal{F}_1(n)$, $b(n) = \mathcal{F}_2(n)$ are as in (18) satisfying (20).

Further, after the substitution $x(n) = z(n) \prod_{i=0}^{n-1} [1 - \mathcal{F}_1(i)]$ equation (18) becomes

$$z(n+1) - z(n) + B(n)z(n-\omega) = 0,$$

where $B(n) = \mathcal{F}_2(n) \prod_{i=n-\omega}^{n} [1 - \mathcal{F}_1(i)]^{-1}$. By (20) for any $\varepsilon > 0$ not exceeding ε defined in Theorem 3.1 we have $\dfrac{mRQ(n)(Y_1-\varepsilon)^m}{(R+(Y_2+\varepsilon)^m)^2} \prod_{i=n-\omega}^{n-1} [1 - \lambda(i)(Y_1+\varepsilon)]^{-1} \le B(n)$ $\le \dfrac{mRQ(n)(Y_2+\varepsilon)^m}{(R+(Y_1-\varepsilon)^m)^2} \prod_{i=n-\omega}^{n-1} [1 - \lambda(i)(Y_2-\varepsilon)]^{-1}$ if n is large enough. Therefore, we deduce $\sum_{n=0}^{\infty} B(n) = \infty$; thus, (34),(35) imply (29),(30), respectively. The reference to Theorem 3.1 completes the proof. □

Remark 2. We can substitute Y_1 by μ_1 and Y_2 by M_1 in the conditions of Theorem 3.2, where μ_1 and M_1 are defined in (13), or by any μ_i, M_i in the sequence. This will give explicit global attractivity tests in terms of coefficients. It is also to be noted that if $Y_1 = Y_2$, then the positive periodic solution is globally attractive.

Global attractivity results are illustrated by the following example.

Example. Consider the autonomous equation

$$y(n+1) = y(n) \exp\left(-\lambda y(n) + p - \frac{Qy^m(n-\omega)}{R+y^m(n-\omega)}\right), \quad n \ge 0, \tag{36}$$

Let $\lambda = 0.4$, $Q = 0.8$, $R = 1$, $m = 3$, $\omega = 4$. First, let us compare conditions of Theorem 3.2 for these values of parameters. Inequalities (33),(34) and (35) can be rewritten in the autonomous case as

$$\frac{mRQY_2^m}{(R+Y_1^m)^2} < \lambda Y_1, \tag{37}$$

$$\frac{mQY_2^m}{(R+Y_1^m)^2} < \frac{1}{\omega}(1-\lambda Y_2)^{\omega+1}, \tag{38}$$

$$\frac{mQY_2^m}{(R+Y_1^m)^2} < \left(\frac{3}{2} + \frac{1}{2(\omega-1)}\right) \frac{1}{\omega+1}(1-\lambda Y_2)^{\omega+1}, \tag{39}$$

respectively. Since $\frac{1}{4} = \frac{1}{\omega} < \left(\frac{3}{2} + \frac{1}{2(\omega-1)}\right)\frac{1}{\omega+1} = \frac{1}{3}$ for given values of parameters, then (39) outperforms (38). Computations demonstrate that (39) also outperforms (37). Numerical estimation gives that for these values of parameters (33), (34) and (35) are satisfied for $p \leq 0.1519, p \leq 0.1520$ and $p \leq 0.1528$, respectively. Let us also mention that iterations can significantly improve the estimates. For instance, for $p = 0.15$ we have $\mu_1 \approx 0.071$, $M_1 \approx 0.667$, while $\mu_{60} \approx 0.30$, $M_{60} \approx 0.33$. Thus the iteration process confirms the global attractivity of the positive equilibrium solution in a wider range of parameters than estimates (37)-(39).

Acknowledgement

The first author was partially supported by NSERC grant. Both authors are grateful to the referee for valuable comments and remarks.

References

1. L. Berezansky and E. Braverman, On exponential dichotomy, Bohl-Perron type theorems and stability of difference equations, *J. Math. Anal. Appl.* **304** (2005), 511–530.
2. L. Berezansky, E. Braverman and E. Liz, Sufficient conditions for the global stability of nonautonomous higher order difference equations, *J. Difference Equ. Appl.* **11** (2005), no. 9, 785–798.
3. A. Cabada, V. Otero-Espinar and D.R. Vivero, Optimal conditions to ensure the stability of periodic solutions of first order difference equations lying between lower and upper solutions, *J. Comput. Appl. Math.* **176** (2005), no. 1, 45–57.
4. E.M. Elabbasy and S. H. Saker, Periodic solutions and oscillation of discrete nonlinear delay population dynamics model with external force, *IMA J. Appl. Math.* **70** (2005), 753–767.
5. L. Erbe, H. Xia and J. S. Yu, Global stability of a linear nonautonomous delay difference equation, *J. Differ. Eqns. Appl.* **1** (1995), 151–161.
6. I. Györi and G. Ladas, Oscillation Theory of Delay Differential Equations, Clarendon Press, Oxford, 1991.
7. I. Györi and M. Pituk, Asymptotic stability in a linear delay difference equation, in *Advances in difference equations* (Proceedings of SICDEA, Veszprém, Hungary, 1995), 295–299, Gordon and Breach, Amsterdam, 1997.
8. V.G. Nazarenko, Influence of delay on auto oscillation in cell population, *Biofisika* **21** (1976), 352–356.

Periodically Forced Rational Difference Equations

Elias Camouzis

American College of Greece, Department of Mathematics,
6 Gravias Street, 15342 Aghia Paraskevi, Athens, Greece
camouzis@acgmail.gr

We study the boundedness nature and the periodic character of solutions, of nonautonomous rational difference equations including Pielou's equation.

1. Introduction

We study the boundedness nature and the periodic character of solutions, of nonautonomous rational difference equations including Pielou's equation.

Consider the periodically forced rational equation

$$x_{n+1} = \frac{x_n}{A_n + B_n x_n + C_n x_{n-1}}, \; n = 0, 1, \ldots \tag{1}$$

with arbitrary positive initial conditions x_{-1} and x_0, and each one of the sequences $\{A_n\}$, $\{B_n\}$, and $\{C_n\}$ is periodic with prime period k, with nonnegative values such that the following hold:

$$\prod_{i=0}^{k-1} A_i > 0 \;\text{ or }\; \sum_{i=0}^{k-1} A_i = 0, \tag{2}$$

$$\prod_{i=0}^{k-1} B_i > 0 \;\text{ or }\; \sum_{i=0}^{k-1} B_i = 0, \tag{3}$$

and

$$\prod_{i=0}^{k-1} C_i > 0 \;\text{ or }\; \sum_{i=0}^{k-1} C_i = 0. \tag{4}$$

Eq. (1) contains the following 7 special cases:

$$x_{n+1} = \frac{x_n}{A_n}, \; n = 0, 1, \ldots \tag{5}$$

$$x_{n+1} = \frac{1}{B_n}, \; n = 0, 1, \ldots \tag{6}$$

$$x_{n+1} = \frac{x_n}{C_n x_{n-1}}, \; n = 0, 1, \ldots \tag{7}$$

$$x_{n+1} = \frac{x_n}{A_n + B_n x_n}, \; n = 0, 1, \ldots \tag{8}$$

$$x_{n+1} = \frac{x_n}{A_n + C_n x_{n-1}}, \; n = 0, 1, \ldots \tag{9}$$

$$x_{n+1} = \frac{x_n}{B_n x_n + C_n x_{n-1}}, \; n = 0, 1, \ldots \tag{10}$$

$$x_{n+1} = \frac{x_n}{A_n + B_n x_n + C_n x_{n-1}}, \; n = 0, 1, \ldots \tag{11}$$

Difference equations with periodic coefficients have been studied by several authors. See, for example, [2]–[24], and the references cited therein. One of the goals of introducing periodic coefficients is to test whether the averages of the resulting periodic solutions are larger, equal, or smaller than the equilibrium values of the associated autonomous equations. See [8] and [9].

2. Periodic Trichotomy of Eq. (1)

The following theorem is a trichotomy result about Eq. (5). The proof of the theorem is straightforward and will be omitted.

Theorem 1.

(i) *Every solution of Eq. (5) decreases to zero when*

$$\prod_{i=0}^{k-1} A_i > 1.$$

(ii) *Every solution of Eq. (5) converges to periodic solution of period-k, when*

$$\prod_{i=0}^{k-1} A_i = 1.$$

(iii) *Every solution of Eq. (5) is unbounded, when*

$$\prod_{i=0}^{k-1} A_i < 1.$$

3. Periodicity Destroys Boundedness

All nontrivial solutions of Eq. (7) with constant coefficients, that is,

$$x_{n+1} = \frac{x_n}{Cx_{n-1}}, \quad n = 0, 1, \ldots \tag{12}$$

are periodic with period six and so bounded. If

$$x_{-1} = \phi \text{ and } x_0 = \psi,$$

the solution of the equation is the six cycle:

$$\phi, \psi, \frac{\psi}{C\phi}, \frac{1}{C^2\phi}, \frac{1}{C^2\psi}, \frac{\phi}{C\psi}, \ldots .$$

It is interesting to note that periodicity may destroy the boundedness of solutions of Eq. (7) as the following theorem states.

Theorem 2. *(See [4] and [6)] Let*

$$C_n = \begin{cases} 1, & \text{if } n = 6k + i \text{ with } i \in \{0, 1, 2, 3, 4\} \\ C, & \text{if } n = 6k + 5 \end{cases}, k = 0, 1, \ldots$$

with $C > 0$. Then every solution of Eq. (7) with initial conditions

$$x_{-1} = x_0 = 1$$

is ***unbounded****, if and only if*

$$C \neq 1.$$

Therefore in Eq. (12), **periodicity may destroy the boundedness of its solutions.** For some results on the asymptotic behavior of Eq. (7), see [2].

Open Problem 1. *Let $\{C_n\}$ be a positive periodic sequence with prime period $p \geq 2$. Obtain necessary and sufficient conditions on p and*

$$C_0, \ldots, C_{p-1}$$

such that every solution of Eq. (7) is bounded.

4. Periodically Forced Pielou's Equation

The equation

$$x_{n+1} = \frac{x_n}{A_n + C_n x_{n-1}}, \quad n = 0, 1, \ldots$$

where each one of the sequences $\{A_n\}$ and $\{C_n\}$ is periodic of prime period k, with positive values, written in the form:

$$x_{n+1} = \frac{\beta_n x_n}{1 + x_{n-1}}, \quad n = 0, 1, \ldots \tag{13}$$

where the sequence $\{\beta_n\}$ is periodic of prime period k, with positive values, is **Pielou's equation** with periodic coefficient (see[23] and[24]). In the special case where $k = 2$, it was established by Kulenovic and Merino in[21] that when

$$\beta_0\beta_1 > 1 \;\; (\textbf{or equivalently} \;\; A_0A_1 < 1)$$

every positive solution of Eq. (13) converges to a period-two solution. Also, Camouzis and Ladas in,[5] established that, when $k \in \{1, 2, \ldots\}$ and

$$\prod_{i=0}^{k-1} \beta_i > 1 \;\; (\textbf{or equivalently} \;\; \prod_{i=0}^{k-1} A_i < 1)$$

every positive solution of Eq. (13) converges to a prime period-k solution.

In the following theorem we will establish that, when $k = 2$ or $k = 3$, the average over k values of a period-k solution

$$\ldots, \bar{x}_0, \ldots, \bar{x}_{k-1}, \ldots$$

of Eq. (13) is smaller than the average over the k positive values of the sequence $\{\beta_n - 1\}$. In other words period-two and period-three are both deleterious for the population.

Theorem 3. *Assume that $k = 2$ or $k = 3$ and that the sequence $\{\beta_n\}$ is periodic with prime period k, with values greater than one. Then*

$$\frac{\sum_{i=0}^{k-1} \bar{x}_i}{k} < \frac{\sum_{i=0}^{k-1} (\beta_i - 1)}{k}. \tag{14}$$

Proof. When $k = 2$, observe that

$$\frac{(\beta_0 - 1) + (\beta_1 - 1)}{2} - \frac{\bar{x}_0 + \bar{x}_1}{2} = \frac{(x_1 - x_0)^2(x_1 + x_0 + 1)}{2x_0x_1},$$

from which the result follows. When $k = 3$, we find that

$$\sum_{i=0}^{2} \beta_i = \frac{\bar{x}_1}{\bar{x}_0} + \frac{\bar{x}_1\bar{x}_2}{\bar{x}_0} + \frac{\bar{x}_2}{\bar{x}_1} + \frac{\bar{x}_2\bar{x}_0}{\bar{x}_1} + \frac{\bar{x}_0}{\bar{x}_2} + \frac{\bar{x}_0\bar{x}_1}{\bar{x}_2}.$$

Due to the fact that

$$\frac{\bar{x}_1}{\bar{x}_0} + \frac{\bar{x}_2}{\bar{x}_1} + \frac{\bar{x}_0}{\bar{x}_2} \geq 3$$

and

$$\frac{\bar{x}_1\bar{x}_2}{\bar{x}_0} + \frac{\bar{x}_2\bar{x}_0}{\bar{x}_1} + \frac{\bar{x}_0\bar{x}_1}{\bar{x}_2} > \bar{x}_0 + \bar{x}_1 + \bar{x}_2$$

the result follows. The proof is complete. □

The result of Theorem 3 is not true, when $k = 4$. Observe that, given

$$\bar{x}_0, \bar{x}_1, \bar{x}_2, \bar{x}_3 \in (0, \infty)$$

there exist

$$\beta_0, \beta_1, \beta_2, \beta_3 \in (1, \infty)$$

such that

$$\ldots, \bar{x}_0, \bar{x}_1, \bar{x}_2, \bar{x}_3, \bar{x}_0, \bar{x}_1, \bar{x}_2, \bar{x}_3, \ldots \tag{15}$$

is a period-four solution of Eq. (13). Indeed, by setting

$$\beta_0 = \frac{\bar{x}_1(1+\bar{x}_3)}{\bar{x}_0}, \quad \beta_1 = \frac{\bar{x}_2(1+\bar{x}_0)}{\bar{x}_1}, \quad \beta_2 = \frac{\bar{x}_3(1+\bar{x}_1)}{\bar{x}_2}, \quad \text{and} \quad \beta_3 = \frac{\bar{x}_0(1+\bar{x}_2)}{\bar{x}_3},$$

(15) is a period-four solution of Eq. (13). In particular, when

$$\beta_0 = 1.09109, \quad \beta_1 = 9174.4, \quad \beta_2 = 11000, \quad \text{and} \quad \beta_3 = 1100,$$

we find that

$$\ldots, 1000, 100000, 109, 10, 1000, 100000, 109, 10, \ldots$$

is a prime period-four solution of Eq. (13). In this particular example

$$\frac{\sum_{i=0}^{3} \bar{x}_i}{4} > \frac{\sum_{i=0}^{3} (\beta_i - 1)}{4}$$

and so period-four might be beneficial for the population, in the sense that the average population might be more in such a periodic environment than it is in a constant environment.

Open Problem 2. *For all values of $k \geq 4$ find necessary and sufficient conditions in terms of the parameters*

$$\beta_0, \ldots, \beta_{k-1}$$

such that

$$\frac{\sum_{i=0}^{k-1} \bar{x}_i}{k} \geq \frac{\sum_{i=0}^{k-1} (\beta_i - 1)}{k}.$$

5. The Beverton-Holt Equation

The equation

$$x_{n+1} = \frac{x_n}{A_n + B_n x_n}, \quad n = 0, 1, \ldots \tag{16}$$

with $A_n = \frac{1}{\mu}$, $B_n = \frac{\mu-1}{\mu K_n}$, $\{K_n\}$ periodic sequence of prime period k, with positive values, and with $\mu > 1$, is the periodic k-Beverton-Holt equation

$$x_{n+1} = \frac{\mu K_n x_n}{K_n + (\mu - 1)x_n}, \quad n = 0, 1, \ldots .$$

Cushing and Henson conjectured in[9] that every positive solution of this equation converges to a period-k solution and also that the average over k values of a period- k solution is smaller than the average of the k values of the sequence $\{K_n\}$.

This conjecture was established by S. Elaydi and R. Sacker in.[12] The convergence result was also established earlier by Clark and Gross in[7] and by Kocic and Ladas in.[17] The following theorem which contains Eq. (16) as a special case has been recently established.

Theorem 4. *(See[3]) Assume that each one of the sequences $\{A_n\}$, $\{B_n\}$, and $\{C_n\}$ is periodic with prime period k, with positive values. Then every positive solution of each one of the following equations:*

$$x_{n+1} = \frac{x_n}{A_n + B_n x_n}, \quad n = 0, 1, \ldots \tag{17}$$

$$x_{n+1} = \frac{x_n}{A_n + C_n x_{n-1}}, \quad n = 0, 1, \ldots \tag{18}$$

$$x_{n+1} = \frac{x_n}{A_n + B_n x_n + C_n x_{n-1}}, \quad n = 0, 1, \ldots \tag{19}$$

converges to a period-k solution when

$$\prod_{i=0}^{k-1} A_i < 1. \tag{20}$$

Remark 1. When

$$\prod_{i=0}^{k-1} A_i \geq 1.$$

every positive solution of each one of the four equations listed in the previous Theorem converges to zero.

The equation with constant coefficients

$$x_{n+1} = \frac{x_n}{A + Bx_n + Cx_{n-1}}, \quad n = 0, 1, \ldots \tag{21}$$

was investigated in [,[17]Kocic and Ladas], where it was shown that when

$$A > 0$$

every positive solution converges to the positive equilibrium. We present here a new, simple, and elegant proof that, every positive solution of Eq. (21) converges to the positive equilibrium. It is an amazing fact that the idea of our proof also extends to the periodically forced Eq. (1) (see[3]).

Theorem 5. *Every positive solution of Eq. (21) converges to a finite limit.*

Proof. When $A \geq 1$, the proof follows from the inequality

$$x_{n+1} \leq \frac{1}{A} x_n.$$

Assume that

$$A < 1$$

and let $\{x_n\}$ be a positive solution of Eq. (21). From Eq. (21) it follows that

$$x_n \leq \frac{1}{B}, \quad \text{for} \quad n \geq 1.$$

We also claim that $\{x_n\}$ is also bounded from below, that is,

$$\liminf_{n \to \infty} x_n > 0. \tag{22}$$

Assume for the sake of contradiction that there exists a sequence of indices $\{n_i\}$ such that

$$x_{n_i+1} \to 0 \quad \text{and} \quad x_{n_i+1} < x_j, \quad \text{for} \quad j < n_i + 1. \tag{23}$$

Then, clearly

$$x_{n_i} \quad \text{and} \quad x_{n_i-1} \to 0.$$

There exists a positive number ϵ, such that

$$\epsilon < \frac{1 - A}{B + C}$$

and that for i sufficiently large

$$x_{n_i}, x_{n_i-1} < \epsilon.$$

Then, eventually

$$x_{n_i+1} = \frac{x_{n_i}}{A + Bx_{n_i} + Cx_{n_i-1}} > \frac{x_{n_i}}{A + (B+C)\epsilon} > x_{n_i}$$

which contradicts (23) and proves (22). Using the change of variables

$$y_n = \frac{1}{x_n},$$

Eq. (21) becomes

$$y_{n+1} = B + Ay_n + \frac{Cy_n}{y_{n-1}}, \quad n = 0, 1, \ldots \,. \tag{24}$$

From this it follows that

$$\frac{y_{n+1}}{y_n} = \frac{B}{y_n} + A + \frac{C}{y_{n-1}}, \quad n = 0, 1, \ldots$$

and so

$$y_{n+1} = B + CA + Ay_n + \frac{CB}{y_{n-1}} + \frac{C^2}{y_{n-2}}, \quad n \geq 0\,. \tag{25}$$

Set

$$S = \limsup_{n\to\infty} y_n \text{ and } I = \liminf_{n\to\infty} y_n. \tag{26}$$

It suffices to show that

$$I = S.$$

Indeed,

$$SI \leq (B + CA)I + ASI + CB + C^2$$

and

$$SI \geq (B + CA)S + ASI + CB + C^2,$$

from which we find that

$$(B + CA)S + CB + C^2 \leq (1 - A)SI \leq (B + CA)I + CB + C^2$$

and so

$$S = I.$$

The proof is complete. □

References

1.
2. R. Abu-Saris and R. DeVault, On the asymptotic behavior of $x_{n+1} = a_n x_n / x_{n-1}$, *J. Math. Anal. Appl.*, Vol. 280, No. 1, pp. 148–154, 2003.
3. E. Camouzis, Global convergence in periodically forced rational equations, *J. Difference. Equ. Appl.*, (2008).
4. E. Camouzis and G. Ladas, When does periodicity destroys boundedness in rational difference equations? *J. Difference. Equ. Appl.*, **12**(2006), 961–979.
5. E. Camouzis and G. Ladas, Periodically forced Pielou's Equation, *J. Math. Anal. Appl.*, (2007).
6. E. Camouzis and G. Ladas, *Dynamics of Third Order Rational Difference Equations; With Open Problems and Conjectures*, Chapman & Hall/CRC Press, (to appear in 2007).
7. M. E. Clark and L. J. Gross, Periodic solutions to nonautonomous difference equations, *Math. Biosciences*, **102**(1990), 105–119.
8. J. M. Cushing, Periodically forced nonlinear systems of difference equations, *J. Difference Equ. Appl.*, **3**(1998), 547–561.
9. J. M. Cushing and S. M. Henson, A periodically forced Beverton-Holt equation, *J. Difference Equ. Appl.* **8**(2002), 1119–1120.
10. R. Devault, V. L. Kocic, and D. Stutson, Global Behavior of solutions of the nonlinear difference equation $x_{n+1} = p_n + \frac{x_{n-1}}{x_n}$, *J. Difference Equ. Appl.*, **11**(2005), 707–719.
11. R. DeVault, G. Ladas, and S. W. Schultz, On the recursive sequence $x_{n+1} = \frac{A}{x_n} + \frac{1}{x_{n-2}}$, *Proc. of the American Mathematical Society* **126**(1998), 3257–3261.
12. S. Elaydi and R. J. Sacker, Global Stability of periodic orbits of nonautonomous difference equations in population biology and the Cushing-Henson conjectures, *Proc. 8th Internat. Conf. Difference Equat. Appl.*, Brno, Czech Republic, 2003.
13. C. H. Gibbons, S. Kalabusic, and C. B. Overdeep, The Dichotomy Character of $x_{n+1} = \frac{\beta_n x_n + \gamma_n x_{n-1}}{A_n + B_n x_n}$ with Period-two Coefficients, (to appear).
14. E. A. Grove, C. M. Kent, and G. Ladas, Boundedness and persitence of the nonautonomous Lyness and Max equations, *J. Difference Equ. Appl.*, **3** (1998), 241–258.
15. E. A. Grove and G. Ladas, Periodicities in Nonlinear Difference Equations, Chapman and Hall/CRC Press, 2005.
16. V. L. Kocic, *A note on the nonautonomous Beverton-Holt model, J. Difference Equ. Appl.*, (to appear).
17. V. L. Kocic and G. Ladas, *Global Asymptotic Behavior of Nonlinear Difference Equations of Higher Order with Applications*, Kluwer Academic Publishers, Dordrecht, 1993.
18. V. L. Kocic, D. Stutson, and G. Arora, *Global Behavior of Solutions of a nonautonomous delay logistic difference equation, J. Difference Equ. Appl.*, **10**, (2004), 1267–1279.

19. R. Kon, A note on attenuant cycles of population models with periodic carrying capacity, *J. Difference Equ. Appl.*, **10**, (2004), 791–793.
20. M. R. S. Kulenović, G. Ladas, and C. B. Overdeep, Open problems and conjectures on the dynamics of $x_{n+1} = p_n + \frac{x_{n-1}}{x_n}$ with a Period-Two Coefficient, *J. Difference Equ. Appl.*, **9**(2003), 1053–1056.
21. M. R. S. Kulenović, and O. Merino, Stability Analysis of Pielou's Equation with Period-Two Coefficient, *J. Difference Equ. Appl.* (to appear).
22. M. R. S. Kulenović and O. Merino, *Convergence to a period-two solution of a class of second order rational difference equations* (to appear).
23. E. C. Pielou, *An Introduction to Mathematical Ecology, Wiley-Interscience,* New York, 1969.
24. E. C. Pielou, *Population and Community Ecology, Gordon & Breach,* New York, 1974.

Chaotic Synchronization of Piecewise Linear Maps

Acilina Caneco

CIMA-UE and Mathematics Unit, DEETC,
Instituto Superior de Engenharia de Lisboa,
Rua Conselheiro Emídio Navarro, 1, 1949-014 Lisboa, Portugal
E-mail: acilina@deetc.isel.ipl.pt

Clara Grácio

Department of Mathematics, Universidade de Évora and CIMA-UE,
Rua Romão Ramalho, 59, 7000-671 Évora, Portugal
E-mail: mgracio@uevora.pt

J. Leonel Rocha

Mathematics Unit, DEQ, Instituto Superior de Engenharia de Lisboa,
Rua Conselheiro Emídio Navarro, 1, 1949-014 Lisboa, Portugal
E-mail: jrocha@deq.isel.ipl.pt

We derive a threshold value for the coupling strength in terms of the topological entropy, to achieve synchronization of two coupled piecewise linear maps, for the unidirectional and for the bidirectional coupling. We prove a result that relates the synchronizability of two m-modal maps with the synchronizability of two conjugated piecewise linear maps. An application to the bidirectional coupling of two identical chaotic Duffing equations is given.

1. Introduction

Synchronization is a process wherein two or more systems starting from slightly different initial conditions would evolve in time, with completely different behaviour, but after some time they adjust a given property of their motion to a common behaviour, due to coupling or forcing. Various types of synchronization have been studied. This includes complete synchronization (CS), phase synchronization (PS), lag synchronization (LS) generalized synchronization (GS), anticipated synchronization (AS), and others [2]. The coupled systems might be identical or different, the coupling might be unidirectional, (master-slave or drive-response), or bidirectional (mutual coupling) and the driving force might be deterministic or stochastic.

In [3], A. Kenfack studied the linear stability of the coupled double-well Duffing oscillators projected on a Poincaré section. In [4], Kyprianidis *et al.* observed numerically the synchronization of two identical single-well Duffing oscillators.

In this work we investigate the unidirectional and bidirectional synchronization of two identical $m+1$ piecewise linear maps and obtain, analytically, the value of the coupling parameter for which the complete synchronization is achieved. Then, we study the relationship between the synchronization of two coupled identical $m-$modal maps and the synchronization of the corresponding conjugated piecewise linear maps. Next, we verify numerically the chaotic synchronization of two identical bidirectionally coupled double-well Duffing oscillators.

2. Main results

Consider a discrete dynamical system $u_{n+1} = f(u_n)$, where $u = (u_1, u_2, ..., u_m)$ is an m-dimensional state vector with f defining a vector field $f : \mathbb{R}^m \to \mathbb{R}^m$. The coupling of two such identical maps $x_{n+1} = f(x_n)$ and $y_{n+1} = f(y_n)$ defines another discrete dynamical system $\varphi : \mathbb{N}_0 \times \mathbb{R}^{2m} \to \mathbb{R}^{2m}$, *i.e.*, $\varphi(0, x, y) = (x, y)$, $\forall (x, y) \in \mathbb{R}^{2m}$ and $\varphi(t+s, x, y) = \varphi(t, \varphi(s, x, y))$, $\forall (x, y) \in \mathbb{R}^{2m}$, $\forall (t, s) \in \mathbb{N}_0^2$.

Denoting by k the coupling parameter, if we consider an unidirectional coupling

$$\begin{cases} x_{n+1} = f(x_n) \\ y_{n+1} = f(y_n) + k\left[f(x_n) - f(y_n)\right] \end{cases}, \tag{1}$$

then $\varphi(n, x, y) = (f(x_n)\ ,\ f(y_n) + k\left[f(x_n) - f(y_n)\right])$. If the coupling is bidirectional

$$\begin{cases} x_{n+1} = f(x_n) - k\left[f(x_n) - f(y_n)\right] \\ y_{n+1} = f(y_n) + k\left[f(x_n) - f(y_n)\right] \end{cases}, \tag{2}$$

then $\varphi(n, x, y) = (f(x_n) + k\left[f(y_n) - f(x_n)\right]\ ,\ f(y_n) + k\left[f(x_n) - f(y_n)\right])$.

To be able to say if the two systems are synchronized we must look to the difference $z_n = y_n - x_n$ and see if this difference converges to zero, as $n \to \infty$. If the coupling is unidirectional then

$$z_{n+1} = (1-k)\left[f(y_n) - f(x_n)\right]. \tag{3}$$

If the coupling is bidirectional then

$$z_{n+1} = (1-2k)\left[f(y_n) - f(x_n)\right]. \tag{4}$$

These two systems are said in complete synchronization if there is an identity between the trajectories of the two systems. In [7] and [8] it is establish that this kind of synchronization can be achieved provided that all the Lyapunov exponents are negative.

2.1. *Synchronization of piecewise linear maps*

Let $I = [a, b] \subseteq \mathbb{R}$ be a compact interval. By definition, a continuous map $f : I \to I$ which is piecewise monotone, *i.e.*, there exist points $a = c_0 < c_1 < \cdots c_m < c_{m+1} = b$ at which f has a local extremum and f is strictly monotone in each of the subintervals $I_0 = [c_0, c_1]$, ..., $I_m = [c_m, c_{m+1}]$, is called a m-modal map. As a particular case, if f is linear in each subinterval I_0, ..., I_m, then f is called a $m + 1$ piecewise linear map.

By theorem 7.4 from Milnor and Thurston [5] and Parry [6] it is known that every m-modal map f: $I = [a, b] \subset \mathbb{R} \to I$, with growth rate s and positive topological entropy $h_{top}(f)$ ($\log s = h_{top}(f)$) is topologically semi-conjugated to a $p + 1$ piecewise linear map T, with $p \leqslant m$, defined on the interval $J = [0, 1]$, with slope $\pm s$ everywhere and $h_{top}(T) = h_{top}(f) = \log s$, *i.e.*, there exist a function h continuous, monotone and onto, $h : I \to J$, such that $T \circ h = h \circ f$. If, in addition, h is a homeomorphism, then f and T are said topologically conjugated.

According to the above statements, we will investigate the synchronization of two identical $p + 1$ piecewise linear maps with slope $\pm s$ everywhere (Theorem 2.1.) and also the synchronization of two identical m-modal maps (Theorem 2.2.).

In what follows we will use the symbols f and k to represent, respectively, the m-modal map and its coupling parameter and the symbols T and c to represent, respectively, the $p + 1$ piecewise linear map and its coupling parameter.

Let $T : J = [a_1, b_1] \subseteq \mathbb{R} \to J$, be a continuous piecewise linear map, *i.e.*, there exist points $a_1 = d_0 < d_1 < \cdots d_p < d_{p+1} = b_1$ such that T is linear in each subintervals $J_i = [d_i, d_{i+1}]$, ($i = 0, ..., p$), with slope $\pm s$ everywhere.

So, the unidirectional coupled system for T is

$$\begin{cases} X_{n+1} = T(X_n) \\ Y_{n+1} = T(Y_n) + c\,[T(X_n) - T(Y_n)] \end{cases}, \tag{5}$$

and the difference $Z_n = Y_n - X_n$ verifies

$$Z_{n+1} = (1 - c)\,[T(Y_n) - T(X_n)]\,. \tag{6}$$

For the bidirectionally coupled system

$$\begin{cases} X_{n+1} = T(X_n) - c\,[T(X_n) - T(Y_n)] \\ Y_{n+1} = T(Y_n) + c\,[T(X_n) - T(Y_n)] \end{cases}, \tag{7}$$

the difference $Z_n = Y_n - X_n$ verifies

$$Z_{n+1} = (1-2c)\,[T(Y_n) - T(X_n)]\,. \tag{8}$$

Theorem 2.1. *Let $T : J \to J$, be a continuous $p+1$ piecewise linear map with slope $\pm s$ everywhere, with $s > 1$. Let $c \in [0,1]$ be the coupling parameter. Then one has:*

(i) The unidirectional coupled system (5) is synchronized if

$$c > \frac{s-1}{s}.$$

(ii) The bidirectional coupled system (7) is synchronized if

$$\frac{s+1}{2s} > c > \frac{s-1}{2s}.$$

Proof. Attending to the fact that T is linear with slope $\pm s$ in each subinterval J_0, ..., J_p, then, the total variation of T is

$$V_{b_1}^{a_1}(T) = \int_{a_1}^{b_1} |T'(t)|\,dt = \sum_{i=0}^{p} \int_{d_i}^{d_{i+1}} |T'(t)|\,dt = s \sum_{i=0}^{p} |d_{i+1} - d_i| = s\,|b_1 - a_1|\,.$$

We have

$$|T(Y_n) - T(X_n)| = \left| \int_{X_n}^{Y_n} T'(t)\;dt \right| \leqslant \int_{X_n}^{Y_n} |T'(t)|\,dt = V_{Y_n}^{X_n}(T) = s\,|Y_n - X_n|\,.$$

Attending to (6), it follows that,

$$|Z_{n+1}| \leqslant |(1-c)\,s|\,|Z_n| \text{ and then } |Z_q| \leqslant |(1-c)\,s|^q\,|Z_0|\,.$$

So, letting $q \to \infty$, we have $\lim_{q\to\infty} |(1-c)\,s|^q\,|Z_0| = 0$, if $|(1-c)\,s| < 1$. The previous arguments shows that, if $c \in [0,1]$ then the unidirectional coupled system (5) is synchronized if $c > \frac{s-1}{s}$.

On the other hand, using the same arguments as before and attending to (8), we have

$$|Z_{n+1}| \leqslant |(1-2c)\,s|\,|Z_n| \text{ and then } |Z_q| \leqslant |(1-2c)\,s|^q\,|Z_0|\,.$$

Thus, considering $q \to \infty$, we have $\lim_{q\to\infty} |(1-2c)\,s|^q\,|Z_0| = 0$, if $|(1-2c)\,s| < 1$. Therefore, we may conclude that, if $c \in [0,1]$ the bidirectional coupled system (8) is synchronized if $\frac{s+1}{2s} > c > \frac{s-1}{2s}$. □

Note that, the bidirectional synchronization occurs at half the value of the coupling parameter for the unidirectional case, as mentioned by Belykh *et al* [1].

2.2. *Conjugacy and synchronization*

In this section our question is to know the relationship between the synchronization of two coupled identical m-modal maps and the synchronization of the two coupled corresponding conjugated $p+1$ piecewise linear maps, with $p \leqslant m$. Consider in the interval J the pseudometric defined by

$$d(x,y) = |h(x) - h(y)|\,.$$

If h is only a semiconjugacy, d is not a metric because one may have $d(x,y) = 0$ for $x \neq y$. Nevertheless, if h is a conjugacy, then the pseudometric d, defined above, is a metric. Two metrics d_1 and d_2 are said to be topologically equivalent if they generate the same topology. A sufficient but not necessary condition for topological equivalence is that for each $x \in I$, there exist constants k_1, $k_2 > 0$ such that, for every point $y \in I$,

$$k_1\ d_1(x,y) \leqslant d_2(x,y) \leqslant k_2\ d_1(x,y).$$

Consider the pseudometric d defined above, $d_2(x,y) = d(x,y)$ and $d_1(x,y) = |x-y|\,.$

Suppose $h : I \to J$ is a bi-Lipschitz map, *i.e.*, $\exists\ N,\ M > 0$, such that,

$$0 < N|x-y| \leqslant |h(x) - h(y)| \leqslant M|x-y|, \quad \forall (x,y) \in I^2. \tag{9}$$

If h is a conjugacy and verifies (9), then the metrics d and $|.|$ are equivalents.

Consider $f : I\,[a,b] \subset \mathbb{R} \to I$ a $m-$modal function with positive entropy. For the unidirectional coupled system given by (1) we have the difference (3). As for the bidirectional coupled system given by (2) we have the difference (4).

As an extension of Theorem 2.1., for the synchronization of piecewise linear maps, we can establish the following result concerning the synchronization of the corresponding semiconjugated piecewise monotone maps.

Theorem 2.2. *Let $f\colon I \to I$, be a continuous and piecewise monotone map with positive topological entropy $h_{top} = \log s$ and $h : I \to J$ a semiconjugacy* *between f and a continuous piecewise linear map $T : J \to J$, with slope $\pm s$ everywhere. If there exist constants N, $M > 0$ satisfying (9), then one has:*

(i) The unidirectional coupled system (1) is synchronized if

$$k > 1 - \frac{N}{M}\frac{1}{s}.$$

(ii) The bidirectional coupled system (2) is synchronized if

$$1 + \frac{N}{M}\frac{1}{2s} > k > 1 - \frac{N}{M}\frac{1}{2s}.$$

Proof. *If f is monotone in the interval $[x, y]$, then T is monotone in the interval $[h(x), h(y)]$, because h is monotone, so*

$$|h(f(x)) - h(f(y))| = |T(h(x)) - T(h(y))| = s\,|h(x) - h(y)|\,.$$

Therefore $d(x, y) = s^{-1}d\,(f(x), f(y))$, if f is monotone in the interval $[x, y]$. If f is not monotone in the interval $[x, y]$, but there exist, points c_i $(i = 1, ..., p-1)$, such that $c_i < c_{i+1}$, $c_i \in [x, y]$ and f is monotone in each subinterval $I_1 = [x = c_0, c_1]$, $I_2 = [c_1, c_2]$,..., $I_p = [c_{p-1}, y = c_p]$, we have

$$\begin{aligned} d(x, y) &= \sum_{j=0}^{p-1} d(c_j, c_{j+1}) \\ &= s^{-1}\sum_{j=0}^{p-1} d\,(f(c_j), f(c_{j+1})) \\ &= s^{-1}\sum_{j=0}^{p-1} |h\,(f(c_j)) - h(f(c_{j+1}))| \\ &\geqslant s^{-1}\,|h\,(f(x)) - h(f(y))| \\ &= s^{-1}d\,(f(x), f(y))\,. \end{aligned}$$

So, we can write $d(x, y) \geqslant s^{-1}d\,(f(x), f(y))$, $\forall x, y \in I$. From (9) and for the unidirectional coupling (3) we have

$$\begin{aligned} d(y_{n+1}, x_{n+1}) &\leqslant M\,|y_{n+1} - x_{n+1}| = M\,|1-k|\,|f(y_n) - f(x_n)| \\ &\leqslant M\,|1-k|\,N^{-1}d\,(f(y_n), f(x_n)) \leqslant M\,|1-k|\,N^{-1}s\;d(y_n, x_n). \end{aligned}$$

It follows that

$$d(y_{n+r}, x_{n+r}) \leqslant M^r\,|1-k|^r\,N^{-r}s^r d(y_n, x_n),$$

so $d(y_{n+r}, x_{n+r}) \underset{r\to\infty}{\to} 0$ *if* $\left|M(1-k)N^{-1}s\right| < 1$. *Then, the coupled system (1) is synchronized if*

$$k > 1 - \frac{N}{M}\frac{1}{s}.$$

For the bidirectional coupling (4) and using the same arguments as before, we also have that

$$d(y_{n+1}, x_{n+1}) \leqslant M\,|1-2k|\,N^{-1}s\; d(y_n, x_n).$$

It follows that $d(y_{n+r}, x_{n+r}) \underset{r\to\infty}{\to} 0$ *if* $\left|M(1-2k)N^{-1}s\right| < 1$. *Then, the coupled system (2) is synchronized if*

$$\frac{1}{2}\left(1 + \frac{N}{M}\frac{1}{s}\right) > k > \frac{1}{2}\left(1 - \frac{N}{M}\frac{1}{s}\right). \qquad \square$$

Denote by k^* the synchronization threshold for (1), *i.e.* the system of piecewise monotone functions synchronizes for $k > k^*$. Denote by c^* the value such that for $c > c^*$ the system of piecewise linear maps (5) is synchronized. Note that

$$N(1-k^*) = M(1-c^*). \tag{10}$$

With the assumptions we made, if the piecewise monotone coupled maps synchronizes, so do the conjugated piecewise linear coupled maps and conversely, if the piecewise linear coupled maps synchronizes, so do the conjugated piecewise monotone coupled maps. In fact, from (9) we have $d(y_n, x_n) \leqslant M\,|y_n - x_n|$, therefore if system (1) synchronizes for $k > k^*$, then system (5) synchronizes for $c > c^*$, because $k^* \geqslant c^*$. On the other hand, we have also from (9), $|y_n - x_n| \leqslant N^{-1}d(y_n, x_n)$, therefore if the system (5) synchronizes for $c > c^*$, then the system (1) synchronizes for $k > k^*$ with k^* verifying (10).

For the bidirectional coupling, we have

$$1 - \frac{1}{s} \leqslant 1 - \frac{N}{M}\frac{1}{s} \leqslant 1 + \frac{N}{M}\frac{1}{s} \leqslant 1 + \frac{1}{s},$$

so the synchronization interval for the piecewise monotone coupled maps is contained in the synchronization interval for the conjugated piecewise linear coupled maps.

3. Duffing oscillators's example and symbolic synchronization

Consider two identical bidirectionally coupled Duffing oscillators, see [3] and references therein

$$\begin{cases} x''(t) = x(t) - x^3(t) - \alpha x'(t) + k\left[y(t) - x(t)\right] + \beta\ Cos(wt) \\ y''(t) = y(t) - y^3(t) - \alpha y'(t) - k\left[y(t) - x(t)\right] + \beta\ Cos(wt) \end{cases} \tag{11}$$

where k is the coupling parameter. We consider parameter values for which each uncoupled ($k = 0$) oscillator exhibits a chaotic behaviour, so if they synchronize, that will be a chaotic synchronization. We did a Poincaré section defined by $y = 0$ and found in the parameter plane (α, β), a region $\mathcal{U}$ where the first return Poincaré map behaves like a unimodal map and a region $\mathcal{B}$ where the first return Poincaré map behaves like a bimodal map. We choose, for example, $w = 1.18$, $x_0 = 0.5$, $x'_0 = -0.3$, $y_0 = 0.9$, $y'_0 = -0.2$ and $\alpha = 0.4$, $\beta = 0.3578$, for the unimodal case and $\alpha = 0.5$, $\beta = 0.719$, for the bimodal case.

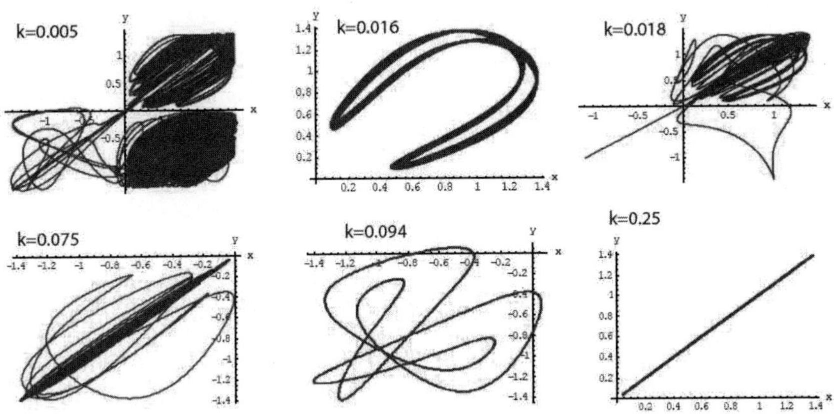

Fig. 1. Evolution of x versus y for the bidirectional coupled Duffing oscillators, for some values of k, in the unimodal case ($\alpha = 0.4, \beta = 0.3578$).

Numerically we can see the evolution of the difference $z = y - x$ with k. The synchronization will occur when $x = y$. See some examples in Fig. 1 for the unimodal case. Although not shown in this figure, the graphics of the difference $y - x$ for k greater then 0.122 are always a diagonal like in the picture for $k = 0.25$, showing that these Poincaré unimodal maps are synchronized. For $\alpha = 0.4$ and $\beta = 0.3578$ we have $h = 0.2406\ldots$, then $s = 1.272...$ If the coupled maps where piecewise linear maps with slope

$s = \pm 1.272$, the synchronization will occurs for $c > c^* = \frac{s-1}{2s} = 0.107$ and we see numerically that these unimodal Poincaré maps for the Duffing equations synchronizes at a little greater value, $k^* \approx 0.122$, so these pictures confirms numerically the above theoretical results, though we cannot guarantee that the semiconjugation between the unimodal and the piecewise linear maps is a conjugation.

4. Conclusions

We obtained explicitly the value c^* of the coupling parameter, such that for $c > c^*$ two piecewise linear maps, unidirectional or bidirectional coupled are synchronized. Moreover we prove that, in certain conditions, the synchronization of two m-modal maps is equivalent to the synchronization of the corresponding conjugated piecewise linear maps, but for different values of the coupling parameter. A numerical application to the coupling of two Duffing equations is given.

Acknowledgments

The authors are thankful to the referee for his authoritative comments and suggestions, which helped so much in improving the presentation of our results. This research has been partially supported by CIMA-UE and ISEL.

References

1. I. Belykh, M. Hasler and V. Belykh, When symmetrization guarantees synchronization in directed networks, *Int. J. Bif. Chaos*, **17(10)** (2007).
2. S. Boccaletti, J. Kurths, G. Osipov, D. L. Valladares and C. S. Zhou, The synchronization of chaotic systems, *Physics Reports* **366** (2002) 1–101.
3. A. Kenfack, Bifurcation structure of two coupled periodically driven double-well Duffing oscillators, *Chaos Sol. Frac.* **15** (2003) 205–218.
4. I. M. Kyprianidis, Ch. Volos, I. N. Stouboulos and J. Hadjidemetriou, Dynamics of two resistively coupled Duffing-type electrical oscillators, *Int. J. Bif. Chaos*, **16** (2006) 1765–1775.
5. J. Milnor and W. Thurston, On iterated maps of the interval I and II, *Lect. Notes in Math*, No. 1342, Springer-Verlag, (1988) 465–563.
6. W. Parry, Symbolic dynamics and transformations of the unit interval, *Trans. Amer. Math. Soc.*, **122,** 368–378, (1964).
7. L. M. Pecora and T. L. Carroll, Synchronization in chaotic systems, *Phys. Rev. Lett.*, **64**, (1990) 821–824.
8. L. M. Pecora and T. L. Carroll, Driving systems with chaotic signals, *Phys. Rev. A*, **44**, (1991) 2374–2383.

On the Iteration of Smooth Maps

M. F. Correia[1], C. C. Ramos[2] and S. Vinagre[3]
Department of Mathematics and CIMA-UE, University of Évora,
Évora, 7000-617, Portugal
[1]*mfac@uevora.pt*, [2]*ccr@uevora.pt and* [3]*smv@uevora.pt*
www.uevora.pt

Iteration of smooth maps appears naturally in the study of continuous difference equations and boundary value problems. Moreover, it is a subject that may be studied by its own interest, generalizing the iteration theory for interval maps. Our study is motivated by the works of A. N. Sharkovsky *et al.*[1,3] , E. Yu. Romanenko *et al.*[2] , S. Vinagre *et al.*[4] and R. Severino *et al.*[5] . We study families of discrete dynamical systems of the type (Ω, f), where Ω is some class of smooth functions, *e.g.*, a sub-class of $C^r(J, \mathbb{R})$, where J is an interval, and f is a smooth map $f : \mathbb{R} \to \mathbb{R}$. The action is given by $\varphi \mapsto f \circ \varphi$. We analyze in particular the case when f is a family of quadratic maps. For this family we analyze the topological behaviour of the system and the parameter dependence on the spectral decomposition of the iterates.

Keywords: Smooth map, difference equations, spectrum, symbolic dynamics, discrete dynamical systems, iteration theory.

1. Introduction

Usually the term iteration of smooth functions means that we have a smooth function f and we study the behaviour of the orbits of points, under iteration of f. In the present paper, we also have a smooth function f. However, here, the term iteration of smooth functions means that the underlying space of the dynamical system is a space of smooth functions, therefore the orbits or trajectories, under iteration of f, are no longer points in an interval but smooth functions. Therefore, we deal with infinite dimension discrete dynamical systems. Iteration of smooth functions appears naturally in different contexts, namely difference equations and boundary value problems, see for example[1–5] . Among boundary value problems for partial differential equations, there are certain classes of problems reducible to difference equations and to other relevant equations. These classes consist mainly on problems for which the representation of the general solution of partial differential equations is known. The reduction to a difference equation with continuous argument, followed by the employment of the properties of the one-dimensional map associated with the difference equation, allows an insight into the properties of chaotic solutions for the original problem.

2. Iterated smooth functions

Let us consider the quadratic family depending on the parameter μ,

$$\begin{aligned} f_\mu : \mathbb{R} &\to \mathbb{R} \\ x &\mapsto 1 - \mu x^2 \end{aligned}$$

with $x_c = 0$ being the critical point. Note that there is an invariant interval which is $[-1, 1]$. Regarding the iteration of the interval map f_μ we have the following: as usual, we assign the symbols L (left) and R (right) to each point x in the subintervals of monotonicity $[-1, 0)$ and $(0, 1]$, respectively, and the symbol C to the critical point $x_c = 0$. The *address* of x, $ad(x)$, is this assignment. Therefore, we get a correspondence between orbits of points and symbolic sequences of the alphabet $\{L, C, R\}$. This correspondence is called the *itinerary* by the map f_μ,

$$it_{f_\mu}(x) = ad\,(x)\,ad\,(f_\mu\,(x))\,ad\,\left(f_\mu^2\,(x)\right)\ldots. \in \{L, C, R\}^{\mathbb{N}}.$$

An *admissible sequence* is a sequence in $\{L, C, R\}^{\mathbb{N}}$ which occurs as an itinerary for some point $x \in [-1, 1]$ and an *admissible word* is some word occurring in an admissible sequence.

The itinerary of the image of the critical point is of particular importance and it is called the *kneading sequence* $\mathcal{K}_\mu = it_{f_\mu}(f_\mu(0)) = K_1K_2... \in \{L, C, R\}^{\mathbb{N}}$.

Let $X = \{x_1, x_2, ..., x_n\} \subset [-1, 1]$. We define $orb_{f_\mu}(X) = \bigcup\limits_{j=0}^{\infty} f_\mu^j(X)$.

Now, consider the following class of smooth functions

$$\mathcal{A} = \{\varphi \in C^\infty([0, 1]) : \varphi'(0) = \varphi'(1) = 0\}.$$

Every function in $\mathcal{A}$ can be written as a linear combination of $\cos(k\pi x)$ (the cosine - Fourier expansion). We also consider the space

$$\mathcal{B} = \left\{\varphi \in \mathcal{A} : \varphi(x) = \sum_{k=0}^{n} c_k \cos(k\pi x), n \in \mathbb{N}\right\},$$

the space of the functions in $\mathcal{A}$ which are finite linear combinations of the $\cos(k\pi x)$. Let T_μ be the operator

$$\begin{aligned} T_\mu : \mathcal{A} &\to \quad \mathcal{A} \\ \varphi &\mapsto f_\mu \circ \varphi. \end{aligned}$$

Note that T_μ is well defined since $(f_\mu \circ \varphi)'(0) = (f_\mu \circ \varphi)'(1) = 0$. We will be interested in the discrete dynamical system $(\mathcal{A}, T_\mu)$. Since f_μ is a polynomial map, we have $T_\mu(\mathcal{B}) \subseteq \mathcal{B}$ and we also consider the dynamical system $(\mathcal{B}, T_\mu)$, where T_μ means $T_{\mu|\mathcal{B}}$. Moreover, if $\phi \in \mathcal{A}$ and $Im(\phi) \subset [-1,1]$ then $Im(T_\mu^k \phi) \subset [-1,1]$ for every $k \in \mathbb{N}$. We will assume, unless otherwise stated, that $Im(\phi) \subset [-1,1]$.

2.1. *Localization of critical points and critical values*

Let $I_{S_0 \dots S_k} \subset [-1,1]$ denote the interval of points x which satisfy

$$ad(x) = S_0,\ ad(f_\mu(x)) = S_1, \dots,\ ad(f_\mu^k(x)) = S_k.$$

Proposition 2.1. *Let $\phi \in \mathcal{A}$ and let $J \subset [0,1]$ be an interval. If $\phi_{|J}(x)$ is a maximal (resp. minimal) value in $I_{RS_1 \dots S_k}$, then $f \circ \phi_{|J}(x)$ is a minimal (resp. maximal) value in $I_{S_1 \dots S_k}$. If $\phi_{|J}(x)$ is a maximal (resp. minimal) value in $I_{LS_1 \dots S_k}$, then $f_\mu \circ \phi_{|J}(x)$ is a maximal (resp. minimal) value in $I_{S_1 \dots S_k}$.*

Proof. First, by the definition of $I_{S_0 \dots S_k}$, we have $f_\mu(I_{S_0 \dots S_k}) = I_{S_1 \dots S_k}$ for any admissible word $S_0 \dots S_k$. Next, since $f_{\mu|I_R}$ is decreasing it reverses the order, therefore maximal (*resp.* minimal) points for $\phi_{|J}$ in some J, subinterval of $[0,1]$, correspond to minimal (*resp.* maximal) points for $f_\mu \circ \phi_{|J}$. The same reasoning, noting that $f_{\mu|I_L}$ is increasing and preserves the order, leads to the claimed result. □

Let $\mathrm{c}(\phi)$ denote the set of the critical points of $\phi \in \mathcal{A}$, $\mathrm{cv}(\phi)$ the set of the critical values of $\phi \in \mathcal{A}$ and $\mathrm{z}(\phi)$ the set of zeros of ϕ. Note that $\mathrm{cv}(\phi) = \phi(\mathrm{c}(\phi))$.

Proposition 2.2. *Let $\phi \in \mathcal{A}$. Then*

$$\mathrm{c}(f_\mu \circ \phi) = \mathrm{z}(\phi) \cup \mathrm{c}(\phi) \text{ and } \mathrm{cv}(f_\mu \circ \phi) = f_\mu(\phi(\mathrm{z}(\phi))) \cup f_\mu(\mathrm{cv}(\phi)).$$

Proof. From the chain rule $(f_\mu \circ \phi)'(x) = f_\mu'(\phi(x))\phi'(x) = 0$. Therefore, either $\phi'(x) = 0$ or $f_\mu'(\phi(x)) = 0$. The first claim follows. The critical values of $f_\mu \circ \phi$ are naturally the images under $f_\mu \circ \phi$ of the critical points of $f_\mu \circ \phi$. Moreover, $f_\mu(\mathrm{cv}(\phi)) = f_\mu \circ \phi(\mathrm{c}(\phi))$ and the result follows. □

A consequence of this result is the following:

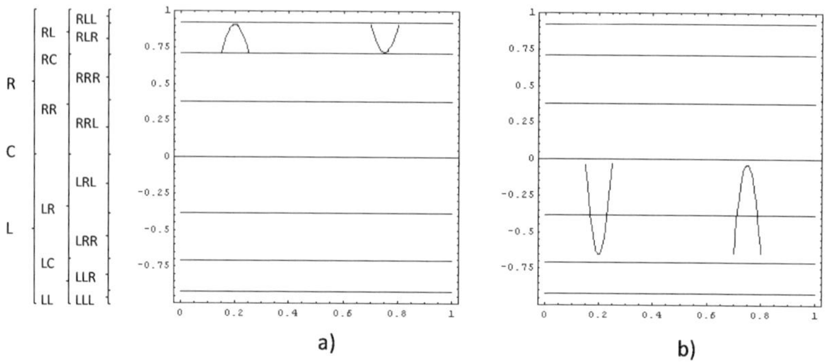

Fig. 1. a) Graph of the restriction of a function ϕ, in a subinterval J, with a maximum (resp. minimum) in the region I_{RLR}. b) Graph of the restriction of $f_\mu \circ \phi$, in the same subinterval J, which has a minimum (resp. maximum) in the region I_{LR}.

Corollary 2.1. *Let* $\phi_0 \in \mathcal{A}$ *and* $\phi_{k+1} = f_\mu \circ \phi_k$. *Then*

$$\mathrm{c}(\phi_{k+1}) = \bigcup_{j=0}^{k} \mathrm{z}(f_\mu^j(\phi_0)) \cup \mathrm{c}(\phi_0)$$

and

$$\mathrm{cv}(\phi_{k+1}) = \bigcup_{j=0}^{k} f_\mu^{k+1-j}\left(\phi_j(\mathrm{z}(\phi_j)) \cup f_\mu^{k+1}(\mathrm{cv}(\phi_0)\right).$$

This last result means that the maxima and minima of the iterates ϕ_k of some initial function ϕ_0 are obtained from the orbits of the maxima and minima of ϕ_0 and the appearance of new critical points corresponds to the appearance of new zeros of $\phi_0, \phi_1, ..., \phi_k$. A similar phenomena is presented in R. Severino *et al.*[5] . In order to illustrate the results given above, consider the Figures 1, 2 and 3.

By the previous results, the critical points of a function $\phi \in \mathcal{A}$ are also critical points of $f_\mu \circ \phi$, that is, $\mathrm{c}(\phi) \subset \mathrm{c}(f_\mu \circ \phi)$. Moreover, if ϕ has a zero then it will be also a critical point of $f_\mu \circ \phi$. Now, let y be a j-pre-image of 0 under the iteration by f_μ. The itinerary of y will be in the form $S_1...S_{j-1}CK_1K_2...$, (recall that $\mathcal{K}_\mu = it_{f_\mu}(f_\mu(0)) = K_1K_2...$) for some admissible word $S_1...S_j \in \{L, R\}^j$. Therefore, if $k \in \mathbb{N}$ is fixed and if we represent the horizontal lines, ordered by the symbolic sequences, corresponding to every j-pre-image of 0, with $0 < j \leq k$, we obtain a procedure to identify all the new critical points and critical values up to the iterate k of any $\phi \in \mathcal{A}$. These new critical points and values will be the

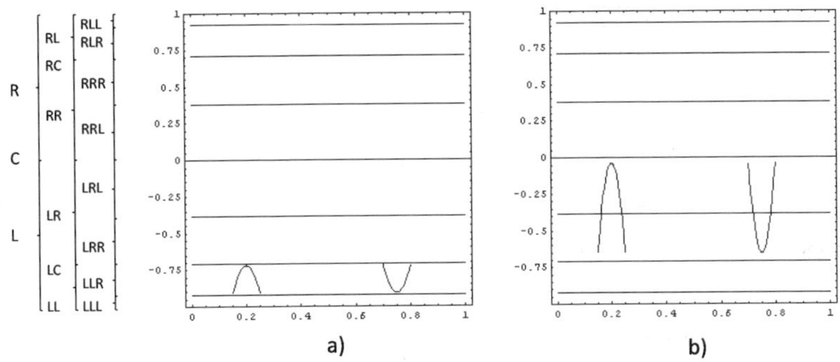

Fig. 2. a) Graph of the restriction of a function ϕ, in a subinterval J, with a maximum (resp. minimum) in the region I_{LLR}. b) Graph of the restriction of $f_\mu \circ \phi$, in the same subinterval J, which has a maximum (resp. minimum) in the region I_{LR}.

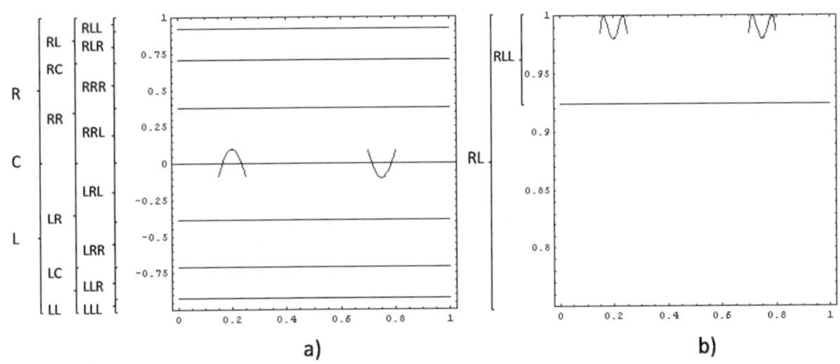

Fig. 3. a) Graph of the restriction of a function ϕ, in a subinterval J, with zeros at some points x_1 and x_2 and a maximum (resp. minimum) in the region I_{RRL} (I_{LRL}). b) Graph of the restriction of a function $f_\mu \circ \phi$, in the same subinterval J, with new maximal values at x_1 and x_2 and a minimum in the region I_{RL} (I_{RL}); note that in b) the vertical scale is changed.

intersection of ϕ with the horizontal lines referred above, see Figures 4 a) and 4 b).

2.2. *Spectrum*

Now, we analyze the spectral evolution under iteration of f_μ and its dependence on the initial condition and on the parameter μ.

Theorem 2.1. *Let* $\phi_0(x) = \sum_{j=0}^{m_0} c_j(0)\cos(j\pi x) \in \mathcal{B}$, *with* $m_0 \in \mathbb{N}$ *and*

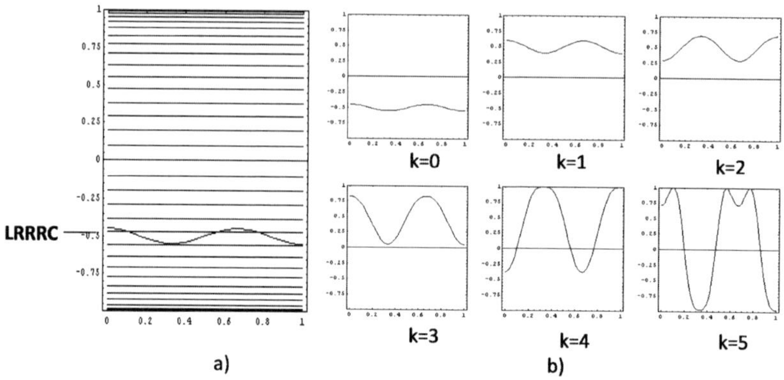

Fig. 4. a) Graph of $\phi_0(x) = a_0 + a_1 \cos(3\pi x)$ with $a_0 = -0.5$, $a_1 = 0.05$ so that $\phi_5 = f_\mu^5 \circ \phi_0$ has three new critical points ($\mu = 2$). b) Graphs of the $k-$iterations under f_μ of $\phi_0(x) = a_0 + a_1 \cos(3\pi x)$, with $\mu = 2$, $a_0 = -0.5$ and $a_1 = 0.05$.

$c_j(0) \in \mathbb{R}$, $j = 0, ..., m_0$. Let $\phi_k = f_\mu^k \circ \phi_0 = \sum_{j=0}^{m_k} c_j(k) \cos(j\pi x)$. Then

$$c_0(k+1) = 1 - \mu c_0(k)^2 - \frac{\mu}{2} \sum_{n=1}^{m_k} c_n(k)^2$$

and

$$c_r(k+1) = -\frac{\mu}{2} \sum_{j=0}^{r} c_{r-j}(k) c_j(k) - \mu \sum_{j=r}^{m_k} c_{j-r}(k) c_j(k),$$

with $m_k = 2^k m_0$.

Proof. Let $\phi_k = \sum_{j=0}^{m_k} c_j(k) \cos(j\pi x) = f_\mu^k \circ \phi_0$. We have $\phi_{k+1} = 1 - \mu (\phi_k)^2 = \sum_{j=0}^{m_{k+1}} c_j(k+1) \cos(j\pi x)$, *i.e.*,

$$1 - \mu \left[c_0(k) + c_1(k) \cos(\pi x) + ... + c_{m_k}(k) \cos(m_k \pi x)\right]^2$$
$$= \sum_{r=0}^{m_{k+1}} c_r(k+1) \cos(r\pi x).$$

Since $\cos(i\pi x)\cos(j\pi x) = \frac{1}{2}\left[\cos((i+j)\pi x) + \cos((i-j)\pi x)\right]$, we have for $r = 0$,

$$c_0(k+1) = 1 - \frac{1}{2} \sum_{\substack{i,j=1,\ldots,m_k: \\ i+j=0}} c_i(k)c_j(k) + \frac{1}{2} \sum_{\substack{i,j=1,\ldots,m_k: \\ |i-j|=0}} c_i(k)c_j(k)$$

$$= 1 - \mu c_0(k)^2 - \frac{\mu}{2} \sum_{n=1}^{m_k} c_n(k)^2$$

and for $r > 0$,

$$c_r(k+1) = -\frac{\mu}{2} \sum_{\substack{i,j=1,\ldots,m_k: \\ i+j=r}} c_i(k)c_j(k) - \frac{\mu}{2} \sum_{\substack{i,j=1,\ldots,m_k: \\ |i-j|=r}} c_i(k)c_j(k)$$

$$= -\frac{\mu}{2} \sum_{j=0}^{r} c_{r-j}(k)c_j(k) - \frac{\mu}{2} \sum_{j=0}^{r} c_{r+j}(k)c_j(k) - \frac{\mu}{2} \sum_{j=r}^{m_k} c_{j-r}(k)c_j(k).$$

Since $\sum_{j=0}^{r} c_{r+j}(k)c_j(k) = \sum_{j=r}^{m_k} c_{j-r}(k)c_j(k)$ the claim follows. □

The Figure 5 illustrates the spectral dependence on the parameter μ.

If $\phi \in \mathcal{B}$ it makes sense to talk about the *maximal harmonic* of ϕ, denoted by $m(\phi)$ (corresponding to a maximal frequency). It is equal to the maximal natural number m so that the coefficient of $\cos(m\pi x)$ in the ϕ cos-expansion is non-zero. We define also $\nu(\phi)$ as the absolute value of this coefficient. Next, we obtain an explicit dependence of $\nu(\phi_k)$ on the parameter μ. Thus, we see $\nu(\phi_k)$ as a function of μ that is $\nu(\phi_k) = \nu(\phi_k, \mu)$.

Proposition 2.3. *Let $\phi \in \mathcal{B}$ and $\phi_k = f_\mu^k \circ \phi$. Then $\nu(\phi_k, \mu)$ is an increasing function on μ for every $k \geq 1$ and for every initial condition ϕ. As a consequence, $\nu(\phi_k, \mu)$ depends monotonically on the topological entropy.*

Proof. Let $\phi \in \mathcal{B}$ and $\phi_k = f_\mu^k \circ \phi$ for some $k \in \mathbb{N}$. From the Theorem 2.1 we have that $m_k = m(\phi_k) = 2^k m(\phi)$ and $\nu(\phi_k, \mu) = |c_{m_k}(k)|$. A simple computation shows that $c_{m_{k+1}}(k+1) = -\frac{\mu}{2} c_{m_k}(k)^2$ and consequently $\nu(\phi_k, \mu) = a\mu^N$, for some real number a and N integer. Therefore, in the considered interval $\mu \in [0, 2]$, the function $\nu(\phi_k, \mu)$ is monotone. Since the topological entropy of f_μ depends monotonically on μ, the result follows.

□

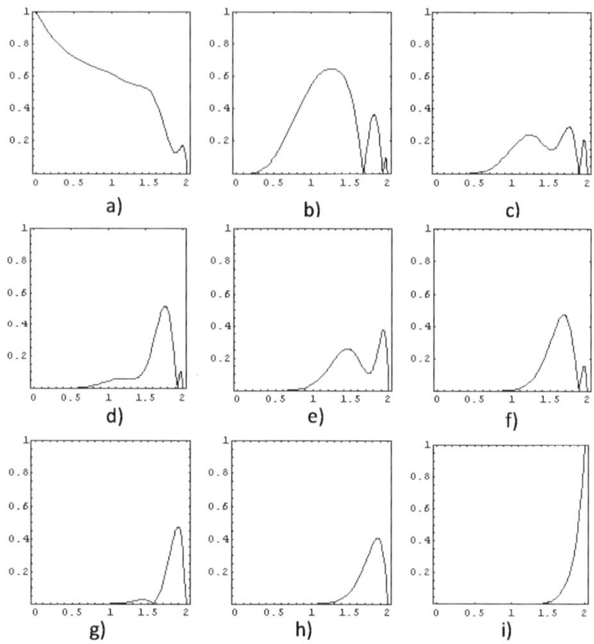

Fig. 5. Graphs of the spectral decomposition $c_j(k)$ of $\phi_k = f_\mu^k$, with $k = 4$, the functions of the parameter μ, with the initial condition $\phi(x) = \cos(\pi x)$. a) $c_0(4)$, b) $c_1(4)$, c) $c_2(4)$, d) $c_3(4)$, e) $c_4(4)$, f) $c_5(4)$, g) $c_6(4)$, h) $c_7(4)$ and i) $c_8(4)$.

Acknowledgements

The first author is supported by Calouste Gulbenkian Foundation. This work has been partially supported by the research center CIMA-UE, FCT-Portugal funding program.

References

1. A. N. Sharkovsky, Yu. S. Maistrenko and E. Yu. Romanenko, Difference equations and their applications, *Kluwer Academic Publishers*, (1993).
2. E. Yu. Romanenko and A.N. Sharkovsky, From boundary value problems to difference equations: a method of investigation of chaotic vibrations, *Internat J. Bifur. Chaos Appl. Sci. Engrg. 9*, **7**, (1999), 1285-1306.
3. A. N. Sharkovsky, Difference equations and boundary value problems, *In "New Progress in Difference Equations", Proceedings of the ICDEA'2001, Taylor and Francis*, (2003), 3-22.
4. S. Vinagre, R. Severino and J. Sousa Ramos, Topological invariants in nonlinear boundary value problems, *Chaos Solitons Fractals*, **25**, (2005), 65-78.
5. R. Severino, A. Sharkovsky, J. Sousa Ramos and S. Vinagre, Symbolic Dynamics in Boundary Value problems, *Grazer Math. Ber*, **346**, (2004), 393-402.

The Comparative Index and the Number of Focal Points for Conjoined Bases of Symplectic Difference Systems

J. V. Elyseeva

Moscow State University of Technology,
Department of Applied Mathematics,
Vadkovskii per. 3a,
101472 Moscow,
Russia
E-mail: elyseeva@mtu-net.ru

In this work we derive the formula which connects the number of focal points in $(0,\ N+1]$ for conjoined bases of symplectic difference systems. The similar result is obtained for the number of focal points in $[0,\ N+1)$. We prove that the number of focal points in $(0,\ N+1]$ and $[0,\ N+1)$ for the principal solutions at $i=0$ and $i=N+1$ coincides. The consideration is based on the new concept of the comparative index.

1. Introduction

We consider the symplectic difference system

$$Y_{i+1} = W_i\, Y_i\,, i=0,1,\ldots,N, \quad W_i = \begin{bmatrix} A_i & B_i \\ C_i & D_i \end{bmatrix}, \; Y_i = \begin{bmatrix} X_i \\ U_i \end{bmatrix}, \tag{1}$$

$$W_i^T J W_i = J,\; J = \begin{bmatrix} 0 & I \\ -I & 0 \end{bmatrix},$$

where W_i, Y_i, J are real partitioned matrices with $n\times n$ blocks, and I, 0 are the identity and zero matrices.

A matrix solution Y_i of (1) is said to be a conjoined basis of (1) if the conditions $Y_i^T\, J\, Y_i = 0$, $\mathrm{rank} Y_i = n$ hold. For two conjoined bases Y_i, $\hat{Y}_i$ of (1) the Wronskian identity [1]

$$w(Y_i, \hat{Y}_i) = w_i = const \tag{2}$$

holds.

According to the definition [1], a conjoined basis of (1) is said to have a focal point in $(i, i+1]$ if the conditions

$$\mathrm{Ker}X_{i+1} \subseteq \mathrm{Ker}X_i,$$
$$X_i X_{i+1}^{\dagger} B_i \geq 0$$

do not hold (here, † denotes the Moore-Penrose inverse [2] of the matrix A, $\mathrm{Ker}A$ denotes the kernel of A, for a symmetric matrix A we write $A \geq 0$ if A is positive semidefinite).

Recent results [4–8] in Sturmian theory of (1) are based on the following concept of the multiplicities of focal points introduced in [3].

Definition 1.1. A conjoined basis Y_i has a focal point of the multiplicity $m_1(i)$ at the point $i+1$ if $m_1(i) = \mathrm{rank} M_i$, where

$$M_i = \left(I - X_{i+1} X_{i+1}^{\dagger}\right) B_i,$$

and this basis has a focal point of the multiplicity $m_2(i)$ in the interval $(i, i+1)$ if $m_2(i) = \mathrm{ind}(T_i^T X_i X_{i+1}^{\dagger} B_i T_i)$, $T_i = I - M_i^{\dagger} M_i$, where indA is the number of negative eigenvalues of a symmetric matrix A. The number of focal points in $(i, i+1]$ is defined by $m(i) = m_1(i) + m_2(i)$.

Consider the reciprocal symplectic system [1]

$$Y_i = W_i^{-1} Y_{i+1},\ W_i^{-1} = \begin{bmatrix} D_i^T & -B_i^T \\ -C_i^T & A_i^T \end{bmatrix}. \tag{3}$$

Observe that Y_i solves (1) if and only if Y_i solves (3). By the definition [1], a conjoined basis of (3), (or (1)) is said to have a focal point in $[i, i+1)$ if the conditions

$$\mathrm{Ker}X_i \subseteq \mathrm{Ker}X_{i+1},$$
$$X_{i+1} X_i^{\dagger} B_i^T \geq 0$$

do not hold. By analogy with Definition 1.1 introduce the number of focal points in $[i, i+1)$ interchanging the roles of i and $i+1$.

Definition 1.2. A conjoined basis Y_i of (3) has a focal point of the multiplicity $m_1^*(i)$ at the point i if

$$m_1^*(i) = \mathrm{rang}\tilde{M}_i,\ \tilde{M}_i = \left(I - X_i X_i^{\dagger}\right) B_i^T,$$

and this basis has a focal point of the multiplicity $m_2^*(i)$ in the interval $(i, i+1)$ if $m_2(i) = \mathrm{ind}(\tilde{T}_i^T X_{i+1} X_i^{\dagger} B_i^T \tilde{T}_i)$, $\tilde{T}_i = I - \tilde{M}_i^{\dagger} \tilde{M}_i$. The number of focal points in $[i, i+1)$ is defined by $m^*(i) = m_1^*(i) + m_2^*(i)$.

Recall that the conjoined basis Y_i^0 of (1) given by the initial conditions $X_0 = 0$, $U_0 = I$ is called the principal solution of (1) at 0. The principal solution of (3) at $N+1$ given by $X_{N+1} = 0$, $U_{N+1} = -I$ is denoted by Y_i^{N+1}. Define the numbers of focal points

$$l(Y_i) = \sum_{i=0}^{N} m(i),\ l^*(Y_i) = \sum_{i=0}^{N} m^*(i) \tag{4}$$

for a conjoined basis Y_i in $(0,\ N+1]$ and $[0,\ N+1)$ respectively.

In the recent papers [4, 8] the following inequality $|l(Y_i) - l(\hat{Y}_i)| \leq n$ is proved for conjoined bases Y_i, $\hat{Y}_i$ of (1). In [7] we derive an equality which connects $l(Y_i)$ and $l(\hat{Y}_i)$ for conjoined bases Y_i, $\hat{Y}_i$ (see also Theorem 3.1 in section 3). Using the similar formula for $l^*(Y_i)$ and $l^*(\hat{Y}_i)$ we prove the following theorem.

Theorem 1.1. *If Y_i^0 and Y_i^{N+1} are the principal solutions of* (1) *at* 0 *and* $N+1$ *respectively, then*

$$l = l(Y_i^0) = l^*(Y_i^{N+1}).$$

The consideration is based on the concept of the comparative index [6] closely related to the concept of the multiplicities of focal points [3].

2. The main properties of the comparative index

In this section we consider $2n \times n$-matrices $Y = \begin{bmatrix} X \\ U \end{bmatrix}$, $\hat{Y} = \begin{bmatrix} \hat{X} \\ \hat{U} \end{bmatrix}$ which obey the conditions

$$Y^T J Y = 0,\ \hat{Y}^T J \hat{Y} = 0,$$
$$\text{rank} Y = \text{rank} \hat{Y} = n.$$

Introduce the matrices

$$w = w(Y, \hat{Y}) = Y^T J \hat{Y} \tag{5}$$
$$w^* = w^*(\hat{Y}, Y) = Y^T J^T \hat{Y}. \tag{6}$$

Then the *comparative index* $\mu(Y, \hat{Y}) = \mu_1(Y, \hat{Y}) + \mu_2(Y, \hat{Y})$ associated with (5) is defined by the following formulas

$$\mu_1(Y, \hat{Y}) = \text{rank}\mathcal{M},\ \mathcal{M} = \left(I - X^\dagger X\right) w, \tag{7}$$
$$\mu_2(Y, \hat{Y}) = \text{ind}\mathcal{P},\ \mathcal{P} = \mathcal{T}^T \left(w^T X^\dagger \hat{X}\right) \mathcal{T},\ \mathcal{T} = I - \mathcal{M}^\dagger \mathcal{M}. \tag{8}$$

Introduce the *reciprocal comparative index* $\mu^*(Y,\hat{Y}) = \mu_1^*(Y,\hat{Y}) + \mu_2^*(Y,\hat{Y})$ (associated with (6)) replacing w by w^* in (7), (8).

Because of $J^T = -J$, $w^*(\hat{Y},Y) = -w(Y,\hat{Y})$ we have that

$$\mu_1^*(Y,\hat{Y}) = \mu_1(Y,\hat{Y}),\ \mu_2^*(Y,\hat{Y}) = \text{ind}(-\mathcal{P}), \tag{9}$$

where $\mathcal{P}$ is defined by (8).

According to Theorem 2.1 in [6], the matrix $\mathcal{M}$ in (7) can be replaced by

$$\tilde{\mathcal{M}} = \left(I - XX^\dagger\right)\hat{X}, \tag{10}$$

the matrix $\mathcal{P}$ in (8) is symmetric and

$$\mathcal{P} = \mathcal{T}^T\left(\hat{X}^T[\hat{Q} - Q]\hat{X}\right)\mathcal{T}$$

for any symmetric Q, $\hat{Q}$ such that $X^TQX = X^TU$, $\hat{X}^T\hat{Q}\hat{X} = \hat{X}^T\hat{U}$.

The main properties of $\mu(Y,\hat{Y})$ are formulated in the following theorem.

Theorem 2.1. *Let matrices $Z,\hat{Z}$ be symplectic and $Z[0\,I]^T = Y$, $\hat{Z}[0\,I]^T = \hat{Y}$, then*

i) for any nonsingular $n \times n$ matrices C_1, C_2 we have

$$\mu_k(YC_1,\hat{Y}C_2) = \mu_k(Y,\hat{Y}),\ k = 1,2,$$

ii) $\mu_k(L_1Y,\, L_1\hat{Y}) = \mu_k(Y,\hat{Y})$, $k = 1,2$, *for any symplectic block lower triangular matrix L_1,*

iii) $\mu_k(Z[0\,I]^T,\hat{Z}[0\,I]^T) = \mu_k^*(Z^{-1}[0\,I]^T, Z^{-1}\hat{Z}[0\,I]^T)$, $k = 1,2$,

iv)

$$\mu_1(Y,\hat{Y}) = \text{rank}\hat{X} - \text{rank}X + \mu_1^*(\hat{Y},Y),$$
$$\mu_2(Y,\hat{Y}) = \mu_2^*(\hat{Y},Y),$$

v) $\mu(Y,\hat{Y}) + \mu(\hat{Y},Y) = \text{rank}(w(Y,\hat{Y}))$,

vi) $\mu(JZ[0\,I]^T, J\hat{Z}[0\,I]^T) = \mu(Z[0\,I]^T,\hat{Z}[0\,I]^T) + \mu(JZ[0\,I]^T,[I\,0]^T) - \mu(J\hat{Z}[0\,I]^T,[I\,0]^T)$.

The proof based on the factorization approach [6] can be found in [7].

Theorem 2.2. *For any symplectic matrices Z, $\hat{Z}$, W the following identity*

$$\mu(WZ[0\,I]^T, W\hat{Z}[0\,I]^T) = \mu(Z[0\,I]^T,\hat{Z}[0\,I]^T) + \mu(WZ[0\,I]^T, W[0\,I]^T) - \mu(W\hat{Z}[0\,I]^T, W[0\,I]^T). \tag{11}$$

holds.

Proof. Note that (11) is proved for the important particular cases $W = L_1$ and $W = J$ (see Theorem 2.1 ii), vi)). Next, using the result (see Proposition 6.2 in [9]) on the factorization of any symplectic matrix in the form $W = JL_1J^TL_2JL_3J^T$, where L_1, L_2, L_3 are symplectic block lower triangular matrices it is possible to prove (11) for any symplectic matrix W. For example, applying sequentially Theorem 2.1 ii), vi), i) we have

$$\begin{aligned}
&\mu(WZ[0\,I]^T, W\hat{Z}[0\,I]^T) = \\
&\quad \mu(JL_1J^TL_2JL_3J^TZ[0\,I]^T, JL_1J^TL_2JL_3J^T\hat{Z}[0\,I]^T) = \\
&\quad \mu(J^TL_2JL_3J^TZ[0\,I]^T, J^TL_2JL_3J^T\hat{Z}[0\,I]^T)+ \\
&\quad \mu(WZ[0\,I]^T, [I\,0]^T) - \mu(W\hat{Z}[0\,I]^T, [I\,0]^T) = \ldots = \\
&\mu(Z[0\,I]^T, \hat{Z}[0\,I]^T) + \{\mu(J^TZ[0\,I]^T, [I\,0]^T) - \mu(J^T\hat{Z}[0\,I]^T, [I\,0]^T)+ \\
&\quad \mu(JL_3J^TZ[0\,I]^T, [I\,0]^T) - \mu(JL_3J^T\hat{Z}[0\,I]^T, [I\,0]^T)+ \\
&\quad \mu(J^TL_2JL_3J^TZ[0\,I]^T, [I\,0]^T) - \mu(J^TL_2JL_3J^T\hat{Z}[0\,I]^T, [I\,0]^T)+ \\
&\quad \mu(WZ[0\,I]^T, [I\,0]^T) - \mu(W\hat{Z}[0\,I]^T, [I\,0]^T)\}. \qquad (12)
\end{aligned}$$

Putting in (12) $\hat{Z} = I$ we derive a similar formula for $\mu(WZ[0\,I]^T, W[0\,I]^T)$. Replacing Z by $\hat{Z}$ we obtain a similar expansion for $\mu(W\hat{Z}[0\,I]^T, W[0\,I]^T)$. Next, it is easy to verify directly that the bracketed expression in (12) coincides with $\mu(WZ[0\,I]^T, W[0\,I]^T) - \mu(W\hat{Z}[0\,I]^T, W[0\,I]^T)$. The proof is completed. □

Note that in Theorem 2.1, Theorem 2.2 we can interchange the role of μ and μ^*. So we have the following theorem for μ^*.

Theorem 2.3. *For any symplectic matrices Z, $\hat{Z}$, W the following identity*

$$\mu^*(Z[0\,I]^T, \hat{Z}[0\,I]^T) = \mu^*(WZ[0\,I]^T, W\hat{Z}[0\,I]^T) + \mu^*(Z[0\,I]^T, W^{-1}[0\,I]^T) - \mu^*(\hat{Z}[0\,I]^T, W^{-1}[0\,I]^T). \qquad (13)$$

holds.

3. Separation results

In this section we apply the concept of the comparative index μ to the oscillation theory of (1). So we connect this concept with the concept of the multiplicities of focal points.

Lemma 3.1. *If Z_i is a symplectic fundamental matrix for* (1) *and* $Y_i = Z_i[0\,I]^T$, *then*

$$m_k(i) = \mu_k(Z_{i+1}[0\,I]^T, W_i[0\,I]^T) = \mu_k^*(Z_{i+1}^{-1}[0\,I]^T, Z_i^{-1}[0\,I]^T), \qquad (14)$$

$$m_k^*(i) = \mu_k^*(Z_i[0\,I]^T, W_i^{-1}[0\,I]^T) = \mu_k(Z_i^{-1}[0\,I]^T, Z_{i+1}^{-1}[0\,I]^T), k = 1, 2, \qquad (15)$$

where $m(i)$ and $m^*(i)$ are the numbers of focal points of Y_i in $(i, i+1]$ and $[i, i+1)$ respectively.

Proof. Note that the proof of (14) is presented in [6] (see Lemmas 2.2, 2.3). Consider the proof of (15). By the definition of $\mu^*(Y, \hat{Y})$ (see section 2) for the case $Y := Y_i$, $\hat{Y} := W_i^{-1}[0\,I]^T = [-B_i,\, A_i]^T$, we have $\mu_1^*(Z_i[0\,I]^T, W_i^{-1}[0\,I]^T) = m_1^*(i)$, where we use (10), (9) for the evaluation of $\mu_1^*(Y, \hat{Y})$. Next, $w^* = Y_i^T J^T W_i^{-1}[0\,I]^T = Y_i^T W_i^T J^T [0\,I]^T = -X_{i+1}^T$, then by (8), $\mu_2^*(Z_i[0\,I]^T, W_i^{-1}[0\,I]^T) = \operatorname{ind}\tilde{T}^T \left(w^{*\,T} X_i^\dagger(-B_i^T)\right) \tilde{T} = m_2^*(i)$, where we use the notation of Definition 1.2. Note that $\mu_k^*(Z_i[0\,I]^T, W_i^{-1}[0\,I]^T) = \mu_k(Z_i^{-1}[0\,I]^T, Z_{i+1}^{-1}[0\,I]^T)$ holds because of Theorem 2.1 iii), where we use additionally that $Z_i Z_{i+1}^{-1} = W_i^{-1}$. The proof is completed. □

Remark 3.1. Note that $m_1(i) = \operatorname{rank}\breve{M}_i$, $\breve{M}_i = (I - X_{i+1}^\dagger X_{i+1}) X_i^T$ by (14), and M_i in Definition 1.1 can be replaced by $\breve{M}_i$ because of Lemma 3.1.

Corollary 3.1. *If $m(i)$ and $m^*(i)$ are the numbers of focal points for a conjoined basis Y_i of* (1) *in $(i, i+1]$ and $[i, i+1)$ respectively, then*

$$m^*(i) - m(i) = \operatorname{rank}(X_{i+1}) - \operatorname{rank}(X_i). \qquad (16)$$

Proof. According to Lemma 3.1, $m(i) = \mu^*(Z_{i+1}^{-1}[0\,I]^T, Z_i^{-1}[0\,I]^T)$, $m^*(i) = \mu(Z_i^{-1}[0\,I]^T, Z_{i+1}^{-1}[0\,I]^T)$, then, by Theorem 2.1 iv), we have

$$\mu(Z_i^{-1}[0\,I]^T, Z_{i+1}^{-1}[0\,I]^T) = \operatorname{rank}(X_{i+1}) - \operatorname{rank}(X_i) + \mu^*(Z_{i+1}^{-1}[0\,I]^T, Z_i^{-1}[0\,I]^T)$$

or (16). □

Corollary 3.2. *Let Y_i, $\hat{Y}_i$ be conjoined bases of* (1) *and* $\mu(i) = \mu(Y_i, \hat{Y}_i)$, $\mu^*(i) = \mu^*(Y_i, \hat{Y}_i)$. *Then*

$$\Delta\mu(i) = m(i) - \hat{m}(i), \; -\Delta\mu^*(i) = m^*(i) - \hat{m}^*(i), \qquad (17)$$

where $m(i)$, $m^*(i)$ and $\hat{m}(i)$, $\hat{m}^*(i)$ are the numbers of focal points for conjoined bases Y_i, $\hat{Y}_i$ of (1) *in $(i, i+1]$ and $[i, i+1)$.*

Proof. For the proof we use Theorems 2.2, 2.3. Put in (11) $Z := Z_i$, $W := W_i$, $\hat{Z} := \hat{Z}_i$, where Z_i, $\hat{Z}_i$ are symplectic fundamental matrices for (1), i.e. $Z_{i+1} = W_i Z_i$, $\hat{Z}_{i+1} = W_i \hat{Z}_i$, and $Y_i = Z_i[0\,I]^T$, $\hat{Y}_i = \hat{Z}_i[0\,I]^T$. Then, according to Lemma 3.1 and (11), we obtain $\mu(Y_{i+1}, \hat{Y}_{i+1}) = \mu(Y_i, \hat{Y}_i) + m(i) - \hat{m}(i)$ or the first equation in (17). Similarly, by Lemma 3.1 and (13) we have $\mu^*(Y_i, \hat{Y}_i) = \mu^*(Y_{i+1}, \hat{Y}_{i+1}) + m^*(i) - \hat{m}^*(i)$ or the second equation in (17). □

Theorem 3.1. *Let Y_i, $\hat{Y}_i$ be conjoined bases of (1), then*

$$l(Y_i) - l(\hat{Y}_i) = \mu(N+1) - \mu(0),\; l^*(Y_i) - l^*(\hat{Y}_i) = \mu^*(0) - \mu^*(N+1), \quad (18)$$

where $l(Y_i)$, $l^*(Y_i)$ defined by (4) are the numbers of focal points for Y_i in $(0,\, N+1]$ and $[0,\, N+1)$.

Proof. Summing the equations in (17) from $i = 0$ to $i = N$ we derive (18). □

Corollary 3.3.

$$|l(Y_i) - l(\hat{Y}_i)| \le \operatorname{rank}(w(Y_i, \hat{Y}_i)) \le n,$$
$$|l^*(Y_i) - l^*(\hat{Y}_i)| \le \operatorname{rank}(w(Y_i, \hat{Y}_i)) \le n, \quad (19)$$

for the Wronskian $w(Y_i, \hat{Y}_i) = w_i$ defined by (2).

Proof. By (18), we obtain $-\mu(0) \le l(Y_i) - l(\hat{Y}_i) \le \mu(N+1)$ because $\mu(i) \ge 0$. Next, by Theorem 2.1 v), we have $\mu(i) \le \operatorname{rank} w_i$, then the first inequality in (19) holds because of the Wronskian identity (2). The proof of the second inequality is similar. □

Proof of Theorem 1.1. Summing the equations in (16) from $i = 0$ to $i = N$ we obtain that

$$l^*(Y_i) - l(Y_i) = \operatorname{rank}(X_{N+1}) - \operatorname{rank}(X_0). \quad (20)$$

Using (18), (20) for the case $Y_i := Y_i^{N+1}$, $\hat{Y}_i := Y_i^0$, where Y_i^{N+1}, Y_i^0 are the principal solutions at $N+1$ and 0 we have

$$l^*(Y_i^{N+1}) - l(Y_i^{N+1}) = -\operatorname{rank}(X_0^{N+1}) \quad (21)$$

by (20), and

$$l(Y_i^{N+1}) - l(Y_i^0) = \mu(Y_{N+1}^{N+1}, Y_{N+1}^0) - \mu(Y_0^{N+1}, Y_0^0) =$$
$$\mu([0\; -I]^T, Y_{N+1}^0) - \mu(Y_0^{N+1}, [0\,I]^T) = \operatorname{rank}(X_{N+1}^0) \quad (22)$$

by (18) and the definition of μ. Note that $\operatorname{rank}(X_0^{N+1}) = \operatorname{rank}(X_{N+1}^0)$ due to the Wronskian identity $w(Y_0^0, Y_0^{N+1}) = -X_0^{N+1} = w(Y_{N+1}^0, Y_{N+1}^{N+1}) = -X_{N+1}^{0\,T}$. Then, summing (21) and (22) we prove Theorem 1.1. □

References

1. M. Bohner and O. Došlý, Disconjugacy and transformations for symplectic systems, *Rocky Mountain J. Math.*, **27** (1997), 707–743.
2. F. Gantmacher, The Theory of Matrices, v.1, Chelsea Publishing Company, New York, 1959.
3. W. Kratz, Discrete Oscillation, *J. Difference Equ. Appl.* **9** (2003), 135–147.
4. O. Došlý and W. Kratz, A Sturmian Separation Theorem for symplectic difference systems, *J. Math. Anal. Appl.* **325** (2007), 333–341.
5. R. Hilscher, V. Ruzickova, Implicit Riccati equations and quadratic functionals for discrete symplectic systems, *Int. J. Difference Equ.* **1** (2006), 135–154.
6. J. Elyseeva, The comparative index for conjoined bases of symplectic difference systems in *Difference equations, Special functions and Orthogonal polinomials*, World Scientific, Singapore, 2007, 168–177.
7. J. Elyseeva, The Comparative index for solutions of symplectic difference systems, *Differ. Equ.*, to appear.
8. M. Bohner, O. Došlý, W. Kratz, Sturmian and spectral theory for discrete symplectic systems, *Trans. Amer. Math. Soc.*, to appear.
9. M. Zelikin, Control theory and optimization I : homogeneous spaces and the Riccati equation in the calculus of variations, Springer, Berlin; New York, 2000.

Conductance and Mixing Time in Discrete Dynamical Systems

Sara Fernandes

Department of Mathematics, Universidade de Évora and CIMA-UE,
Rua Romão Ramalho, 59, 7000-671 Évora, Portugal
E-mail: saf@uevora.pt

Sumitha Jayachandran

Department of Mathematics, Universidade de Évora,
Rua Romão Ramalho, 59, 7000-671 Évora, Portugal
E-mail: sumithajaya@rediffmail.com

We have introduced the notion of conductance in discrete dynamical systems using the known results from graph theory applied to systems arising from the iteration of continuous functions. The conductance allowed differentiating several systems with the same topological entropy, characterizing them from the point of view of the ability of the system to go out from a small subset of the state space. There are several other definitions of conductance and the results differ from one to another. Our goal is to understand the meaning of each one concerning the dynamical behaviour in connection with the decay of correlations and mixing time. Our results are supported by computational techniques using symbolic dynamics, and the tree-structure of the unimodal and bimodal maps.

Keywords: Conductance, mixing time, decay correlations, kneading theory.

AMS Subject Classification 37A25, 37A35, 37E05

1. Introduction

The problem of classification of dynamical systems depends on the context in which it is applied. It can have a topological, differentiable, algebraic or other character. In each context there are parameters, which remain constant in equivalent systems (isomorphic) and we call invariants.

The classification is achieved only when we know all invariants required, as is the case of Bernoulli systems, where the entropy is a complete invariant. However this situation is uncommon, there are few examples in the theory where it is known.

In the last years we have been studying discrete dynamical systems, arising from the iterates of a Markov real function in the interval $f : I \rightarrow I$,

$$x_{k+1} = f\left(x_k\right), \; x_0 \in I.$$

Such systems can be viewed as topological Markov chains, where the state space corresponds to the intervals of the Markov partition and the transition matrix is obtained from the following:

$$\left(A_f\right)_{ij} := \begin{cases} 1 & \text{if} \quad f\left(I_i\right) \supseteq I_j \\ 0 & \text{otherwise.} \end{cases}$$

It is known that the topological entropy of the system is related to the spectral radius of A, but it is not a complete invariant for we have families of maps with constant topological entropy, but with different dynamics. We have presented a study using the second eigenvalue (in magnitude) of the transition matrix A, that allowed us to classify a certain family of bimodal maps with constant topological entropy in a binary ordered tree of trajectories.[3] Nevertheless it was showed that this tree possessed branches corresponding to different dynamics, but with the same spectral invariants (first and second eigenvalue of A).[4] This situation led us to the research of different kind of parameters to accomplish the task we have proposed: to find the relation of the topological parameters of the bimodal maps we were iterating, with the dynamics of the system itself.[5,7] With this proposal we have introduced the notion of conductance and discrete laplacian in the context of discrete dynamical systems and we have used these concepts in several families of m-modal maps to characterize the dynamics.

In this work we return to the first invariant we have treated before, the mixing time of a discrete dynamical system, and we show the relationship with the conductance in families of unimodal and bimodal maps. We present a numerical study that shows that the conductance, in all four versions of it, presents an effective measure of the mixing time.

All investigations are done considering the kneading theory and the symbolic dynamics as the main tool to produce concrete results.

2. Settings

Consider the discrete dynamical system of the iterates of a function in the interval $(I, f, \mathcal{B}, \mu)$, where $(I, \mathcal{B}, \mu)$ is the probabilistic measure space, $I = [0, 1]$, $\mathcal{B}$ to be the Borel sets in I, μ the Lebesgue measure and $f : I \rightarrow I$ is a non-singular and measure preserving transformation, i.e., $\mu\left(f^{-1}\left(B\right)\right) = \mu\left(B\right)$, for all $B \in \mathcal{B}$.

Definition 2.1. The map $f : I \to I$ is called mixing if and only if

$$\lim_{t\to\infty} \mu\left(B \cap f^{-t}(C)\right) = \mu(B)\,\mu(C) \qquad \text{for all } B, C \in \mathcal{B}$$

where $f^{-t}(C) = \{x \in I : f^t(x) \in C\}$. If we consider points x in $B \cap f^{-t}(C)$, when $t \to \infty$, the measure of the set of these points is just $\mu(B)\,\mu(C)$. It means that any set $C \in \mathcal{B}$ under the action of f becomes asymptotically independent of a fixed set $B \in \mathcal{B}$.

Obviously the mixing property depends of the measure μ, and sometimes we say that the system is (f, μ)-mixing meaning that the map f is mixing with respect to μ.

If f is a Markov function, there is a unique Markov chain associated with it, via the Parry measure, see,[10] which is induced by the Perron eigenvectors of the matrix A. This Markov chain is characterized by a stochastic matrix P which depends only on the matrix A and the referred eigenvectors. This matrix P is irreducible, that is,

$$\text{for all } i, j, \text{ there exists } t, \text{ such that } \left(P^t\right)_{ij} > 0$$

and aperiodic, that is

$$\text{for all } i, \ \gcd\{t : \left(P^t\right)_{ii} > 0\} = 1$$

then there exists a (unique) stationary distribution π, such that $\pi P = \pi$ and moreover, the system is ergodic and this distribution, π, also called the equilibrium, verifies the mixing condition

$$\lim_{t\to\infty} \left(P^t\right)_{ij} = \pi_j \text{ for all } i, j.$$

Here the probability distributions in the time step t, $p^{(t)} = p^{(0)} P^t$ are given by a row vector $1 \times n$, as well as the stationarity π. The entry P_{ij} of the Markov matrix indicates the probability of transition from the state i to the state j in one step.

In several practical models we put the question of knowing the convergence speed, concretely, some times is useful to know how long the chain must evolves to attempt some given proximity of the equilibrium. Intuitively the convergence is faster if there are no bottlenecks. That means that there are no small subsets of the state space which retains the system. These subsets are detected by the conductance of the discrete dynamical system as defined in references 5,6.

Define distance of $p^{(t)}$ from π as the l^1 distance $\|\cdot\|_{l^1(\pi)}$, which is defined[9] by

$$\left\|p^{(t)} - \pi\right\|_{l^1(\pi)} = \sum_{i=1}^{n} \left|p_i^{(t)} - \pi_i\right|$$

For a given precision $\varepsilon > 0$ we define the mixing time as follows

$$\tau(\varepsilon) = \min\{t : \left\|p^{(t')} - \pi\right\|_{l^1(\pi)} \leq \varepsilon, \text{ for all } t' \geq t\} \tag{1}$$

There are other distances,[9] namely the l^p distances or the χ^2-distance which give rise to another mixing times, nevertheless in this work we have adopted the l^1-distance.

3. Conductance

Bezrukov[1] in a survey on isoperimetric problems on graphs have presented four different versions of isoperimetric numbers on graphs. They were established to determine the edge expansion/congestion of the graph and so to prove rapid mixing. This edge expansion, in a weighted version, is also called conductance of the graph.

All these definitions were setting for a simple, non-weighted, connected graph $G = (V, E)$, with $|V| = n$ vertices. Denote by $\deg(i)$ the degree of the vertex i and define, for $U \subset V$,

$$Vol(U) = \sum_{i \in U} \deg(i),$$
$$\theta(U) = \{(i, j) \in E : i \in U,\ j \notin U\}.$$

With these settings let us introduce the four definitions of conductance in a graph

$$i_1(G) = \min_{\emptyset \neq U \subset V} \frac{|\theta_G(U)|}{\min\{|U|,\ n - |U|\}};\ i_2(G) = \min_{\emptyset \neq U \subset V} \frac{|\theta_G(U)|}{\min\{Vol(U), Vol(\overline{U})\}};$$

$$i_3(G) = \min_{\emptyset \neq U \subset V} \frac{|\theta_G(U)|\ n}{|U| . |\overline{U}|};\qquad i_4(G) = \min_{\emptyset \neq U \subset V} \frac{|\theta_G(U)|}{|U| .\ \log\left(\frac{n}{|U|}\right)}.$$

In Ref. 6 we have considered the adaptation of these four definitions as definitions of conductance of a discrete dynamical system using a similar adaptation from Bollobás[2] to the case of a random walk in a weighted graph.

$$\Phi_1 = \min_{\emptyset \neq U \subset V} \frac{\sum\limits_{i\in U, j\in \overline{U}} \pi_i P_{ij}}{\min\{\sum\limits_{i\in U} \pi_i, \sum\limits_{i\in \overline{U}} \pi_i\}};$$

$$\Phi_2 = \min_{\emptyset \neq U \subset V} \frac{\sum\limits_{i\in U, j\in \overline{U}} \pi_i P_{ij}}{\min\{\sum\limits_{i\in U, j\in V} (P_{ij}+P_{ji}), \sum\limits_{i\in \overline{U}, j\in V} (P_{ij}+P_{ji})\}};$$

$$\Phi_3 = \min_{\emptyset \neq U \subset V} \frac{\sum\limits_{i\in U, j\in \overline{U}} \pi_i P_{ij}}{\sum\limits_{i\in U} \pi_i . \sum\limits_{i\in \overline{U}} \pi_i};$$

$$\Phi_4 = \min_{\emptyset \neq U \subset V} \frac{\sum\limits_{i\in U, j\in \overline{U}} \pi_i P_{ij}}{\sum\limits_{i\in U} \pi_i . \log\left(\dfrac{1}{\sum\limits_{i\in U} \pi_i}\right)}.$$

As we have said before these definitions of conductance were used to characterize families of discrete dynamical systems,[6] namely those with the same topological entropy. We have presented an example with the same topological entropy and also the same conductance Φ_1. This example was the start point to the research of other invariants that led us to the present study.

4. Unimodal and bimodal applications

Return to the discrete dynamical system. We have considered f in the families of unimodal and bimodal applications. These maps are continuous piecewise monotone maps in the interval with one and two critical points. It is known[8] that they are topologically semiconjugated to piecewise linear maps with constant slope $\pm s$ and so it is sufficiently wide to analyze the family of piecewise linear maps to have a general outlook. We have made a strong use of symbolic dynamics in the choice of families of maps as it is done in Ref. 4. We identify each map by the itinerary (or itineraries) of his critical point (or points) given by the sequence of the letters of the alphabet, also called the kneading sequences. In the unimodal case we use the alphabet $\{L, C, R\}$ and in the bimodal case $\{L, A, M, B, R\}$.

Basically we "walk" in the complete ordered tree of unimodal maps, where the trajectories of the critical points are disposed according to the period (each period corresponds to one floor) and to the topological entropy (in each floor), see Fig. 1. Furthermore, if we are in one floor and we "go" to the next floor going to the left makes the topological entropy decreasing

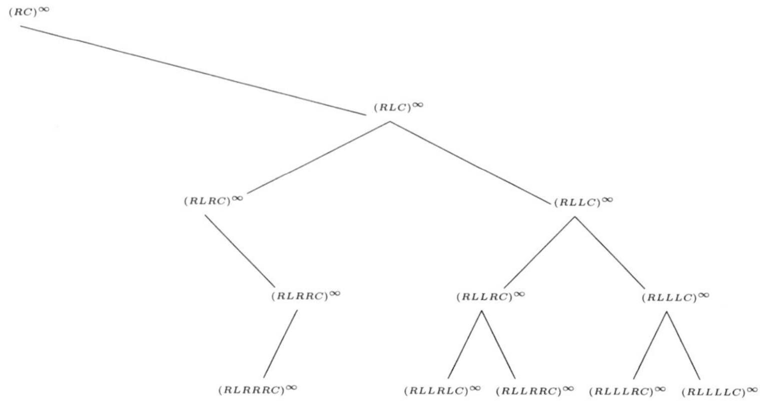

Fig. 1. Kneading sequences tree of the unimodal maps up to period 6.

and to the right it increases. In the bimodal case, we need two itineraries to identify each map, unless the trajectory of one critical point, A or B, falls in the trajectory of the other, as is the case in the considered family below.

We present the results for conductance (all four possibilities) in comparison with the mixing time, computed by the formula (1), where we have taken $\varepsilon = 0.001$ in all chosen families. We present, in the left, the evolution of both quantities with the topological entropy and, in the right, the evolution with the number of states in the state space (number of intervals in the Markov partition, or, the length of the Markov matrix).

Results concerning the unimodal family, $(CRLLLRLR^k)^\infty$

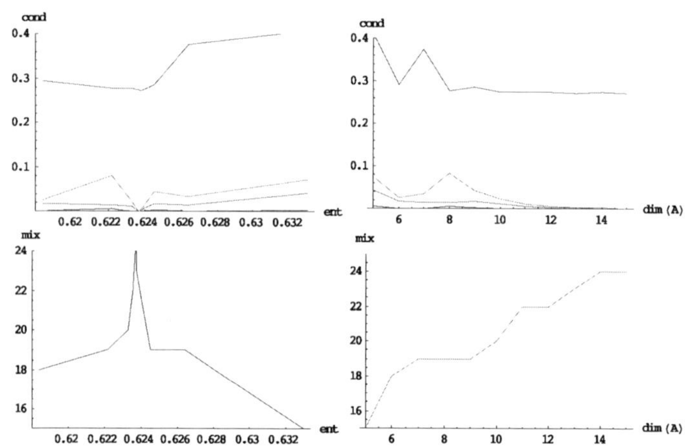

Another unimodal family, $(CRLR^K)^\infty$

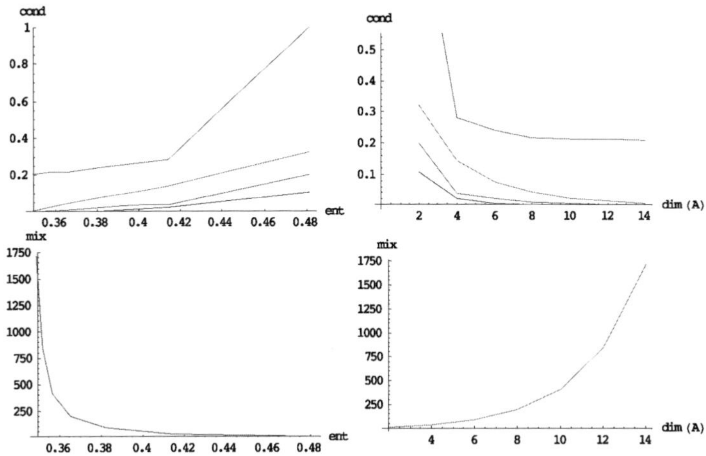

Next we have considered all admissible unimodal maps up to period 8. In the Figures below are plotted the variation of conductance and mixing time with the topological entropy.

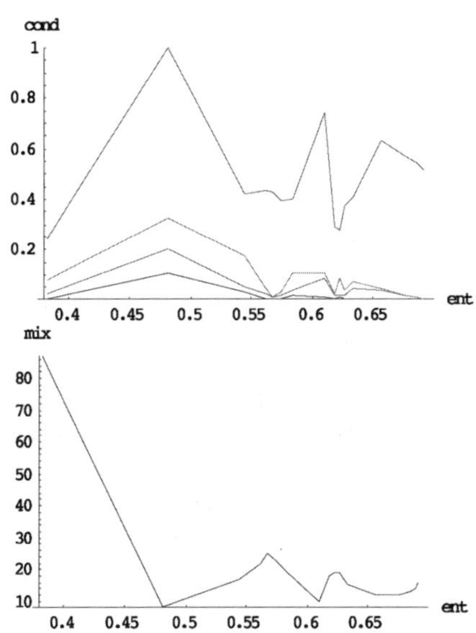

Here is a bimodal family, $AR\left(L^{k}BL\right)^{\infty}$, which has constant topological entropy, $\log 2$. In this family we have constant $\Phi_1 = 0.5$, but different $\Phi_{2,3,4}$. In the next Figure we have plotted just the evolution of conductance and mixing time with the number of possible states.

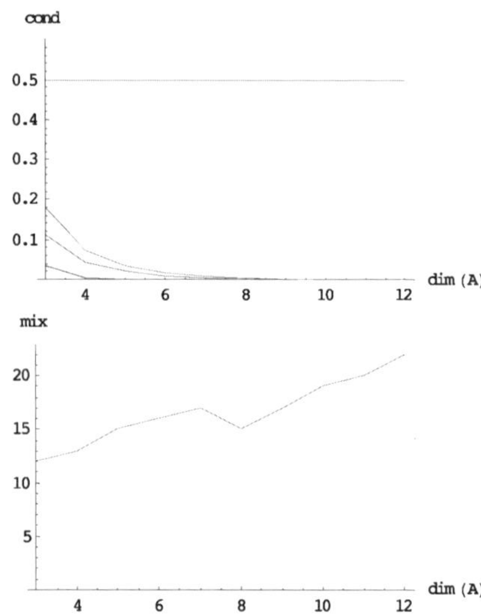

The relation of the conductance with the mixing time shows, as expected, that the existence of funnels is detected by all definitions of conductance and implies a slower convergence to the stationarity.

References

1. S. L. Bezrukov, Edge Isoperimetric Problems on Graphs, *Theoretical Computer Science* Vol. 307, Issue3 (2003) 473–492.
2. B. Bollobás, Volume Estimates and Rapid Mixing, *Flavours of Geometry*, MSRI Publications, Vol. 31, (1997).
3. S. Fernandes,J. Sousa Ramos, Second smaller zero of kneading determinant, *Proceedings of the conference ICDEA2003,* 8th International Conference on Difference Equations and Applications, July 28 - August 1, 2003, Brno, Czech Republic.
4. S. Fernandes,J. Sousa Ramos, Second eigenvalue of transition matrix associated to iterated maps. *Chaos, Solitons & Fractals* Volume 31, Issue 2, (2007), 316–326.
5. S. Fernandes, J. Sousa Ramos, Conductance, Laplacian and mixing rate in discrete dynamical systems, *Nonlinear Dynamics*, 44 (2006) No. 2, 117–126.

6. S. Jayachandran, S. Fernandes, C.Grácio, C. Ramos, Conductance in digraphs arising from discrete dynamical systems (submitted).
7. C.Grácio and J. Sousa Ramos, Spectrum of the laplacian on hyperbolic surfaces (to appear).
8. Milnor, J., Thurston, W., On the Iterated Maps of the Interval, *Dynamical Systems: Proc. Univ. of Maryland 1986–1987, Lecture Notes in Mathematics,* **1342** (Springer, Berlin, New York), 465–563 (1988).
9. R. Montenegro, Vertex and edge expansion properties for rapid mixing ,Random Structures & Algorithms, Vol. 26, Issue 1-2 (2005), *Proceedings of the Eleventh International Conference "Random Structures and Algorithms"*, 52–68.
10. P. Walters, An Introduction to Ergodic Theory, Springer 2007.

Pulsating Equilibria: Stability through Migration*

Takao Fujimoto

Fukuoka University,
Fukuoka, 814-0180, Japan
E-mail: takao@econ.fukuoka-u.ac.jp

Mingzhe Li

Fukuoka University,
Fukuoka, 814-0180, Japan
E-mail: lmz@fukuoka-u.ac.jp

Pascal Mossay

University of Alicante,
Alicante, 03080, Spain
E-mail: mossay@merlin.fae.ac.es

This paper is to present a model of spatial equilibrium using a nonlinear generalization of Markov-chain type model, and to show the dynamic stability of a unique equilibrium. Even at an equilibrium, people continue to migrate among regions as well as among agent-types, and yet their overall distribution remain unchanged. The model is also adapted to suggest a theory of traffic distribution in a city.

Keywords: Indecomposability, Nonlinear Positive Mappings, Primitivity, Spatial Equilibrium, Stability, Traffic Network.

1. Introduction

Scarf [16] made it clear that in Walrasian general equilibrium models, dynamic stability may not be guaranteed under the tatonnement adjustment process[a]. And so, Sonnenschein [17] introduced migration of resources among industries to establish stability. Then, Mossay ([12] and [13]) considered stability allowing for migration of consumers. In both Sonnenschein's

*Thanks are due to the referee for the comments, which are useful to make this article more readable.

[a]For an interesting recent contribution, see Mukherji [14].

and Mossay's model, prices are still the key variables, though quantities are adjusted through migration of agents on a circumferential territory. In this paper, we present a model in which the distributions of people among regions as well as agent-types are the key variable, and establish the dynamic stability through migration of agents. Our model is a variant of nonlinear Markov-chain model with a different interpretation put on the transition probability matrix.

In Section 2, we explain our model, and show its dynamic stability in Section 3. The following Section 4 contains detailed consideration on some conditions under which our assumption of primitivity of the transition probability matrix is satisfied. An interpretation of each condition is also presented. Section 5 gives an application of our model to a theory of traffic distribution in an area. The final section includes some remarks.

2. Model

In our model of an economy, there are $n (n \geq 1)$ regions among which people migrate. There exist also $m (m \geq 1)$ types of people. These types may represent a producer of a particular commodity, or a transporter of a commodity from a region to another, or an employee in an industry with a specific taste[b], or a person who is a producer and consumer at the same time. All the people need not be rational in the ordinary sense, and some types of people are allowed to be irrational so long as that kind of irrationality persists through time and so they behave the same way consistently under the same environment. Given the distribution of people, $x \in D \equiv (R_+^{m \times n} - \{0\})$ among the n regions and the m types at the beginning of a period, the supply of and demand for various commodities in each region for the period are determined: we assume there are $k (k \geq 1)$ kinds of goods and services. (The symbol $R_+^{m \times n}$ stands for the nonnegative orthant of the Euclidean space of dimension $(m \times n)$.) After observing these supplies and demands, some people migrate, at the end of the period, to another region depending upon their own decision process.[c]

People may also change their types. Changes in types of people, with intention or not, naturally involve migration of resources among industries.

[b]Various tastes may be represented by different utility functions people have adopted.
[c]People certainly have many reasons to migrate other than the inequalities between supply and demand. For example, some wish to move out of a crowded region, while others move in to look for a job. The point is that those reasons are definable solely by the present distribution of people.

Besides, producers shifts to a different industry together with their assets.[d] Concerning how people migrate among regions as well as among types, we adopt a Markov-chain type transition coefficient matrix, $T(x)$, whose size is $(m \times n) \times (m \times n)$, and the $(i,\, j)$ entry is denoted by $t_{ij}(x)$, showing its dependence on the distribution of x. We have just mentioned a transition *coefficient* matrix, not a *probability* one. In our model, we may regard the transition among regions and types as *deterministic*, and each column-sum of coefficients needs not be unity, thus allowing for expansion or contraction of our economy. We define

$$F(x) \equiv T(x) \cdot x.$$

Also define $x(k)$ to be the distribution of people at a period k, then the dynamics of our model is expressed by the following equation.

$$x(k+1) = F(x(k)) \equiv T(x(k)) \cdot x(k) \quad \text{for } k = 0, 1, 2, \ldots,$$

with $x(0)$ being the initial distribution.

We assume the following.

Assumption A1. The mapping $T(x)$ does not depend on periods in a direct way, i.e., the process is homogeneous.

Assumption A2. Each $t_{ij}(x)$ is continuous and homogeneous of degree zero with respect to $x \in D$, and $t_{ij}(x) \geq 0$ for all $x \in D$. Moreover, $F(x)$ is monotone, i.e., $F(y) \geq F(x)$ if $y \geq x$.

Thus, the mapping $T(x)$ is nonnegative as a matrix. A simple case where the assumption A2 is verified is that in which each element of $F(x)$ is nonnegative, monotone, homogeneous of degree one and continuously differentiable with respect to $x \in D$: we can apply Euler's theorem on homogeneous functions.

Assumption A3. If $t_{ij}(x) > 0$ at some $x \in D$, it satisfies $t_{ij}(x) > 0$ at all $x \in D$.

This assumption allows us to judge the primitivity of the matrix T only by its sign pattern at an arbitrary x.

Assumption A4. The matrix $T(x)$ is primitive at an arbitrary $x \in D$, i.e., there exists a positive integer k such that $T^k(x) \equiv \{T(x)\}^k \gg 0$ for any $x \in D$.

[d] When assets get larger or smaller through investment for an agent, we regard this agent has changed types. Thus, a difference in assets yields that in types. To have a finite number of types, however, each commodity should be indivisible at a certain unit.

Here, the inequality sign $\gg 0$ means that every element of the matrix on the LHS is positive.

3. Stability as Strong Ergodicity

Let us consider the normalized process starting from an initial vector $x(0) \in D$. That is, we define the following normalized map

$$G(x) \equiv \frac{F(x)}{\|F(x)\|} ,$$

where $\|\cdot\|$ is any given norm on $R^{m\times n}$, and examine the vector sequence

$$S(x(0)) \equiv \{x(0),\ G(x(0)),\ G^2(x(0)),\ \ldots\}.$$

We can prove the strong ergodicity of our model, that is,

Theorem 3.1. *Starting from an arbitrary $x(0) \in D$, the sequence $S(x(0))$ converges to a unique vector $x^* \gg 0$.*

Proof. The mapping $T(x) \cdot x$ is from D into D, continuous and homogeneous, and some power of $T(x)$ is strictly increasing, i.e., $T^k(x) \gg 0$ by the assumption A4. Thus, we can apply the main result of Kohlberg [7] or Corollary 1 in Fujimoto and Krause (p.106 of [4]) to have the stated theorem. □

The stability here is the strong ergodicity of the normalized process, i.e., we have a directional stability or a ray-stability, though the vector sequence $\{x(0),\ F(x(0)),\ F^2(x(0)),\ \ldots\}$ itself may continue to expand or shrink. Besides, at a unique vector x^*, it repeats time after time under $G(x)$, and yet migration of people still takes place in each period. Such a phenomenon cannot be observed in Sonnenschein's model [17] or Mossay's [13].

Some words are in order about the prices of goods and services. In our process explained above, prices are pushed away to the background, and thus, go through a non-Walrasian adjustment process. The prices can, however, be thought of as changing based on Walrasian rules: the price of a commodity in a region rises when there is an excess demand for it in that region, and falls when an excess supply is observed there. So, given the initial price vector $p(0) \in R_+^{k\times n}$, this vector gets adjusted as the distribution x is transformed. Certainly it is awkward if the prices continue to vary even after the distribution arrives at the unique equilibrium x^*. All we have to assume is that at this equilibrium, there is no excess demand for each commodity in every region. Or put simply, when there is excess demand for some commodity, the distribution of agents will change in the next period: the degree of irrationality is limited.

4. Primitivity

Now we had better examine some conditions under which our assumption of primitivity A4 is met. To do this, we consider the transition coefficient matrix partitioned region by region.

$$T(x) \equiv \begin{pmatrix} T_{11} & T_{12} & \cdots & T_{1n} \\ T_{21} & T_{22} & \cdots & T_{2n} \\ \vdots & \vdots & \ddots & \vdots \\ T_{n1} & T_{n2} & \cdots & T_{nn} \end{pmatrix}.$$

Each $T_{k\ell}$, or more precisely $T_{k\ell}(x)$, shows the $m \times m$ transition coefficient matrix among the agent-types, more precisely, the coefficient $t_{(k\ell)\,ij}(x)$ means the transition coefficient from the type j in the region ℓ to the type i in the region k. (As we have explained above, agents not only move among regions, but also may expand or shrink while moving, e.g., children join parents' company.) Almost needless to say, when we have every $T_{k\ell} \gg 0$, then our operator satisfies $T \gg 0$ without any iteration: it is primitive. Actually, however, we know there can be many zeros in $T_{k\ell}$, especially in the case where $k \neq \ell$ and when two regions are geographically far away.

Now a helpful and powerful proposition is that when a nonnegative square matrix is indecomposable(or irreducible) and has at least one positive element in the diagonal, it is primitive. (Concerning indecomposability (or irreducibility), the reader is referred to Bauer [1], [2] and Nikaido [10], [11]). This proposition is obvious thanks to a characterization of indecomposable matrices due to Frobenius [3], and explicitly stated in Nikaido (Theorem 8.2, p.117 of [9]).

Therefore, first we make

Assumption P1. Each T_{kk} is indecomposable for $k = 1, 2, \ldots, n$.

Assumption P2. There is at least one positive entry in the diagonal of the whole transition coefficient matrix $T(x)$.

The assumption P1 amounts to saying that in each region, we cannot divide the types of people into two groups between which no flow of people is observed in either one of the two directions even when there exist a positive number of people in each type. The assumption P2 is much weaker than requiring that each T_{kk} has at least one positive element. When we assume that each T_{kk} has at least one positive element, it becomes primitive because of the assumption P1. So, if $T_{k\ell} = 0$ for $k \neq \ell$, and if $t_{(kk)\,ij}(x)$ depends only upon the distribution of agents within region k for all k, then within each region we have strong ergodicity. This result may not be so interesting simply because there is no migration among different regions.

Then, as was shown in Frobenius [3], square matrices of the following sign pattern are indecomposable.

$$T \equiv \begin{pmatrix} 0 & 0 & 0 & 0 & 1 \\ 1 & 0 & 0 & 0 & 0 \\ 0 & 1 & 0 & 0 & 0 \\ \vdots & \vdots & \cdots & \ddots & \vdots \\ 0 & 0 & \cdots & 1 & 0 \end{pmatrix}. \tag{1}$$

We can establish

Theorem 4.1. *Suppose the regions are suitably reordered so that there is at least one positive element in every* $T_{k\ell}$*, where* $k = \ell + 1$*, for* $\ell = 1, 2, \ldots, n$.[e] *Given the assumptions P1 and P2, the matrix* $T(x)$ *is primitive.*

Proof. Since the assumption P2 is postulated, all we have to show is the indecomposability of $T(x)$, or simply T. Let us prove this by reduction ad absurdum, and suppose to the contrary: T is decomposable. Then by some permutation of the columns as well as rows, we should be able to transform T to a form, something like

$$\begin{pmatrix} * & * & * & * & * \\ * & * & * & * & * \\ 0 & 0 & * & * & * \\ 0 & 0 & * & * & * \\ 0 & 0 & * & * & * \end{pmatrix}. \tag{2}$$

Namely, the elements of a south-west corner are all zero. By this zero pattern, the whole region-type combinations are divided into two groups. We know that each region cannot be set apart by this division, because every T_{kk} is indecomposable by the assumption P1. Thus, the division simply

[e]When $\ell = n$, we set $k = 1$.

classifies the regions into two groups. If this division creates the sign pattern (2) above, it implies we can also transform a matrix of the sign pattern (1) to the one with the pattern (2), which is decomposable, leading to a contradiction. □

The requirement in Theorem 4.1 is not so demanding, allowing for the existence of many zeros. What is required is that after an appropriate reordering the regions, there is a circular flow of people from region i to region $(i+1)$.[f] It is unnecessary for all the types in region i migrate. At least one type in region i is assumed to move to region $(i+1)$.

It is important to note that all the regional matrices T_{ii} should be square matrices, but can have different sizes. In other words, there can be region-specific types, or some types cannot exist in certain regions. The same proof in the above applies. (See Nikaido, Theorem 8.2, p.117 of [9]).

5. Traffic Distribution

It may be interesting to notice that our model above can be used to produce a model of traffic distribution in a city or a country, and to prove the existence of a unique equilibrium and its dynamic stability, i.e., strong ergodicity. Let us consider a network of roads in a city area, and a finite number of modes to move on roads: walking, riding on a bicycle, in a private car, in a public bus, or by underground etc.

Now we make the following definition.

Definition 5.1. A path is a connected route from one node (node of departure) to another (node of destination) which is normally used by people when they commute to work or go shopping.

This path serves as a 'region' in the above model of economy. Then, what work as 'types' are any combinations of modes of traffic available in respective paths, and which combination completes the journey along the path. So, the next definition is:

Definition 5.2. A method of transportation (simply method) is a particular combination of modes available in respective paths, and which combination brings passengers from the departure node to the destination node of a particular path.

[f]When $i = n$, the receiving region is 1, not $(n+1)$.

In some paths, a set of specific methods of transportation may not be available, thus making the sizes of the matrices, T_{kk}, distinct from each other. By the remark at the end of the previous section, however, the proposition on stability remains valid, so long as the primitivity of $T(x)$ is guaranteed.

The adjustment process goes as follows. In the initial period, the distribution of people among various paths as well as methods is given. People, after observing the current distribution of traffic, change their path as well as method of transportation. One path contains more than one road segment (like a road along one block), hence we need to sum up the amounts of traffic of all the paths when we wish to know the traffic of a particular road segment. In our model, some people may flow in, for example, by judging this city is less crowded than others, while others may move out, thinking the city is too much congested, thus rendering the total population expanding or shrinking.

Assumption P1 is now to be interpreted that in each path, we cannot divide the methods of transportation into two groups between which no flow of people is observed in either one of the two directions even when there exist a positive number of people in each method. The meaning of Assumption P2 is that there is at least one person (or one group of persons) who sticks to a particular path and a particular method available within the path. The supposition in Theorem 4.1 requires that we should be able to reorder the paths so that there is a positive fraction of people who shift from path i to $(i+1)$[g].

When applying to a real problem, we have to limit the numbers of paths and of modes. To estimate a particular transition coefficient, we may employ the following function:

$$\begin{aligned} t_{(k\ell)\,ij}(x) &= a_{(k\ell)\,ij} + b_{(k\ell)\,ij} \cdot \frac{x_{(k)\,i}}{\sum_{p=1}^{m} x_{(k)\,p}} + c_{(k\ell)\,ij} \cdot \frac{x_{(\ell)\,j}}{\sum_{p=1}^{m} x_{(\ell)\,p}} \\ &\quad + d_{(k\ell)\,ij} \cdot \frac{x_{k\,(i)}}{\sum_{q=1}^{n} x_{q\,(i)}} + e_{(k\ell)\,ij} \cdot \frac{x_{\ell\,(j)}}{\sum_{q=1}^{n} x_{q\,(j)}} . \end{aligned}$$

Here, the symbol $x_{(k)\,i}$ stands for the number of people who is in region k and of type i, with the bracket (k) meaning the index inside is fixed while summing up: other symbols are used in a similar manner. Certainly

[g]When $i = n$, the receiving path is 1, not $(n+1)$.

one may also add some quadratic terms such as

$$\beta_{(k\ell)\,ij} \cdot \left(\frac{x_{(k)\,i}}{\sum_{p=1}^{m} x_{(k)\,p}} \right)^2 .$$

6. Remarks

In this final section, we give several brief remarks.

(1) In our model of moving equilibria, prices are relegated to a supplementary position. In this sense, our adjustment process may be called of a Marshallian type.

(2) Even in the state of equilibrium, people are likely to continue to migrate among the regions and the types. Net migrations are, however, absent. In Sonnenschein [17] and Mossay [13], there is no flow of people or firms in an equilibrium.

(3) In our model, the accumulation (or decumulation) of (indivisible) assets in the possession of individuals is allowed for as migrations among types.

(4) When the transition coefficient matrix depends on the current prices, or when people change their taste, the process becomes inhomogeneous, and we may have only weak ergodicity in place of strong one. See Fujimoto and Krause [6].

(5) In the literature of traffic distribution, the nature of equilibrium and how to find out equilibrium states have been discussed. See, e.g., Wardrop [18] and Patriksson [15]. The stability of an equilibrium through the adjustment by individual behaviour has not been dealt with.

References

1. Brauer, A.: On the Theorems of Perron and Frobenius on nonnegative matrices, in *Studies in Mathematical Analysis and Related Topics*, Gilberg, Solomon, and others(eds.), Stanford University Press. 1962, 48-55.
2. Brauer, A.: On the Characteristic Roots of Non-negative Matrices, in *Recent Advances in Matrix Theory*, H.Schneider (ed.), University of Wisconsin Press: Madison, 1964, 3-38.
3. Frobenius, F. G.: Über Matrizen aus nicht negativen Elementen, *Sitzungsberichte Kgl. Preussischen Akademie der Wissenschaften* (1912), 456-477.
4. Fujimoto, T., Krause, U.: Strong Ergodicity for Strictly Increasing Nonlinear Operators. *Linear Algebra and Its Applications* 71(1984), 101-112.
5. Fujimoto, T., Krause, U.: Stable Inhomogeneous Iterations of Nonlinear Positive Operators on Banach Spaces. *SIAM Journal on Mathematical Analysis* 25(1994), 1195-1202.

6. Fujimoto, T., Krause, U.: Primitivity and Irreducibility of Nonlinear Positive Mappings in a Finite-Dimensional Space. *Fukuoka University Review of Economics* 51(2006), 1-8.
7. Kohlberg, E.: The Perron-Frobenius Theorem without Additivity. *Journal of Mathematical Economics* 10(1982), 299-303.
8. Krause, U.: *Positive Discrete Dynamical Systems. Theory, Models, and Applications.* (Manuscript, 2004).
9. Nikaido, H.: *Linear Algebra for Economics,* (in Japanese), Tokyo: Baifuhkan, 1961.
10. Nikaido, H.: *Convex Structures and Economic Theory.* New York: Academic Press, 1968.
11. Nikaido, H.: *Introduction to Sets and Mappings in Modern Economics.* New York: Academic Press, 1970.
12. Mossay, P.: Global Trade and Local Migration, mimeo, PhD Dissertation, Chapter 2, University of Louvain, 2001.
13. Mossay, P.: Stability of Spatial Adjustments across Local Exchange Economies, *Regional Science and Urban Economics*, 36 (2006), 431-449.
14. Mukherji, A.: The Stability of a Competitive Economy: A Reconsideration, *International Journal of Economic Theory*, 4 (2008), 317-336.
15. Patriksson, M.: *The Traffic Assignment Problem: Models and Methods.* Utrecht: VSP, 1994.
16. Scarf, H.: Some Examples of Global Instability of Competitive Equilibrium, *International Economic Review* 1(1960), 157-172.
17. Sonnenschein, H.: Price Dynamics Based on the Adjustment of Firms', *American Economic Review* 72(1982), 1088-1096.
18. Wardrop, J.G.: Some Theoretical Aspects of Road Traffic Research, *Engineers Part II* (1952), 325-378.

Nonlinear Prediction in Complex Systems Using the Ruelle-Takens Embedding

R. Gonçalves

Faculdade de Engenharia, Universidade do Porto
4200-465, Porto, Portugal
E-mail: rjasg@fe.up.pt

A. A. Pinto

Departamento de Matemática, Universidade do Minho
4710 - 057, Braga, Portugal
E-mail: aapinto@math.uminho.pt

We exploit ideas of nonlinear dynamics in a complex non-deterministic dynamical setting. Our object of study is the observed riverflow time series of the Portuguese Paiva river whose water is used for public supply. The Ruelle-Takens delay embedding of the daily riverflow time series revealed an intermittent dynamical behavior due to precipitation occurrence. The laminar phase occurs in the absence of rainfall. The nearest neighbor method of prediction revealed good predictability in the laminar regime, but we warn that this method is misleading in the presence of rain. We present some new insights between the quality of the prediction in the laminar regime, the embedding dimension, and the number of nearest neighbors considered.

Keywords: Nonlinear dynamics; Ruelle-Takens embedding; BHP distribution; river flow prediction.

1. Introduction

A direct link between the real world and deterministic dynamical systems theory is the analysis of real systems time series in terms of nonlinear dynamics with noise. Advances have been made to exploit ideas of dynamical systems theory in cases where the system is not necessarily deterministic but it displays a structure not captured by classical stochastic methods. The application of dynamical systems methods to data found a firm ground on the works of Ruelle-Takens[19] and Sauer, Yorke and Casdagli ([17]). The motivation for applying methods of deterministic dynamics in riverflow time series lies in the natural tendency of river systems to present recurrent behavior (see[3,7,13,15,18]). We do a dynamical analysis starting with a Takens delay coordinates reconstruction of the daily flow series which indicates the intermittent character of Paiva river dynamical system. This intermittent dynamical behavior is characterized by a laminar and an irregular phase. The laminar phase occurs in the absence of rainfall and the irregular phase occurs under the action of rain. Hence, the forcing of the dynamical system is not of a deterministic type because rainfall is stochastic.

We present some new insights between the quality of the prediction, the embedding dimension, and the number of nearest neighbors considered (see[14]). We use a neighborhood of the current runoff whose radius is essentially proportional to the value of the current runoff, and we study the influence of the tuning of the constant of proportionality in the quality of the runoff predictors. The nearest neighbor method of prediction reveals good predictability in the laminar regime. We compute the mean of the relative predicted decays, for different regimes and embedding dimensions, that characterize the bias of the runoff predictors. The nearest neighbor method of prediction does not exhibit good predictability in the irregular phase indicating the stochastic predominance over the deterministic in the irregular phase. Since most of the data is laminar, we warn that the use of nonlinear deterministic prediction methods can be just misleading when both dynamical regimes are considered.

In the laminar regime, the nearest neighbor method of prediction indicates that the dynamics can be approximated by a one to three dimensional dynamical system. The prediction results revealed that it is essential to know the current runoff to predict future values. In,[4] we use these results to reconstruct an approximation of the one-dimensional dynamics of the runoff using different prediction estimators. In,[5] we discovered that the normalized empirical distributions of the relative first difference, for some runoff regimes, exhibit a good fit to the distribution BHP.[1,2,6] Furthermore, the empirical distribution of the relative first difference computed using the nearest neighbors predictor is close to the BHP distribution. Hence, the nearest neighbors predictor captures the BHP distribution as the essential randomness of the data. Like that, we have described the stochastic dynamics of the laminar regime.

2. Data analysis

The most relevant data for this work consist of the time series of mean daily runoff of the Paiva river, measured at Fragas da Torre in the North of Portugal.[a]

Table 1. Principal descriptive statistics for the river Paiva runoff data set.

Statistic	Mean	Median	Skewness	Kurtosis	Maximum	Minimum
Value	0.25	20.73	5.66	45.98	920.0	0.06

Note: [a] Values in (m^3/s) where applicable.

The sample period runs from 1st of October of 1946 to 30th of September of 1999 for a total of 19358 observations. The Paiva river has a small basin of about $700Km^2$ with both smooth and rocky bed, and reacting rapidly to rainfall. The total drought does not occur during the observation period. Hence, the numerical problems[18] associated to the predominance of zero values in data are absent here.

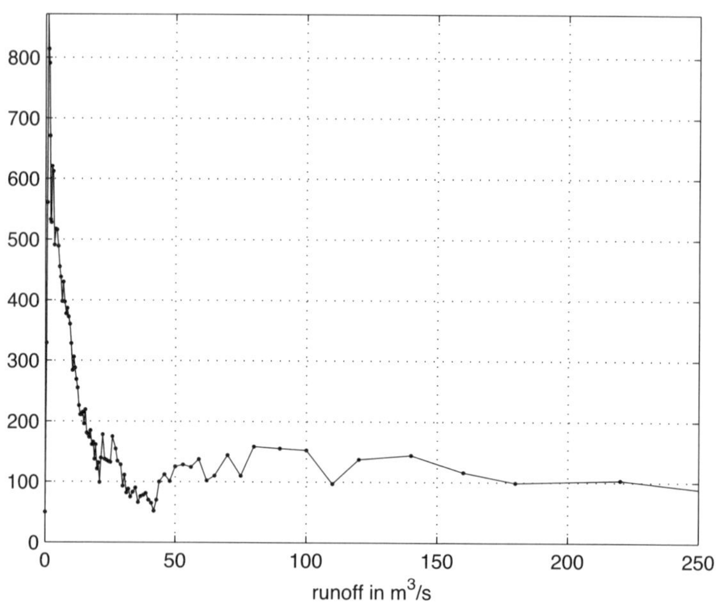

Fig. 1. Histogram of the mean daily runoff series of Paiva river.

[a]The data is available for download in the *Instituto Nacional da Água* webpage`http://www.inag.pt`

The daily river flow descriptive statistics, (see Tab. 1 and Fig. 1) shows the strong asymmetry of the data. The Paiva basin does not have regulators such as dams or glaciers.

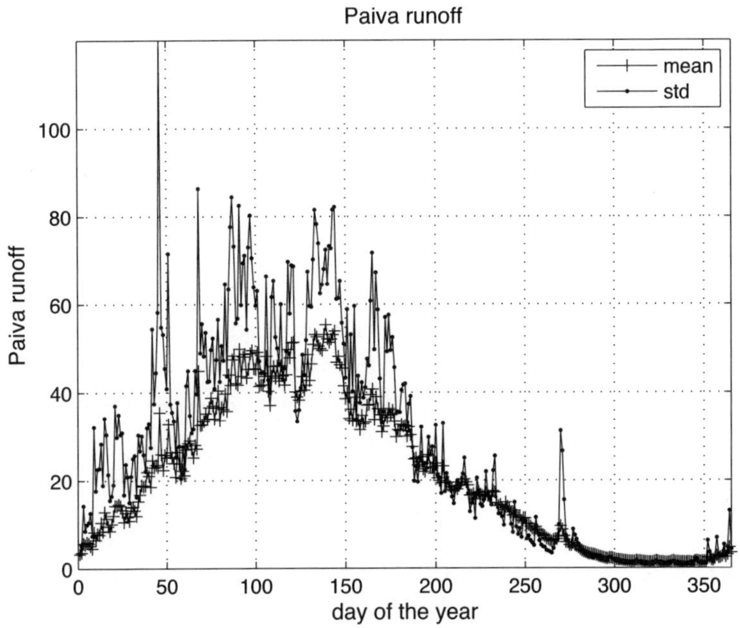

Fig. 2. Mean and standard deviation for each day of the year.

The average and the standard deviation for each day of the year (see Fig. 2) [b] show a strong statistical irregularity for each day of the year that increases with the runoff value.

3. Intermittent dynamics of Paiva

The dynamic characterization includes invariant estimation and in this direction we do a correlation-integral analysis for all the data and then we consider only the runoffs less than $20m^3/s$ which represents about 75% of the data corresponding mainly to the laminar phase, i. e., periods without

[b]The 29th of February of each year were deleted.

rain. The Correlation Integral of a system is by definition the probability of finding a fraction of points observed of a set of data is given by Eq. (1).

$$C_N^{(m)}(\varepsilon) = \frac{2}{N(N-1)} \sum_{1 \le i < j \le N} \Theta\left(\varepsilon - \left\|X^i - X^j\right\|\right) \tag{1}$$

where $X^t = (X_t, X_{t+1}, \ldots, X_{t+m-1})$ is a reconstructed vector which elements are values of the time series, $\{X_t\}_{t=1}^N$, N is the number of data points of the series, Θ the Heaviside function, ε the neighborhood radius and m the embedding dimension of the reconstructed phase space. The sample CI is a statistic used in the correlation dimension estimation, it was proposed by.[8] The sample CI is a statistic used in the the observed fraction of reconstruction vectors (RV) at a distance smaller than ε. The sum, (1), is computed for a set of distances, $\varepsilon_1, \ldots, \varepsilon_n$ evenly spaced on a logarithmic scale. A *scaling range* is said to exists if for such a range of values the sample correlation integral behaves like a power law. In practice, there is a cut-off on the radius size due to data size restrictions. Defining $d(N, \varepsilon) = \partial \ln C_N^{(m)}(\varepsilon)/\partial \ln \varepsilon$, we have that

$$D_C = \lim_{\varepsilon \to 0^+} \lim_{N \to \infty} d(N, \varepsilon) \tag{2}$$

Hence, $d(N, \varepsilon)$ is the slope of the CI curve for a certain range, and D_C is then the estimate of the correlation dimension. In Fig. 3, we present the correlation integral slopes. We can distinguish three different behaviors in the correlation-integral curve for different ranges of the radius, ε. For the runoff values larger than $30m^3/s$ no scaling range exists. For the runoffs in the interval $[5 - 30m^3/s]$ there is a scaling range which point towards a one-dimensional attractor. This dimension is not fractal and indicates that the behavior of riverflow for that range is close to that of a curve. This show us the existence in the reconstructed phase-space of a one-dimensional manifold to which all the laminar phase orbits are close, i. e. the orbits are mainly contained in a small neighborhood of a one-dimensional curve.

4. Nonlinear prediction

Several authors used nonlinear prediction methods for river flow data locally in the phase space,[10,11,13,16] and[7] among others. In this work, we use a different version of the nearest neighbors method proposed by[12] to predict the next day runoff. In our approach the predicted value is the average of the phase-space images of the neighbors defined as below. Other authors,[13]

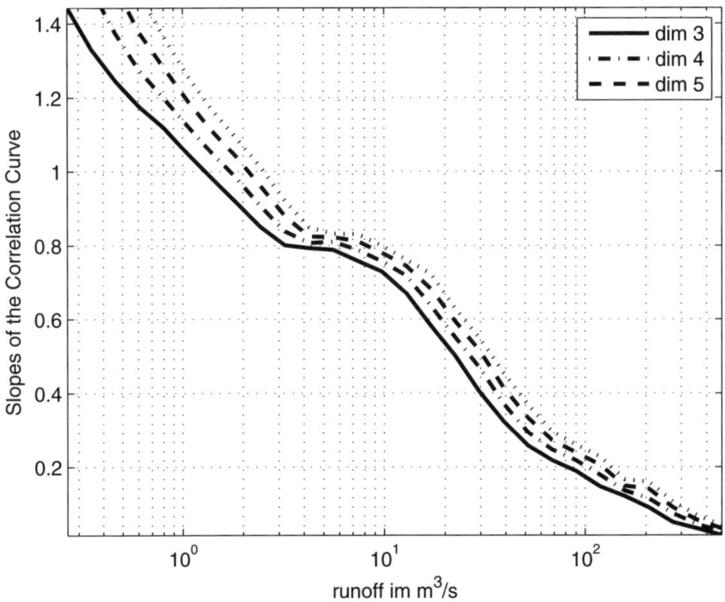

Fig. 3. Slopes of the sample correlation integral curve of the Paiva river data and for several embedding dimensions.

reported that predictors based on the phase space average give better results than other local linear functions. Taking into account the findings of the former section, we started by considering small embeddings and a time delay of one day. Since our goal is to predict the runoff value during the laminar regime (absence of rain) we will filter appropriately the reconstruction vectors. Hence given the dimension, m, of the Ruelle-Takens embedding we consider only the reconstruction vectors $\mathbf{X}_t = (X_t, X_{t+1}, \cdots, X_{t+m-1})$ satisfying the following *δ-relative non-increasing rule*, see Eq. (3)

$$X_{t+i-1} \leq X_{t+i}(1+\delta),\ 1 \leq i \leq m \tag{3}$$

where δ is a fixed positive value. Therefore the total numbers $T_m(X_t)$ of filtered reconstruction vectors depends on the runoff value of X_t and on the embedded dimension m considered. In this line of reasoning, it only makes sense to consider reconstruction vectors whose future points to a non-increasing runoff. As a rule of thumb we choose a 15% of increasing tolerance for runoffs lower than $4m^3/s$ and 5% for runoffs above that threshold. This distinction was based on the observed effects of measurement error.

Table 2. Percentages P_i of the neighbors in each neighbors set RV_1 - RV_8.

RV_i	1	2	3	4	5	6	7	8
Percentage (%)	0.25	0.5	1.0	5.0	10.0	12.5	15.0	20

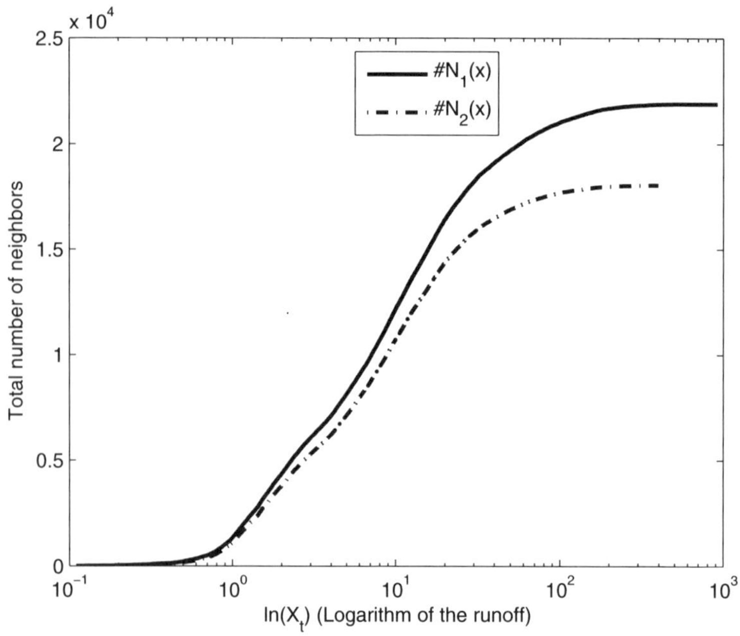

Fig. 4. T_m vs the Logarithm of the runoff value for embedding dimension 1, upper curve, and embedding dimension 3, lower curve.

In this analysis, instead of using all the $T_{i,m}(X_t)$ neighbors within a fixed radius, we use a fixed proportion P_i, , see Tab. 2, of the closest neighbors of $\mathbf{X}_t$ such that, $\#N_{i,m}(X_t) = P_i * \#T_{i,m}(X_t)$. In Figure 4, we present the curves N_1 and N_3. The resulting curves show that there is a plateau for values in the range of $3.2m^3/s$ to $3.6m^3/s$ that occurs for various embedding dimensions.

The Fig. 5 gives us the variation of the relative distance of the predicted value, using the close neighbors method with respect to the RV's, runoff

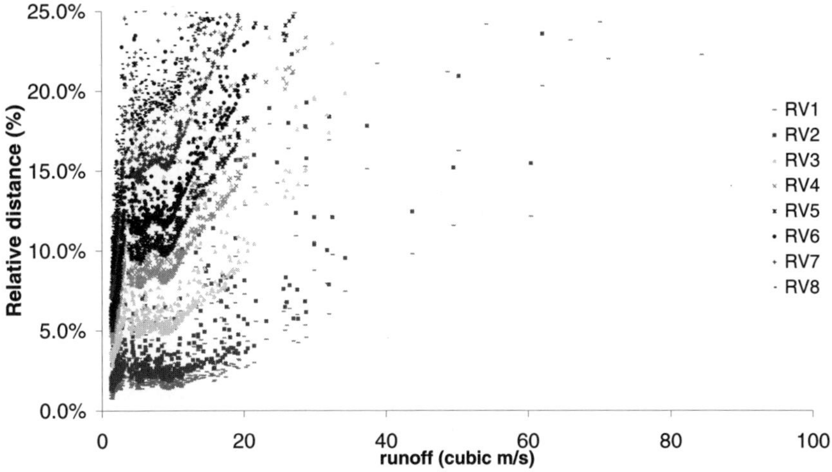

Fig. 5. Relative distance (%) of RV's as function of the last coordinate for different neighborhoods and for dimension 3.

regimes and embedding dimensions. The relative distance of neighbors in the phase space increases with the runoff regime and also with the embedding dimension. Moreover, the mean for several regimes and number of neighbors (RV_i) considered approximately increases linearly with m.

5. Conclusions

A dynamical analysis of the Paiva river data was performed using the Ruelle-Takens method of dynamical reconstruction. We concluded that the Paiva river is an intermittent system. The laminar phase takes place in the absence of rainfall and the irregular phase occurred under the action of rain. The nearest neighbor method of prediction revealed good predictability in the laminar regime. However, since 75% of data is laminar, the use of nonlinear deterministic prediction methods can be just misleading when both dynamical regimes are considered. We studied the dependence of the nearest neighbors runoff predictor on the embedding dimension and on the relative average distance of the nearest neighbors with respect to the runoff value. The prediction results revealed that it is essential to know the current runoff to predict future values. We noticed small improvements in prediction when the former two runoffs are used.

References

1. S. T. Bramwell, J. Y. Fortin, P. C. W. Holdsworth, S. Peysson, J. F. Pinton, B. Portelli & M. Sellitto *Phys. Rev* **E 63** 041106 (2001).
2. K. Dahlstedt & H. J. Jensen *Physica* **A 348** 596-610 (2005).
3. R. Gonçalves, A. A. Pinto & F. Calheiros in *Differential equations, Chaos and Variational problems*, (Birkhäuser, Staicu, Vasile (Ed.), 231, 2007).
4. R. Gonçalves, A. A. Pinto & N. Stollenwerk *submitted.*
5. R. Gonçalves, H. Ferreira, A. A. Pinto & N. Stollenwerk, *Journal of Difference Equations and Applications*, 1-10, (2009). (To be published)
6. R. Gonçalves & A. A. Pinto *Journal of Difference Equations and Applications*, 1-10, (2009). (To be published)
7. S. Islam & B. Sivakumar *Advances in Water Resources*, **25**, 179, (2002).
8. P. Grassberger & I. Procaccia, *Physica A*, **9**, 189, (1983).
9. A. W. Jayawardena & A. Gurung, *Journal of Hydrology*, **228**, 242, (2000).
10. A. W. Jayawardena & F. Lai *Journal of Hydrology*, **153**, 23, (1994).
11. A. Jayawardena, W. Li & P. Xu *Journal of Hydrology*, **258**, 40, (2002).
12. H. Kantz & T. Schreiber, *Nonlinear Time Series Analysis*, (Cambridge University Press, 1997).
13. Q. Liu, S. Islam, I. Rodriguez-Iturbe & Y. Le *Advances in Water Resources*, **210**, 463, (1998).
14. E. Lorenz, *Journal of Atmospheric Sciences*, **20**, 636, (1969).
15. L. Porporato & L. Ridolfi, *Water Resources Research*, 1353, **33**, (1997).
16. A. Porporato & L. Ridolfi *Journal of Hydrology*, **248**, 109, (2001).
17. T. Sauer, J. Yorke & M. Casdagli, *Journal of Statistical Physics*, **65**, 579 (1991).
18. B. Sivakumar *Journal of Hydrology*, **227** (2000).
19. F. Takens, in *Lecture Notes in Mathematics*, **898**, (Springer, Rand D. A., Young L. Editors, 1980).

Reid Roundabout Theorems for Time Scale Symplectic Systems

R. Hilscher*

Department of Mathematics and Statistics,
Faculty of Science, Masaryk University,
Kotlářská 2,
CZ-61137 Brno, Czech Republic
E-mail: hilscher@math.muni.cz

V. Zeidan

Department of Mathematics, Michigan State University,
East Lansing, MI 48824-1027, USA,
E-mail: zeidan@math.msu.edu

In this paper we survey Reid roundabout theorems for time scale symplectic systems $(\mathcal{S})$. These theorems list equivalent conditions for the positivity and nonnegativity of the quadratic functional $\mathcal{F}$ associated with $(\mathcal{S})$. The Reid roundabout theorems in this paper do not impose any normality assumption. We also show that Jacobi systems for nonlinear time scale control problems naturally lead to time scale symplectic systems, and that such a system consists of the Hamiltonian equations corresponding to the weak maximum principle for the quadratic functional $\mathcal{F}$.

Keywords: Time scale; Riccati equation; Quadratic functional; Positivity; Nonnegativity; Normality; Controllability; Conjoined basis.

1. Introduction

Let $\mathbb{T}$ be a time scale, i.e., a nonempty closed subset of $\mathbb{R}$, and denote the time scale interval by $[a,b]_{\mathbb{T}} := [a,b] \cap \mathbb{T}$. In this paper we discuss the positivity and nonnegativity of the time scale quadratic functional

$$\mathcal{F}(x,u) := \begin{pmatrix} x(a) \\ x(b) \end{pmatrix}^T \Gamma \begin{pmatrix} x(a) \\ x(b) \end{pmatrix} + \mathcal{F}_0(x,u), \qquad (1)$$

*Corresponding author.

where

$$\mathcal{F}_0(x,u) := \int_a^b \{x^T \mathcal{C}^T (I+\mu\mathcal{A})\, x + 2\mu\, x^T \mathcal{C}^T \mathcal{B} u + u^T (I+\mu\mathcal{D})^T \mathcal{B} u\}(t)\, \Delta t,$$

subject to admissible pairs (x,u), i.e., $x \in \mathrm{C}^1_{\mathrm{prd}}$ (piecewise rd-continuously delta-differentiable functions) and $u \in \mathrm{C}_{\mathrm{prd}}$ (piecewise rd-continuous functions) such that

$$x^\Delta(t) = \mathcal{A}(t)\, x(t) + \mathcal{B}(t)\, u(t), \quad \text{for all } t \in [a,\rho(b)]_{\mathbb{T}}$$

and satisfying the boundary conditions

$$\mathcal{M} \begin{pmatrix} x(a) \\ x(b) \end{pmatrix} = 0. \tag{2}$$

Such functionals are closely related to the associated time scale symplectic system

$$x^\Delta = \mathcal{A}(t)\, x + \mathcal{B}(t)\, u, \quad u^\Delta = \mathcal{C}(t)\, x + \mathcal{D}(t)\, u. \tag{$\mathcal{S}$}$$

Here the coefficients $\mathcal{A}, \mathcal{B}, \mathcal{C}, \mathcal{D} \in \mathrm{C}_{\mathrm{prd}}[a,\rho(b)]_{\mathbb{T}}$ are real $n \times n$ matrix functions such that the $2n \times 2n$ matrix $\mathcal{S} := \left(\begin{smallmatrix} \mathcal{A} & \mathcal{B} \\ \mathcal{C} & \mathcal{D} \end{smallmatrix}\right)$ satisfies the identity

$$\mathcal{S}^T(t)\, \mathcal{J} + \mathcal{J}\mathcal{S}(t) + \mu(t)\, \mathcal{S}^T(t)\, \mathcal{J}\mathcal{S}(t) = 0 \quad \text{on } [a,\rho(b)]_{\mathbb{T}}, \tag{3}$$

$\mathcal{M}$ is a real $2n \times 2n$ projection, Γ is a real $2n \times 2n$ matrix such that $\Gamma = (I - \mathcal{M})\, \Gamma\, (I - \mathcal{M})$, and $\mathcal{J} := \left(\begin{smallmatrix} 0 & I \\ -I & 0 \end{smallmatrix}\right)$ a $2n \times 2n$ skew-symmetric matrix.

The main goal of this paper is to present Reid roundabout theorems for the system $(\mathcal{S})$, as well as to relate these systems to nonlinear control problems on time scales. Reid roundabout theorems list a number of conditions which are equivalent to the positivity and nonnegativity of the quadratic functional $\mathcal{F}$.

In the continuous time case, the system $(\mathcal{S})$ reduces to the classical linear Hamiltonian system

$$x' = A(t)\, x + B(t)\, u, \quad u' = C(t)\, x - A^T(t)\, u. \tag{H$_c$}$$

It was W. T. Reid in [31, Theorem VII.5.1] who provided a complete characterization of the positivity of the quadratic functional corresponding to system (H_c). See also related more recent results by the second author in [34–36], and by W. Kratz in [29,30]. Such a result was later called, by Reid's student C. D. Ahlbrandt in [1–3], a Reid roundabout theorem. Discrete Reid roundabout theorems were proven under various degree of generality by several authors, see the references in [11, pg. 130] and [5,12,13,19,21,23] and the recent thesis [32]. For the time scale setting and under a certain normality assumption on the system $(\mathcal{S})$, a Reid roundabout theorem was established in [11, Theorem 1], see also [8, Theorem 10.52], for the positivity of $\mathcal{F}_0$ with zero endpoints, including conditions on focal points of the principal solution of $(\mathcal{S})$, a certain conjoined basis (X, U) of $(\mathcal{S})$ with $X(t)$ invertible on $[a, b]_{\mathbb{T}}$, the explicit Riccati matrix equation involving the Riccati operator

$$R[Q](t) := Q^{\Delta} - [\mathcal{C}(t) + \mathcal{D}(t)\, Q] + Q^{\sigma}[\mathcal{A}(t) + \mathcal{B}(t)\, Q],$$

and generalized zeros of vector solutions of $(\mathcal{S})$. For a subclass of time scale symplectic systems consisting of linear Hamiltonian systems of the form

$$\eta^{\Delta} = A(t)\, \eta^{\sigma} + B(t)\, q, \quad q^{\Delta} = C(t)\, \eta^{\sigma} - A^{T}(t)\, q, \qquad (\mathrm{H}^{\sigma})$$

where $I - \mu(t)\, A(t)$ is invertible and $B(t)$ and $C(t)$ are symmetric on $[a, \rho(b)]_{\mathbb{T}}$, and under a normality assumption, a Riccati *inequality* for the positivity of $\mathcal{F}_0$ with zero endpoints was included in [16, Theorem 3.1]. Moreover, in the special case of the time scale calculus of variations (which is automatically normal) and under the corresponding time scale strengthened Legendre condition, additional conditions in terms of the nonexistence of conjugate points are derived in [24, Theorems 5.1 and 6.1] for the positivity and nonnegativity of $\mathcal{F}$ with zero right endpoint, and in terms of the coercivity of $\mathcal{F}$ with general endpoints in [25, Theorem 4.1].

Recently in [22], the authors initiated the study of time scale symplectic systems without any normality assumption and derived characterizations of the positivity and nonnegativity of $\mathcal{F}$ in this general setting in terms of a natural conjoined basis (X_a, U_a) of $(\mathcal{S})$. The focus of this paper is to present Reid roundabout theorems for both the positivity and nonnegativity of $\mathcal{F}$ *without normality,* including the results on the natural conjoined basis from [22] and several other conditions obtained in separate publications, such as the Riccati inequality from [26], implicit Riccati equations from [27], and perturbed quadratic functionals from [14,15].

The motivation for the study of time scale symplectic systems originated in the unification of the continuous time linear Hamiltonian systems (H_c) and discrete symplectic systems. However, it is known in these two special cases that these systems are Jacobi systems for nonlinear calculus of variations and control problems, see e.g. [7,9,17,18,33,37,39]. This fact is also known for the time scale calculus of variations problems, see [4,20]. On the other hand, this question remained open for the time scale control setting. In this paper we fill this gap and show that time scale symplectic systems $(\mathcal{S})$ play the same role for general control problems on time scales. In particular, by using a recently obtained time scale weak maximum principle [28], we prove that nonlinear time scale control problems lead to time scale linear Hamiltonian systems (in fact, two kinds of linear Hamiltonian systems, one with the shift in η as in (H^σ) and one with the shift in q, depending on whether the original control problem has or does not have the shift in x), which naturally possess a symplectic structure and hence, can be embedded into time scale symplectic systems. This fact highlights the theory of time scale symplectic systems as an ultimate field for second order optimality conditions for such variational problems. Furthermore, we also prove that the system $(\mathcal{S})$ is indeed the Euler-Lagrange system for the functional $\mathcal{F}$.

The paper is divided as follows. In Section 2 we present basic notions related to time scale symplectic systems. Reid roundabout theorems for positive and nonnegative definite quadratic functionals with separated endpoints are given in Section 3. In this section we also discuss Reid roundabout theorems for functionals with jointly varying endpoints, although we do not state such results explicitly. Finally, Section 4 is devoted to the connection of time scale symplectic systems with control problems on time scales.

2. Time scale symplectic systems

We refer to [6] for the elementary topics of the time scale calculus, and to [8] for the basic concepts of the time scale symplectic systems. Alternatively, the reader may consult [22], since it is the main reference for abnormal time scale symplectic systems. In particular, $\sigma(t)$ and $\rho(t)$ denote the forward and backward jump operators, $\mu(t) := \sigma(t) - t$ is the graininess, $f^\Delta(t)$ is the time scale delta-derivative, and $\int_a^b f(t)\,\Delta t$ is the time scale delta-integral. We also write $f^\sigma(t)$ for $f\big(\sigma(t)\big)$.

The property defined by (3) is translated to be the following equivalent conditions in terms of the coefficients

$$\left.\begin{aligned} &\mathcal{C}^T(I+\mu\mathcal{A}) \text{ and } \mathcal{B}^T(I+\mu\mathcal{D}) \text{ are symmetric, and}\\ &\mathcal{A}^T+\mathcal{D}+\mu\,(\mathcal{A}^T\mathcal{D}-\mathcal{C}^T\mathcal{B})=0. \end{aligned}\right\} \tag{4}$$

It follows that the matrix $I + \mu(t)\,\mathcal{S}(t)$ is symplectic, hence invertible, on $[a, \rho(b)]_{\mathbb{T}}$, so that solutions of $(\mathcal{S})$ are uniquely determined by their initial values at any point $t_0 \in [a, b]_{\mathbb{T}}$.

Expanding the delta-derivatives in $(\mathcal{S})$ with the formula $\mu(t)\,f^\Delta(t) =$

$f^\sigma(t) - f(t)$, we see that solutions of $(\mathcal{S})$ satisfy the identities

$$x^\sigma = (I + \mu\mathcal{A})\, x + \mu\mathcal{B}u, \quad u^\sigma = \mu\,\mathcal{C}x + (I + \mu\mathcal{D})\, u, \tag{5}$$

$$x = (I + \mu\mathcal{D}^T)\, x^\sigma - \mu\mathcal{B}^T u^\sigma, \quad u = -\mu\,\mathcal{C}^T x^\sigma + (I + \mu\mathcal{A}^T)\, u^\sigma. \tag{6}$$

System $(\mathcal{S})$ can be written in the equivalent (adjoint) form

$$x^\Delta = -\mathcal{D}^T(t)\, x^\sigma + \mathcal{B}^T(t)\, u^\sigma, \quad u^\Delta = \mathcal{C}^T(t)\, x^\sigma - \mathcal{A}^T(t)\, u^\sigma. \tag{7}$$

This can be seen from the coefficient identities (4) for the right-dense points t and from the equations in (6) for the right-scattered points t.

In this paper we study time scale quadratic functionals with general endpoints as in (2) and with separated endpoints. In the latter case we have $\Gamma = \text{diag}\{\Gamma_a, \Gamma_b\}$ and $\mathcal{M} = \text{diag}\{\mathcal{M}_a, \mathcal{M}_b\}$, where the $n \times n$ matrices Γ_a and Γ_b are symmetric, $\mathcal{M}_a$ and $\mathcal{M}_b$ are projections, and $\Gamma_a = (I - \mathcal{M}_a)\, \Gamma_a\, (I - \mathcal{M}_a)$ and $\Gamma_b = (I - \mathcal{M}_b)\, \Gamma_b\, (I - \mathcal{M}_b)$. In this case we consider the quadratic functional

$$\mathcal{F}(x, u) := x^T(a)\, \Gamma_a\, x(a) + x^T(b)\, \Gamma_b\, x(b) + \mathcal{F}_0(x, u) \tag{8}$$

over admissible pairs (x, u) with separated endpoints

$$\mathcal{M}_a\, x(a) = 0, \quad \mathcal{M}_b\, x(b) = 0. \tag{9}$$

When $\mathcal{M}_a = \mathcal{M}_b = I$ and $\Gamma_a = \Gamma_b = 0$, we say that the quadratic functional $\mathcal{F} = \mathcal{F}_0$ has *zero endpoints.*

Let us recall some necessary terminology and notation related to the time scale symplectic systems. We shall always denote the $2n \times n$ matrix solutions of $(\mathcal{S})$, typically (X, U), by capital letters. A solution (X, U) of $(\mathcal{S})$ is said to be *conjoined basis* if $X^T(t)\, U(t)$ is symmetric and $\text{rank}\,\big(X^T(t)\ U^T(t)\big) = n$ at some (and hence at any) point $t \in [a, b]_{\mathbb{T}}$. The *natural conjoined basis* of $(\mathcal{S})$, denoted by (X_a, U_a), is the conjoined basis satisfying the initial conditions $X_a(a) = I - \mathcal{M}_a$ and $U_a(a) = \Gamma_a + \mathcal{M}_a$. In the case of zero initial endpoint, i.e., when $I - \mathcal{M}_a = 0 = \Gamma_a$, the natural conjoined basis reduces to the *principal solution* $(\hat{X}, \hat{U})$ which starts with the initial values $\hat{X}(a) = 0$ and $\hat{U}(a) = I$.

Following [30] and [22], a matrix valued function $X(t)$ has *piecewise constant kernel on* $[a, b]_{\mathbb{T}}$ if there are points $\{t_k\}_{k=0}^m \subseteq [a, b]_{\mathbb{T}}$ with $a = t_0 < t_1 < \cdots < t_{m-1} < t_m = b$ such that

$$\text{Ker}\, X(t) \text{ is constant for all } t \in (t_{k-1}, t_k)_{\mathbb{T}},\ k = 1, \ldots m. \tag{10}$$

Condition (10) is void on the intervals $(t_{k-1}, t_k)_{\mathbb{T}}$ where $t_k = \sigma(t_{k-1})$. A conjoined basis (X, U) of $(\mathcal{S})$ has *no generalized focal points* in the interval

$(a,b]_{\mathbb{T}}$ if $\operatorname{Ker} X(t) \subseteq \operatorname{Ker} X(\tau)$ for all $t, \tau \in [a,b]_{\mathbb{T}}$, $\tau \le t$, and

$$P(t) := X(t)\,[X^{\sigma}(t)]^{\dagger}\,\mathcal{B}(t) \ge 0 \quad \text{for all } t \in [a, \rho(b)]_{\mathbb{T}}. \tag{11}$$

These conditions are called the *kernel condition* and the *P-condition*, respectively. In (11) and elsewhere in this paper, the dagger denotes the Moore-Penrose generalized inverse of the indicated matrix. Moreover, we shall use the following $n \times n$ matrices M and T, defined via a given conjoined basis (X, U),

$$M(t) := \big\{ [I - X^{\sigma}(X^{\sigma})^{\dagger}]\,\mathcal{B} \big\}(t), \quad T(t) := I - M^{\dagger}(t)\,M(t), \tag{12}$$

and the symmetric $n \times n$ matrix $\mathcal{P}$, defined via a symmetric matrix Q,

$$\mathcal{P}(t) := \big\{\mathcal{B} + \mu\,(\mathcal{D}^T - \mathcal{B}^T Q^{\sigma})\,\mathcal{B}\big\}(t).$$

The quadratic functional $\mathcal{F}$ in (1), resp. in (8), is *nonnegative* and we write $\mathcal{F} \ge 0$, if $\mathcal{F}(x,u) \ge 0$ for all admissible pairs (x,u) satisfying the boundary conditions (2), resp. (9). The quadratic functional $\mathcal{F}$ is *positive* and we write $\mathcal{F} > 0$, if $\mathcal{F}(x,u) > 0$ for all admissible (x,u) satisfying (2), resp. (9), and $x \not\equiv 0$ on $[a,b]_{\mathbb{T}}$. For brevity, we will simply say that $\mathcal{F} \ge 0$ or $\mathcal{F} > 0$ *over* the corresponding boundary conditions without repeating the admissibility requirement on (x,u).

3. Reid roundabout theorems

In this section we present Reid roundabout theorems for the quadratic functional $\mathcal{F}$ in (8). Our first result is a Reid roundabout theorem regarding the positive definiteness of $\mathcal{F}$.

Theorem 3.1 ($\mathcal{F} > 0$, separated endpoints). *The following conditions are equivalent.*

(i) The quadratic functional $\mathcal{F}$ in (8) is positive over (9) and $x \not\equiv 0$.

(ii) The natural conjoined basis (X_a, U_a) has no generalized focal points in $(a,b]_{\mathbb{T}}$ and satisfies

$$U_a(b)\,X_a^{\dagger}(b) + \Gamma_b > 0 \quad \text{on } \operatorname{Ker} \mathcal{M}_b \cap \operatorname{Im} X_a(b).$$

(iii) There exists a conjoined basis (X,U) of $(\mathcal{S})$ with no generalized focal points in $(a,b]_{\mathbb{T}}$ such that $X(t)$ is invertible for all $t \in [a,b]_{\mathbb{T}}$ and satisfying

$$X^T(a)\,[\,\Gamma_a X(a) - U(a)\,] > 0 \quad \text{on } \operatorname{Ker} \mathcal{M}_a X(a), \tag{13}$$

$$X^T(b)\,[\,\Gamma_b X(b) + U(b)\,] > 0 \quad \text{on } \operatorname{Ker} \mathcal{M}_b X(b). \tag{14}$$

(iv) *There exists a symmetric solution* $Q(t)$ *on* $[a,b]_{\mathbb{T}}$ *of the explicit Riccati equation*

$$R[Q](t) = 0 \quad \textit{on } [a,\rho(b)]_{\mathbb{T}},$$

such that for all $t \in [a,\rho(b)]_{\mathbb{T}}$

$$I + \mu(t)\,[\mathcal{A}(t) + \mathcal{B}(t)\,Q(t)\,] \textit{ is invertible,} \tag{15}$$

$$\{I + \mu(t)\,[\mathcal{A}(t) + \mathcal{B}(t)\,Q(t)\,]\}^{-1}\mathcal{B}(t) \geq 0, \tag{16}$$

and satisfying the initial and final endpoint inequalities

$$\Gamma_a - Q(a) > 0 \quad \textit{on } \operatorname{Ker}\mathcal{M}_a, \tag{17}$$

$$\Gamma_b + Q(b) > 0 \quad \textit{on } \operatorname{Ker}\mathcal{M}_b. \tag{18}$$

(v) *The system*

$$X^{\Delta} = \mathcal{A}(t)\,X + \mathcal{B}(t)\,U, \quad (X^{\sigma})^T\{U^{\Delta} - \mathcal{C}(t)\,X - \mathcal{D}(t)\,U\} \leq 0,$$

$t \in [a,\rho(b)]_{\mathbb{T}}$, *has a solution* (X,U) *on* $[a,b]_{\mathbb{T}}$ *such that* $X^T(t)\,U(t)$ *is symmetric and* $X(t)$ *is invertible for all* $t \in [a,b]_{\mathbb{T}}$, $P(t) = \{X(X^{\sigma})^{-1}\mathcal{B}\}(t) \geq 0$ *on* $[a,\rho(b)]_{\mathbb{T}}$, *and satisfying the endpoint inequalities* (13) *and* (14).

(vi) *The Riccati inequality*

$$R[Q](t)\,\{I + \mu(t)\,[\mathcal{A}(t) + \mathcal{B}(t)\,Q)]\,\}^{-1} \leq 0, \quad t \in [a,\rho(b)]_{\mathbb{T}},$$

has a symmetric solution $Q(t)$ *on* $[a,b]_{\mathbb{T}}$ *satisfying conditions* (15) *and* (16) *and the endpoint inequalities* (17) *and* (18).

(vii) *There exist* $\alpha > 0$ *and* $\beta > 0$ *such that*

$$\mathcal{F}(x,u) + \alpha\,\|\mathcal{M}_a\,x(a)\|^2 + \beta\,\|\mathcal{M}_b\,x(b)\|^2 > 0 \quad \textit{over} \quad x \not\equiv 0.$$

Moreover, if $X_{\mathrm{a}}^{\dagger}(\cdot)$ *is continuous on* $(a,b]_{\mathbb{T}}$, *then each of the conditions (i)–(vii) is equivalent to the following Riccati type condition.*

(viii) *The matrix* $X_{\mathrm{a}}^{\dagger}(\cdot)$ *has piecewise constant kernel on* $[a,b]_{\mathbb{T}}$ *and there exists a symmetric* $n \times n$ *matrix function* $Q(t)$ *on* $[a,b]_{\mathbb{T}}$ *such that* $Q \in \mathrm{C}^1_{\mathrm{prd}}(a,b]_{\mathbb{T}}$ *and satisfying*

- *the time scale implicit Riccati equation*

$$R[Q](t)\,X_{\mathrm{a}}(t) = 0 \quad \textit{on } (a,\rho(b)]_{\mathbb{T}}, \tag{19}$$

and the equation in (19) *holds also at* $t = a$ *if* a *is right-scattered,*

- *the initial condition*

$$Q(a) = \Gamma_a \quad \textit{if } a \textit{ is right-scattered,} \tag{20}$$

$$(I - \mathcal{M}_a) \lim_{t \to a^+} Q(t)\, X_a(t) = \Gamma_a \quad \textit{if } a \textit{ is right-dense,} \tag{21}$$

- *the final endpoint inequality*

$$Q(b) + \Gamma_b > 0 \quad \textit{on } \operatorname{Ker} \mathcal{M}_b \cap \operatorname{Im} X_a(b),$$

- *and the $\mathcal{P}$-condition*

$$\mathcal{P}(t) \geq 0 \quad \textit{on } [a, \rho(b)]_{\mathbb{T}}.$$

Remark 3.1. (i) Note that in [15, Theorem 4.2] there are several more perturbation conditions similar to the one in (vii) of Theorem 3.1. In these conditions, one may choose to keep only one out of the two boundary conditions from (9) or even combine them together in a certain way.

(ii) When the endpoints are zero and under a normality assumption on the intervals $[a, s]_{\mathbb{T}}$ with s being a dense point, the equivalence of conditions (i)–(iv) in Theorem 3.1 is known from [11, Theorem 1], see also [8, Theorem 10.52]. Furthermore, under the same normality assumption but for the special case of time scale linear Hamiltonian system (H^σ), the equivalence of conditions (i), (v), and (vi) in Theorem 3.1 is known in [16, Theorem 3.1].

(iii) When the initial endpoint is zero, i.e. when $\mathcal{M}_a = I$ and $\Gamma_a = 0$, a modified perturbation condition (in a sense that only the initial endpoint is perturbed) similar to (vii) in Theorem 3.1 is shown to be equivalent the positivity of $\mathcal{F}$ in [26, Theorem 5.1]. For both endpoints being zero this latter result is known also in [14, Result 1.3].

(iv) In the time scale calculus of variations setting and for the zero right endpoint case, another condition equivalent to the positivity of $\mathcal{F}$, that is to condition (i) of Theorem 3.1, is known in [24, Theorem 6.1]. In particular, it is the nonexistence of conjugate points in the interval $(a, b]_{\mathbb{T}}$. Moreover, the coercivity and positivity of $\mathcal{F}$ are shown to be equivalent in [25, Theorem 4.1] for the general time scale calculus of variations case.

Proof of Theorem 3.1. The equivalence "(i) $\Leftrightarrow$ (ii)" is proven in [22, Theorem 4.1] via generalized Picone identity and the results on piecewise constant kernel of X_a. The equivalence "(i) $\Leftrightarrow$ (iii) $\Leftrightarrow$ (iv)" is established in [26, Theorem 6.1] by means of a transformation of the quadratic functional $\mathcal{F}$ to a functional on the extended time scale $[a-1, b]_{\mathbb{T}}$ with zero left endpoint $x(a-1) = 0$ and by a perturbation of the initial conditions at $t = a - 1$. The equivalence "(i) $\Leftrightarrow$ (v) $\Leftrightarrow$ (vi)" is proven in [26, Theorem 7.1] by using a Sturmian comparison theorem for time scale symplectic systems. The equivalence "(i) $\Leftrightarrow$ (vii)" is shown in [15, Theorem 4.2] via a transformation of the functional $\mathcal{F}$ into a functional on the extended time scale $[a-1, b+2]_{\mathbb{T}}$ with zero endpoints $x(a-1) = 0 = x(b+2)$ and via the application of the zero endpoints result from [15, Theorem 3.5]. Finally, the equivalence "(i) $\Leftrightarrow$ (viii)" is established in [27, Theorem 4.15] via a generalized Picone type identity and results on piecewise constant kernel of X_a. □

Next we present a Reid roundabout theorem regarding the nonnegativity of $\mathcal{F}$.

Theorem 3.2 ($\mathcal{F} \geq 0$, separated endpoints). *The following conditions are equivalent.*

(i) The quadratic functional $\mathcal{F}$ in (8) is nonnegative over (9).

(ii) The natural conjoined basis (X_a, U_a) has piecewise constant kernel on $[a, b]_{\mathbb{T}}$ and satisfies the image condition

$$x(t) \in \operatorname{Im} X_a(t) \quad \textit{for all } t \in [a, b]_{\mathbb{T}} \tag{22}$$

and for all (x, u) admissible and satisfying (9), the P-condition

$$T(t)\, P_a(t)\, T(t) \geq 0 \quad \textit{for all } t \in [a, \rho(b)],$$

where the matrices $P_a(t)$ and $T(t)$ are defined in (11) and (12) through (X_a, U_a), and the final endpoint inequality

$$U_a(b)\, X_a^{\dagger}(b) + \Gamma_b \geq 0 \quad \textit{on } \operatorname{Ker} \mathcal{M}_b \cap \operatorname{Im} X_a(b).$$

(iii) There exist $\alpha > 0$ and $\beta > 0$ such that

$$\mathcal{F}(x, u) + \alpha \left\|\mathcal{M}_a\, x(a)\right\|^2 + \beta \left\|\mathcal{M}_b\, x(b)\right\|^2 \geq 0$$
$$\textit{over} \quad \mathcal{M}_b\, x(b) - \bar{Z} \mathcal{M}_a\, x(a) \in \operatorname{Im} Z,$$

where the $n \times n$ matrices Z and $\bar{Z}$ are defined by

$$Z := (\Gamma_b + \mathcal{M}_b)\, X_a(b) + (I - \mathcal{M}_b)\, U_a(b),$$
$$\bar{Z} := (\Gamma_b + \mathcal{M}_b)\, \bar{X}(b) + (I - \mathcal{M}_b)\, \bar{U}(b),$$

and where $(\bar{X}, \bar{U})$ is the conjoined basis of $(\mathcal{S})$ starting with the initial values $\bar{X}(a) = \mathcal{M}_a$ and $\bar{U}(a) = \mathcal{M}_a - I$.

Moreover, if $X_a^{\dagger}(\cdot)$ is continuous on $(a, b)_{\mathbb{T}}$, then each of the conditions (i)–(iii) is equivalent to any of the following Riccati type conditions.

(iv) The matrix $X_a(\cdot)$ has piecewise constant kernel on $[a,b]_{\mathbb{T}}$ and there exists a symmetric $n \times n$ matrix function $Q(t)$ on $[a,b]_{\mathbb{T}}$ such that $Q \in \mathrm{C}^1_{\mathrm{prd}}(a,b)_{\mathbb{T}}$ and satisfying

- *the time scale implicit Riccati equation*

$$[X_a^\sigma(t)]^T R[Q](t)\, X_a(t) = 0 \quad \text{on } (a,\rho(b))_{\mathbb{T}}, \tag{23}$$

  and the equation in (23) holds also at $t=a$ if a is right-scattered, and at $t=\rho(b)$ if b is left-scattered,
- *the initial condition (20) and (21),*
- *the final endpoint inequality*

$$Q(b) + \Gamma_b \geq 0 \quad \text{on } \operatorname{Ker} \mathcal{M}_b \cap \operatorname{Im} X_a(b) \tag{24}$$

  if b is left-scattered, or

$$X_a^T(b) \lim_{t \to b^-} [\Gamma_b + Q(t)]\, X_a(t) \geq 0 \quad \text{on } \operatorname{Ker} \mathcal{M}_b\, X_a(b) \tag{25}$$

  if b is left-dense,
- *the $\mathcal{P}$-condition*

$$T(t)\, \mathcal{P}(t)\, T(t) \geq 0 \quad \text{on } [a,\rho(b)]_{\mathbb{T}}, \tag{26}$$

  where the matrix $T(t)$ is defined in (12) through (X_a, U_a),
- *and for any admissible (x,u) with (9) we have the image condition (22) and*

$$\mu(t)\, [I - T(t)]\, [u(t) - Q(t)\, x(t)] = 0 \quad \text{on } [a,\rho(b)]_{\mathbb{T}}.$$

(v) The matrix $X_a(\cdot)$ has piecewise constant kernel on $[a,b]_{\mathbb{T}}$ and there exists a symmetric $n \times n$ matrix function $Q(t)$ on $[a,b]_{\mathbb{T}}$ such that $Q \in \mathrm{C}^1_{\mathrm{prd}}(a,b)_{\mathbb{T}}$ and satisfying

- *the initial condition (20) and (21),*
- *the final endpoint condition (24) and (25),*
- *for all points $t \in [a,b]_{\mathbb{T}}$ which are right-scattered or left-scattered Q and (X_a, U_a) satisfy the identity*

$$Q(t)\, X_a(t) = U_a(t)\, X_a^\dagger(t)\, X_a(t), \tag{27}$$

- *the $\mathcal{P}$-condition (26), where the matrix $T(t)$ is defined in (12) through (X_a, U_a),*
- *and for any admissible (x,u) with (9) we have*

  (a) the Riccati type identity

$$[x^\sigma(t)]^T R[Q](t)\, x(t) = 0 \quad \text{on } (a,\rho(b))_{\mathbb{T}}, \tag{28}$$

and the equation in (28) *holds also at* $t = a$ *if* a *is right-scattered and at* $t = \rho(b)$ *if* b *is left-scattered,*

(b) the image condition at b*:*

$$x(b) \in \operatorname{Im} X_a(b) \quad \textit{if } b \textit{ is left-dense.}$$

Remark 3.2. (i) Note that in [15, Theorem 4.1] there are several more perturbation conditions similar to the one in (iii) of Theorem 3.2. In these conditions, one may choose to keep only one out of the two boundary conditions from (9) or even combine them together in a certain way.

(ii) For the case of both endpoints being zero, a perturbation condition similar to (iii) of Theorem 3.2 (in a sense that only the initial endpoint is perturbed) was established in [14, Theorem 2.1].

(iii) For the time scale calculus of variations case with the zero right endpoint, a conjugate point condition equivalent to the nonnegativity of $\mathcal{F}$ is known from [24, Theorem 5.1]. In particular, this condition reads as the nonexistence of conjugate points in the interval $(a, b)_{\mathbb{T}}$ and b is not strictly conjugate to a if b is left-scattered.

(iv) The difference between the implicit Riccati equation conditions (iv) and (v) in Theorem 3.2 lies mainly in the presence or absence of the image condition (22). While condition (iv) contains the whole image condition (22), in condition (v) it is taken only at $t = b$ if b is left-dense, and it is then "compensated" by using (27).

Proof of Theorem 3.2. The equivalence "(i) $\Leftrightarrow$ (ii)" is established in [22, Theorem 4.2] via generalized Picone identity and the results on piecewise constant kernel of X_a. The equivalence "(i) $\Leftrightarrow$ (iii)" is proven in [15, Theorem 4.1] by transforming the functional $\mathcal{F}$ into a functional on the extended time scale $[a-1, b+2]_{\mathbb{T}}$ with zero endpoints $x(a-1) = 0 = x(b+2)$ and applying the zero endpoints result from [15, Theorem 3.1]. The equivalence "(i) $\Leftrightarrow$ (iv)" is shown in [27, Theorem 4.1] by using a generalized Picone type identity and results on piecewise constant kernel of X_a. Finally, the equivalence "(i) $\Leftrightarrow$ (v)" is established in [27, Theorem 4.9] by the combination of the method for proving condition (iv) with the results on reachable sets from [22, Lemma 5.2]. $\square$

Next we wish to discuss Reid roundabout theorems for the quadratic functional $\mathcal{F}$ in (1) with jointly varying endpoints (2). As it is now well known, such a functional can be transformed into a functional with separated endpoints in the double dimension $2n$, that is, with $2n \times 2n$ coefficient matrices, see e.g. [26, Section 4] or [19, Section 6]. Namely, the conditions on the natural conjoined basis (X_a, U_a) are replaced by conditions on the principal solution $(\hat{X}, \hat{U})$ and the augmented matrices

$$X_*(t) := \begin{pmatrix} 0 & I \\ \hat{X}(t) & \bar{X}(t) \end{pmatrix}, \quad U_*(t) := \begin{pmatrix} -I & 0 \\ \hat{U}(t) & \bar{U}(t) \end{pmatrix},$$

where $(\bar{X}, \bar{U})$ is the solution of $(\mathcal{S})$ given by the initial conditions $\bar{X}(a) = I$ and $\bar{U}(a) = 0$. The main difficulty in obtaining the results for the jointly varying endpoints case was the abnormality of the augmented time scale symplectic system resulting from this transformation. However, once we have Reid roundabout theorems for the separated endpoints case without any normality assumption (Theorems 3.1 and 3.2), the parallel results for the jointly varying endpoints can be easily derived. We shall not list these Reid roundabout theorem explicitly, but we comment on the literature where the corresponding partial results appeared.

Remark 3.3. More precisely, we can formulate a Reid roundabout theorem for the positivity of $\mathcal{F}$ with joint endpoints in a parallel way to Theorem 3.1 with the following references:

(i) The functional $\mathcal{F}$ in (1) is positive definite over (2) and $x \not\equiv 0$.
(ii) The condition on no generalized focal points for the principal solution $(\hat{X}, \hat{U})$ was derived in [26, Theorem 4.1].
(iii) The existence of an augmented conjoined basis (X_*, U_*) with invertible $X_*(t)$ on $[a, b]_{\mathbb{T}}$ and appropriate augmented boundary conditions was established in [26, Theorem 6.2].
(iv) The augmented explicit Riccati equation condition is also proven in [26, Theorem 6.2].
(vii) The perturbation result for the jointly varying endpoints case can be found in [15, Theorem 5.3].
(viii) The implicit Riccati equation condition with an augmented final endpoint inequality is derived in [27, Theorem 5.5].

Conditions (v) and (vi) corresponding to the Riccati inequality for the jointly varying endpoints case were never stated in the literature explicitly.

Remark 3.4. Similarly, we can formulate a Reid roundabout theorem for the nonnegativity of $\mathcal{F}$ with joint endpoints in a parallel way to Theorem 3.2 with the following references:

(i) The functional $\mathcal{F}$ in (1) is nonnegative over (2).

(ii) The image condition for the principal solution $(\hat{X}, \hat{U})$, involving also the conjoined basis $(\bar{X}, \bar{U})$, and an augmented final endpoint inequality was derived in [26, Theorem 4.2].

(iii) The perturbation result for the jointly varying endpoints case can be found in [15, Theorem 5.2].

(iv) The implicit Riccati equation condition with an image condition on $[a, b]_{\mathbb{T}}$ and with an augmented final endpoint inequality is derived in [27, Theorem 5.1].

(v) The implicit Riccati equation condition involving an identity of the form (27) for the principal solution $(\hat{X}, \hat{U})$ instead of the natural conjoined basis (X_a, U_a), with an image condition at $t = b$ only if b is left-dense, and with an augmented final endpoint inequality is established in [27, Theorem 5.3].

4. Jacobi systems for control problems

The Jacobi equation or Jacobi system is the (linear) Euler-Lagrange system for the quadratic functional arising as the second variation of the original variational problem. In the time scale *calculus of variations* setting, see [20,24], the corresponding Jacobi equation is of the form

$$\left[R(t)\, \eta^{\Delta} + Q^T(t)\, \eta^{\sigma}\right]^{\Delta} = P(t)\, \eta^{\sigma} + Q(t)\, \eta^{\Delta} \tag{J}$$

with symmetric matrices $P(t)$ and $R(t)$. Upon setting $q := R(t)\, \eta^{\Delta} + Q^T(t)\, \eta^{\sigma}$ and assuming that $R(t)$ and $R(t) + \mu(t)\, Q^T(t)$ are invertible on $[a, \rho(b)]_{\mathbb{T}}$, equation (J) can be written as the linear Hamiltonian system (H^{σ}) with

$$\begin{aligned} &A(t) := -R^{-1}(t)\, Q^T(t), \quad C(t) := P(t) - Q(t)\, R^{-1}(t)\, Q^T(t), \\ &B(t) := R^{-1}(t). \end{aligned} \tag{29}$$

Note that, under the given assumptions, the matrix $I - \mu(t)\, A(t) = R^{-1}(t)\, [R(t) + \mu(t)\, Q^T(t)]$ is indeed invertible and that $B(t)$ and $C(t)$ are symmetric. Thus, Jacobi equation (J) is a special case of the time scale symplectic system ($\mathcal{S}$), see also Proposition 4.1 below. In this section we show in a similar way that linear Hamiltonian systems arising from time scale *control problems* lead naturally to time scale symplectic systems. This work is motivated by its continuous and discrete counterparts in [38, Section 6] and [19, Section 2].

4.1. *Control problem with shift in the state variable*

Consider the time scale control problem

$$\text{minimize} \quad \mathcal{J}(x,u) := K\big(x(a), x(b)\big) + \int_a^b L\big(t, x^\sigma(t), u(t)\big)\,\Delta t, \qquad (\mathrm{C}^\sigma)$$

subject to $x \in \mathrm{C}^1_{\mathrm{prd}}[a,b]_{\mathbb{T}}$ and $u \in \mathrm{C}_{\mathrm{prd}}[a,\rho(b)]_{\mathbb{T}}$ satisfying

$$x^\Delta(t) = f\big(t, x^\sigma(t), u(t)\big), \quad t \in [a,\rho(b)]_{\mathbb{T}},$$

$$\psi\big(t, u(t)\big) = 0, \quad t \in [a,\rho(b)]_{\mathbb{T}}, \qquad (30)$$

$$\varphi\big(x(a), x(b)\big) = 0, \qquad (31)$$

where the data satisfy

$$\begin{aligned}
&L : [a,\rho(b)]_{\mathbb{T}} \times \mathbb{R}^n \times \mathbb{R}^m \to \mathbb{R}, \quad m \le n, && K : \mathbb{R}^{2n} \to \mathbb{R},\\
&f : [a,\rho(b)]_{\mathbb{T}} \times \mathbb{R}^n \times \mathbb{R}^m \to \mathbb{R}^n, && \varphi : \mathbb{R}^{2n} \to \mathbb{R}^r, \quad r \le 2n,\\
&\psi : [a,\rho(b)]_{\mathbb{T}} \times \mathbb{R}^m \to \mathbb{R}^k, \quad k \le m.
\end{aligned}$$

It is shown in [28, Section 7] that the second variation at a normal feasible pair $(\bar{x}, \bar{u})$ in the direction (η, v) takes the form of the quadratic functional

$$\mathcal{J}''(\bar{x},\bar{u};\eta,v) :=$$
$$\begin{pmatrix}\eta(a)\\ \eta(b)\end{pmatrix}^T \Gamma \begin{pmatrix}\eta(a)\\ \eta(b)\end{pmatrix} + \int_a^b \big\{(\eta^\sigma)^T P\, \eta^\sigma + 2\,(\eta^\sigma)^T Q\, v + v^T R\, v\big\}(t)\,\Delta t,$$

subject to $\eta \in \mathrm{C}^1_{\mathrm{prd}}[a,b]_{\mathbb{T}}$ and $v \in \mathrm{C}_{\mathrm{prd}}[a,\rho(b)]_{\mathbb{T}}$ satisfying

$$\eta^\Delta(t) = \bar{\mathcal{A}}(t)\,\eta^\sigma(t) + \bar{\mathcal{B}}(t)\,v(t), \quad t \in [a,\rho(b)]_{\mathbb{T}}, \quad v \in \mathcal{T}. \qquad (32)$$

$$M \begin{pmatrix}\eta(a)\\ \eta(b)\end{pmatrix} = 0, \qquad (33)$$

The quantities appearing in the functional $\mathcal{J}''(\bar{x},\bar{u};\cdot,\cdot)$ and in (32)–(33) are defined, for some vector $\bar{\gamma} \in \mathbb{R}^r$ and functions $\bar{\lambda} : [a,\rho(b)]_{\mathbb{T}} \to \mathbb{R}^k$, $\bar{\lambda} \in \mathrm{C}_{\mathrm{prd}}$, and $\bar{p} : [a,b]_{\mathbb{T}} \to \mathbb{R}^n$, $\bar{p} \in \mathrm{C}^1_{\mathrm{prd}}$, as follows:

$$\begin{aligned}
&\bar{\mathcal{A}}(t) := f_x(t), && P(t) := L_{xx}(t) + \bar{p}^T(t)\, f_{xx}(t),\\
&\bar{\mathcal{B}}(t) := f_u(t), && Q(t) := L_{xu}(t) + \bar{p}^T(t)\, f_{xu}(t),\\
&M := \nabla\varphi\big(\bar{x}(a), \bar{x}(b)\big), && R(t) := L_{uu}(t) + \bar{p}^T(t)\, f_{uu}(t) + \bar{\lambda}^T(t)\,\psi_{uu}(t),\\
&N(t) := \nabla_u \psi\big(t, \bar{u}(t)\big), && \Gamma := \nabla^2 K\big(\bar{x}(a), \bar{x}(b)\big) + \bar{\gamma}^T \nabla^2 \varphi\big(\bar{x}(a), \bar{x}(b)\big),
\end{aligned} \qquad (34)$$

and the tangent space $\mathcal{T}$ of tangent function is

$$\mathcal{T} := \big\{v(\cdot) \in \mathrm{C}_{\mathrm{prd}}[a,\rho(b)]_{\mathbb{T}} : \; N(t)\,v(t) = 0 \text{ for all } t \in [a,\rho(b)]_{\mathbb{T}}\big\}.$$

The first and second order partial derivatives of L, f, and ψ are evaluated at $\big(t, \bar{x}^\sigma(t), \bar{u}(t)\big)$ and $\big(t, \bar{u}(t)\big)$, respectively.

We assume that the matrices M and $N(t)$, $t \in [a, \rho(b)]_{\mathbb{T}}$, have full rank, and that the linear system (32) is M-controllable over $\mathcal{T}$, see [28, Definition 4.2]. Moreover, we denote by $Y \in \mathrm{C}_{\mathrm{prd}}$ the $m \times (m-k)$ matrix function whose columns form, for each $t \in [a, \rho(b)]_{\mathbb{T}}$, an orthonormal basis for $\operatorname{Ker} N(t)$. Hence, every tangent function $v \in \mathcal{T}$ is of the form $v(t) = Y(t)\, w(t)$ with some $w : [a, \rho(b)]_{\mathbb{T}} \to \mathbb{R}^{m-k}$, $w \in \mathrm{C}_{\mathrm{prd}}$. Then, by [28, Theorem 7.2], a necessary condition for $(\bar{x}, \bar{u})$ being a weak local minimum for problem (C^σ) is that the quadratic functional $\mathcal{J}''(\bar{x}, \bar{u}; \eta, v)$ is nonnegative for all (η, v) satisfying (32) and (33). As it is common, (η, v) satisfying (32) will be called *admissible*. Consequently, the application of the weak maximum principle [28, Theorem 6.1] to the functional $\mathcal{J}''(\bar{x}, \bar{u}; \cdot, \cdot)$ yields the existence of a quadruple $\lambda_0 = 1$, $\omega \in \mathbb{R}^r$, $\lambda \in \mathrm{C}_{\mathrm{prd}}[a, \rho(b)]_{\mathbb{T}}$, and $q \in \mathrm{C}^1_{\mathrm{prd}}[a, b]_{\mathbb{T}}$ (changing the sign of $q(\cdot)$) such that for $t \in [a, \rho(b)]_{\mathbb{T}}$

$$q^\Delta(t) = -\bar{\mathcal{A}}^T(t)\, q(t) + P(t)\, \eta^\sigma(t) + Q(t)\, v(t), \tag{35}$$

$$-\bar{\mathcal{B}}^T(t)\, q(t) + Q^T(t)\, \eta^\sigma(t) + R(t)\, v(t) + N^T(t)\, \lambda(t) = 0, \tag{36}$$

$$\begin{pmatrix} q(a) \\ -q(b) \end{pmatrix} = M^T \omega + \Gamma \begin{pmatrix} \eta(a) \\ \eta(b) \end{pmatrix}, \tag{37}$$

where $v(t) = Y(t)\, w(t)$ as mentioned above. If $Y^T(t)\, R(t)\, Y(t)$ is invertible on $[a, \rho(b)]_{\mathbb{T}}$, then we may solve equation (36) for $w(t)$, hence also for $v(t)$, and then

$$v(t) = Z(t)\, \big[\bar{\mathcal{B}}^T(t)\, q(t) - Q^T(t)\, \eta^\sigma(t)\big], \tag{38}$$

where

$$Z(t) := Y(t)\, \big[Y^T(t)\, R(t)\, Y(t)\big]^{-1} Y^T(t). \tag{39}$$

By inserting (38) into equations (32) and (35), it follows that (η, q) satisfies the linear Hamiltonian system (H^σ) with

$$\begin{aligned} A(t) &:= \bar{\mathcal{A}}(t) - \bar{\mathcal{B}}(t)\, Z(t)\, Q^T(t), \quad C(t) := P(t) - Q(t)\, Z(t)\, Q^T(t), \\ B(t) &:= \bar{\mathcal{B}}(t)\, Z(t)\, \bar{\mathcal{B}}^T(t). \end{aligned} \tag{40}$$

These equations reduce to equations (29) for the calculus of variations case where $\bar{\mathcal{A}}(t) \equiv 0$, $\bar{\mathcal{B}}(t) \equiv I$, and $Y(t) \equiv I$. Then the following property was shown in [10], where the time scale linear Hamiltonian systems were introduced.

Proposition 4.1 (Hamiltonian to symplectic). *Let $A(t)$, $B(t)$, and $C(t)$ be defined by the formulas in (40). Under the natural solvability assumption that*

$$I - \mu(t)\,A(t) \quad \text{is invertible for all } t \in [a, \rho(b)]_{\mathbb{T}},$$

the linear Hamiltonian system (H^σ) *is a special case of the time scale symplectic system* ($\mathcal{S}$) *with*

$$\begin{aligned} \mathcal{A}(t) &:= D(t)\,A(t), & \mathcal{C}(t) &:= C(t)\,D(t), \\ \mathcal{B}(t) &:= D(t)\,B(t), & \mathcal{D}(t) &:= \mu(t)\,C(t)\,D(t)\,B(t) - A^T(t), \end{aligned}$$

and where $D(t) := [I - \mu(t)\,A(t)]^{-1}$.

4.2. *Control problem without shift in the state variable*

On the other hand, starting with the time scale control problem which does not have the shift in x, i.e., with

$$\text{minimize} \quad \mathcal{J}(x,u) := K\big(x(a), x(b)\big) + \int_a^b L\big(t, x(t), u(t)\big)\,\Delta t, \tag{C}$$

subject to $x \in \mathrm{C}^1_{\mathrm{prd}}[a,b]_{\mathbb{T}}$ and $u \in \mathrm{C}_{\mathrm{prd}}[a,\rho(b)]_{\mathbb{T}}$ satisfying the constraints (30) and (31) and with the dynamics

$$x^\Delta(t) = f\big(t, x(t), u(t)\big), \quad t \in [a, \rho(b)]_{\mathbb{T}},$$

leads, by [28, Section 9], to the second variation

$$\mathcal{J}''(\bar{x}, \bar{u}; \eta, v) :=$$

$$\begin{pmatrix} \eta(a) \\ \eta(b) \end{pmatrix}^T \Gamma \begin{pmatrix} \eta(a) \\ \eta(b) \end{pmatrix} + \int_a^b \{\eta^T P\,\eta + 2\,\eta^T Q\,v + v^T R\,v\}(t)\,\Delta t,$$

subject to $\eta \in \mathrm{C}^1_{\mathrm{prd}}[a,b]_{\mathbb{T}}$ and $v \in \mathrm{C}_{\mathrm{prd}}[a,\rho(b)]_{\mathbb{T}}$ satisfying the endpoint constraint (33) and

$$\eta^\Delta(t) = \tilde{\mathcal{A}}(t)\,\eta(t) + \tilde{\mathcal{B}}(t)\,v(t), \quad t \in [a, \rho(b)]_{\mathbb{T}}, \quad v \in \mathcal{T}. \tag{41}$$

Here, for some vector $\bar{\gamma} \in \mathbb{R}^r$ and functions $\bar{\lambda} : [a, \rho(b)]_{\mathbb{T}} \to \mathbb{R}^k$, $\bar{\lambda} \in \mathrm{C}_{\mathrm{prd}}$, and $\bar{p} : [a,b]_{\mathbb{T}} \to \mathbb{R}^n$, $\bar{p} \in \mathrm{C}^1_{\mathrm{prd}}$, the coefficients $\tilde{\mathcal{A}}(t)$, $\tilde{\mathcal{B}}(t)$, $P(t)$, $Q(t)$, $R(t)$ are defined in a similar manner as the corresponding coefficients for the problem (C^σ) in equation (34), but with the first and second order partial derivatives of L and f evaluated at $\big(t, \bar{x}(t), \bar{u}(t)\big)$ instead of at $\big(t, \bar{x}^\sigma(t), \bar{u}(t)\big)$, and with $\bar{p}^\sigma(t)$ instead of $\bar{p}(t)$.

Under a corresponding M-controllability assumption on the linear system (41), the weak maximum principle [28, Theorem 9.4] applied to the functional $\mathcal{J}''(\bar{x}, \bar{u}; \cdot, \cdot)$ yields the existence of a quadruple $\lambda_0 = 1$, $\omega \in \mathbb{R}^r$, $\lambda \in \mathrm{C}_{\mathrm{prd}}[a, \rho(b)]_{\mathbb{T}}$, and $q \in \mathrm{C}^1_{\mathrm{prd}}[a, b]_{\mathbb{T}}$ (changing the sign of $q(\cdot)$) such that for $t \in [a, \rho(b)]_{\mathbb{T}}$

$$q^{\Delta}(t) = -\tilde{\mathcal{A}}^T(t)\, q^{\sigma}(t) + P(t)\, \eta(t) + Q(t)\, v(t), \tag{42}$$

$$-\tilde{\mathcal{B}}^T(t)\, q^{\sigma}(t) + Q^T(t)\, \eta(t) + R(t)\, v(t) + N^T(t)\, \lambda(t) = 0, \tag{43}$$

and satisfying the transversality condition (37). Similarly to the situation in Subsection 4.1, if $Y^T(t)\, R(t)\, Y(t)$ is invertible on $[a, \rho(b)]_{\mathbb{T}}$, then we may solve equation (43) for $v(t)$, i.e.,

$$v(t) = Z(t)\, [\tilde{\mathcal{B}}^T(t)\, q^{\sigma}(t) - Q^T(t)\, \eta(t)], \tag{44}$$

where the matrix $Z(t)$ is defined in (39). Upon inserting formula (44) into equations (41) and (42), it follows that (η, q) satisfies the linear Hamiltonian system

$$\eta^{\Delta} = A(t)\, \eta + B(t)\, q^{\sigma}, \quad q^{\Delta} = C(t)\, \eta - A^T(t)\, q^{\sigma}, \tag{H}$$

where the coefficients are defined by

$$\begin{aligned} A(t) &:= \tilde{\mathcal{A}}(t) - \tilde{\mathcal{B}}(t)\, Z(t)\, Q^T(t), \quad C(t) := P(t) - Q(t)\, Z(t)\, Q^T(t), \\ B(t) &:= \tilde{\mathcal{B}}(t)\, Z(t)\, \tilde{\mathcal{B}}^T(t). \end{aligned} \tag{45}$$

Then we can easily prove the following.

Proposition 4.2 (Hamiltonian to symplectic). *Let $A(t)$, $B(t)$, and $C(t)$ be defined by the formulas in (45). Under the natural solvability assumption that*

$$I + \mu(t)\, A(t) \quad \textit{is invertible for all } t \in [a, \rho(b)]_{\mathbb{T}},$$

the linear Hamiltonian system (H) *is a special case of the time scale symplectic system* $(\mathcal{S})$ *with*

$$\begin{aligned} \mathcal{A}(t) &:= \mu(t)\, B(t)\, D(t)\, C(t) + A(t), & \mathcal{C}(t) &:= D(t)\, C(t), \\ \mathcal{B}(t) &:= B(t)\, D(t), & \mathcal{D}(t) &:= -A^T(t)\, D(t), \end{aligned} \tag{46}$$

and where $D(t) := [I + \mu(t)\, A^T(t)]^{-1}$.

Proof. Using the formula $\mu q^{\Delta} = q^{\sigma} - q$, we can solve the second equation in (H) for q^{σ} in terms of η and q. Putting this into both equations in (H) yields the coefficients in (46). Then it is a straightforward calculation that the corresponding matrix $\mathcal{S}(t)$ satisfies identity (3), thus defining a time scale symplectic system. □

In conclusion, no matter whether the original time scale control problem (and hence, also a time scale calculus of variations problem) contains the shift in the state variable x or does not, the resulting Jacobi system is a time scale symplectic system of the form $(\mathcal{S})$. Note also that in the continuous time case both linear Hamiltonian systems (H^{σ}) and (H), as well as the system $(\mathcal{S})$, reduce to the same differential system (H_c).

4.3. *Euler-Lagrange system for functional $\mathcal{F}$*

Next we show that the system $(\mathcal{S})$ is indeed the Euler-Lagrange (or Jacobi) system for the quadratic functional $\mathcal{F}$ given in (1). Consider the variational problem

$$\text{minimize} \quad \mathcal{F}(x,u) \tag{VP}$$

subject to admissible (x,u) satisfying

$$M\begin{pmatrix} x(a) \\ x(b) \end{pmatrix} = 0. \tag{47}$$

We assume that $M \in \mathbb{R}^{r\times 2n}$, $r \leq 2n$, and that M has full rank r.

Remark 4.1. In (2) we considered the boundary conditions for $\mathcal{F}$ in terms of a projection matrix $\mathcal{M}$. Obviously, if $\operatorname{rank}\mathcal{M} < 2n$, then we may simply disregard in (2) the linearly dependent equations and obtain the equivalent constraint (47). On the other hand, starting with M as in (47), then we can set $\mathcal{M} := M^T(MM^T)^{-1}M$, which is a projection with the property that $\mathcal{M}\alpha = 0$ if and only if $M\alpha = 0$. In this case we can also redefine the matrix Γ to be $(I-\mathcal{M})\,\Gamma\,(I-\mathcal{M})$, because for $\alpha \in \operatorname{Ker} M = \operatorname{Ker}\mathcal{M}$ we have $\alpha^T\Gamma\,\alpha = \alpha^T(I-\mathcal{M})\,\Gamma\,(I-\mathcal{M})\,\alpha$.

Next we consider the boundary value problem

$$(\mathcal{S}), \quad M\begin{pmatrix} x(a) \\ x(b) \end{pmatrix} = 0, \quad \begin{pmatrix} u(a) \\ -u(b) \end{pmatrix} = \Gamma\begin{pmatrix} x(a) \\ x(b) \end{pmatrix} + M^T\gamma, \tag{BVP}$$

for some $\gamma \in \mathbb{R}^r$, where the system $(\mathcal{S})$ is defined in Section 1. The system $(\mathcal{S})$ is called *M-normal* if the only solution $u(\cdot)$ of

$$u^{\Delta} = \mathcal{D}(t)\,u, \quad \mathcal{B}(t)\,u = 0, \quad t \in [a,\rho(b)]_{\mathbb{T}}, \quad \begin{pmatrix} u(a) \\ -u(b) \end{pmatrix} = M^T\gamma,$$

for some $\gamma \in \mathbb{R}^r$, is the trivial solution $u(\cdot) \equiv 0$. Equivalently, the M-normality of the system $(\mathcal{S})$ means that if (x,u) solves (BVP) with $x(\cdot) \equiv 0$, then also $u(\cdot) \equiv 0$.

In the formulation of the next result we need the first variation of the functional $\mathcal{F}$ at an admissible pair (x, u) in the direction (ξ, w). By [28, Section 9], the first variation takes the form

$$\mathcal{F}'(x,u;\xi,w) = \begin{pmatrix} x(a) \\ x(b) \end{pmatrix}^T \Gamma \begin{pmatrix} \xi(a) \\ \xi(b) \end{pmatrix}$$

$$+ \int_a^b \{x^T \mathcal{C}^T (I + \mu \mathcal{A})\, \xi + \mu\, x^T \mathcal{C}^T \mathcal{B} w + \mu\, u^T \mathcal{B}^T \mathcal{C}\, \xi + u^T (I + \mu \mathcal{D})^T \mathcal{B} w\}(t)\, \Delta t$$

over admissible (ξ, w) satisfying

$$M \begin{pmatrix} \xi(a) \\ \xi(b) \end{pmatrix} = 0. \tag{48}$$

Again, the pair (ξ, w) is called admissible if $\xi \in \mathrm{C}^1_{\mathrm{prd}}[a,b]_{\mathbb{T}}$, $w \in \mathrm{C}_{\mathrm{prd}}[a, \rho(b)]_{\mathbb{T}}$, and $\xi^\Delta(t) = \mathcal{A}(t)\, \xi(t) + \mathcal{B}(t)\, w(t)$ for $t \in [a, \rho(b)]_{\mathbb{T}}$.

The discrete version of the following result can be found in [19, Theorem 4]. In particular, condition (ii) in the statement below says that the time scale symplectic system $(\mathcal{S})$ is the Euler-Lagrange system for the quadratic functional $\mathcal{F}$.

Proposition 4.3 (Euler-Lagrange system for $(\mathcal{S})$). *The following implications relate the variational problem* (VP) *and the boundary value problem* (BVP).

(i) Let $(\bar{x}, \bar{u})$ solve (BVP) *and let $\mathcal{F} \geq 0$ over* (47)*. Then $\mathcal{F}'(\bar{x}, \bar{u}; \cdot, \cdot) = 0$ and $(\bar{x}, \bar{u})$ solves* (VP).

(ii) Conversely, if $(\bar{x}, \bar{u})$ solves (VP)*, then $\mathcal{F}'(\bar{x}, \bar{u}; \cdot, \cdot) = 0$, $\mathcal{F} \geq 0$ over* (47)*, and there exist $\lambda_0 \geq 0$, $\gamma \in \mathbb{R}^r$, and $\bar{p} \in \mathrm{C}^1_{\mathrm{prd}}[a,b]_{\mathbb{T}}$ such that the pair $(\lambda_0 \bar{x}, \bar{p})$ solves* (BVP)*. In addition, if the system $(\mathcal{S})$ is M-normal, then $\lambda_0 = 1$, and γ and $\bar{p}(\cdot)$ are unique.*

Proof. "(i)" Suppose that $(\bar{x}, \bar{u})$ solves (BVP) and $\mathcal{F} \geq 0$. Then, by [28, Theorem 9.6], $\mathcal{F}'(\bar{x}, \bar{u}; \xi, w) = 0$ for any admissible (ξ, w) with (47). Moreover, for any admissible (x, u) the pair $(\xi, w) := (x, u) - (\bar{x}, \bar{u})$ is also admissible, it satisfies (48), and

$$\mathcal{F}(x,u) - \mathcal{F}(\bar{x}, \bar{u}) = 2\, \mathcal{F}'(\bar{x}, \bar{u}; \xi, w) + \mathcal{F}(\xi, w) = \mathcal{F}(\xi, w) \geq 0.$$

Hence, $(\bar{x}, \bar{u})$ solves (VP).

"(ii)" Conversely, let $(\bar{x}, \bar{u})$ solve (VP), i.e., $(\bar{x}, \bar{u})$ is admissible, (47) holds, and for every other such pair (x, u) we have $\mathcal{F}(\bar{x}, \bar{u}) \leq \mathcal{F}(x, u)$. Then, by [28, Theorems 9.6 and 9.7], $\mathcal{F}'(\bar{x}, \bar{u}; \xi, w) = 0$ and $2\, \mathcal{F}(\xi, w) =$

$\mathcal{F}''(\bar{x},\bar{u};\xi,w)\geq 0$ for any admissible (ξ,w) satisfying (48). Furthermore, by the time scale weak maximum principle [28, Theorem 9.4], there is $\lambda_0\geq 0$, $\gamma\in\mathbb{R}^r$, and $\bar{p}\in \mathrm{C}^1_{\mathrm{prd}}[a,b]_{\mathbb{T}}$ (changing the sign of $\bar{p}(\cdot)$) such that (suppressing the argument t in the following calculations) on $[a,\rho(b)]_{\mathbb{T}}$

$$\bar{p}^{\Delta} = -\mathcal{A}^T\,\bar{p}^{\sigma} + \lambda_0\left[\mathcal{C}^T\,(I+\mu\,\mathcal{A})\,\bar{x} + \mu\,\mathcal{C}^T\mathcal{B}\bar{u}\right], \tag{49}$$

$$-\mathcal{B}^T\bar{p}^{\sigma} + \lambda_0\left[\mu\,\mathcal{B}^T\mathcal{C}\bar{x} + (I+\mu\mathcal{D}^T)\,\mathcal{B}\bar{u}\right] = 0, \tag{50}$$

$$\begin{pmatrix}\bar{p}(a)\\ -\bar{p}(b)\end{pmatrix} = \lambda_0\,\Gamma\begin{pmatrix}\bar{x}(a)\\ \bar{x}(b)\end{pmatrix} + M^T\gamma. \tag{51}$$

The admissibility of $(\bar{x},\bar{u})$ yields from (49), via the first identity in (5), the equation

$$\bar{p}^{\Delta} = \lambda_0\,\mathcal{C}^T\bar{x}^{\sigma} - \mathcal{A}^T\,\bar{p}^{\sigma}\quad \text{on } [a,\rho(b)]_{\mathbb{T}}, \tag{52}$$

while from (50) it gives on $[a,\rho(b)]_{\mathbb{T}}$

$$\lambda_0\,(\bar{x}^{\Delta} - \mathcal{A}\bar{x} - \mathcal{B}\bar{u}) + \lambda_0\left[\mu\,\mathcal{B}^T\mathcal{C}\bar{x} + (I+\mu\mathcal{D}^T)\,\mathcal{B}\bar{u}\right] = \mathcal{B}^T\bar{p}^{\sigma}.$$

By using the first identity in (5) once more and by the third equation in the coefficient identities (4), it follows that

$$\lambda_0\,\bar{x}^{\Delta} = \lambda_0\,(\mathcal{A}\bar{x} - \mu\,\mathcal{B}^T\mathcal{C}\bar{x} - \mu\,\mathcal{D}^T\mathcal{B}\bar{u}) + \mathcal{B}^T\bar{p}^{\sigma} = -\lambda_0\,\mathcal{D}^T\bar{x}^{\sigma} + \mathcal{B}^T\bar{p}^{\sigma}. \tag{53}$$

Therefore, equations (52) and (53) show that the pair $(\lambda_0\bar{x},\bar{p})$ satisfies the adjoint time scale symplectic system (7). Consequently, this pair satisfies $(\mathcal{S})$ and hence, with the transversality condition (51), the pair $(\lambda_0\bar{x},\bar{p})$ solves (BVP). Finally, under the M-normality, $\lambda_0 = 1$ and γ and $\bar{p}(\cdot)$ are unique, as it follows from the time scale weak maximum principle in [28, Theorem 9.4]. □

Acknowledgments

The first author was supported by the Grant Agency of the Academy of Sciences of the Czech Republic under grant KJB100190701, by the Czech Grant Agency under grant 201/07/0145, and by the research projects MSM 0021622409 and ME 891 (program KONTAKT) of the Ministry of Education, Youth, and Sports of the Czech Republic. He wishes to thank the Michigan State University for the hospitality provided while conducting a part of this research project. The second author was supported by the National Science Foundation under grant DMS – 0707789.

References

1. C. D. Ahlbrandt, Discrete variational inequalities, in: "General Inequalities 6", Proceedings of the 6th International Conference on General Inequalities (Oberwolfach, 1990), W. Walter, editor, pp. 93–107, Birkhäuser, Basel, 1992.
2. C. D. Ahlbrandt, Equivalence of discrete Euler equations and discrete Hamiltonian systems, *J. Math. Anal. Appl.* **180** (1993), 498–517.
3. C. D. Ahlbrandt, S. L. Clark, J. W. Hooker, W. T. Patula, A discrete interpretation of Reid's roundabout theorem for generalized differential systems, *Comput. Math. Appl.* **28** (1994), no. 1–3, 11–21.
4. M. Bohner, Calculus of variations on time scales, *Dynam. Systems Appl.* **13** (2004), no. 3–4, 339–349.
5. M. Bohner, O. Došlý, W. Kratz, Positive semidefiniteness of discrete quadratic functionals, *Proc. Edinburgh Math. Soc.* **46** (2003), 627–636.
6. M. Bohner, A. Peterson, *Dynamic Equations on Time Scales. An Introduction with Applications,* Birkhäuser, Boston, 2001.
7. A. E. Bryson, Y. C. Ho, *Applied Optimal Control. Optimization, Estimation, and Control,* Hemisphere Publishing Corp., Washington, 1975.
8. O. Došlý, S. Hilger, R. Hilscher, Symplectic dynamic systems, in: "Advances in Dynamic Equations on Time Scales", M. Bohner and A. Peterson, editors, pp. 293–334, Birkhäuser, Boston, 2003.
9. M. R. Hestenes, *Calculus of Variations and Optimal Control Theory,* John Wiley & Sons, New York, 1966.
10. R. Hilscher, Linear Hamiltonian systems on time scales: positivity of quadratic functionals, *Math. Comput. Modelling* **32** (2000), no. 5–6, 507–527.
11. R. Hilscher, Reid roundabout theorem for symplectic dynamic systems on time scales, *Appl. Math. Optim.* **43** (2001), no. 2, 129–146.
12. R. Hilscher, V. Růžičková, Riccati inequality and other results for discrete symplectic systems, *J. Math. Anal. Appl.* **322** (2006), no. 2, 1083–1098.
13. R. Hilscher, V. Růžičková, Implicit Riccati equations and quadratic functionals for discrete symplectic systems, *Internat. J. Difference Equ.* **1** (2006), no. 1, 135–154.
14. R. Hilscher, V. Růžičková, Perturbation of nonnegative time scale quadratic functionals, in: "Difference Equations, Special Functions, and Orthogonal Polynomials", Proceedings of the International Conference (Munich, 2005), S. Elaydi *et al.*, editors, pp. 266–275, World Scientific, London, 2007.
15. R. Hilscher, V. Růžičková, Perturbation of time scale quadratic functionals with variable endpoints, *Adv. Dyn. Syst. Appl.* **2** (2007), no. 2, 207–224.
16. R. Hilscher, P. Řehák, Riccati inequality, disconjugacy, and reciprocity principle for linear Hamiltonian dynamic systems, *Dynam. Systems. Appl.* **12** (2003), no. 1–2, 171–189.
17. R. Hilscher, V. Zeidan, Discrete optimal control: the accessory problem and necessary optimality conditions, *J. Math. Anal. Appl.* **243** (2000), no. 2, 429–452.
18. R. Hilscher, V. Zeidan, Second order sufficiency criteria for a discrete optimal control problem, *J. Difference Equ. Appl.* **8** (2002), no. 6, 573–602.

19. R. Hilscher, V. Zeidan, Symplectic difference systems: variable stepsize discretization and discrete quadratic functionals, *Linear Algebra Appl.* **367** (2003), 67–104.
20. R. Hilscher, V. Zeidan, Calculus of variations on time scales: weak local piecewise C^1_{rd} solutions with variable endpoints, *J. Math. Anal. Appl.* **289** (2004), no. 1, 143–166.
21. R. Hilscher, V. Zeidan, Nonnegativity and positivity of a quadratic functional in the discrete calculus of variations: A survey, *J. Difference Equ. Appl.* **11** (2005), no. 9, 857–875.
22. R. Hilscher, V. Zeidan, Time scale symplectic systems without normality, *J. Differential Equations* **230** (2006), no. 1, 140–173.
23. R. Hilscher, V. Zeidan, Coupled intervals for discrete symplectic systems, *Linear Algebra Appl.* **419** (2006), no. 2–3, 750–764.
24. R. Hilscher, V. Zeidan, Legendre, Jacobi, and Riccati type conditions for time scale variational problem with application, *Dynam. Systems Appl.* **16** (2007), no. 3, 451–480.
25. R. Hilscher, V. Zeidan, Time scale embedding theorem and coercivity of quadratic functionals, *Analysis (Munich)* **28** (2008), no. 1, 1–28.
26. R. Hilscher, V. Zeidan, Applications of time scale symplectic systems without normality, *J. Math. Anal. Appl.* **340** (2008), no. 1, 451–465.
27. R. Hilscher, V. Zeidan, Riccati equations for abnormal time scale quadratic functionals, *J. Differential Equations* **244** (2008), no. 6, 1410–1447.
28. R. Hilscher, V. Zeidan, Weak maximum principle and accessory problem for control problems on time scales, *Nonlinear Anal.*, to appear.
29. W. Kratz, *Quadratic Functionals in Variational Analysis and Control Theory,* Akademie Verlag, Berlin, 1995.
30. W. Kratz, Definiteness of quadratic functionals, *Analysis (Munich)* **23** (2003), no. 2, 163–183.
31. W. T. Reid, *Ordinary Differential Equations,* Wiley, New York, 1971.
32. V. Růžičková, *Discrete Symplectic Systems and Definiteness of Quadratic Functionals,* PhD dissertation. Masaryk University, Brno, 2006.
33. G. Stefani, P. Zezza, Constrained regular *LQ*-control problems, *SIAM J. Control Optim.* **35** (1997), no. 3, 876–900.
34. V. Zeidan, Sufficiency criteria via focal points and via coupled points, *SIAM J. Control Optim.* **30** (1992), no. 1, 82–98.
35. V. Zeidan, Sufficient conditions for variational problems with variable endpoints: coupled points, *Appl. Math. Optim.* **27** (1993), no. 2, 191–209.
36. V. Zeidan, Nonnegativity and positivity of a quadratic functional, *Dynam. Systems Appl.* **8** (1999), no. 3–4, 571–588.
37. V. Zeidan, New second-order optimality conditions for variational problems with C^2-Hamiltonians, *SIAM J. Control Optim.* **40** (2001), no. 2, 577–609.
38. V. Zeidan, P. Zezza, The conjugate point condition for smooth control sets, *J. Math. Anal. Appl.* **132** (1988), no. 2, 572–589.
39. P. Zezza, The Jacobi condition in optimal control, in: "Control Theory, Stochastic Analysis and Applications", S. Chen, J. Yong, editors, pp. 137–149, World Scientific Publishing, River Edge, NJ, 1991.

The Global Properties of a Two-Dimensional Competing Species Model Exhibiting Mixed Competition

P. J. Johnson* and M. E. Burke

Department of Mathematics and Statistics,
University of Limerick, Ireland
**E-mail: patrick.johnson@ul.ie*
www.ul.ie

In this paper we consider a two species population model based on the discretisation of the original Lotka-Volterra competition equations. We analyse the global dynamic properties of the resulting two-dimensional noninvertible dynamical system in the case where the interspecific competition is considered to be "mixed". The main results of this paper are derived from the study of some global bifurcations that change the structure of the attractors and their basins. These bifurcations are investigated using the method of Critical Curves.

Keywords: Critical curves; Global bifurcations; Noninvertible maps; Competing species.

1. Introduction

In this paper we will show how the global dynamics of a biological model can be analysed through a study of global bifurcations that cause qualitative changes in the attractors and their basins of attraction, as some parameters of the model are varied. In particular, we consider a two-dimensional quadratic map derived from the Lotka-Volterra competing species population model which has very close analogies to the well-known one-dimensional standard logistic map. A previous investigation of the dynamics of this map has been given for a range of parameter values which show both characteristic one-dimensional and two-dimensional behaviours.[1] However, this model has been largely ignored by other researchers due to the charge of biological infeasibility. Once the population sizes grow beyond a certain point then at the next time step at least one of the populations becomes negative. This is biologically meaningless and thus represents the main drawback of this model. However the model is extremely useful in demonstrating the effects of global bifurcations. In this paper these bifurcations are analysed using the method of critical curves. This is a technique for investigating the global properties of noninvertible two-dimensional maps[2,3] and is used extensively in this paper when analysing the global properties of the map. Using critical curves the map has previously been studied as a particular model of Cournot duopoly, though for a restricted set of parameter values.[4,5]

The discrete-time model analysed in this paper can easily be deduced from the Lotka-Volterra two species competition model which was originally expressed as a system of differential equations.[6] This model permits either the coexistence of both species or the extinction of one or other of the competing species. It is possible, by discretisation and rescaling, to derive from the original Lotka-Volterra model a discrete time two-dimensional quadratic map $T : \mathbb{R}^2_+ \to \mathbb{R}^2_+$. Denoting the normalised populations of the two species at time k by x_k and y_k respectively, the map determines the time evolution of the two species according to $(x_{k+1}, y_{k+1}) = T(x_k, y_k)$ or

$$\begin{aligned} x_{k+1} &= ax_k(1 - x_k - sy_k) \\ y_{k+1} &= by_k(1 - y_k - tx_k), \end{aligned} \tag{1}$$

where a and b are the intrinsic growth rates, and s and t the (interspecific) competition parameters, of species x and y respectively. Per head of species population, the size of the competition parameter determines whether intraspecific competition is more important (parameter < 1) or whether interspecific competition dominates (parameter > 1). This paper deals with the case where one of the interspecific competition parameters is greater than 1 and the other interspecific competition parameter is less than 1. This subcase shall be referred to as *mixed competition*. In this scenario it is possible to exhibit permanence,[7] bistability[8] or dominance.[9] Permanence occurs when both species coexist whereas dominance is seen when of one species always drives the other species to extinction. Bistability occurs where only one of the species survives but unlike dominance which species survives depends on the initial population densities.

2. General Properties of the Quadratic Map

The evolution of the dynamical system is obtained by the iteration of the two-dimensional map T :

$$\begin{aligned} x' &= ax(1 - x - sy), \\ y' &= by(1 - y - tx), \end{aligned} \tag{2}$$

where a and b are greater than 1, s and t are positive parameters and $'$ denotes the unit-time advancement operator. Since this map (2) is basically a two-dimensional version of the standard logistic map with interactions permitted, there are limitations on the values of the parameters to ensure that the populations remain nonnegative.

The map (2) is defined only for nonnegative values of x and y that lie below the lines $1 - x - sy = 0$ and $1 - y - tx = 0$. More specifically, the *Live Region*(LR) of the map can be defined as follows:

$$LR = \{(x, y);\, x \geq 0,\, y \geq 0,\, 1 - x - sy \geq 0,\, 1 - y - tx \geq 0\}.$$

In the case of mixed competition only one of the lines $1 - x - sy = 0$ and $1 - y - tx = 0$ is active in forming the boundary of LR. Therefore, LR is triangular in shape.

If $(x, y) \in \mathbb{R}^2_+ \setminus LR$, then the next iterate will produce negative values for at least one of the species, which is biologically meaningless. Furthermore, it is important to realise, that under certain conditions an initial value whose starting point lies in LR, may enter into the region $\mathbb{R}^2_+ \setminus LR$ after a finite number of iterations of the map.[10] We shall call a point (x_0, y_0) a *feasible point* if its full trajectory[a] remains in LR. We will refer to such a trajectory as a *feasible trajectory*.[11] Finally, we denote by $\mathcal{F}$ the *feasible set*, defined as the set of feasible points.

There are three axial fixed points of the map T:

$$E_x = \left(\frac{a-1}{a}, 0\right), \quad O = (0, 0), \quad E_y = \left(0, \frac{b-1}{b}\right).$$

The fixed points E_x and E_y are related to the non-zero fixed points of the one-dimensional quadratic maps that govern the dynamics restricted to the invariant axes. There is also another fixed point given by the solution of the equations

$$\begin{aligned} a - ax - asy - 1 &= 0, \\ b - by - btx - 1 &= 0. \end{aligned}$$

We will refer to this off-axis fixed point as E_*. We will concentrate on the situation where O is always an unstable node ($a, b > 1$) as otherwise it will be attracting to trajectories in its local neighbourhood.

The case of mixed competition can be split into two subcases; subcase 1 is when $st < 1$ and subcase 2 is when $st > 1$. In subcase 1, by fixing the growth parameters between 1 and 3 the fixed points E_x and E_y will both be saddle points and the off-axis fixed point E_* will be stable. When the growth parameters are increased beyond 3 the fixed points E_x and

[a]The sequence $(T^k(x_0, y_0))_{k\in\mathbb{N}}$ of values calculated by iterating the map T.

E_y become completely unstable and trajectories starting on the axes are now attracted to the higher period stable points embedded in the axes. For increased values of the parameters the off-axis fixed point, E_*, will be replaced by higher period points. It should be noted too that it is possible for the fixed point E_* to be located outside of LR.[10]

In subcase 2, the fixed point E_* is located outside of LR for most of the choices of parameters and so all generic trajectories starting within LR are attracted to one of the attractors on the coordinate axes. It is possible to configure the map (2) so that the fixed point E_* is located in LR. E_* will be a saddle point in this case and its stable set acts as a separatrix between the attractors embedded in the invariant coordinate axes. These results are obtained through a standard study of the local stability of the fixed points.[10]

2.1. *Critical Curves*

Map (2)[b] belongs to the class of noninvertible maps because solving for (x, y) in terms of a given (x', y') in (2) will typically return no solution or more than one solution. Noninvertible biological maps[12,13] have previously been studied using the method of critical curves.

As the point (x', y') varies in the plane $\mathbb{R}^2$ the number of preimages of (x', y'), changes. Pairs of real preimages appear or disappear as the point (x', y') crosses the boundaries separating regions whose points have a different number of preimages. Such boundaries are usually characterised by the presence of two coincident or merging preimages. This leads to the definition of the *critical curves*. The critical curve of rank-1, denoted by LC (from **L**igne **C**ritique), is defined as the locus of points having two, or more, coincident preimages, located on a set called LC_{-1} (the *curve of merging preimages*). Arcs of LC separate the plane into regions characterised by a different number of real preimages. A region of the plane is labelled "a Z_n region" when its points possess n distinct preimages.

For a continuously differentiable map T the curve of merging first rank preimages LC_{-1} belongs to the set :

$$LC_{-1} \subseteq \left\{(x, y) \in \mathbb{R}^2 \mid \det(DT) = 0\right\}$$

where DT is the Jacobian of T. Also note that LC is the rank-1 image of LC_{-1} under T, i.e. $LC = T(LC_{-1})$.

---

[b]This map (2) has previously been analysed using the method of critical curves; In Bischi and Naimzada (1999)[4] for the parameter values $st = 1/4$, and in Bischi et al. (1998)[5] for $s = t = 1/2$ and $a = b$.

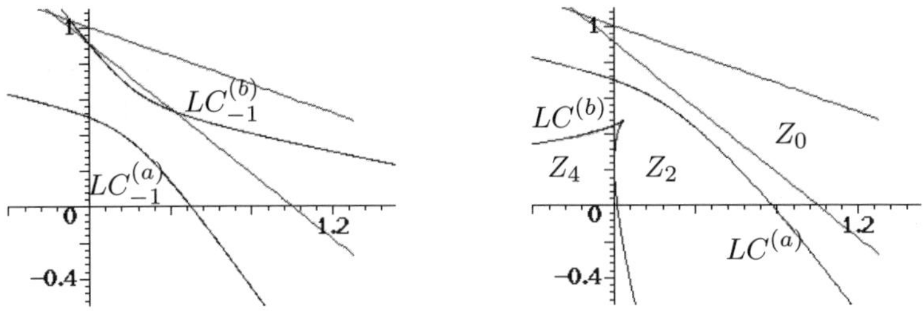

(a) (b)

Fig. 1. (a) Critical curve LC_{-1} in the case when $st < 1$. (b) Critical curve LC in the case when $st < 1$.

In the case where $st < 1$, the locus of points for which $\det(DT) = 0$ is given by the union of two branches of a hyperbola, denoted by $LC_{-1}^{(a)}$ and $LC_{-1}^{(b)}$ in Fig. 1(a). Also LC is the union of two branches, denoted by $LC^{(a)}$ $= \mathrm{T}(LC_{-1}^{(a)})$ and $LC^{(b)} = \mathrm{T}(LC_{-1}^{(b)})$ (Fig. 1(b)): $LC^{(a)}$ separates the region Z_0, whose points have no preimages, from the region Z_2, whose points have two distinct rank-1 preimages, and $LC^{(b)}$ separates the region Z_2 from Z_4, whose points have four distinct rank-1 preimages.

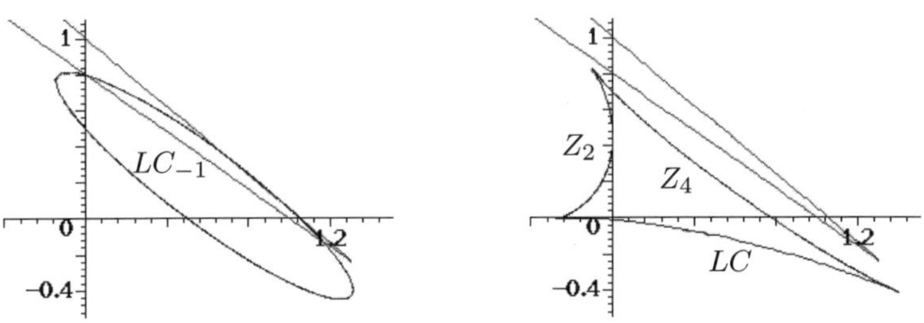

(a) (b)

Fig. 2. (a) Critical curve LC_{-1} in the case when $st > 1$. (b) Critical curve LC in the case when $st > 1$.

In the case where $st > 1$ LC_{-1} is represented by an ellipse (Fig. 2(a)). In this case LC is represented by a closed curve presenting three-cusp points as shown in Fig. 2(b). LC here bounds a region Z_4 surrounded by an unbounded region Z_2.

3. Global Bifurcations

We know from Mira et al. (1996)[3] that global bifurcations are characterised by a contact between the boundary, denoted $\partial\mathcal{F}$, of the feasible set and the arcs of the critical curves LC.

In the case when $st < 1$ the possible biological outcomes from the system are dominance and permanence. Dominance results in all feasible trajectories being attracted to one of the coordinate axis whereas permanence results in an attractor in the interior of LR which leads to more interesting behaviour when analysing global bifurcations.[14]

Recall that if $s > 1$ then the line $1 - x - sy = 0$ forms the internal boundary of LR whereas if $t > 1$ the line $1 - y - tx = 0$ forms the boundary. We shall look at this phenomenon for the case where $s > 1$ although similar results can be found in the case where $t > 1$. Fix the parameters b, s and t and vary the parameter a. As a is increased the critical curve branch, $LC^{(a)}$ which separates Z_0 from Z_2 moves upward and outward (See Fig. 3(a) & (b)). The exact value of $a(= \Phi)$ at which this contact bifurcation occurs is found from the solution of the following expression,

$$(2s[\frac{a}{b} + t] - [4 - a(1 - s)])^2 = [4 - a(1 - s)]^2 - 4s[4t + a(s - t)].$$

A detailed derivation of Φ can be found in Appendix C of.[8]

For $a > \Phi$, immediately after the bifurcation, a segment of $1 - x - sy = 0$ enters the region Z_2 so that the region R_1, bounded completely by $LC^{(a)}$ and $1 - x - sy = 0$ now has two preimages (see Fig. 3(b)). These two preimages merge along $LC_{-1}^{(a)}$ and form a *hole* (or *lake*) inside $\mathcal{F}$.[3] This has the effect of changing the set $\mathcal{F}$ from simply connected to multiply connected (or connected with holes).[3] This can be seen clearly in Fig. 3(c), where the hole H_0 is the preimage of the region R_1. H_0 belongs to the set of points that generate non-feasible trajectories because the points of H_0 are mapped into R_1, outside of LR, and therefore go negative in x on their next iteration. This is only the first of infinitely many preimages of R_1 which form a sequence of smaller and smaller holes within $\mathcal{F}$.

As the parameter a is further increased (yet still remaining below 4), $LC^{(a)}$ moves upwards and outward and the region R_1 enlarges and, consequently, all its preimages enlarge and become infinitely more pronounced. When the parameter a is increased beyond 4, the critical curve $LC^{(a)}$ no longer forms a closed region with the boundary of the feasible region $\mathcal{F}$. This adjustment has the effect of changing the structure of the boundary of $\mathcal{F}$ from smooth to fractal (Fig. 3(d)). Thus it can be concluded that the transformation of the region $\mathcal{F}$ from a simply connected set into a multiply

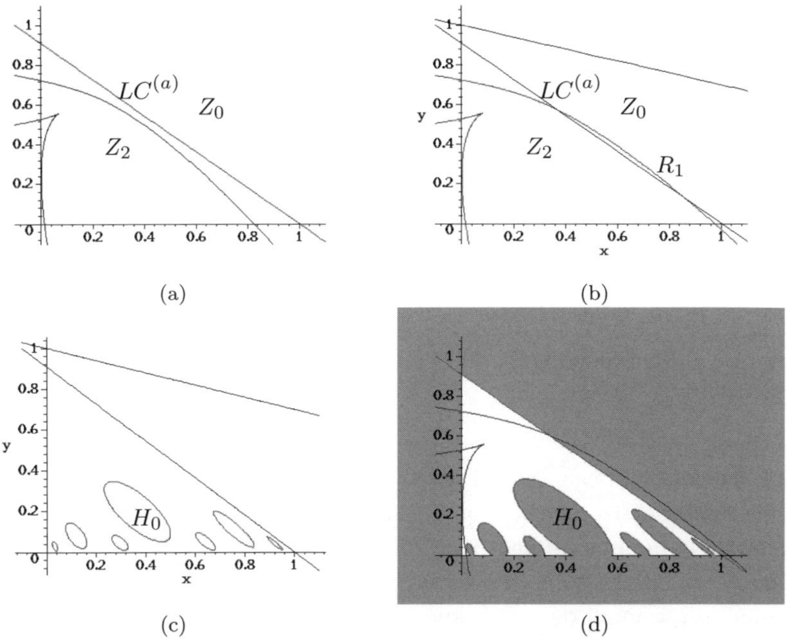

Fig. 3. (a) LC inside $\mathcal{F}$ prior to contact. (b) Crossing between LC and $\partial\mathcal{F}$. (c) The resulting hole H_0 inside $\mathcal{F}$. (d) The boundary of $\mathcal{F}$ is now fractal. The Basin of Infinity is represented by the gray region.

connected set occurs because of a global (or non-classical) bifurcation due to contact between $\partial\mathcal{F}$ and branches of the critical set LC.

The attractor, $\mathcal{A}$, existing inside $\mathcal{F}$ changes its structure for increasing values of the parameters a and b when $st < 1$. For values of a and b less than 3 the attractor is the stable fixed point E_*. As a and b increase above the value 3, E_* loses stability through a flip (or period doubling) bifurcation at which point E_* becomes a saddle point, and an attracting cycle of period 2 is created near it. If we now fix b at some value greater than 3, and further increase a, the cycle of period 2 undergoes a flip bifurcation at which point an attracting cycle of period 4 is created, which becomes the unique attractor inside $\mathcal{F}$. Almost all points inside $\mathcal{F}$ have trajectories that converge to the 4-cycle attractor, so that $\mathcal{F}$ can be identified with the basin of attraction of the attractor for all practical purposes.

These flip bifurcations are followed by a sequence of further period doublings, which creates a sequence of attracting cycles of period 2^n followed

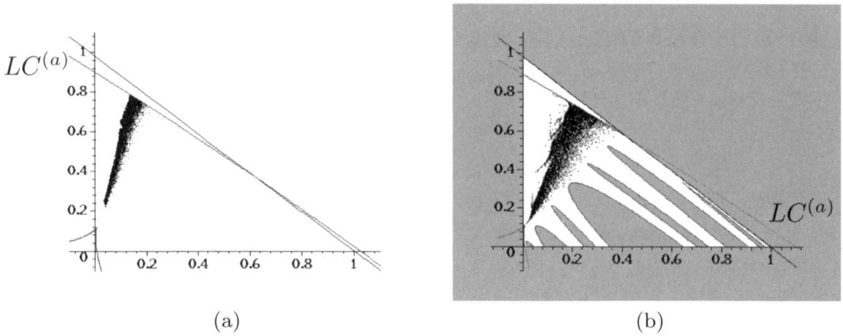

Fig. 4. (a) Chaotic attractor in the interior of $\mathcal{F}$. (b) Just prior to the contact between the attractor $\mathcal{A}$ and $\partial\mathcal{F}$.

by the creation of chaotic attractors (Fig. 4(a)). The numerical simulations show that the size of the chaotic attractor increases as a increases, and with the parameter values that were used in Fig. 4(b) the chaotic set is extremely close to a contact with the boundary of its basin. This contact bifurcation, when it occurs, is known as the *final bifurcation* and causes the destruction of the attractor $\mathcal{A}$.[3] After this type of contact bifurcation, any initial point generates non-feasible trajectories, that is, the growth model does not generate a stable feasible evolution of the population system.

In the case when $st > 1$ it is possible for the system to exhibit all three biological outcomes. In the case of dominance there is no internal fixed point in LR and all internal trajectories end up on one of the coordinate axes. In the case of bistability the internal fixed point is a saddle and again all internal trajectories are attracted to the coordinate axes. When permanence occurs in the system in this case there is no internal fixed point and all trajectories are attracted to a higher period stable point in the interior of LR.[14] We will now focus on the attractor embedded in the y-axis to demonstrate some unusual behaviour in the case when $st > 1$.

Prior to the contact between $1 - x - sy = 0$ and the segment of LC in LR, the attractor on the y-axis exists completely below LC. When the parameters are chosen so that $sb > 4$, LC intersects the y-axis at a value greater than $\frac{1}{s}$ and hence a portion of the attractor embedded in the axis now exists in the interval $(\frac{1}{s}, 1]$ which is above the line $1 - x - sy = 0$. Trajectories will continue to be attracted to the axial attractor but in this instance they will exit LR en route to the attractor. Although this is not a typical contact bifurcation (change in the structure of the attractor or holes within the basin of attraction) it does occur due to a contact between

LC and the boundary of the feasible set. Even though the attractor still exists after this "quasi-final bifurcation", and for trajectories starting on the axis is still viable, all points starting in the interior of LR that are attracted to this attractor will now generate infeasible trajectories en route to the attractor. Therefore for parameter values $sb > 4$ (y-axis) and $ta > 4$ (x-axis) a contact bifurcation occurs that results in infeasible trajectories for points originating in the interior of LR.

4. Conclusions

In this paper, an investigation of the global properties of a two-dimensional competing species model displaying mixed competition is made using the method of critical curves. Critical curves allow us to thoroughly analyse the changing structure of the system as the parameters vary. A general study of the properties of the attractors and of their basins in the case of mixed competition is carried out. The biological relevance of the different cases is presented to highlight the significance of the analysis carried out. In contrast to the two species continuous-time Lotka-Volterra model with mixed competition which results in dominance by one species (e.g.[15]), this discrete-time model is capable of exhibiting permanence, bi-stability or dominance depending on the parameter values.

References

1. D. Blackmore, J. Chen, J. Perez and M. Savescu, *Chaos, Solitons & Fractals* **12**, 2553 (2001).
2. C. Mira, *Chaotic Dynamics* (Singapore: World Scientific, 1987).
3. C. Mira, L. Gardini, A. Barugola and J. C. Cathala, *Chaotic Dynamics in Two-Dimensional Noninvertible Maps* (Singapore: World Scientific, 1996).
4. G. I. Bischi and A. Naimzada, *Advances in Dynamics Games and Applications* **5**, 361 (1999).
5. G. I. Bischi, L. Stefanini and L. Gardini, *Mathematics and Computers in Simulation* **44**, 559 (1998).
6. V. Volterra, *Journal du Conseil International pour l'Exploration de la Mer* **3**, 3 (1928).
7. J. Hofbauer, V. Hutson and W. Jansen, *Journal of Mathematical Biology* **25**, 553 (1987).
8. P. J. Johnson, An investigation of permanence and exclusion in a two-dimensional discrete time competing species model, PhD thesis, University of Limerick, 2006.
9. G. Hardin, *Science* **131**, 1292 (1960).
10. P. J. Johnson and M. E. Burke, *Discrete and Continuous Dynamical Systems - Series B* **10**, 109 (2008).

11. G. I. Bischi, L. Gardini and M. Kopel, *Journal Economic Dynamics & Control* **24**, 855 (2000).
12. R. López-Ruiz and D. Fournier-Prunaret, *Mathematical Biosciences and Engineering* **1**, 307 (2004).
13. R. López-Ruiz and D. Fournier-Prunaret, *Chaos, Solitons & Fractals* **24**, 85 (2005).
14. P. J. Johnson and M. E. Burke, Permanence in a two-dimensional discrete-time competing species model with bounded population densities, (2009).
15. F. Brauer and C. Castillo-Chávez, *Mathematical Models in Population Biology and Epidemiology* (Springer, 2001).

Ordinal Pattern Distributions in Dynamical Systems

K. Keller* and M. Sinn

Institute of Mathematics, University of Lübeck,
Wallstraße 40, D-23560 Lübeck, Germany
**E-mail: keller@math.uni-luebeck.de*

In this work we discuss the structure of ordinal pattern distributions obtained from orbits of dynamical systems. In particular, we consider the extreme cases of systems with a singular pattern distribution and of realizing each ordinal pattern of any order, respectively. Finally, we review results relating the Kolmogorov-Sinai entropy and the topological entropy of one-dimensional dynamical systems to the richness of the underlying ordinal pattern distribution.

Keywords: Ordinal time series analysis; Ordinal pattern; Permutation entropy; Topological Permutation entropy.

1. Introduction

Ordinal time series analysis (see Bandt[4], Keller et al.[7]) is a new approach to the investigation of long and complex time series. From the modelling viewpoint, the time series considered are either realizations of stochastic processes or orbits of dynamical systems. Here we concentrate to the second case. The central subjects are ordinal patterns describing the up and down in orbits.

Definition 1.1. For $d \in \mathbb{N} = \{1, 2, 3, \ldots\}$ we call $\mathcal{I}_d := \times_{l=1}^{d} \{0, 1, \ldots, l\}$ the *set of ordinal patterns of order* d. Further, let $\mathcal{X} \subseteq \mathbb{R}$ and let $f : \mathcal{X} \hookleftarrow$. By the *ordinal pattern of order* $d \in \mathbb{N}$ *realized by* $x \in \mathcal{X}$ we understand the sequence $\mathbf{i}_d = \mathbf{i}_d(x) = (i_1, i_2, \ldots, i_d) \in \mathcal{I}_d$ defined by

$$i_l = \#\{r \in \{0, 1, \ldots, l-1\} \mid f^{\circ(d-l)}(x) \geq f^{\circ(d-r)}(x)\}$$

for $l = 1, 2, \ldots, d$, where $f^{\circ n}(x)$ with $n \in \mathbb{N}$ denotes the n-th iterate of a point $x \in \mathcal{X}$ with respect to f.

If $d \in \mathbb{N}$, $x \in \mathcal{X}$ and $\mathbf{i}_{d+1}(x) = (i_1, i_2, \ldots, i_d, i_{d+1})$, then obviously $\mathbf{i}_d(f(x)) = (i_1, i_2, \ldots, i_d)$. The relationship between $\mathbf{i}_{d+1}(x)$ and

$\mathbf{i}_d(x) = (i'_1, i'_2, \ldots, i'_d)$ is more complicated. As shown in Keller et al.,[7] the $i'_1, i'_2, \ldots, i'_d$ are recursively given by

$$i'_l = \begin{cases} i_{l+1} \text{ if } \sum_{k=1}^{l+1} i_k - \sum_{k=1}^{l-1} i'_k \leq l \\ i_{l+1} - 1 \text{ else} \end{cases}. \tag{1}$$

Here note that originally the ordinal pattern of order d realized by $x \in \mathcal{X}$ was defined as the permutation $\pi_d(x) = (r_0, r_1, \ldots, r_d)$ of the set $\{0, 1, \ldots, d\}$ satisfying

$$f^{\circ\,(d-r_0)}(x) \geq f^{\circ\,(d-r_1)}(x) \geq \ldots \geq f^{\circ\,(d-r_{d-1})}(x) \geq f^{\circ\,(d-r_d)}(x) \tag{2}$$

and

$$r_{l-1} > r_l, \text{ if } f^{\circ\,(d-r_{l-1})}(x) = f^{\circ\,(d-r_l)}(x). \tag{3}$$

(The determination of $\pi_d(x)$ from $\mathbf{i}_d(x)$ and vice versa is simple, but not relevant here.) In the permutation language the ordinal pattern of order d realized by x can be immediately seen from the ordinal pattern of order $d+k$ realized by x for k in $\mathbb{N}$. Namely $\pi_d(x)$ is obtained from $\pi_{d+k}(x) = (r_0, r_1, \ldots, r_{d+k})$ by omitting the entries $0, 1, \ldots, k$ in $(r_0, r_1, \ldots, r_{d+k})$ and subtracting k from the remaining entries. In order to get $\pi_d(f^{\circ k}(x))$ from $\pi_{d+k}(x) = (r_0, r_1, \ldots, r_{d+k})$, one simply has to omit the entries $d+1, d+2, \ldots, d+k$ in $(r_0, r_1, \ldots, r_{d+k})$.

2. Ordinal patterns under iteration

Prerequisites and notions. Let $\mathcal{X} \subseteq \mathbb{R}$ and let $\mathcal{E}$ be a σ-algebra on $\mathcal{X}$. We will consider measurable maps $f : (\mathcal{X}, \mathcal{E}) \hookleftarrow$ with the property that

$$\{x \in \mathcal{X} \,|\, x < f^{\circ n}(x)\} \in \mathcal{E} \text{ for all } n \in \mathbb{N}. \tag{4}$$

Clearly, this property is satisfied in the case $\mathcal{E} = \mathbb{B}(\mathcal{X})$, where $\mathbb{B}(\mathcal{X})$ is the Borel-σ-algebra on $\mathcal{X}$. A probability measure κ on $\mathcal{E}$ is said to be f-*invariant* if $\kappa(f^{-1}(E)) = \kappa(E)$ for all $E \in \mathcal{E}$. Further, let for $(i_l)_{l=1}^d \in \mathcal{I}_d$ with $d \in \mathbb{N}$

$$E_{(i_l)_{l=1}^d} = \{x \in \mathcal{X} \,|\, x \text{ realizes } (i_l)_{l=1}^d\}.$$

Property (4) ensures that $E_{(i_l)_{l=1}^d} \in \mathcal{E}$. Finally, let $\mathbb{N}_0 = \{0, 1, 2, \ldots\}$.

Infinite patterns and the backward map. Ordinal patterns of order d realized by $\mathcal{X}$ classify orbits $(x, f(x), \ldots, f^{\circ d-1}(x), z = f^{\circ d}(x))$ starting at some point $x \in \mathcal{X}$. From the viewpoint of prediction, for example, it is however better to see them as determined from an orbit landing at z. The higher the order d of the ordinal pattern is, the more information on the 'past' of z is given, where higher order patterns are 'extensions' of lower order ones. Therefore, it is natural to consider the set $\mathcal{I}_\infty = \times_{l=1}^{\infty}\{0,1,\ldots,l\}$ of *infinite ordinal patterns*. There is a special dynamics on $\mathcal{I}_\infty$ which we describe now.

First of all, we equip $\mathcal{I}_\infty$ with the topology generated by the set of *elementary cylinder sets*

$$A_{(i_l)_{l=1}^d} = \{i_1\} \times \{i_2\} \times \ldots \times \{i_d\} \times \mathop{\times}_{l=d+1}^{\infty} \{0,1,\ldots,l\}$$

given for $(i_l)_{l=1}^d \in \mathcal{I}_d$ with $d \in \mathbb{N}$. Note that $\mathcal{I}_\infty$ is compact with respect to this topology. The backward map describes the intrinsic structure of ordinal patterns of arbitrary order under iteration (compare text below Def. 1.1).

Definition 2.1. The map $B_\infty : \mathcal{I}_\infty \hookleftarrow$ with $B_\infty((i_l)_{l=1}^\infty) = (i'_l)_{l=1}^\infty$, where $i'_1, i'_2, \ldots$ are recursively defined by

$$i'_l = \begin{cases} i_{l+1} \text{ if } \sum_{k=1}^{l+1} i_k - \sum_{k=1}^{l-1} i'_k \leq l \\ i_{l+1} - 1 \text{ else} \end{cases},$$

is called the *backward map*.

As shown in Keller et al.,[7] the map $B_\infty : \mathcal{I}_\infty \hookleftarrow$ is continuous, hence Borel-measurable. Note that it plays a similar role as the shift-map on $\{1,2,\ldots,n\}^\infty$ in symbolic dynamics for a symbolization decomposing the state space into n pieces.

3. Distribution of ordinal patterns

Given a map f on a measurable space $(\mathcal{X}, \mathcal{E})$ with (4) and an invariant probability measure κ, we are interested in the distribution of ordinal patterns realized by $\mathcal{X}$, which is well-defined by the invariance of κ. This distribution can be considered as the unique measure $\mu = \mu_\kappa$ on the space $(\mathcal{I}_\infty, \mathbb{B}(\mathcal{I}_\infty))$ of infinite ordinal patterns defined by

$$\mu(A_{(i_l)_{l=1}^d}) = \kappa(E_{(i_l)_{l=1}^d}) \text{ for all } d \in \mathbb{N} \text{ and } (i_l)_{l=1}^n \in \mathcal{I}_d.$$

Note that the probability of (finite) ordinal patterns $(i_l)_{l=1}^n \in \mathcal{I}_d$ is here directly addressed by the measure of the elementary cylinder sets $A_{(i_l)_{l=1}^d}$. One easily sees that μ is B_∞-invariant (see Keller et al.[7]).

The richness of the ordinal pattern distribution μ characterizes somehow the complexity of a dynamical system. Let us first describe the case that μ is singular.

Theorem 3.1. *Let $\mathcal{X} \subset \mathbb{R}$, let $\mathcal{E}$ be a σ-algebra on $\mathcal{X}$, and let $f : (\mathcal{X}, \mathcal{E}) \hookleftarrow$ be a measurable map. Further, let κ be an f-invariant probability measure on $(\mathcal{X}, \mathcal{E})$ satisfying (4). Then for $\mu = \mu_\kappa$ the following statements are equivalent:*

(i) μ *is singular.*
(ii) There exist $p \in \mathbb{N}$ *and* $\mathbf{i} \in \mathcal{I}_\infty$ *with* $\mu(\{\mathbf{i}\}) > 0$ *and* $B_\infty^{\circ p}(\mathbf{i}) = \mathbf{i}$.
(iii) There exists a set $E \in \mathcal{E}$ *whose elements realize the same ordinal patterns for each given order and some* $p \in \mathbb{N}$ *with* $f^{\circ p}(E) \subset E$ *and* $\kappa(E) > 0$.

Proof. (i) $\Longrightarrow$ (ii): If μ is singular, then fix some $\mathbf{i} \in \mathcal{I}_\infty$ with $\mu(\{\mathbf{i}\}) > 0$. Invariancy of μ implies $\mu(\{\mathbf{i}\}) = \mu(B_\infty^{-1}(\mathbf{i})) = \mu(B_\infty^{-\circ 2}(\mathbf{i})) = \mu(B_\infty^{-\circ 3}(\mathbf{i})) = \ldots$. Since μ is finite, $B_\infty^{-\circ k}(\mathbf{i}) \cap B_\infty^{-\circ m}(\mathbf{i}) \neq \emptyset$ for some $k, m \in \mathbb{N}_0$ with $k < m$, thus $\mathbf{i} \in B_\infty^{-\circ p}(\{\mathbf{i}\})$ for $p = m - k$. This shows (ii).

(ii) $\Longrightarrow$ (iii): Assuming that (ii) is satisfied with $\mathbf{i} = (i_l)_{l=1}^\infty$, let E_k for $k = 1, 2, \ldots$ be the set of all $x \in \mathcal{X}$ realizing the ordinal pattern $(i_l)_{l=1}^{kp}$. Then $E_1 \supset E_2 \supset E_3 \supset \ldots$ and $f^{\circ p}(E_{k+1}) \subset E_k$ for all $k \in \mathbb{N}$. Therefore, for $E := \bigcap_{k=1}^\infty E_k$ one easily obtains $f^{\circ p}(E) \subset E$. By the construction of μ it holds $\kappa(E_1) \geq \kappa(E_2) \geq \kappa(E_3) \geq \ldots \geq \mu(\{\mathbf{i}\}) > 0$, hence $\kappa(E) > 0$. Since the ordinal pattern of order kp realized by $x \in E$ is equal to $(i_l)_{l=1}^{kp}$ for each $x \in E$, the ordinal patterns realized in E coincide for any given order.

(iii) $\Longrightarrow$ (i): Given an E as described in (iii), let $(i_1^{(d)}, i_2^{(d)}, \ldots, i_d^{(d)})$ denote the ordinal pattern of order $d \in \mathbb{N}$ realized by E. Note that $(i_1^{((k+1)p)}, i_2^{((k+1)p)}, \ldots, i_{kp}^{((k+1)p)}) = (i_1^{(kp)}, i_2^{(kp)}, \ldots, i_{kp}^{(kp)})$ for each $k \in \mathbb{N}$. Hence, there exists some $\mathbf{i} \in \mathcal{I}_\infty$ such that $\bigcap_{k=1}^\infty A_{(i_l^{(kp)})_{l=1}^{kp}} = \mathbf{i}$. Since $\mu(A_{(i_l^{(kp)})_{l=1}^{kp}}) \geq \kappa(E) > 0$ for each $k \in \mathbb{N}$, it follows that $\mu(\{\mathbf{i}\}) > 0$, that is, μ is singular. $\square$

Note the part '(i) $\Longrightarrow$ (ii) of Theorem 3.1 shows that the singular part of $\mu = \mu_\kappa$ is concentrated on a set of points being periodic with respect to the map B_∞.

If ordinal patterns are obtained from a real-valued stochastic process with independent identically distributed continuous random variables, all ordinal patterns of a given order have the same probability. The distribution describing this case adequately is the probability measure Λ on $(\mathcal{I}_\infty, \mathbb{B}(\mathcal{I}_\infty))$ - the equidistribution - defined by

$$\Lambda(A_{(i_l)_{l=1}^d}) = \frac{1}{(d+1)!} \text{ for all } d \in \mathbb{N} \text{ and } (i_l)_{l=1}^n \in \mathcal{I}_d.$$

Let us consider the ordinal pattern distributions relative to this equidistri-

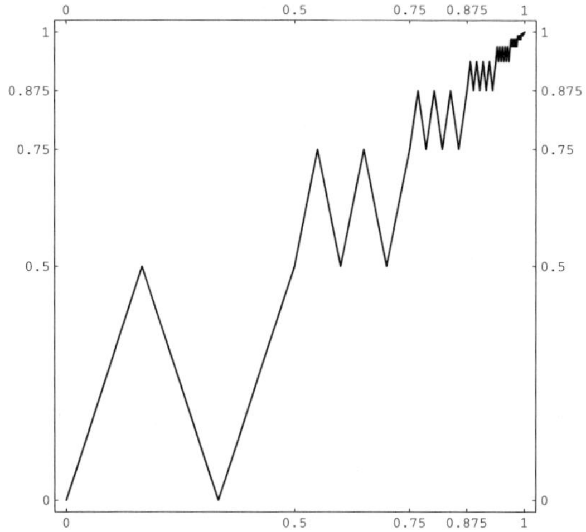

Fig. 1. Map realizing all ordinal patterns

bution. First of all, as shown in Keller et al.,[7] Λ is B_∞-invariant and B_∞ is strongly mixing for Λ. The latter means that

$$\lim_{s\to\infty} \Lambda(B_\infty^{-\circ s}(A_1) \cap A_2) = \Lambda(A_1)\Lambda(A_2) \text{ for all } A_1, A_2 \in \mathbb{B}(\mathcal{I}_\infty),$$

with the consequence of ergodicity of B_∞ for Λ, that is, $\Lambda(B_\infty^{-1}(A) \,\Delta\, A) = 0$ for $A \in \mathbb{B}(\mathcal{I}_\infty)$ implies $\Lambda(A) = 0$ or $\Lambda(A) = 1$ (compare Walters[11]).

The following statement related to ergodicity of B_∞ for Λ and also shown in Keller et al.[7] roughly says that the distribution of ordinal patterns is either very thin or very fat. The *support* C of a probability measure μ on $(\mathcal{I}_\infty, \mathbb{B}(\mathcal{I}_\infty))$ is the intersection of all closed sets D with $\mu(D) = 1$. It holds $\mu(C) = 1$.

Theorem 3.2. *If μ is a B_∞-invariant measure on $(\mathcal{I}_\infty, \mathbb{B}(\mathcal{I}_\infty))$, then for the support C of μ either $\Lambda(C) = 0$ or $\Lambda(C) = 1$.*

Let us give a continuous map f on the interval $[0,1]$ for which the Lebesgue measure λ is f-invariant and μ_λ is concentrated on $\mathcal{I}_\infty$.

Example 3.1. For all $n \in \mathbb{N}$ consider the map $f_n : [0,1] \hookleftarrow$ defined by

$$f_n(x) = \begin{cases} (2n+1)x - k \text{ for } x \in [\frac{k}{2n+1}, \frac{k+1}{2n+1}[;\ k = 0, 2, \ldots, 2n \\ 1 + k - (2n+1)x \text{ for } x \in [\frac{k}{2n+1}, \frac{k+1}{2n+1}[;\ k = 1, 3, \ldots, 2n-1 \\ 1 \text{ for } x = 1 \end{cases}.$$

We want to show that with respect to f_n each ordinal pattern of order $2n$ is realized by the points of a non-empty open subset of $[0,1]$. Let $J_k =]\frac{k}{2n+1}, \frac{k+1}{2n+1}[$ for $k = 0, 1, 2, \ldots, 2n$. Given a permutation

$$\pi = \begin{pmatrix} 0 & 1 & 2 & \ldots & 2n \\ \pi(0) & \pi(1) & \pi(2) & \ldots & \pi(2n) \end{pmatrix}$$

of $\{0, 1, \ldots, 2n\}$, fix successively $x_{2n} \in J_{\pi^{-1}(0)}, x_{2n-1} \in J_{\pi^{-1}(1)}, x_{2n-2} \in J_{\pi^{-1}(2)}, \ldots, x_0 \in J_{\pi^{-1}(2n)}$ with $f_n(x_{l-1}) = x_l$ for all $l = 1, 2, \ldots, 2n$. This is possible since each of the intervals J_k is mapped onto $]0,1[$. One easily sees that x_0 realizes the ordinal pattern π and that there exists a neighborhood of x_0 consisting of points realizing the same ordinal pattern.

Now for $n \in \mathbb{N}_0$ let $a_n = 1 - 2^{-n}$. Then clearly $[0,1[$ is the disjoint union of the intervals $I_n = [a_{n-1}, a_n[; n \in \mathbb{N}$ having length 2^{-n}. Let $f : \mathcal{X} = [0,1] \hookleftarrow$ be defined by

$$f(x) = \begin{cases} a_{n-1} + 2^{-n} f_n(2^n(x - a_{n-1})) & \text{for } x \in I_n \\ 1 & \text{for } x = 1\,. \end{cases}$$

f is continuous and maps each of intervals $[a_{n-1}, a_n]$ onto itself 'like as small copy of f_n' (see Fig. 3). This shows that f realizes each ordinal pattern of arbitrary order on a non-empty open set. It can easily be seen that the Lebesgue measure is f-invariant.

4. Permutation entropy and topological permutation entropy

With the exception of a few number of extreme cases, the cardinality of ordinal patterns of order d realized by a map f is increasing for d. The rate of this increasing can be used for quantifying the complexity of a given map. This more generally leads to the concepts of permutation entropy and topological permutation entropy introduced by Bandt et al.:[6]

Definition 4.1. For $\mathcal{X} \subseteq \mathbb{R}$ and $f : \mathcal{X} \hookleftarrow$ let $\mathcal{X}_d$ be the set of points x with different iterates $x, f(x), f^{\circ 2}(x), \ldots, f^{\circ d}(x)$. Further, let

$$h^*_{0,d} = \ln(\#\{\mathbf{i} \in \mathcal{I}_d \mid E_{\mathbf{i}} \cap \mathcal{X}_d \neq \emptyset\}).$$

If $\mathcal{E}$ is a σ-algebra with (4), f is measurable and κ is an f-invariant measure on $(\mathcal{X}, \mathcal{E})$, let

$$h^*_{\kappa,d} = -\sum_{\mathbf{i} \in \mathcal{I}_d} \kappa(E_{\mathbf{i}} \cap \mathcal{X}_d) \ln \kappa(E_{\mathbf{i}} \cap \mathcal{X}_d).$$

Then $h_0^* = \limsup_{d\to\infty} \frac{h_{0,d}^*}{d}$ and $h_\kappa^* = \limsup_{d\to\infty} \frac{h_{\kappa,d}^*}{d}$ is called *topological permutation entropy* of f and *permutation entropy* of f with respect to κ, respectively.

Remark 4.1. Note that the exclusion of points with some coinciding iterates in the definition of $h_{0,d}^*$ and $h_{\kappa,d}^*$ seems to be not substantial in the most cases. Thus for the 'normal' cases $h_\kappa^* = \limsup_{d\to\infty} \frac{-\sum_{\mathfrak{i}\in\mathcal{I}_d} \mu(A_\mathfrak{i}) \ln \mu(A_\mathfrak{i})}{d}$ for $\mu = \mu_\kappa$, had been used to define permutation entropy. This definition would extend to a quantity related to a B_∞-invariant measure μ on $(\mathcal{I}_\infty, \mathbb{B}(\mathcal{I}_\infty))$.

In the case that all ordinal patterns are realized by a map f as in Example 3.1, the topological permutation entropy is infinite. Since according to Stirling's formulae $\lim_{n\to\infty} \frac{\ln n!}{(n+\frac{1}{2})\ln n - n + \frac{1}{2}\ln(2\pi)} = 1$, it holds

$$\lim_{d\to\infty} \frac{h_{0,d}^*/d}{\ln d} = 1,$$

in particular, the topological permutation entropy is infinite (see also Amigó et al.[3]). Bandt et al.[6] have shown the following interesting statement:

Theorem 4.1. *Let $f : I \hookleftarrow$ be a piecewise monotone interval map, i.e. a continuous map on a compact interval I decomposing into finitely many subintervals where the map is monotone. Then the following holds:*

(i) The topological entropy of f is equal to h_0^*.
(ii) For each f-invariant measure κ on $(I, \mathbb{B}(I)$ the Kolmogorov-Sinai entropy of f is equal to h_κ^*.

Note that for piecewise monotone interval maps the permutation entropy is determined by ordinal patterns in the periodic orbits (see Misiurewicz[9]). Since piecewise monotone interval maps have finite topological entropy (see Misiurewicz and Szlenk,[10] there must be non-realizable ordinal patterns (see Keller et al.[7], Amigó et al.[3]), hence the following is valid (Keller et al.[7]):

Corollary 4.1. *If $f : I \hookleftarrow$ is a piecewise monotone interval map and κ an f-invariant measure on $(I, \mathbb{B}(I))$, then for the support C of μ_κ it holds $\Lambda(C) = 0$.*

By the above result, for ergodic piecewise monotone interval maps the Kolmogorov-Sinai can principally be estimated using the empirical permutation entropy of a part of an orbit. Unfortunately, there is not much known about for which maps (i) resp. (ii) of Theorem 4.1 is valid. In this connection, note that Misiurewicz[9] has given an example of an interval map with zero topological entropy but non-zero permutation entropy.

In order to generalize Theorem 4.1 in a certain sense, Amigó et al.[1] and Amigó and M. B. Kennel[2] have given relaxed definitions of permutation entropy and topological permutation entropy. Roughly speaking, they mimic the Kolmogorov-Sinai entropy and topological entropy by considering permutation entropy and topological entropy of a coarse-graining and then letting the diameter of pieces go to zero. For ergodic and expanding interval maps they get (i) and (ii) of Theorem 4.1, repectively, for the relaxed definitions. Note that from the practical viewpoint, the definitions are not too different from the original ones, because each computer uses finite precision. It is an advantage of the approaches of Amigo and coauthors that their approach also includes multidimensional interval maps.

References

1. J. M. Amigó, M. B. Kennel and L. Kocarev, The permutation entropy rate equals the metric entropy rate for ergodic information sources and ergodic dynamical systems, *Physica D* 210 (2005), 77 – 95.
2. J. M. Amigó, M. B. Kennel. Topological Permutation Entropy. *Physica D* 231 (2007), 137 – 142.
3. J. M. Amigó, S. Zambrano, M.A.F. Sanjuán, True and false forbidden patterns in deterministic and random dynamics, *Europhys. Lett.* 79 (2007), 50001.
4. C. Bandt, Ordinal time series analysis, *Ecological modelling* 182 (2005), 229 – 238.
5. C. Bandt, B. Pompe, Permutation entropy: A natural complexity measure for time series, *Phys. Rev. Lett.* 88 (2002), 174102.
6. C. Bandt, G. Keller, B. Pompe, Entropy of interval maps via permutations, *Nonlinearity* 15 (2002), 1595 – 1602.
7. K. Keller, J. Emonds, M. Sinn, Time series from the ordinal viewpoint, *Stochastics and Dynamics* 2 (2007), 247 – 272.
8. K. Keller, H. Lauffer, M. Sinn, Ordinal analysis of EEG time series, *Chaos and Complexity Letters* 2 (2007), 247 – 258.
9. M. Misiurewicz, Permutations and topological entropy for interval maps, *Nonlinearity* 16 (2003), 971 – 976.
10. M. Misiurewicz, W. Szlenk, Entropy of piecewise monotone mappings. *Stud. Math.* 67 (1980), 45 – 63.
11. P. Walters, *An Introduction to Ergodic Theory*, Springer, New York, 1982.

Discretized Pantograph Equation with a Forcing Term: Note on Asymptotic Estimate

Petr Kundrát

Institute of Mathematics, FME BUT, Technická 2,
Brno, 616 69, Czech republic
E-mail: kundrat@fme.vutbr.cz

The paper deals with the asymptotic estimate of solutions of a difference equation, which arises as a discretization of pantograph equation with a forcing term. The term with delayed argument is approximated via linear interpolation between the closest mesh points. The derived asymptotic estimate is compared with the estimate corresponding to the continuous counterpart.

Keywords: Asymptotic estimate, difference equation, MSC: 39A11, 34K25.

1. Introduction

The main goal of the article is to formulate an asymptotic estimate of solution of the difference equation

$$x_{n+1} - x_n = -ahx_n + bh((1-r_n)x_{\lfloor \lambda n \rfloor} + r_n x_{\lfloor \lambda n \rfloor + 1}) + hf_n, \qquad (1)$$

where $a > 0$, $b \neq 0$, $0 < \lambda < 1$ are reals, $|b|/a < 1$, $r_n := \lambda n - \lfloor \lambda n \rfloor$, $n = 0, 1, 2, \ldots$, $h > 0$ is the stepsize and $\lfloor * \rfloor$ means the floor function of the argument $*$. The term f_n is a given sequence of reals. The investigation of this difference equation was motivated in the connection to the numerical discretization of

$$\dot{x}(t) = -ax(t) + bx(\lambda t) + f(t), \qquad t \in I := [0, \infty) \qquad (2)$$

via Euler method, where the term with the proportional delay is approximated via linear interpolation. Really, if $t_n := nh$ and $f_n := f(t_n)$, then $x_n \approx x(t_n)$, $n = 0, 1, 2, \ldots$. Equation (2) (in the vector form and without the forcing term) is in the scientific literature referred to as a pantograph equation. This name comes from an application on British railways,[1] where the motion of electric locomotive's pantograph along a trolley wire has been studied. There was many publications dealing with the properties of pantograph equation and its modifications. Let us recall e.g.[2–6] The discrete cases have been studied e.g. in[7–10] and many others.

2. Preliminaries

In this section we mention asymptotic estimate of solution of Eq. (2) and formulate some assumptions, which will be considered to derive the asymptotic estimate of solution in the discrete case (1).

In the continuous case, solutions of auxiliary functional equation

$$a\psi(t) = |b|\psi(\lambda t), \tag{3}$$

are often utilized to derive the asymptotic estimate of solutions of the pantograph equation. Dealing with the discrete case, instead of the above functional equation (3) we similarly consider the functional inequalities

$$a\rho(t) \geq |b|\rho(\lfloor \lambda t/h \rfloor h), \qquad t \in I. \tag{4}$$

and

$$a\rho(t) \geq |b| \left[(1 - r(t))\rho(\lfloor \lambda t/h \rfloor h) + r(t)\rho((\lfloor \lambda t/h \rfloor + 1)h)\right], \qquad t \in I, \tag{5}$$

where $r(t) = \lambda t/h - \lfloor \lambda t/h \rfloor$.

Remark 2.1. We can easily verify that if $|b|/a < 1$, then $\rho(t) = (t + h/(1-\lambda))^\alpha$, $\alpha = \log(|b|/a)/\log \lambda^{-1}$ is a positive continuous and decreasing solution of (4) such that $\rho(t+\tilde{h}) - \rho(t)$ is nondecreasing on I for arbitrary real $0 < \tilde{h} \leq h$. Moreover, the decreasing solution of (4) is also a solution of (5) provided $|b|/a < 1$, as one can see from the relation

$$\begin{aligned} a\rho(t) &\geq |b|\rho(\lfloor \tfrac{\lambda t}{h} \rfloor h) \\ &\geq |b| \left((1 - r(t))\rho(\lfloor \tfrac{\lambda t}{h} \rfloor h) + r(t)\rho((\lfloor \tfrac{\lambda t}{h} \rfloor + 1)h)\right), \quad t \in I. \end{aligned}$$

In the end of this section we recall the relevant asymptotic estimate for the delay differential equation (2), which was derived in.[11]

Theorem 2.1. *Let $a > 0$, $f(t) \in C^1([t_0, \infty))$ and $f(t) = O(t^\alpha)$, $\dot{f}(t) = O(t^{\alpha-1})$ as $t \to \infty$, where $\alpha = \log(|b|/a)/\log \lambda^{-1}$. Then the asymptotic estimate*

$$x(t) = O(t^\alpha \log t) \qquad \text{as} \quad t \to \infty \tag{6}$$

holds for any solution $x(t)$ of (2).

The goal of the next section is to find conditions under which we obtain the same asymptotic estimate also in the corresponding discrete case.

3. Main result

Theorem 3.1. *Let x_n be a solution of equation (1), where $0 < ah < 1$, $b \neq 0$, $|b|/a < 1$ and let $f_n = O(n^\alpha)$, where $\alpha = \log(|b|/a)/\log(\lambda^{-1})$. Then $x_n = O(n^\alpha \log n)$ as $n \to \infty$.*

Proof. Let $\tilde{a}$ be such that $\tilde{a}^h = 1 - ah$. Then we can rewrite the difference equation (1) as

$$x_{n+1} = \tilde{a}^h x_n + bh((1 - r_n)x_{\lfloor \lambda n \rfloor} + r_n x_{\lfloor \lambda n \rfloor + 1}) + hf_n, \quad n = 1, 2, 3, \ldots. \tag{7}$$

We introduce the substitution $y_n = x_n/\rho_n$ in (7) to obtain

$$\rho_{n+1} y_{n+1} = \tilde{a}^h \rho_n y_n + bh((1-r_n)\rho_{\lfloor \lambda n \rfloor} y_{\lfloor \lambda n \rfloor} + r_n \rho_{\lfloor \lambda n \rfloor + 1} y_{\lfloor \lambda n \rfloor + 1}) + hf_n, \tag{8}$$

where $\rho_n = \rho(t_n)$, $n = 1, 2, 3$. Function $\rho(t)$ satisfies the inequality (5) with the properties mentioned in Remark 2.1. Our aim is to show that the estimate $y_n = O(\log n)$ as $n \to \infty$ holds for every solution y_n of Eq. (8). Multiplying Eq. (8) by $1/\tilde{a}^{t_{n+1}}$ we get

$$\frac{\rho_{n+1} y_{n+1}}{\tilde{a}^{t_{n+1}}} = \frac{\rho_n y_n}{\tilde{a}^{t_n}} + \frac{bh}{\tilde{a}^{t_{n+1}}}((1-r_n)\rho_{\lfloor \lambda n \rfloor} y_{\lfloor \lambda n \rfloor} + r_n \rho_{\lfloor \lambda n \rfloor + 1} y_{\lfloor \lambda n \rfloor + 1}) + \frac{hf_n}{\tilde{a}^{t_{n+1}}},$$

i.e.,

$$\Delta\left(\frac{\rho_n y_n}{\tilde{a}^{t_n}}\right) = \frac{bh}{\tilde{a}^{t_{n+1}}}((1 - r_n)\rho_{\lfloor \lambda n \rfloor} y_{\lfloor \lambda n \rfloor} + r_n \rho_{\lfloor \lambda n \rfloor + 1} y_{\lfloor \lambda n \rfloor + 1}) + \frac{hf_n}{\tilde{a}^{t_{n+1}}}. \tag{9}$$

Now we construct a sequence of intervals $I_m, m = 0, 1, 2, \ldots$ such that $\cup_{m=0}^{\infty} I_m = I$ and the function λt is mapping I_{m+1} onto I_m for all $m = 1, 2, \ldots$: We put $T_{-1} = 0$ and $T_m = \lambda^{-m} h$, $m = 0, 1, 2, \ldots$. Then we set $I_m := [T_{m-1}, T_m]$.

Let us take any $t_v \in I_{m+1}$, $m = 1, 2, \ldots$. We define nonnegative integers $k_m(t_v) := \lfloor (t_v - T_m)/h \rfloor$. Denote $t_u := t_v - k_m(t_v)h - h$. Summing the equation (9) from t_u to t_{v-1}, we get

$$y_v = \frac{\rho_u \tilde{a}^{t_v}}{\rho_v \tilde{a}^{t_u}} y_u + \frac{\tilde{a}^{t_v}}{\rho_v} \sum_{s=u}^{v-1} \frac{h}{\tilde{a}^{t_{s+1}}}(b((1-r_s)\rho_{\lfloor \lambda s \rfloor} y_{\lfloor \lambda s \rfloor} + r_s \rho_{\lfloor \lambda s \rfloor + 1} y_{\lfloor \lambda s \rfloor + 1}) + f_s).$$

Furthermore, we denote $M_m := \sup\left\{|y_i|, \quad t_i \in \bigcup_{j=0}^{m} I_j\right\}$. In accordance with (5) and the assumption on f_n we obtain

$$|y_v| \leq \frac{\rho_u \tilde{a}^{t_v}}{\rho_v \tilde{a}^{t_u}} M_m + \frac{\tilde{a}^{t_v}}{\rho_v} \sum_{s=u}^{v-1} \frac{(1 - \tilde{a}^h)\rho_s}{\tilde{a}^{t_{s+1}}} M_m + \frac{\tilde{a}^{t_v}}{\rho_v} \sum_{s=u}^{v-1} \frac{hK\rho_s}{\tilde{a}^{t_{s+1}}},$$

where $K > 0$ is a suitable real constant. Using the relation $\frac{1-\tilde{a}^h}{\tilde{a}^{t_{s+1}}} = \Delta\left(\frac{1}{\tilde{a}}\right)^{t_s}$ we get

$$|y_v| \leq \frac{\rho_u \tilde{a}^{t_v}}{\rho_v \tilde{a}^{t_u}} M_m + \left(M_m + \frac{hK}{1-\tilde{a}^h}\right) \frac{\tilde{a}^{t_v}}{\rho_v} \sum_{s=u}^{v-1} \Delta \left(\frac{1}{\tilde{a}}\right)^{t_s} \rho_s,$$

and summing by parts we obtain

$$\begin{aligned} |y_v| &\leq \frac{\rho_u \tilde{a}^{t_v}}{\rho_v \tilde{a}^{t_u}} M_m + \left(M_m + \frac{hK}{1-\tilde{a}^h}\right) \frac{\tilde{a}^{t_v}}{\rho_v} \left(\frac{\rho_v}{\tilde{a}^{t_v}} - \frac{\rho_u}{\tilde{a}^{t_u}} - \sum_{s=u}^{v-1} \left(\frac{1}{\tilde{a}}\right)^{t_{s+1}} \Delta\rho_s\right) \\ &= \left(M_m + \frac{hK}{1-\tilde{a}^h}\right)\left(1 - \frac{\tilde{a}^{t_v}}{\rho_v} \sum_{s=u}^{v-1} \left(\frac{1}{\tilde{a}}\right)^{t_{s+1}} \Delta\rho_s\right) - \frac{hK}{1-\tilde{a}^h} \frac{\rho_u \tilde{a}^{t_v}}{\rho_v \tilde{a}^{t_u}} \leq \\ &\leq \left(M_m + \frac{hK}{1-\tilde{a}^h}\right)\left(1 - \frac{\tilde{a}^{t_v}}{\rho_v} \sum_{s=u}^{v-1} \left(\frac{1}{\tilde{a}}\right)^{t_{s+1}} \Delta\rho_s\right). \end{aligned} \tag{10}$$

Now, we can take $\rho = (t + h/(1-\lambda))^\alpha$, $\alpha = \frac{\log \frac{|b|}{a}}{\log \lambda^{-1}}$ as a decreasing solution of (5). Let us recall that $\Delta\rho$ is nondecreasing on I (see Remark 2.1). Then from (10) we obtain

$$\begin{aligned} |y_v| &\leq \left(M_m + \frac{hK}{1-\tilde{a}^h}\right)\left\{1 + \frac{\rho_u - \rho_{u+1}}{\rho_v} \sum_{s=u}^{v-1} \left(\frac{\tilde{a}^{t_v}}{\tilde{a}^{t_{s+1}}}\right)\right\} \\ &\leq \left(M_m + \frac{hK}{1-\tilde{a}^h}\right)\left\{1 + \frac{\rho_u - \rho_{u+1}}{\rho_v}\xi\right\} \\ &\leq \left(M_m + \frac{hK}{1-\tilde{a}^h}\right)\left\{1 + \xi\frac{-\Delta\rho(T_m - h)}{\rho(T_{m+1})}\right\}, \end{aligned}$$

where $\xi := 1/(1-\tilde{a}^h)$. The repeated application of this procedure yields

$$|y_v| \leq \left(M_1 + \frac{mhK}{1-\tilde{a}^h}\right) \prod_{j=1}^{m} \left(1 + \xi\frac{-\Delta\rho(T_j - h)}{\rho(T_{j+1})}\right),$$

i.e.,

$$M_{m+1} \leq \left(M_1 + \frac{mhK}{1-\tilde{a}^h}\right) \prod_{j=1}^{m} \left(1 + \xi\frac{-\Delta\rho(T_j - h)}{\rho(T_{j+1})}\right). \tag{11}$$

Substituting the solution $\rho = (t + h/(1-\lambda))^\alpha$ and utilizing the binomial theorem it could be shown that $\sum_{j=1}^{\infty} \xi\frac{-\Delta\rho(T_j - h)}{\rho(T_{j+1})}$ is convergent as $n \to \infty$. Hence, the product in (11) converges as $m \to \infty$ as well and M_m fulfills the asymptotic estimate $M_m \leq O(m)$ as $m \to \infty$. Since $t_v \in I_{m+1}$, we have $m \leq \log(t_v/h)/\log\lambda^{-1} \leq m+1$. Hence, the asymptotic estimate $y_n = O(\log n)$ as $n \to \infty$ holds. □

4. Examples and final remarks

Remark 4.1. The assumption on the stepsize h ($h < 1/a$) enables us to preserve the correlation of asymptotic estimates of discrete and continuous case. In other words, the estimates of solutions in the discrete case and the continuous case are expressed via the same function, resp. sequence (provided the stepsize h is sufficiently small).

Example 4.1. Consider the initial problem:

$$\dot{x}(t) = -9x(t) + \frac{1}{3}x(t/3) + \frac{2}{(t+1)^3}, \quad x(0) = 1, \quad t \in [0, \infty). \tag{12}$$

In accordance with (6) we get the asymptotic estimate

$$x(t) = O(t^{-3} \log t) \qquad \text{as} \quad t \to \infty.$$

In the corresponding discrete case we consider formula (1) in the form

$$x_{n+1} = (1 - 9h)x_n + \frac{h}{3}((1 - r_n)x_{\lfloor n/3 \rfloor} + r_n x_{\lfloor n/3 \rfloor + 1}) + \frac{2h}{(nh + 1)^3}, \tag{13}$$

where $x_0 = 1$, $r_n = n/3 - \lfloor n/3 \rfloor$. Then according to Theorem 3.1 we get $x_n = O(n^{-3} \log n)$ as $n \to \infty$ provided $h < 1/a$. If we violate the condition on stepsize, this asymptotic formula is not valid. Indeed, if we set $h = 1 > 1/a$, then the corresponding discrete equation

$$x_{n+1} = -9x_n + \frac{1}{3}\left((1 - r_n)x_{\lfloor n/3 \rfloor} + r_n x_{\lfloor n/3 \rfloor + 1}\right) + \frac{2}{(n+1)^3}$$

admits unbounded solutions as $n \to \infty$ (one can see that the relation $|x_n| \leq |x_{n+1}|$ holds for $n = 0, 1, 2, \ldots$ provided $t_0 = 0$ and $x_0 = 1$).

Remark 4.2. The above mentioned result follows the previous investigation in,[12] where Eq. (2) was discretized via Euler method, but with a piecewise constant approximation of the delayed term. Although we have derived the result only for $|b|/a < 1$, the same result can be derived also for the case $|b|/a \geq 1$ applying analogous proof technique.

Acknowledgments

Published results were acquired using the subsidization of the Ministry of Education, Youth and Sports of the Czech Republic, research plan MSM 0021630518 "Simulation modelling of mechatronic systems".

References

1. J. R. Ockendon and A. B. Tayler, The Dynamics of a Current Collection System for an Electric Locomotive, in *Proc. Roy. Soc. London Ser. A* **322** (1971), pp. 447–468.
2. J. Čermák, The Asymptotics of Solutions for a Class of Delay Differential Equations, *Rocky Mountain J. Mathematics* **33** (2003), pp. 775–786.
3. M. L. Heard, A Change of Variables for Functional Differential Equations, *J. Differential Equations* **18** (1975), pp. 1–10.
4. A. Iserles, Exact and Discretized Stability of the Pantograph Equation, *Appl. Numer. Math.* **24** (1997) 295–308.
5. Y. Liu, Asymptotic Behaviour of Functional-Differential Equations with Proportional Time Delays, *European J. Appl. Math.* **7** (1996), pp. 11–30.
6. T. Kato and J. B. McLeod, The Functional Differential Equation $y'(x) = ay(\lambda x) + by(x)$, *Bull. Amer. Math. Soc.* **77** (1971), pp. 891–937.
7. I. Györi and M. Pituk, Comparison Theorems and Asymptotic Equilibrium for Delay Differential and Difference Equations, *Dynam. Systems Appl.* **5** (1996), pp. 277–302.
8. P. Kundrát, Asymptotic Properties of the Discretized Pantograph Equation, *Studia Univ. Babeş-Bolyai, Mathematica,* **L** 1 (2005), pp. 77–84.
9. G. Makay and J. Terjéki, On the Asymptotic Behaviour of the Pantograph Equations, *Electron. J. Qual. Theory Differ. Equ.* **2** (1998), pp. 1–12.
10. H. Péics, On the Asymptotic Behaviour of Difference Equations with Continuous Arguments, it Ser A. Math. Anal. **9** 2 (2002), pp. 257–273.
11. J. Čermák, A Change of Variables in the Asymptotic Theory of Differential Equations with Unbounded Delays, *J. Computational and Appl. Math.* **143** (2002), pp. 81–93.
12. P. Kundrát, Asymptotic Estimation of the Discrete Pantograph Equation with a Forcing Term, *Folia FSN Universitatis Masarykianae Brunensis, Mathematica* **16** (2007), pp. 101–106.

General Theory of Linear Functional Equations in the Space of Strictly Monotonic Functions

Jitka Laitochová

Department of Mathematics, Palacký University,
Olomouc, 771 40, Czech Republic
E-mail: jitka.laitochova@upol.cz
www.upol.cz

The classical theory of k-th order linear functional and difference equations is obtained as a special case of the theory developed here for the k-th order functional equation model which generalizes the first order model $f \circ \phi - Af = B$. The function ϕ belongs to a special space S of continuous strictly monotonic functions equipped with a group multiplication $\circ$. The equation's domain can be a finite interval, a half-line or the real line.

Announced here are the main theorems on solution structure for k-th order homogeneous linear functional equations in S.

Keywords: Linear functional equation, homogeneous structure, space of continuous strictly monotonic functions, group multiplication.

1. Introduction

Studied here is the general theory for generalized k-th order homogeneous linear functional and difference equations in the space S of strictly monotonic functions of the form

$$a_k(x)\left(f \circ \phi^k\right)(x) + \cdots + a_1(x)\left(f \circ \phi^1\right)(x) + a_0(x)\left(f \circ \phi^0\right)(x) = 0. \quad (*)$$

Particular and general solutions of these equations are defined and fundamental theorems on solution structure are stated. Linearly dependent and independent functions in S are introduced so the fundamental system of solutions of equation (*) can be defined and used in the new theory, which is analogous to the classical theory of functional and difference equations [1-5].

Based on this general theory of linear functional and difference equations are the results of the author's papers [6-8], where also applications appear. The first paper [6] uses the theory developed here to solve generalized Abel functional equations in S. The second paper [7] applies the theory to solve equation (*) with constant coefficients using roots of the characteristic equation and a continuous solution of the associated Abel functional equation $(\alpha \circ \phi)(x) = X(x+1) \circ \alpha(x)$. In the third paper [8], iterative solution formulas are given for the first order linear equation $(f \circ \phi)(x) = Af(x) + B$ to produce approximate values for the solution $f(x)$ and for computer assisted visualization of solutions. The two cases are considered there: when A, B are constant and when A, B depend on x. Existence-uniqueness for kth order equation (*) appears in [8], pages 114-116.

Preliminaries

Notation: Let $\mathcal{J} = (-\infty, \infty)$, $\mathbf{Z}$ the set of integers, $\mathbf{R}$ the set of real numbers, $\bar{\mathbf{R}} = \langle -\infty, \infty \rangle$ and $C_0(\mathcal{J})$ the set of continuous functions on the interval $\mathcal{J}$.

Definition 1.1. Let $a, b \in \bar{\mathbf{R}}$, $a < b$. The set of all functions f which satisfy $f \in C_0(\mathcal{J})$ and f maps one-to-one the interval $\mathcal{J}$ on the interval (a, b) will be denoted by symbol S and called a *space of strictly monotonic functions.*

Definition 1.2. An arbitrarily chosen increasing function $X \in S$ will be called a *canonical function* in S. The inverse to the canonical function X will be denoted by X^*.

Definition 1.3. Let $\alpha, \beta \in S$. Let X^* be the inverse to the canonical function $X \in S$. The composite function

$$\gamma(x) = \alpha(X^*(\beta(x)))$$

will be called a *product* of functions $\alpha, \beta \in S$ in the class S and we write $\gamma = \alpha \circ \beta$. The product γ is defined on $\mathcal{J}$.

Note. The set S, together with the operation $\circ$, forms a non-commutative group [6]. The neutral element is X, the canonical function, and the inverse of α is $\hat{\alpha}(x) = X(\alpha^*(X(x))$, where α^* is the inverse to function α.

Definition 1.4. Let $X \in S$ be the canonical function. Let $\phi \in S$. The *iterations* of a function ϕ in S are functions in S defined by

$$\phi^0(x) = X(x),$$

$$\phi^{n+1}(x) = (\phi \circ \phi^n)(x), \quad x \in \mathcal{J}, \quad n = 0, 1, 2, \ldots,$$

$$\phi^{n-1}(x) = \left(\phi^{-1} \circ \phi^n\right)(x), \quad x \in \mathcal{J}, \quad n = 0, -1, -2, \ldots,$$

where $\phi^{-1} = \hat{\phi}$, defined by $\hat{\phi}(x) = X(\phi^*(X(x))$, is the inverse element to the element ϕ in S according to the multiplication $\circ$.

Definition 1.5. Let $\phi \in S$. A continuous function $p = p(x)$ on $\mathcal{J}$ satisfying $(p \circ \phi)(x) = p(x)$ for $x \in \mathcal{J}$ is called *automorphic over* ϕ.

An example is any 1-periodic continuous function p on $\mathcal{J}$ and $\phi(x) = x + 1$. For instance, define a 1-periodic function p by $p(x) = \sin(\pi x)$ on $0 \le x \le 1$. Then $p(x) = p(x+1) = (p \circ \phi)(x)$ for all x. Constant functions are automorphic over ϕ, for any ϕ.

Symbol U_ϕ denotes the set of all automorphic functions p over ϕ which satisfy $|p(x)| > 0$ for $x \in \mathcal{J}$, together with the zero function.

Assume hereafter $\phi \in S$, $X \in S$ is a canonical function, $\phi(x) > X(x), x \in \mathcal{J}$. Let $x_0 \in \mathcal{J}$ denote an arbitrary point of $\mathcal{J}$ and let X^* be the inverse to X. Define point iterates $\{x_\mu\}$ and intervals $\{j_\mu\}$ by

$$x_\mu = X^*(\phi^\mu(x_0)), \quad j_\mu = \langle x_\mu, x_{\mu+1}), \quad \mu \in \mathbf{Z}.$$

2. Linear Functional Equations in S

Consider a linear homogeneous functional equation of the k-th order in S of the form

$$a_k(x)\left(f \circ \phi^k\right)(x) + \cdots + a_1(x)\left(f \circ \phi^1\right)(x) + a_0(x)\left(f \circ \phi^0\right)(x) = 0, \quad (1)$$

where $a_i \in C_0(\mathcal{J}), i = 1, 2, \ldots, k$. We look for a solution $f \in C_0(\mathcal{J})$ which satisfies equation (1) identically on $\mathcal{J}$. A function f which is a solution of (1) is called a *particular solution* of (1). **Assume hereafter** $a_k \equiv 1$.

Note. Difference equations can use $\phi(x) = x + 1, X(x) = x$. For fixed x_0, solutions are just real sequences $\{f(x_\mu)\}_{\mu=0}^{\infty}$. See [8].

Theorem 2.1. *If functions f_1 and f_2 are solutions of the given equation (1), then also their sum $f_1 + f_2$ is a solution of equation (1).*

Theorem 2.2. *If a function f is a solution of (1) and $c \in \mathbf{R}$ an arbitrary constant, then also the product cf is a solution of equation (1).*

Theorem 2.3. *If a function f is a solution of (1) and $C \in U_\phi$ an automorphic function, then also $C(x)f(x)$ is a solution of equation (1).*

Proof. To prove the third result, substitute the product into (1):

$$\begin{aligned} &a_k(x)\left(C \circ \phi^k\right)(x)\left(f \circ \phi^k\right)(x) + \cdots + a_0(x)\left(C \circ \phi^0\right)(x)\left(f \circ \phi^0\right)(x) \\ &= C(x)\left[a_k(x)\left(f \circ \phi^k\right)(x) + \cdots + a_0(x)\left(f \circ \phi^0\right)(x)\right] = 0, \end{aligned}$$

because $\left(C\circ\phi^i\right)(x) = C(x)$ for $i = 1, 2, \ldots, k$. □

Theorem 2.4. *If functions $f_1, \ldots, f_k$ are particular solutions of equation (1) and $C_i \in U_\phi$ are arbitrary automorphic functions, then also*

$$C_1(x)f_1(x) + C_2(x)f_2(x) + \cdots + C_k(x)f_k(x) \tag{2}$$

is a solution of equation (1) on $\mathcal{J}$.

Proof. Apply Theorem 2.1 and Theorem 2.3. □

Definition 2.1. Expression (2) is called a *general solution* of (1) if every solution f of (1) is uniquely represented by (2). The *trivial solution* $f = 0$ is uniquely represented by $C_1 \equiv \ldots \equiv C_k \equiv 0$.

Definition 2.2. Let $f = f(x)$ be a solution of equation (1). Functions $f_0, f_1, \ldots, f_{k-1}$ defined by $f_i(x) = f(x), x \in j_i, i = 0, 1, \ldots, k-1$, are called *initial conditions* of the solution f.

3. Linear Dependence and Fundamental Solutions

Definition 3.1. Functions $f_i = f_i(x), i = 1, \ldots, k$, $x \in \mathcal{J}$ are called *linearly dependent* over U_Φ on $\mathcal{J}$, if there are automorphic functions $C_i \in U_\Phi$, $i = 1, \ldots, k$, at least one of which is different from zero, such that for all $x \in \mathcal{J}$

$$C_1(x)f_1(x) + C_2(x)f_2(x) + \cdots + C_k(x)f_k(x) = 0. \tag{3}$$

In particular, for some index i, $|C_i(x)| > 0$ for $x \in \mathcal{J}$, because a function in U_ϕ is either the zero function or else it never takes the value zero.

Linear independence over U_ϕ on $\mathcal{J}$ means that formula (3) holds for all $x \in \mathcal{J}$ only if $C_1 \equiv C_2 \equiv \ldots \equiv C_k \equiv 0$.

Definition 3.2. Let functions $f_i \in C_0(\mathcal{J}), i = 1, \ldots, k$, be given. For $x \in \mathcal{J}$, define $D_0 = \det(F_0)$ where F_0 is the matrix

$$F_0(x) = \begin{pmatrix} \left(f_1\circ\phi^0\right)(x) & \left(f_2\circ\phi^0\right)(x) & \cdots & \left(f_k\circ\phi^0\right)(x) \\ \left(f_1\circ\phi^1\right)(x) & \left(f_2\circ\phi^1\right)(x) & \cdots & \left(f_k\circ\phi^1\right)(x) \\ \vdots & \vdots & & \vdots \\ \left(f_1\circ\phi^{k-1}\right)(x) & \left(f_2\circ\phi^{k-1}\right)(x) & \cdots & \left(f_k\circ\phi^{k-1}\right)(x) \end{pmatrix}.$$

Lemma 3.1. *For $x \in \mathcal{J}$ we have the identity*

$$\left(D_0\circ\phi\right)(x) = a_0(x)D_0(x).$$

Proof. Replace each entry of the last row in determinant $(D_0 \circ \phi)(x)$ by its equivalent expression obtained from the functional equation. A simplification is made by adding combinations of the determinant rows to the last row, until entry i in the last row consists of a single term $-a_0(x)\left(f_i \circ \phi^0\right)(x)$. The common factor of the last row is then $-a_0(x)$. Swapping the last row to the first row implies

$$(D_0 \circ \phi)(x) = (-1)(-a_0(x))D_0(x).$$

The identity is verified. □

Lemma 3.2. *The inequality $|D_0(x)| > 0$ holds for all $x \in \mathcal{J}$ if and only if $|a_0(x)| > 0$ for $x \in \mathcal{J}$ and $|D_0(x)| > 0$ on j_0.*

Proof. Assume $|D_0(x)| > 0$ on $\mathcal{J}$. The previous lemma implies that $a_0(x)D_0(x) = (D_0 \circ \phi)(x)$ is nonzero, therefore $a_0(x) \neq 0$. Because j_0 is contained in $\mathcal{J}$, then it follows that $|D_0(x)| > 0$ on j_0.

Assume $|a_0(x)| > 0$ for $x \in \mathcal{J}$ and $|D_0(x)| > 0$ for $x \in j_0$. Then each $x \in j_1$ can be written $x = X^*(\phi(u))$ for some $u \in j_0$. The previous lemma implies

$$|D_0(x)| = |\,(D_0 \circ \phi)(u)| = |a_0(u)D_0(u)| > 0.$$

The process can be repeated to show that $|D_0(x)| > 0$ for $x \in j_\mu$, $\mu > 0$. Similarly, if $x \in j_{-1}$, then $u = X^*(\phi(x))$ belongs to j_0. The previous lemma implies

$$|a_0(x)D_0(x)| = |D_0(u)| > 0.$$

The inequality implies $|D_0(x)| > 0$ for all $x \in j_{-1}$. The process can be repeated to show that $|D_0(x)| > 0$ for $x \in j_\mu$, $\mu < 0$. Because $\mathcal{J}$ is the union of all j_μ, then $|D_0(x)| > 0$ for $x \in \mathcal{J}$. □

Theorem 3.1. *If functions $f_i, i = 1, \ldots, k$, are linearly dependent over U_ϕ on $\mathcal{J}$, then the determinant D_0 is identically zero on $\mathcal{J}$.*

Proof. Let functions $f_i, i = 1, \ldots, k$, be linearly dependent over U_ϕ on $\mathcal{J}$. Then identity (3) holds, where $C_i \in U_\phi$, $i = 1, 2, \ldots, k$, and for some index i, $|C_i(x)| > 0$ on $\mathcal{J}$. Sample (3) at iterates x, $\phi(x)$, $\phi^2(x)$, ..., to obtain a $k \times k$ linear homogeneous system of algebraic equations with coefficient matrix $F_0(x)$ and variable list $C_1(x)$, ..., $C_k(x)$. Because the algebraic equations have a nontrivial solution, then the determinant of coefficients $D_0(x)$ must vanish, by standard linear algebra theorems. □

Definition 3.3. A system of k solutions $f_i \in C_0(\mathcal{J}), i = 1, \ldots, k$ of (1) is called **fundamental** provided $|D_0| \equiv |\det(F_0)| > 0$ on $\mathcal{J}$. Such a system is called a **fundamental system of solutions** and the matrix F_0 is called a **fundamental matrix**.

Theorem 3.2. *If equation (1) has a fundamental system of solutions f_1, $\ldots$, f_k, then these solutions are linearly independent over U_ϕ on $\mathcal{J}$.*

Proof. Assume

$$C_1(x)f_1(x) + C_2(x)f_2(x) + \cdots + C_k(x)f_k(x) = 0.$$

Create $k-1$ more sampled equations by replacing x by iterates $\phi(x)$, $\phi^2(x)$, $\ldots$, to obtain a $k \times k$ system in variables $C_1(x)$, $\ldots$, $C_k(x)$ with coefficient matrix F_0. Because F_0 is invertible, then each C_i has to be zero on $\mathcal{J}$. □

Theorem 3.3. *If equation (1) has a fundamental system of solutions f_1, $\ldots$, f_k, then a general solution of equation (1) is given by*

$$\begin{aligned} f(x) = C_1(x)f_1(x) + C_2(x)f_2(x) + \cdots + C_k(x)f_k(x), \\ C_i \in U_\phi, \quad i = 1, 2, \ldots, k. \end{aligned} \tag{4}$$

Proof. Assume that besides solutions f_1, $\ldots$, f_k we are given an additional solution f. We will construct automorphic functions satisfying (4).

Sample k times the relation $C_1(x)f_1(x) + \cdots + C_k(x)f_k(x) = f(x)$ at iterates x, $\phi(x)$, $\phi^2(x)$, $\ldots$, to get a $k \times k$ nonhomogeneous system of equations for the symbols $C_1(x)$, $\ldots$, $C_k(x)$. The coefficient matrix for this system is the invertible matrix $F_0(x)$. On interval $\mathcal{J}$, Cramer's rule applies to solve for the symbols $C_i(x)$, as quotients of determinants. We must verify that the functions C_i defined in this manner are indeed continuous on $\mathcal{J}$ and that $(C_i \circ \phi)(x) = C_i(x)$ holds on $\mathcal{J}$.

To verify continuity of C_i, examine its determinant quotient to see that both numerator and denominator are continuous and the denominator D_0 never takes the value zero.

We verify now identity $(C_i \circ \phi)(x) = C_i(x)$. The fraction $(C_i \circ \phi)(x)$ has denominator $(D_0 \circ \phi)(x)$. The numerator of C_i is a determinant D_1 and $D_1 \circ \phi$ is the numerator of $C_i \circ \phi$. Proof details for $(D_0 \circ \phi)(x) = a_0(x)D_0(x)$ apply to give the relation $(D_1 \circ \phi)(x) = a_0(x)D_1(x)$. A previous lemma implies $a_0(x) \neq 0$, therefore the common factor $a_0(x)$ cancels in the quotient, giving the identity

$$(C_i \circ \phi)(x) = C_i(x).$$

Then the C_i are automorphic on $\mathcal{J}$ and

$$C_1(x)f_1(x) + \cdots + C_k(x)f_k(x) = f(x) \text{ on } \mathcal{J}.$$

Because each solution f on $\mathcal{J}$ is expressed as a linear combination of the fundamental system of solutions, with automorphic coefficients C_i, then (4) is satisfied. □

Theorem 3.4. *Given a fundamental system of k solutions $f_1, \ldots, f_k$ of equation (1) and any other solution f_{k+1}, then the appended list $f_1, \ldots, f_{k+1}$ is linearly dependent over U_ϕ.*

Proof. By Theorem 3.3, f_{k+1} is a linear combination of $f_1, f_2, \ldots, f_k$ with coefficients in U_ϕ. □

Theorem 3.5. *If two homogeneous linear equations of k-th order*

$$\begin{aligned}
\left(f \circ \phi^k\right)(x) + p_{k-1}(x)\left(f \circ \phi^{k-1}\right)(x) + \cdots + p_0(x)\left(f \circ \phi^0\right)(x) = 0,\\
\left(f \circ \phi^k\right)(x) + q_{k-1}(x)\left(f \circ \phi^{k-1}\right)(x) + \cdots + q_0(x)\left(f \circ \phi^0\right)(x) = 0,
\end{aligned}$$

have the same fundamental system of solutions $f_1, \ldots, f_k$, then the equations are identical, that is,

$$p_i(x) = q_i(x), \quad i = 0, \ldots, k-1.$$

Proof. Re-write the functional equations for $f_1, \ldots, f_k$ as a system of equations

$$\begin{aligned}
p_{k-1}(x)\left(f_1 \circ \phi^{k-1}\right)(x) + \cdots + p_0(x)\left(f_1 \circ \phi^0\right)(x) &= -\left(f_1 \circ \phi^k\right)(x)\\
\vdots \quad \cdots \quad \vdots \qquad\qquad & \qquad\quad \vdots\\
p_{k-1}(x)\left(f_k \circ \phi^{k-1}\right)(x) + \cdots + p_0(x)\left(f_k \circ \phi^0\right)(x) &= -\left(f_k \circ \phi^k\right)(x)
\end{aligned}$$

Solve for the variables $p_0(x), \ldots, p_{k-1}(x)$ with Cramer's rule. This proves uniqueness of the coefficients. □

Theorem 3.6. *Given a fundamental system of solutions $f_1, \ldots, f_k$ of equation (1) on $\mathcal{J}$, then the corresponding linear homogeneous functional equation in S is*

$$\begin{vmatrix}
\left(f_1 \circ \phi^0\right)(x) & \cdots & \left(f_k \circ \phi^0\right)(x) & \left(f \circ \phi^0\right)(x)\\
\left(f_1 \circ \phi^1\right)(x) & \cdots & \left(f_k \circ \phi^1\right)(x) & \left(f \circ \phi^1\right)(x)\\
\vdots & \vdots & \vdots & \vdots\\
\left(f_1 \circ \phi^k\right)(x) & \cdots & \left(f_k \circ \phi^k\right)(x) & \left(f \circ \phi^k\right)(x)
\end{vmatrix} = 0, \quad x \in \mathcal{J}. \tag{5}$$

Proof. First show that (5) defines a k-th order linear functional equation (1) with fundamental set of solutions $f_1, \ldots, f_k$. Then use Theorem 3.5. □

4. Linear Nonhomogeneous Functional Equation

Consider the nonhomogeneous equation

$$a_k(x)\left(f \circ \phi^k\right)(x) + \cdots + a_0(x)\left(f \circ \phi^0\right)(x) = Q(x) \tag{6}$$

where assumptions replicate those for (1) and $Q(x)$ is continuous. We state without proof the basic result.

Theorem 4.1. *Assume (1) has a fundamental set of solutions $f_1, \ldots, f_k$ over U_ϕ on $\mathcal{J}$. Each solution of linear nonhomogeneous functional equation (6) is the sum of a particular solution $\bar{f}(x)$ of nonhomogeneous equation (6) and the general solution of (1), that is, $f(x)$ in (6) is given by*

$$f(x) = \bar{f}(x) + C_1(x)f_1(x) + C_2(x)f_2(x) + \cdots + C_k(x)f_k(x).$$

References

1. J. Aczél, J. Dhombres, *Functional Equations in Several Variables* (Cambridge, 1989), ISBN 0 521 35276 2.
2. D. R. Curtis, *Relations between the Gramian, the Wronskian and a third determinant connected with the problem of linear independence*, Bull. Amer. Math. Soc. 17(1910), 462-467.
3. S. N. Elaydi, *An Introduction to Difference Equations* (Springer, New York, Berlin, Heidelberg, 1999).
4. A. O. Gelfond, *Differenzenrechnung* (VEB Deutscher Verlag der Wissenschaften, Berlin, 1958).
5. M. Kuczma, *Functional equations in a single variable* (PWN, Warszawa, 1968).
6. J. Laitochová, *Group iteration for Abel's functional equation*, Nonlinear Analysis: Hybrid systems, Vol 1, issue 1, March 2007, 95-102.
7. Laitochová, J., *A remark on k-th order linear functional equations with constant coefficients*, Hindawi Publishing Corporation (Cairo, New York, 2006), Advances in Difference Equations, vol. 2006, Article ID 72615, 8 pages.
8. J. Laitochová, *Linear difference operators in the space of strictly monotonic functions*, Tatra Mt. Math. Publ. 38(2007), 111-121.

A Dynamical System for a Long Term Economic Model*

Rafael Luís† and Henrique Oliveira‡

† *Center for Mathematical Analysis, Geometry, and Dynamical Systems, Instituto Superior Tecnico, Technical University of Lisbon*
‡ *Department of Mathematics, Instituto Superior Tecnico, Technical University of Lisbon*
† *E-mail: rafaelluis@netmadeira.com*
‡ *E-mail: holiv@math.ist.utl.pt*

The Marx model for the profit rate r depending on the exploitation rate e and on the organic composition of the capital k is studied using a dynamical approach. Supposing both $e(t)$ and $k(t)$ are continuous functions of time we derive a law for $r(t)$ in the long term. Depending upon the hypothesis set on the growth of $k(t)$ and $e(t)$ in the long term, $r(t)$ can fall to zero or remain constant. This last case contradicts the classical hypothesis of Marx stating that the profit rate must decrease in the long term. Introducing a discrete dynamical system in the model and, supposing that both k and e depend on the profit rate of the previous cycle, we get a discrete dynamical system for r, $r_{n+1} = f_a(r_n)$, which is a family of unimodal maps depending on the parameter a. In this map we can have a fixed point when a is small and, when we increase a, we get a cascade of period doubling bifurcations leading to chaos. When a is very big, the system has again periodic stable orbits of period five and, finally, period three.

Keywords: Marx Model, Profit rate, Exploitation rate, Organic composition of the capital, Chaos, Topological entropy, Lyapunov exponent.

1. Introduction

We find in [8] a well known equation from economic theory. In this work we will deduct the model and study it in terms of dynamical systems theory.

The equation quoted above describes the behaviour of the profit rate r, with the exploitation rate e and with the organic composition of the capital k, in the Marx model. The study of this equation in a dynamical system perspective seems to be very interesting, because it relates the magnitudes

*This work is part of the first author's Ph.D. dissertation.

r, e and k by the equation

$$r = \frac{e}{1+k}. \tag{1}$$

Our main goal is to use this equation as the starting point of a study in which we may model $e(t)$, $k(t)$ and $r(t)$ as variables depending on t, using the formalism of discrete dynamical systems, which we present in the final section.

In the first section we present a pre-analysis considering that $e(t)$ and $k(t)$ do not depend on one another and that they vary continuously with time. This analysis leads us to conclude that, in this simple situation, the profit rate can decrease or even increase with time and that we aren't in the presence of a mathematical indetermination. Something that has been always motif of discussion in economic theory.

In the third section we approach the problem by making a study of some concrete models for profit rates using several assumptions. We use an asymptotic perspective. We conclude that we aren't in the presence of any indetermination. Thus we may infer the behaviour of the profit rate in the long term.

In the fourth section we introduce in the model a iterative approach: the organic composition of the capital and the exploitation rate are now dependent on the same variable (profit rate) in the previous cycle. We now have discrete variables and the system starts reacts to what happened before. In this section we will also introduce a concrete model, that, like many other possible models, may describe the dynamical situation in a realistic way. We prove that, in certain cases, this approach may lead to stability situations with well defined limits, may lead to periodic orbits (where the profit rate repeats itself from p to p units of time) or may lead to chaos where, in this case, it is impossible to predict the evolution of the system. In this last case, the system becomes strongly dependent on the initial conditions.

In the last section using a discrete dynamical system we study the profit rate with the resources of dynamical system theory. We computed the Lyapunov exponents and the topological entropy. We established the existence of chaos when a is large.

In this work we prove that the Marx model is open to several mathematical approaches and, depending on which approach we use, the conclusion may go from decrease, to stability, to periodicity and even to chaos of the profit rate.

It's wrong to say that this subject is closed by saying that it leads to an indetermination. Quite to the contrary, this model is still open and may be treatable mathematically with profitable economic interpretations.

2. Study of the profit rate as a function of time

In this section we study the profit rate, considering that the variables do not depend on the other variables. We also assume that all the variables are continuous functions of the time t.

To get the relation (1) let's remember that, according to Marx

$$\text{profit rate} = \frac{\text{obtained profit}}{\text{invested capital}},$$

where

$$\text{invested capital} = \text{constant capital} + \text{human capital}.$$

We also know that

$$\text{exploitation rate} = \frac{\text{obtained profit}}{\text{human capital}}$$

and

$$\text{organic composition of the capital} = \frac{\text{constant capital}}{\text{human capital}}.$$

Simplifying the relation for the profit rate we obtain

$$\text{profit rate} = \frac{\text{exploitation rate}}{1 + \text{organic composition of the capital}}.$$

To study the profit rate r, we suppose that this quantity varies with the time t according to equation (1). Both $e(t)$ and $k(t)$ are positive and continuous functions which depend on t,

$$e(t) \in C^0\left(\mathbb{R}_0^+\right) : \mathbb{R}_0^+ \longmapsto \mathbb{R}_0^+ \text{ and } k(t) \in C^0\left(\mathbb{R}_0^+\right) : \mathbb{R}_0^+ \longmapsto \mathbb{R}_0^+.$$

Our assumptions are simple: we suppose that the exploitation rate is a limited function that may or may not be monotonous. We assume that the organic composition of the capital increases unlimitedly, because it accounts the technological innovation which, according to [8], is a magnitude that will grow unbounded due to the never stopping invention and technological innovation of mankind. The exploitation rate will always be limited by physical factors that cannot be surpassed due to human limitation.

It's easy to show that, if $e(t)$ is a positive and bounded function, with

$$e(t) < A, A \in \mathbb{R}^+, \tag{2}$$

and $k(t)$ is a positive increasing and unbounded function, in other words,

$$\forall B \in \mathbb{R}^+ \exists t_1 \in \mathbb{R}^+ : t > t_1 \Rightarrow k(t) > B, \tag{3}$$

then the profit rate tends to zero,

$$\lim_{t \to +\infty} \frac{e(t)}{1 + k(t)} < \lim_{t \to +\infty} \frac{A}{1 + k(t)} = 0$$

because we can take $k(t)$ as big as we want.

These assumptions are justified because there is no physical limit for technological innovation, for invention and for investment. However there exists a physical limitation for the exploitation rate inherent to the human condition.

Example 1. Let $e(t) = \frac{t}{1+t}$ and $k(t) = \sqrt{t}$. We know that $0 \leq \frac{t}{1+t} = 1 - \frac{1}{1+t} < 1$ and $\lim_{t \to +\infty} \sqrt{t} = +\infty$. The profit rate decreases slowly to zero, in long term, after an initial growth.

A similar idea leads to the same conclusions if the exploitation rate grows but in an inferior rate when compared to the organic composition of the capital. This means that the exploitation rate wouldn't be bounded by any human factors. Both this magnitudes would tend to infinity. In this case the growth of the exploitation rate is slower than the growth of the organic composition of the capital, which implies that the limit obtained is also zero.

Example 2. Let $\alpha, \delta > 0$, $e(t) = t^\alpha$ and $k(t) = t^{\alpha+\delta}$. In this case the growth of the exploitation rate and the growth of the organic composition of the capital are unlimited, but the growth of the exploitation rate is smaller than the growth of the organic composition of the capital, because $t^\alpha < t^{\alpha+\delta}, \forall t \in \mathbb{R}^+$. Since $r(t) = \frac{t^\alpha}{1+t^{\alpha+\delta}} = \frac{1}{t^\delta + \frac{1}{t^\alpha}} \underset{t \to +\infty}{\longrightarrow} 0$, the profit rate decreases to zero.

Example 3. Consider that the exploitation rate increases at the same rate as the organic composition of the capital, i.e., $e(t) = k(t), \forall t \in \mathbb{R}^+$. The model for the profit rate is now $r(t) = \frac{1}{1+(e(t))^{-1}} \underset{t \to +\infty}{\longrightarrow} 1$, for any $e(t)$. In other words, the profit rate tends to a constant value after a few cycles.

In an extreme situation, it's easy to verify that if the exploitation rate increases faster than the organic composition of the capital, then the profit rate grows indefinitely, because $\frac{e(t)}{1+k(t)} \underset{t \to +\infty}{\longrightarrow} +\infty$.

From an economic point of view, the situations of the examples 1 and 2 make more sense, because the exploitation rate can not grow indefinitely.

3. Study of concrete models for the profit rate

In this section we will make the analytic study of the variation of the profit rate r as a function that depends on the exploitation rate e and on the organic composition of the capital k.

3.1. $e(t)$ *and* $k(t)$ *grow in a linear way*

A first model, not necessarily realistic, but often used theoretically as a first approach when the time scale is not very large, is the usual linear model.

We consider the hypothesis that $e(t)$ and $k(t)$ obey to a linear laws

$$e(t) = At + B \text{ and } k(t) = Ct + d,$$

where t is the time and $A, B, C,\ D \in \mathbb{R}^{+}$. When the system is in the initial position, $t = 0$, we have the initial conditions

$$e(0) = e_0 = B \text{ and } k(0) = k_0 = D.$$

The values of A and C are, respectively, the rate of change of $e(t)$ and $k(t)$ which respect to t, that is, the derivatives of $e(t)$ and $k(t)$ which respect to time are

$$e'(t) = A \text{ and } k'(t) = C.$$

What happens in the asymptotic situation? We need to study

$$r_{final} = \lim_{t \to +\infty} \frac{At + B}{Ct + D + 1} = \frac{A}{B} \neq 0.$$

We conclude that in the case of a linear growth of $e(t)$ and $k(t)$, the value of $r(t)$ tends to a constant value different from zero.

3.2. $e(t)$ *and* $k(t)$ *grow with power model*

A second possible model happens when we assume that the exploitation rate and the organic composition of the capital obey a power law.

We consider the hypothesis that $e(t)$ and $k(t)$ obey

$$e(t) = At^{\alpha} + B \text{ and } k(t) = Ct^{\beta} + D,$$

where $A, B, C,\ D,\ \alpha,\ \beta \in \mathbb{R}^{+}$.

When the time t is in the initial instant, $t = 0$, we have the initial conditions

$$e(0) = e_0 = B \text{ and } k(0) = k_0 = D.$$

The derivatives of $e(t)$ and $k(t)$ in order to t are given by

$$e'(t) = \alpha A t^{\alpha-1} \text{ and } k'(t) = \beta C t^{\beta-1}.$$

In an asymptotic situation, when t grows to infinity, we have the limit

$$r_{final} = \lim_{t\to+\infty} \frac{At^{\alpha} + B}{Ct^{\beta} + D + 1}.$$

We have three possibilities:

(1) If $\alpha > \beta$, then $r_{final} = +\infty$ and the profit rate grows to infinity. This situation seems to be unrealistic, because it means that the exploitation rate would grow in a more accelerate rhythm than the organic composition of the capital.
(2) If $\alpha = \beta$, then $r_{final} = \frac{A}{C}$. Therefore if $e(t)$ and $k(t)$ both grow with the same exponent, then the value of $r(t)$ tends to a constant, different from zero. This situation is identical, in terms of final result, to the linear situation.
(3) If $\alpha < \beta$, then $r_{final} = 0$. We conclude that the profit rate would have as limit the zero value. This situation seems to be the most realist, because it means that the exploitation rate would grow at a pace less accelerated than the organic composition of the capital.

4. A discrete dynamical system

In this section we will construct a discrete dynamical model for the profit rate, considering that the profit rate for the current cycle is obtained using the exploitation rate and the organic composition of the capital as functions of the profit rate from the time unit immediately before.

4.1. *General model*

In the economic world of today, any decision, is conditioned by what happened immediately before, that is, we can suppose that the exploitation rate and the organic composition of the capital depend on the profit rate obtained in the time instant immediately before. The time unit in this case can be, in general, a year. Therefore $n+1$ represents the year after the current year n. The variable in the year $n+1$ will be a function of the variable of the year n. It's obvious that we can consider another time unit, for example a month or a decade, depending of the rate of change in the system.

Therefore the magnitudes in the $(n+1)$th iteration are functions that depend on the nth iteration. We consider this functions continuous and differentiable. We have

$$e_{n+1} = E\left(r_n\right), k_{n+1} = K\left(r_n\right). \tag{4}$$

In this case the profit rate depends also on the profit rate of the year before. By replacing the relations (4) in (1), we obtain

$$r_{n+1} = \frac{E\left(r_n\right)}{1 + K\left(r_n\right)}. \tag{5}$$

This equation represents a one-dimensional discrete system. If K depends on r_n in an increasing way and, E is a limited function of r_n, then the ideas presented in the early sections apply and the profit rate decreases also to zero. We will have always a more complex situations when E and K adjust themselves to the profit rate step by step. We believe that, this situation will be very interesting in the medium term (ten, twenty, thirty and more years). However the long term situation, from our point of view, will always lead us to a growth of K and a limitation of E, which will originate a decrease of the profit rate in the long term.

4.2. *Model with a discrete dynamical system*

We will now introduce an idealised dynamical system so we can understand what happens, if we consider the economic universe as a whole. We are not interested in considering a specific country or company.

The model that we propose is based on the following assumptions:

(1) We suppose that we haven't had any losses, i.e., when we consider the economy as a whole, we are supposing that a negative profit rate doesn't exist. In fact, we think that in all of the economic activity there exists a positive balance (possibly zero) of the profits. On the other hand, the profit rate can't be unlimited because that contradicts the fact that the total quantity of money in the planet is finite.

(2) When the profit rate is low, the exploitation tends to be high. In general, when the companies are created or when they face financial difficulties it is very common for the workers to make an extra effort to help the company to obtain or regain satisfactory results. Consequently we will consider that the exploitation rate is high when the profit is low.
On the other hand, when the profit rate is high, the pressure on workers tends to decrease and therefore the exploitation rate tends to decrease.

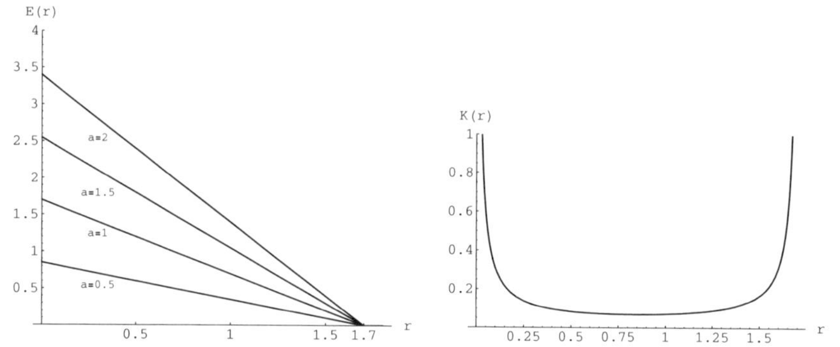

Fig. 1. In the left plot we represent the variation of the exploitation rate when $a = 0.5, 1, 1.5$ and 2 while in the right plot we represent an instance of the organic composition of the capital with an increase of the profits.

We will consider here when the profit rate is very large the exploitation rate is zero. So, we consider the following linear model for the exploitation rate

$$E(r_n) = \begin{cases} a(b - r_n) & \text{if } r_n \in [0, b] \\ 0 & \text{if } r_n > b \end{cases}, \tag{6}$$

based on what was mentioned above, where a is a positive real value and b is an adaptable parameter that adjusts to the economic situation. Therefore, for the study of a concrete model, let's choose $b = 1.7$. Having the previous assumptions in mind, we could have chosen another value. It is clear that this value can be adaptable to the economic information available. On the other hand, the real parameter a has strongly influence in the exploitation rate when profits are low, but when profits are high it will have a low influence. In Fig. 1 we can see some examples.

(3) If the economic system has a low profit rate, the trend will be to incorporate more capital (invest) and, on the other hand, to decrease the human capital, through dismissals, that will lead to an increase of the organic composition of the capital. If the profit is very high, the trend will be to reinvest in the capital: human capital (technical formation) and constant capital (technological innovation). A model that describes this reality is given by a capital function K depending on the profit rate of the previous cycle, that is,

$$K(r_n) = \begin{cases} \frac{0.05}{(r_n + 0.001)(b - r_n)} & \text{if } r_n \in [0, b] \\ 0 & \text{if } r_n > b \end{cases}. \tag{7}$$

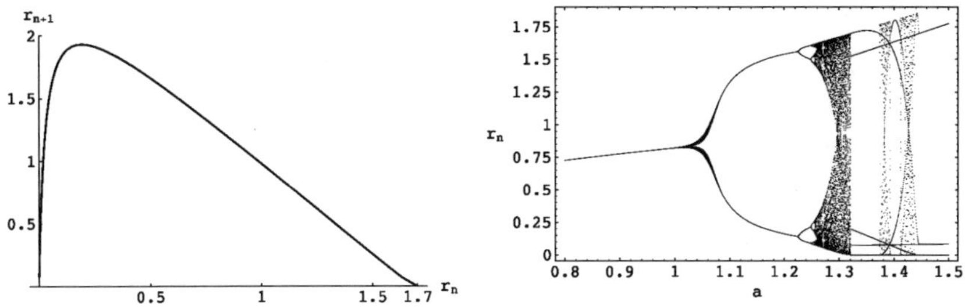

Fig. 2. In the left plot we represent the profit rate for $a = 1.5$. In the right plot we can see the bifurcation diagram for the profit rate as a function of the parameter a.

The value of b in our model was considered equal to 1.7. The value 0.001 is used to give us a reasonable value for the capital when the profit is low, and the value 0.05 will originate an equilibrium in the organic composition of the capital, between low profit and high profit. We emphasize the fact that when the profit is high, i.e., for values near to 1.7, the organic composition of the capital as a function of the profit, grows unlimitedly, which is acceptable because the profit is always bounded. We can see in Fig. 1 a representation of the function $K(r)$.

By replacing the relations (6) and (7) in (5), we obtain for the profit rate

$$r_{n+1} = \begin{cases} \frac{a(b-r_n)^2(r_n+0.001)}{(r_n+0.001)(b-r_n)+0.05} & \text{if } r_n \in [0, b] \\ 0 & \text{if } r_n > b \end{cases} \tag{8}$$

In Fig. 2 we see an instance of the plot for the profit rate defined in 8.

For values $0 < a < 1.0690451$ this iteration has a stable fixed point. For example, for $a = 0.75$, the profit rate tends to 0.70004. When $1.0690451 < a < 1.2267617$ we do not have a stable fixed point, but oscillations between two points. For example for $a = 1.1$ the profit rate oscillates between 0.376535 and 1.32338. If the parameter increases more, we would have an even more complex situation, with period doubling until the system turns completely unpredictable. For $1.3204 < a < 1.37291$, r_n has orbits of period five. When the parameter a grows even more, we have again the aperiodic situation until we fall on a period three zone (for values of a greater than 1.44607), that is, the profit rate has a triennial repetition. In Fig. 2 we represent the bifurcation diagram for the profit rate as a function of the parameter a.

5. Chaos on the profit rate

In this section we make a qualitative study of the profit rate using mathematical tools from dynamical system theory, such as the Lyapunov exponents and topological entropy. We use these concepts to establish that there is chaos in the profit rate when we increase the parameter a.

Let's consider the function $f_a : \mathbb{R}_0^+ \longmapsto \mathbb{R}_0^+$ discussed in section 4.2,

$$f_a(r) = \frac{a(1.7-r)^2(r+0.001)}{(r+0.001)(1.7-r)+0.05} \tag{9}$$

and let $a_l \approx 1,321$. If $a_l < a$, then there exists a real interval $J =]x_1, x_2[$ such that for all $r \in J$ we have $f(r) > 1.7$.

In our model this means that if the profit rate r_n in a specific year n belongs to J then in the next year we have a null profit. For example, for $a = 1.35$ we have $x_1 = 0.0687709$ and $x_2 = 0.468683$.

To calculate the critical point of the function, that we will represent by c, it is necessary to solve the equation $f_a'(r) = 0$. The derivative of the function defined in (9) is

$$f_a'(r) = -a\frac{(r-1.75779)(r-1.7)(r-0.191897)(r+0.251688)}{(r-1.7289)^2(r+0.0299033)^2}. \tag{10}$$

So $c = 0.191897$ is the only critical point of f_a, for all $a > 0$. This means that if in a specific year the profit rate is 0.191897, then in the next year we will necessarily reach the maximum possible profit.

The fixed points of the function are determined through the relation $f_a(x) = x, \forall x \in D_{f_a}$. For example, for $a = 1$ we have $f_1(0.821574) = 0.821574$. The value of $f_a^2(x) = f_a(f_a(x))$ is called the second iterate of x under f_a (the first iterate of x under f_a is $f_a(x)$ and $f_a^0(x) = x$). More generally, $f_a^k(x) = f_a\left(f_a^{k-1}(x)\right)$ is the kth iterate of x under f. So a point $x_0 \in D_f$ is said to be k periodic if, after k iterations under f, it returns to the initial value x_0. In other words, the period is the minimum k such that $f_a^k(x_0) = x_0$. The set

$$O_{f_a}(x_0) = \left\{x_0, x_1 = f_a(x_0), x_2 = f_a^2(x_0), ..., x_{k-1} = f_a^{k-1}(x_0)\right\}$$

is called the forward periodic orbit of x_0. When $f_a^k(x_0) = x_0$ (with k minimum) we say that the orbit is k periodic or a k-cycle. If the orbit

$$O_{f_a}(x_0) = \{x_0, x_1, x_2, ..., x_k, ...\}$$

isn't periodic, then $\#(O_{f_a}) = \infty$.

It is easy to verify that 0.00586137 is a 3-cycle under the iteration $f_{1.5}$, because

$$f_{1.5}^{3}(0.00586137) = f_{1.5}(f_{1.5}(0.479346)) = f_{1.5}(1.68711) = 0.00586137.$$

The orbits can be stable or unstable. According to [2] a k-cycle is stable if $|\left(f^{k}(x_0)\right)'| < 1$ and is unstable when $|\left(f^{k}(x_0)\right)'| > 1$. Note that by the chain rule $|\left(f^{k}(x_0)\right)'|$ is equal to

$$|f'(x_0)\,f'(x_1)\ldots f'(x_{k-1})| = \prod_{n=0}^{k-1} |f'(x_n)|.$$

In the 3-cycle determined above, we have $f_{1.5}'(1.68711) = -0.769278$, $f_{1.5}'(0.00586137) = 56.1711$ and $f_{1.5}'(0.479346) = -1.21476$, so

$$|f_{1.5}'(1.68711) \times f_{1.5}'(0.00586137) \times f_{1.5}'(0.479346)| > 1,$$

and therefore the 3-cycle is unstable.

The number of orbits is not bounded, while the number of stable orbits is bounded, as we will see.

The Schwarzian derivative, Sf_a, of a function f_a at x is defined

$$Sf_a(x) = \frac{f_a'''(x)}{f_a'(x)} - \frac{3}{2}\left(\frac{f_a''(x)}{f_a'(x)}\right)^2.$$

According to Singer's theorem [7] if f_a is a C^3 map in a closed interval $J = [x_1, x_2] \in D_{f_a}$, such that $f_a(x_1) = f_a(x_2) = x_1$, $Sf_a(x) < 0, \forall x \in J\backslash\{P_c\}$ and if f_a has n critical points in J, where $\{P_c\}$ is a set of the critical points of f_a, then for $\forall k \in \mathbb{Z}^+$, f_a has at most n attracting periodic orbits with k period. If we have $f_a(x_1) > x_1$ and $f_a(x_2) = x_1$ or $f_a(x_1) = x_1$ and $f_a(x_2) > x_1$ or yet $f_a(x_1) = f_a(x_2) > x_1$, then we have at most $n+1$ attracting periodic orbits with k period (we include the orbit of x_1 or x_2 according to the case). In the case of $f_a(x_1), f_a(x_2) > x_1$ and $f_a(x_1) \neq f_a(x_2)$, then we will have at most $n+2$ attracting periodic orbits with k period (we include the orbits of x_1 and x_2).

In our example, we have $f_a \in C^3$ in $J = [0, 1.7]$ and $f_a(0) = 0.0558994a > 0$, $a > 0$ and $f_a(1.7) = 0$. $c = 0.191897$ is the only critical point of f_a in J. Once that $Sf_a(x) < 0$, $\forall x \in J\backslash\{c\}$, then $\forall k > 1$, f_a has at most two attracting period orbits with k period.

The Lyapunov exponent is a mathematical indicator of the exponential degree of the velocity at which two arbitrary nearby orbits split apart as the

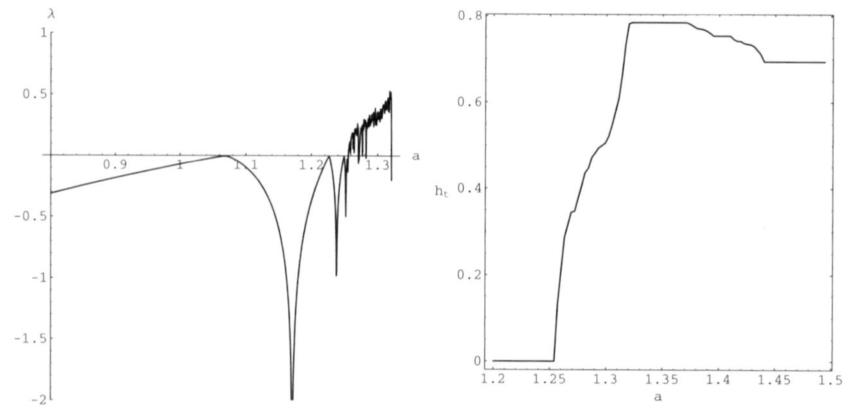

Fig. 3. In the left plot we represent the Lyapunov exponents λ of the function f_a as a function of the parameter a while in the right plot we have the curve of the topological entropy h_t.

number of iterations increases. The Lyapunov exponent $\lambda(x_0)$ for a point x_0 can be defined by the formula

$$\lambda(x_0) = \limsup_{n \to +\infty} \left(\frac{1}{n} \sum_{k=0}^{n-1} \log |f^{'}(x_k)| \right) \tag{11}$$

where $x_k = f_a^k(x_0)$. To calculate with a numeric algorithm the value of the exponent we use

$$\lambda(x_0) = \lim_{n \to +\infty} \left(\frac{1}{n} \sum_{k=0}^{n-1} \log |f^{'}(x_k)| \right) \tag{12}$$

that is equivalent to

$$\lambda(x_0) = \log \left(\lim_{n \to +\infty} \sqrt[n]{|(f^n(x_0))^{'}|} \right).$$

If the absolute value of $f^{'}(x_k)$ is greater than one, then the Lyapunov exponent is positive, which implies a sensitive dependence on the initial conditions. In Fig. 3 we represent, in the (a, λ)-plane, the progress of the Lyapunov exponents of the function f_a.

Another important topological invariant is the topological entropy. To introduce this concept we will use the kneading theory [5]. For this it is necessary to define the growth number s for unimodal maps (a function that has only one critical point). We use the concept of lap number $l(f_a^n)$, i.e., the number of maximal intervals of monotonicity of f_a^n (f_a^n is piecewise-monotone).

The growth number s can be obtained by the relation

$$s(f_a) = \lim_{n \to +\infty} \left(l\left(f_a^n\right)\right)^{\frac{1}{n}}. \tag{13}$$

When the growth of the laps is small (polynomial growth with the number of iterates) we do not have chaos. When the growth of the lap number is exponential we have chaos. This happens when the growth number is greater than 1.

For the turning point $c = 0.191897$, we will define the parity function by

$$\varepsilon(x) = \begin{cases} 1 & \text{if } x < c \\ 0 & \text{if } x = c \\ -1 & \text{if } x > c \end{cases} \tag{14}$$

The kneading determinant is a formal series in t given by

$$Z(t,a) = 1 + \sum_{n=1}^{\infty} \left(\prod_{j=1}^{n} \varepsilon\left(f_a^n(c)\right) t^n \right) \tag{15}$$

In case of periodic orbits of c, $Z(t,a)$ is a polynomial of degree $(n-1)$. The inverse of the least root of $Z(t,a)$ in $[0,1]$ is the growth number of f_a (Milnor-Thurston theorem [5]). The topological entropy h_t is given by the relation $h_t = \log_2(s)$.

Our function is unimodal (more precisely, f_a increases in $[0,c]$ and decreases in $[c,1.7]$). For $a = 1.32$, the first terms of the kneading determinant are

$$1 - t - t^2 + t^3 - t^4 + t^5 - t^6 + t^7 - t^8 + t^9 - t^{10} + t^{11} - t^{12} - t^{13} + \ldots$$

The smallest real root of this polynomial belongs to $[0,1]$ is approximately equal to 0.580692. The topological entropy therefore is given by

$$h_t = \log_2\left(0.580692^{-1}\right) \approx 0.78415.$$

In Fig. 3 we see the evolution of the topological entropy values for our model.

The fact that the topological entropy increases and is greater than zero (associated to the fact that the Lyapunov exponent is positive) means that the dynamical system becomes more complex when the parameter a increases. For values of $a > 1.32099$ the model starts to exhibit chaos, something that results clearly in the bifurcation diagram at the end of the first classical duplication period process. We can see clearly in this diagram an aperiodic band with no stability. From the economic point of view this situation would also result in a huge complexity and instability of the system. This situation happens when the exploitation rate is very high and the profit rate is very low. When $a > 1.372$ the h_t starts to decreases. This happens because some orbits fall outside the interval $[0, b = 1.7]$ where the profit rate is zero. So we have less complexity.

In this model, trying to correct low profit at the cost of high exploitation rate leads inevitably to instability and chaos, both from the mathematical point of view (mathematical definition of chaos) and from the common sense point of view (economic chaos). For values of $a > 1.32099$ our model stops being realistic, because in those circumstances, after a high profit, we will have a null profit. If it was possible to introduce an exploitation rate so high that we would be in this situation, that would mean that the system would react by presenting a reasonable profit rate from five to five or three to three units of time (values of a near to 1.35, in the first case, and higher than 1.446, in the second case). This situation of apparent stability inside chaos coexists with infinity unstable orbits of all periods (as is guaranteed by the Sharkovsky theorem [6]).

In economy the exploitation rate must be kept below a certain critical value, which in our model is $a < 1.069$. This maintains the stability of the system in a constant profit rate along the economic cycles.

6. Conclusion

The problem of the decreasing of the profit rate with time has been the subject of multiple discussions in the last 150 years.

In this purely mathematical study about this subject, we used the equation that arises from value theory, a purely static equation, that only gives us the fixed relation between the profit rate, the exploitation rate and the organic composition of the capital. That equation is our starting point for a several dynamical considerations.

It's obvious that we can't infer conclusions about the profit rate evolution if we regard this equation as a relation between constants. This is what has been done in the past.

Our goal was to construct an evolutive model in time, that starts in this equation and give it a dynamical interpretation.

In sections 2 and 3 we used fundamental assumptions the fact that the exploitation rate can not grow indefinitely, duo to the limitations of all the economic processes that always depends on human factors. In these sections we concluded that: either the profit rate always decreases or, at least, it tends to a constant value. This happens if the profit rate and the organic composition of the capital obey the same evolution law, which isn't absurd from the economic point of view because they are strongly related.

Although mathematically possible, that assumption which arises from Marx's time, that "the profit rate has a decreasing tendency" has not been proved and it is not a theorem. As always in mathematics, everything depends on the assumptions.

In the final sections we introduced a unidimensional dynamical system, more realistic that blind previsions based only on immutable laws, or in linear approaches, that are unreal, in the last case, in a long term.

In this model we obtain either stability or chaotic behaviour, but whatever the order of the magnitudes of a, when we use this last kind of discrete dynamical modulation, the profit rate doesn't necessarily fall to zero.

References

1. K. M. Brooks, H. Bruin, *Topics From One-Dimensional Dynamics*, London Mathematical Society, Student Texts 62, Cambridge University Press, 2004.
2. S. Elaydi, *Discrete Chaos*, Chapman and Hall/CRC, First Edition, 2000.
3. J. Guckenheimer, *On bifurcation of maps of the interval*, Invent. Math., Vol. 39 , pp. 165–178, 1977.
4. T. Li, J. Yorke, *Period Three Implies Chaos*, Am. Math. Monthly, Vol. 82, pp. 985–992, 1975.
5. J. Milnor, W. Thurston, *On iterated maps of the interval*, Dynamical Systems: Proc. Univ. of Maryland, Lecture Notes in Mathematics Vol. 1342, pp. 465–563, 1986–1987.
6. A. Sharkovsky, *Coexistence of cycles of a continuous mapping of the line into itself*, Ukrain. Math. Zh., Vol. 16, pp. 61–71, 1964.
7. D. Singer, *Stable orbits and bifurcation of maps of the interval*, Siam J. Appl. Math., Vol. 35, No. 2, pp. 260–267, 1978.
8. G. Statter, *Tendência decrescente da taxa de lucro*, Preprint, ISCTE, 2003.

Numerical Range, Numerical Radii and the Dynamics of a Rational Function

Helena Melo

Department of Mathematics, Universidade dos Açores, Campus de Ponta Delgada, 9501-801, Ponta Delgada, Portugal E-mail: hmelo@uac.pt

João Cabral

Department of Mathematics, Universidade dos Açores, Campus de Ponta Delgada, 9501-801, Ponta Delgada, Portugal E-mail: jcabral@uac.pt

Sometimes we obtain attractive results when associating facts to simple elements. The goal of this work is to introduce a possible alternative in the study of the dynamics of rational maps. In this study we use the family of maps $f(x) = \frac{x^2-a}{x^2-b}$, making some associations with the matrix $A = \begin{pmatrix} 1 & -a \\ 1 & -b \end{pmatrix}$ of its coefficients. Calculating the numerical range $W(A)$, the numerical radii $r(A)$ and $\widehat{r}(A)$, the boundary of the numerical range $\partial W(A)$, powers and iterations, we found relations very interesting, specially with the entropy of this maps.

1. Introduction

The main goal of this article is to present an alternative tool to study the dynamics of a real rational function, using results from the Numerical Range Theory, and has two main parts. The first, comprising Sections 2-3, is concerned to the adequation of the Numerical Range Theory to the data provided from rational maps. As described in Milnor [8], each map f, in the space Rat_2, can be expressed as a ratio

$$f(z) = \frac{p(z)}{q(z)} = \frac{a_0 z^2 + a_1 z + a_2}{b_0 z^2 + b_1 z + b_2},$$

where a_0 and b_0 are not both zero and $p(z)$, $q(z)$ have no common root. Milnor [8] states that we can obtain a roughly description of the topology of this space Rat_2 that can be identified with the Zariski open subset of *complex projective 5-space* consisting of all points

$$(a_0 : a_1 : a_2 : b_0 : b_1 : b_2) \in CP^5,$$

for which the *resultant*

$$res(p,q) = \det \begin{pmatrix} a_0 & a_1 & a_2 & 0 \\ 0 & a_0 & a_1 & a_2 \\ b_0 & b_1 & b_2 & 0 \\ 0 & b_0 & b_1 & b_2 \end{pmatrix}$$

is non-zero. Taking $z = x+i0$ and $a_0 = 1$, $b_0 = 1$, $a_2 = -a+i0$, $b_2 = -b+i0$, $a_1 = 0$, $b_1 = 0$, with x, a, b real numbers, we obtain

$$B = \begin{pmatrix} 1 & 0 & -a & 0 \\ 0 & 1 & 0 & -a \\ 1 & 0 & -b & 0 \\ 0 & 1 & 0 & -b \end{pmatrix},$$

$res(p,q) = \det B$. This matrix B is associated to the real rational map $f(x) = \left(x^2 - a\right) / \left(x^2 - b\right)$. The map f will be the one that we will use in our results associated to the matrix B.

The second part of this article, comprising Sections 4-5, shows how can we apply the Numerical Range Theory to the dynamics of the map f, establishing the relation between some partitions of an ellipse Ω, and the symbolic space generated by the partition of the domain of f in real intervals. Moreover, we launch a conjecture that could be a path to generate, in the future, an extension of the usual symbolic space applied to rational maps, allowing us to describe much better the dynamics of this maps.

2. Numerical Range Theory

The classical numerical range of a square matrix M_n, with complex numbers elements, is the set $W(M_n) = \{u^* M_n u,\ u \in S(\mathbb{C}^n)\}$, with $S(\mathbb{C}^n)$ the unit sphere, u is a vector in $\mathbb{C}^n$ and u^* is the transpose conjugate of u. The numerical range $W(M_n)$ also can be defined as the the image of the *Rayleight quotient* $R_{M_n}(u) = u^* M_n u / u^* u$, $u \neq 0$. The set $W(M_n)$ is closed and limited, and it is also a subset of the Gaussian $\mathbb{C}$ plane. Toeplitz [10] and Hausdorff [2] proved that $W(M_n)$ is a convex region. From Kippenhahn [5], the boundary of the numerical range, $\partial W(M_n)$ is a piecewise algebraic curve. In the particular case of a square matrix M_2, with eigenvalues λ_1, λ_2, $W(M_2)$ is a subset limited by an ellipse with foci in λ_1 and λ_2, result known as Elliptical Range Theorem, see Li [6].

Proposition 1. *If* $H_{M_n} = (M_n + M_n^*)/2$ *and* $S_{M_n} = (M_n - M_n^*)/2$ *are the Hermitian and skew-Hermitian parts of* M_n*, respectively, then* $\mathrm{Re}(W(M_n)) = W(H_{M_n})$ *and* $\mathrm{Im}(W(M_n)) = W(S_{M_n})$.

Proof. The proof can be found in Melo [7]. □

Theorem 1. *To every complex matrix* $M_n = H_{M_n} + S_{M_n}$ *through the equation* $k_{M_n}(\alpha_1, \alpha_2, \alpha_3) \equiv \det(\alpha_1 H_{M_n} - i\alpha_2 S_{M_n} + \alpha_3 I_n) = 0$ *is associated a curve of class n in homogeneous line coordinates in the complex plane. The convex hull of this curve is the numerical range of the matrix* M_n.

Proof. Adapting the proof in Kippenhahn [5] we have the desired result□

3. Merging $f(x)$ in $W(M_n)$

Hwa-Long Gau [1] states that we can obtain from a 4×4 matrix an elliptical numerical range, thus we could use B as defined in section 1, but we can simplify our results if we use a smaller matrix, A_2, trough a result in linear algebra. The new matrix $A_2 = \begin{pmatrix} 1 & -a \\ 1 & -b \end{pmatrix}$ will produce equivalent results as the obtained from B.

Lemma 1. *The matrix B is unitary decomposable in*

$$\begin{pmatrix} A_2 & \mathbb{O}_2 \\ \mathbb{O}_2 & A_2 \end{pmatrix}$$

a block diagonal matrix.

Proof. In order to prove this result it is sufficient to find an unitary matrix E such that

$$E^* B E = \begin{pmatrix} A_2 & \mathbb{O}_2 \\ \mathbb{O}_2 & A_2 \end{pmatrix}.$$

With some computation we can see that

$$E = \begin{pmatrix} 1 & 0 & 0 & 0 \\ 0 & 0 & 1 & 0 \\ 0 & 1 & 0 & 0 \\ 0 & 0 & 0 & 1 \end{pmatrix}.$$

□

Proposition 2. $W(A_2) = W(B)$.

Proof. By lemma 1 $E^*BE = A_2 \oplus A_2$ and using the properties of numerical range we have

$$W(E^*BE) = \textit{convex hull}\{W(A_2) \cup W(A_2)\} = \textit{convex hull}\{W(A_2)\}.$$

The *convex hull* of a convex set is itself, then $W(E^*BE) = W(A_2)$. But the numerical range of B is invariant under unitary transformations, see Kippenhahn [5], so $W(B) = W(A_2)$. □

In our study we use $f(x) = (x^2 - a)/(x^2 - b)$, with $a > 0$, $b > 0$ and $a > b$. Such map takes all real axis with exceptions $\pm\sqrt{b}$ on $(-\infty, 1) \cup [\frac{a}{b}, +\infty)$. With $\Lambda = \mathbb{R}\backslash\{\pm\sqrt{b}\} \times (-\infty, 1) \cup [\frac{a}{b}, +\infty)$ we can define the graphic of f, *graph*(f), as the pair $(x, f(x)) \in \Lambda$ and $\theta : \mathbb{R}\backslash\{\pm\sqrt{b}\} \longrightarrow \Lambda$.

Definition 1. Let $C = \left\{v \in \mathbb{C}^2 : v = (x, if(x)\right\}$ and

$$\Psi = \left\{z \in \mathbb{C} : z = \frac{v^* A_2 v}{v^* v}, v \neq 0\right\}.$$

We define $V : \Lambda \longrightarrow C$ as $(x, f(x)) \longmapsto (x, if(x))$ and $\Xi : C \longrightarrow \Psi$.

By definition 1 the image of $(x, f(x))$ is $z = (v^* A_2 v)/v^* v$ by $\Xi \circ V$.

Proposition 3. $\Psi \subset W(A_2)$

Proof. By the definition of Ψ and $W(A_2)$, using *Rayleight quotient*, the result follows. □

From proposition 3 we know that Ψ is a subset of $W(A_2) \subset \mathbb{C}$, and using definition 1, we can calculate the elements $z \in \Psi$, and as they were defined, they will become a function of x. After some calculations we have

$$z(x) = \frac{-a^2 b + (2a + b)\,bx^2 - 3bx^4 + x^6 + i\,(a+1)\left(-abx + (b+a)x^3 - x^5\right)}{a^2 + (b^2 - 2a)\,x^2 + (1 - 2b)\,x^4 + x^6}.$$

So, the $z \in \Psi$ is a function such that $z(x) = g(x) + ih(x)$, with $x \in \mathbb{R}$. Some elementary calculus show us that $g(x)$ and $h(x)$ are real rational continuous functions in $\mathbb{R}$, therefore $z(x)$ is continuous in $\mathbb{C}$. We call some attention to the fact that $z(\pm\sqrt{b})$ and $z(\infty)$ exists and are well defined in $\mathbb{C}$.

Observation 1. We have $z(x+1) = z(x)$ for

$$x = \frac{1}{2}\left(-1 \pm \sqrt{1 + 6a - 2b \pm 2\sqrt{4a + 9a^2 - 10ab + b^2}}\right).$$

Let x_2 and x_{12} be the values where $f(x) = 0$. If we calculate $z(0, a/b)$; $z(x_2, 0)$; $z(x_{12}, 0)$; $z(x, x)$; $z(x, -x)$ we obtain four different points of Ψ. With some elementary algebra we calculated the ellipse that contain this four points. This ellipse is

$$\Omega = \left\{ \frac{\left(x - \frac{1-b}{2}\right)^2}{\left(\frac{1+b}{2}\right)^2} + \frac{y^2}{\left(\frac{1+a}{2}\right)^2} = 1,\, (x,\, y) \in \mathbb{R}^2,\, a > b, a > 0, b > 0 \right\},$$

with $1 + b$ and $1 + a$ the minor and major axis length of Ω, respectively.

Moreover, when we use all points $(x,\, f(x))$ they will fall in Ω under transformation by z.

Lemma 2. *Let $z(x) = g(x) + ih(x)$, then the pair $(\mathrm{Re}(z), \mathrm{Im}(z))$ satisfies Ω.*

Proof. We obtain this result replacing in the equation of Ω, x by $\mathrm{Re}(z)$ and y by $\Im(z)$. □

Proposition 4. *If $S_i \in \Psi$ there are, at least one x_i such that $z(x_i) = S_i$.*

Proof. Since $z(x) = g(x) + ih(x)$ is a continuous function in $\mathbb{C}$ and by the lemma 2 the result follows. □

Then we conclude that Ψ can be represented by the ellipse with equation Ω.

Since Ω is constructed in the space $\mathbb{R}^2$ and this space is isomorphic to $\mathbb{C}$, when we refer to an element $z \in \Omega$ it can understood has a vector in $\mathbb{R}^2$ or a complex number in the plane $\mathbb{C}$.

There are relations between the functions f and g that we can observe, described in the following lemmas. The proofs are omitted because they result from straight calculus.

Lemma 3. *If x_0 is a zero of $f(x)$, then $g(x_0)$ is a relative maximum of $g(x)$.*

Lemma 4. *If x_0 is a relative minimum of $f(x)$ or x_0 is a discontinuity value of $f(x)$, then $g(x_0)$ is a relative minimum of $g(x)$.*

There are similar relations between $h(x)$ and $f(x)$.

Follows some results relating $f(x)$ to Ω.

Lemma 5. *Let $f(x_0) = \pm x_0$, then $\Xi \circ V(x_0, f(x_0))$ is vertex of Ω.*

Proof. If $f(x_0) = x_0$, by V we have $(x_0,\, ix_0)$. So

$$\Xi((x_0,\, ix_0)) = \frac{\begin{pmatrix} x_0 & -ix_0 \end{pmatrix}\begin{pmatrix} 1 & -a \\ 1 & -b \end{pmatrix}\begin{pmatrix} x_0 \\ ix_0 \end{pmatrix}}{\begin{pmatrix} x_0 & -ix_0 \end{pmatrix}\begin{pmatrix} x_0 \\ ix_0 \end{pmatrix}} = \frac{1-b}{2} - i\frac{1+a}{2}$$

And if we look at the equation of Ω, we see that $\left(\frac{1-b}{2}, -\frac{1+a}{2}\right)$ is a vertex of Ω.

If $f(x_0) = -x_0$ we obtain another vertex of Ω in a similar way, which is $\left(\frac{1-b}{2}, \frac{1+a}{2}\right)$. □

Lemma 6. *The discontinuities of $f(x)$ and the values where $f(x)$ has a minimum are transformed by $\Xi \circ V \circ \theta$ in the vertex $(-b, 0)$ of Ω, and the roots of $f(x)$ and the ∞ are transformed by $\Xi \circ V \circ \theta$ in the vertex $(1, 0)$ of Ω.*

Proof. Since $z(x) = g(x) + ih(x) = v^* A_2 v / v^* v$, $v = (x, if(x))$ is a continuous function in $\mathbb{C}$, we have $z(\pm\sqrt{b}) = -b$ and $z(\infty) = 1$. □

4. Partitions of Ω

Let x_1, x_2, x_5, x_6, x_7, x_8, x_9, x_{12}, x_{13} be the solutions of $g'(x) = 0$. By lemma 4, and lemma 3, and considering the order of real axis, we will have x_2 and x_{12} as zeros of $f(x)$; x_5 and x_9 as the discontinuities of $f(x)$ and $x_7 = 0$. All this values have image from $z(x)$, lemma 2, including the infinity, being related by $z(x_2) = z(x_{12}) = z(\infty)$ and equal to vertex $(1, 0)$ in Ω, see lemma 6, $z(x_5) = z(x_9) = z(0)$ and equal to vertex $(-b, 0)$ in Ω. Related to the real axis, $z(x_1)$ is symmetric to $z(x_{13})$ and $z(x_6)$ is symmetric to $z(x_8)$ in Ω. Where are the missing x_3, x_4, x_{10}, x_{11} ? They will be the values such that $z(x_1) = z(x_{11})$, $z(x_6) = z(x_{10})$, $z(x_8) = z(x_4)$ and $z(x_{13}) = z(x_3)$.

Using this special values x_i, $i = 1, .., 13$, with order $x_i < x_{i+1}$, we can define a partition function pa, as

$$pa(x) \underset{x\in\mathbb{R}}{=} \begin{cases} I_1, & if\ x < x_1 \\ I_i, & if\ x_{i-1} < x < x_i \text{ with } 2 \le i \le 13 \\ I_{14}, & if\ x > x_{13} \end{cases} .$$

Now we will create partitions in Ω using the images $z(x_i)$, $i = 1, .., 13$ in Ω. Here, we ask attention for one particular aspect of z, see proposition 4. Some intervals I_i will be transformed in the same arc of Ω. The only thing that will distinguish them is the orientation and the origin of its end points.

Definition 2. Let $S_i = z(x_i)$, we define $arc(S_i, S_{i+1})$ as the arc of Ω starting at S_i and ending at S_{i+1}, with counterclockwise orientation.

We define pa_Ω, a partition function, as:

$$pa_\Omega(w) \underset{w\in\mathbb{C}}{=} \begin{cases} J_1, & if\ w \in\ arc(z(\infty),\ z(x_1)) \\ J_i, & if\ w \in\ arc(z(x_{i-1}),\ z(x_i)) \text{ with } 2 \le i \le 13 \\ J_{14}, & if\ w \in\ arc(z(x_{14}),\ z(\infty)) \end{cases}.$$

The functions $pa(x)$ and $pa_\Omega(w)$ are related by

$$z(I_i) = \begin{cases} J_1 & if\ i = 1 \\ -J_i & if\ 2 \le i \le 6 \\ J_i & if\ i = 7 \text{ or } i = 8 \\ -J_i & if\ 9 \le i \le 13 \\ J_{14} & if\ i = 14 \end{cases},$$

thus we can build a matrix T_{14} of the transformation $pa_\Omega(z(pa))$,

$$T_{14} = \begin{bmatrix} N_7 & \mathbb{O}_7 \\ \mathbb{O}_7 & N_7 \end{bmatrix},$$

with

$$N_7 = \begin{bmatrix} 1 & 0 & 0 & 0 & 0 & 0 & 0 \\ 0 & -1 & 0 & 0 & 0 & 0 & 0 \\ 0 & 0 & -1 & 0 & 0 & 0 & 0 \\ 0 & 0 & 0 & -1 & 0 & 0 & 0 \\ 0 & 0 & 0 & 0 & -1 & 0 & 0 \\ 0 & 0 & 0 & 0 & 0 & -1 & 0 \\ 0 & 0 & 0 & 0 & 0 & 0 & 1 \end{bmatrix}.$$

It is easy to see that $T_{14}.T_{14} = I_{14}$, $det(T_{14}) = 1$, and it is an involutary matrix.

5. Dynamics of $f(x)$

Now we have a new tool to study the dynamics of f using a symbolic space. Using Ω to study the behavior of f we will have the same advantages that we would have when studying the behavior of second degree polynomials functions in the unit circle.

If we define a symbolic space using the partitions created by the function pa in the real axis we will have the problem of dealing with the discontinuities of the function f and the infinity itself. So, profiting that z is a continuous complex function in $\mathbb{C} \cup \{\infty\}$ this problem will vanish.

We can build two distinct symbolic spaces. The first will be the classical association between the intervals produced by pa in the real axis, see Milnor [8] for further reference, using the domain of the function f, and considering an alphabet $\mathcal{A}$ with designations I_i for each interval, we will have a symbolic space $\Sigma_c = \mathcal{A}^{\mathbb{N}}$. The second will be constructed as we consider the alphabet $\mathcal{B} = \{J_1, ..., J_{14}\}$, and the set $\Sigma = \mathcal{B}^{\mathbb{N}}$ of symbolic sequences on the elements of $\mathcal{B}$, introducing the map $spa : \mathbb{R} \cup \{\infty\} \longrightarrow \mathcal{B}$.

Conjecture 1. *The symbolic dynamics of f does not change if we use Σ instead of Σ_c.*

Both spaces are connected by the transformation matrix T_{14} and doing some calculus in matrix algebra, since this matrix is an involutary matrix, we could get the result. All computations in our work points in that direction. But we are still working in a suitable proof of this result. Moreover, Σ will work as an extension of Σ_c.

It means that we can identify the periodic orbits in the same values of a and b as we use both spaces Σ and Σ_c. For example for the values $a = 4.01$ and $b = 2.5$ the critical orbit of f is periodic in both spaces, such as all the others values of periodicity found in our research. But they are many new sequences in Σ that needs more work to full understood slightly changes caused by the obliteration of ∞. We are in the way.

This work, when finished, will imply a sequential work in the kneading theory and further study of the entropy of rational real maps.

References

1. Hwa-Long Gau, *Elliptic Numerical Ranges of 4x4 Matrices*, Taiwanese Journal of Mathematics, Vol. 10, n°1, pp. 117–128, 2006.
2. F. Hausdorff, *Der Wertvorrat einer Bilinearform*, Math Z., Vol. 3, pp. 314–316, 1919.
3. E. A. Jonckheere, F. Ahmad and E. Guthin, *Differential Topology of Numerical Range*, Linear Algebra and its Applications, 279, pp. 227–254, 1998.
4. M. Joswig, B. Straub, *On the numerical range map*, J. Austral.Math. Societ., series A, 65, pp. 267–283, 1998.
5. R. Kippenhahn, *On the Numerical Range of a Matrix*, Translation from the german by Paul F. Zachlin and Michiel E. Hochstenbach, 1951.
6. Chi-Kwong Li, *A simple proof of the elliptical range theorem*, proceedings of the American Mathematical Society, Vol. 124, n°7, 1996.

7. H. Melo, *Geração computacional de campos de valores e raios numéricos,* Ph. D. Thesis, Universidade dos Açores, 1999.
8. J. Milnor. *Geometry and Dynamics of Quadratic Rational Maps*, with an appendix by Milnor and Tan Lei, A. K. Peters, Ltd, 1993.
9. P. J. Psarrakos, M. J. Tsatsomeros, *Numerical Range: (in) Matrix Nutshell*, Part 1, Math-Notes, Washigthon State University, Vol. 45 - n°2, pp. 151–165, 2003.
10. O. Toeplitz, *Das algebraische Analogon zu einem Satze von Fejér*, Math. Z., Vol. 2, pp. 187–197, 1918.

On First-Order Nonlinear Difference Equations of Neutral Type

Małgorzata Migda

Institute of Mathematics, Poznań University of Technology, Piotrowo 3A, 60-965 Poznań, Poland
email: mmigda@math.put.poznan.pl

Janusz Migda

Faculty of Mathematics and Computer Science, A. Mickiewicz University, Umultowska 87, 61-614 Poznań, Poland
email: migda@amu.edu.pl

We consider the difference equations of the form

$$\Delta(x_n - p_n x_{n-\tau}) = \pm q_n f(x_{n-\sigma}, x_{n+1-\sigma}, \ldots, x_n).$$

We classify the set of possible nonoscillatory solutions of the above equations according to their asymptotic behavior. Some oscillation results for equation with maximum function are also obtained.

Keywords: Neutral type difference equation, asymptotic behavior, nonoscillation, oscillation.

1. Introduction

In this paper we consider the nonlinear difference equations of neutral type of the form

$$\Delta(x_n - p_n x_{n-\tau}) = \delta q_n f(x_{n-\sigma}, x_{n+1-\sigma}, \ldots, x_n) \qquad (E\delta)$$

where $n \in N(n_0)$, $N(n_0) = \{n_0, n_0 + 1, \ldots\}$, n_0 is fixed in $N = \{1, 2, \ldots\}$, $\delta = \pm 1$, τ is a positive integer, σ is nonnegative integer, $p, q : N(n_0) \longrightarrow R_+$, $f : R^{\sigma+1} \to R$. We assume that f is a nondecreasing function and if $u_1, u_2, \ldots, u_{\sigma+1}$ are positive (negative), then $f(u_1, u_2, \ldots, u_{\sigma+1})$ is also positive (negative).

By a solution of equation $(E\delta)$ we mean a real sequence (x_n) which satisfies equation $(E\delta)$ for all sufficiently large n. A solution is said to be nonoscillatory if it is eventually positive or eventually negative. Otherwise it is said to be oscillatory.

In section 2, we classify the set of possible nonoscillatory solutions of equations $(E\delta)$ according to their asymptotic behavior. In section 3, we study the following neutral type difference equation

$$\Delta(x_n - p_n x_{n-\tau}) = -q_n \max_{s\in[n-\sigma,n]} x_s, \quad n = 0, 1, \dots \tag{e}$$

which is a special case of equation $(E-)$. We obtain, under the hypothesis $\sum_{j=1}^{\infty} q_j = \infty$, sufficient conditions for the oscillation of all solutions of equation (e). The unstable difference equation

$$\Delta(x_n - p_n x_{n-\tau}) = q_n \max_{s\in[n-\sigma,n]} x_s, \quad n = 0, 1, \dots$$

has been studied in [7]. Functional equations involving the maximum function are important since they are often met in the applications, for instance in the theory of automatic control [8]. Some qualitative properties of these equations can be found in [2-5], [7], [9]. The higher order difference equations of type (e) have been studied in [5]. The authors obtained in that paper some oscillatory and asymptotic behaviors under the hypothesis $\sum_{j=1}^{\infty} q_j < \infty$.

2. Classifications of Nonoscillatory Solutions

We begin by analysing the asymptotic behavior of possible nonoscillatory solutions of equations $(E\delta)$. Let (x_n) be a nonoscillatory solution of $(E\delta)$. Put

$$z_n = x_n - p_n x_{n-\tau}. \tag{1}$$

Then, by $(E\delta)$, (Δz_n) is eventually of one sign, so (z_n) is eventually of constant sign, also. Therefore, either

$$x_n z_n > 0 \tag{2}$$

or

$$x_n z_n < 0 \tag{3}$$

for all sufficiently large n. Let us denote by $\mathcal{N}^+$ [or $\mathcal{N}^-$] the set of all nonoscillatory solutions x of equation $(E\delta)$ such that (2) [or (3)] is satisfied.

Let us introduce the following sets:

$$\begin{aligned}
\mathcal{N}_0^+ &= \{x \in \mathcal{N}^+ : \lim_{n \longrightarrow \infty} z_n = 0\},\\
\mathcal{N}_c^+ &= \{x \in \mathcal{N}^+ : \lim_{n \longrightarrow \infty} |z_n| \in (0, \infty)\},\\
\mathcal{N}_\infty^+ &= \{x \in \mathcal{N}^+ : \lim_{n \longrightarrow \infty} |z_n| = \infty\},\\
\mathcal{N}_0^- &= \{x \in \mathcal{N}^- : \lim_{n \longrightarrow \infty} z_n = 0\},\\
\mathcal{N}_c^- &= \{x \in \mathcal{N}^- : \lim_{n \longrightarrow \infty} |z_n| \in (0, \infty)\},\\
\mathcal{N}_\infty^- &= \{x \in \mathcal{N}^- : \lim_{n \longrightarrow \infty} |z_n| = \infty\}.
\end{aligned}$$

From (2) and $(E+)$ it follows that the sequence (z_n) is positive and increasing for large n or it is negative and decreasing, eventually. Hence, for equation $(E+)$ we have $\mathcal{N}_0^+ = \emptyset$. Similarly, from (3) we get $\mathcal{N}_\infty^- = \emptyset$ for equation $(E+)$.
Analogously for equation $(E-)$ we have $\mathcal{N}_\infty^+ = \emptyset$ and $\mathcal{N}_0^- = \emptyset$. Therefore we have the following decomposition of the set $\mathcal{N}$ of all nonoscillatory solutions of equations $(E\delta)$:

$$\begin{aligned}
\mathcal{N} &= \mathcal{N}_c^+ \cup \mathcal{N}_0^+ \cup \mathcal{N}_c^- \cup \mathcal{N}_\infty^- \quad \text{for} \quad \delta = -1\\
\mathcal{N} &= \mathcal{N}_\infty^+ \cup \mathcal{N}_c^+ \cup \mathcal{N}_0^- \cup \mathcal{N}_c^- \quad \text{for} \quad \delta = 1.
\end{aligned} \tag{4}$$

We will consider the following two cases:

(I) $0 < p_n \leq 1, \quad \text{for} \quad n \geq n_0;$
(II) $1 \leq p_n \leq \lambda < \infty, \quad \text{for} \quad n \geq n_0.$

For simplicity, the equation $(E\delta)$ subject to the case (I) [case (II)] will be referred to as equation $(E\delta, I)$ $[(E\delta, II)]$. The simple lemma below indicate that restrictions upon (p_n) may force some of the classes appearing in (4) to be empty.

Lemma 1. *(see [6]) Suppose*

$$0 < p_n \leq 1 \quad \textit{for} \quad n \geq n_0.$$

Let x be a nonoscillatory solution of the inequality

$$x_n(x_n - p_n x_{n-\tau}) < 0$$

for sufficiently large n. Then x is bounded. If, moreover,

$$0 < p_n \leq \lambda < 1, \quad \textit{for} \quad n \geq n_0$$

for some positive constant λ, then $\lim_{n \longrightarrow \infty} x_n = 0$.

From Lemma 1 it follows that

$$\mathcal{N}_{\infty}^{-} = \emptyset \qquad \text{for} \qquad (E-, I).$$

By imposing suitable restrictions on (q_n) we show that some other solution classes appearing in (4) are empty.

Lemma 2. *Suppose that*

$$\sum_{j=n_0}^{\infty} q_j = \infty. \tag{5}$$

Then the inequality

$$(\Delta u_n + q_n f(u_{n-\sigma}, ..., u_n))\, sgn\, u_n \le 0$$

has no nonoscillatory solution (u_n) *such that* $\liminf_{n \longrightarrow \infty} |u_n| > 0$, *and the inequality*

$$(\Delta u_n - q_n f(u_{n-\sigma}, ..., u_n))\, sgn\, u_n \ge 0$$

has no nonoscillatory solution (u_n)*such that* $\limsup_{n \longrightarrow \infty} |u_n| < \infty$.

The proof of Lemma 2 is similar to the proof of Lemma 3 in [6] and will be omitted.

Lemma 3.

(a) If $x \in \mathcal{N}^+$ *for (E+) then the sequence* z *defined by (1) satisfies*

$$(\Delta z_n - q_n f(z_{n-\sigma}, ..., z_n))\, sgn\, z_n \ge 0 \text{ for all large } n.$$

(b) If $x \in \mathcal{N}^-$ *for (E+,I) then the sequence* $y_n = -z_n$ *with* z *given by (1) satisfies*

$$(\Delta y_n + q_n f(y_{n+\tau-\sigma}, ..., y_{n+\tau}))\ sgn\, y_n \le 0 \text{ for all large } n.$$

(c) If $x \in \mathcal{N}^-$ *for (E+,II) then the sequence* $w_n = -\frac{z_n}{\lambda}$ *with* z *given by (1) satisfies*

$$\left(\Delta w_n + \frac{1}{\lambda} q_n f(w_{n+\tau-\sigma}, ..., w_{n+\tau})\right)\ sgn\, w_n \le 0$$

for all large n.

(d) If $x \in \mathcal{N}^+$ *for (E−) then the sequence* z *satisfies*

$$(\Delta z_n + q_n f(z_{n-\sigma}, ..., z_n))\, sgn\, z_n \le 0 \text{ for all large } n.$$

(e) *If* $x \in \mathcal{N}^-$ *for* $(E-, I)$ *then* $y_n = -z_n$ *with* z *given by (1) satisfies*

$$(\Delta y_n - q_n f(y_{n+\tau-\sigma}, ..., y_{n+\tau}))\, sgn\, y_n \geq 0 \ \textit{ for all large } \ n.$$

(f) *If* $x \in \mathcal{N}^-$ *for* $(E-, II)$ *then* $w_n = -\frac{z_n}{\lambda}$ *with* z *given by (1) satisfies*

$$\left(\Delta w_n - \frac{1}{\lambda} q_n f(w_{n+\tau-\sigma}, ..., w_{n+\tau})\right) sgn\, w_n \geq 0$$

for all large n.

The proof of Lemma 3 is similar to that of Proposition 1 in [6] and will be omitted. From the above lemmas we get that if $\sum_{j=n_0}^{\infty} q_j = \infty$ is satisfied then

$$\mathcal{N}_c^+ = \emptyset \quad \text{and} \quad \mathcal{N}_c^- = \emptyset \quad \text{for} \quad (E\delta).$$

Furthermore, the condition $\sum_{j=n_0}^{\infty} q_j = \infty$ also ensures that $\mathcal{N}_0^+ = \emptyset$ for $(E-, II)$.

Summarizing the above observations we conclude that under the condition $\sum_{j=n_0}^{\infty} q_j = \infty$ the classification (4) of the set $\mathcal{N}$ of all nonoscillatory solutions of equations $(E\delta)$ is as follows:

$$\mathcal{N} = \mathcal{N}_\infty^+ \cup \mathcal{N}_0^- \quad \text{for} \quad (E+, I)$$
$$\mathcal{N} = \mathcal{N}_\infty^+ \cup \mathcal{N}_0^- \quad \text{for} \quad (E+, II)$$

and

$$\mathcal{N} = \mathcal{N}_0^+ \quad \text{for} \quad (E-, I)$$
$$\mathcal{N} = \mathcal{N}_\infty^- \quad \text{for} \quad (E-, II).$$

3. Oscillation Results

Our aim in this section is to establish conditions for the oscillation of all solutions of equation (e). We will need the following lemmas.

Lemma 4. *Assume* (q_n) *is a positive real sequence and* k *is a positive integer. If*

$$\lim_{n \longrightarrow \infty} \inf \sum_{i=n-k}^{n-1} q_i > \left(\frac{k}{k+1}\right)^{k+1},$$

then

(i) the difference inequality

$$\Delta x_n + q_n x_{n-k} \le 0$$

has no eventually positive solution;

(ii) the difference inequality

$$\Delta x_n + q_n x_{n-k} \ge 0$$

has no eventually negative solution.

This lemma follows from Theorem 7.6.1 in [1].

Lemma 5. *Assume (q_n) is a positive real sequence and l is a positive integer. If*

$$\lim_{n \longrightarrow \infty} \inf \sum_{i=n+1}^{n+l-1} q_i > \left(\frac{l}{l+1}\right)^{l+1},$$

then

(i) the difference inequality

$$\Delta x_n - q_n x_{n+l} \ge 0$$

has no eventually positive solution;

(ii) the difference inequality

$$\Delta x_n - q_n x_{n+l} \le 0$$

has no eventually negative solution.

The above lemma one can obtain by modification of Theorem 7.5.2 in [1].

From Section 2 we know, that the only possible nonoscillatory solutions of equation (e) are of the type

$$\mathcal{N} = \mathcal{N}_0^+ \qquad \text{for} \quad 0 < p_n \le 1$$
$$\mathcal{N} = \mathcal{N}_\infty^- \qquad \text{for} \quad p_n \ge 1.$$

Therefore we get the following result.

Lemma 6. *Let $\sum\limits_{j=n_0}^{\infty} q_j = \infty$. Then we have:*

(i) if $0 < p_n \le 1$ and (x_n) is an eventually positive solution of (e), then the sequence (z_n) is eventually decreasing, $z_n > 0$ eventually and $\lim\limits_{n \longrightarrow \infty} z_n = 0$;

(ii) if $0 < p_n \leq 1$ *and* (x_n) *is an eventually negative solution of (e), then the sequence* (z_n) *is eventually increasing,* $z_n < 0$ *eventually and* $\lim_{n \longrightarrow \infty} z_n = 0$*;*

(iii) if $1 \leq p_n \leq \lambda < \infty$ *and* (x_n) *is an eventually positive solution of (e), then the sequence* (z_n) *is eventually decreasing,* $z_n < 0$ *eventually and* $\lim_{n \longrightarrow \infty} z(n) = -\infty$*;*

(iv) if $1 \leq p_n \leq \lambda < \infty$ *and* (x_n) *is an eventually negative solution of (e), then the sequence* (z_n) *is eventually increasing,* $z_n > 0$ *eventually and* $\lim_{n \longrightarrow \infty} z_n = \infty$.

As an immediate corollary of Lemma 6 we get the following theorem.

Theorem 1. *Let* $\sum_{j=n_0}^{\infty} q_j = \infty$ *and* $p_n \equiv 1$*. Then every solution of equation (e) is oscillatory.*

Theorem 2. *Let* $\sum_{j=n_0}^{\infty} q_j = \infty, \quad 0 < p_n \leq 1, \quad \sigma > \tau$ *and*

$$\liminf_{n \longrightarrow \infty} \sum_{i=n-\tau}^{n-1} q_i \left(\min_{s \in [i-\sigma, i]} p_s \right) > \left(\frac{\tau}{\tau+1} \right)^{\tau+1}. \tag{6}$$

Then every solution of equation (e) is oscillatory.

Proof. Assume, for the sake of contradiction, that equation (e) has a nonoscillatory solution (x_n). Let $x_n > 0$ eventually. From Lemma 6 (i) it follows that the sequence (z_n) is eventually decreasing and positive. Then, by (1) we have $z_n < x_n$ and

$$\max_{s \in [n-\sigma, n]} z_s < \max_{s \in [n-\sigma, n]} x_s.$$

Therefore from (e) we obtain

$$\Delta z_n + q_n \max_{s \in [n-\sigma, n]} z_s < 0.$$

Since (z_n) is eventually decreasing, for sufficiently large n, we have

$$z_{n-\tau} \leq z_{n-\sigma} = \max_{s \in [n-\sigma, n]} z_s.$$

Hence

$$\Delta z_n + q_n z_{n-\tau} \leq 0.$$

But from (6) and Lemma 4 it follows that the above inequality has no eventually positive solution which is a contradiction.

Now, let $x_n < 0$ eventually. From Lemma 6 (ii) it follows that the sequence (z_n) is eventually increasing and negative. Then, from (1) we have

$$x_n < p_n x_{n-\tau} < p_n z_{n-\tau}$$

and

$$\max_{s\in[n-\sigma,n]} x_s < \max_{s\in[n-\sigma,n]} (p_s z_{s-\tau}).$$

As (z_n) is increasing it is $z_{s-\tau} \le z_{n-\tau}$ for every $s \in [n-\sigma, n]$. Since all p_s are positive then $p_s z_{s-\tau} \le p_s z_{n-\tau}$ for every $s \in [n-\sigma, n]$. Therefore

$$\max_{s\in[n-\sigma,n]} x_s \le \max_{s\in[n-\sigma,n]} (p_s z_{s-\tau}) \le \max_{s\in[n-\sigma,n]} (p_s z_{n-\tau}).$$

As $z_{n-\tau}$ is negative and p_s are positive then

$$\max_{s\in[n-\sigma,n]} (p_s z_{n-\tau}) \le \left(\min_{s\in[n-\sigma,n]} p_s\right) z_{n-\tau}.$$

Hence, by (e) we get

$$0 = \Delta z_n + q_n \max_{s\in[n-\sigma,n]} x_s \le \Delta z_n + q_n \left(\min_{s\in[n-\sigma,n]} p_s\right) z_{n-\tau}.$$

But from (6) and Lemma 4 it follows that the above inequality has no eventually negative solution which is a contradiction. This completes the proof.

Theorem 3. *Let* $\sum\limits_{j=n_0}^{\infty} q_j = \infty, \quad 1 \le p_n \le \lambda < \infty, \quad \tau > \sigma$ *and*

$$\liminf_{n\longrightarrow\infty} \sum_{i=n+1}^{n+\tau-\sigma-1} \frac{q_i}{\max\limits_{s\in[i-\sigma,i]} p_{s+\tau}} > \left(\frac{\tau-\sigma}{\tau-\sigma+1}\right)^{\tau-\sigma+1}. \tag{7}$$

Then every solution of equation (e) is oscillatory.

Proof. Assume, for the sake of contradiction, that equation (e) has a nonoscillatory solution (x_n). Let $x_n > 0$ eventually. From Lemma 6 (iii) it follows that the sequence (z_n) is eventually decreasing and negative. In view of (1) we have

$$z_n > -p_n x_{n-\tau},$$

$$x_n > -\frac{z_{n+\tau}}{p_{n+\tau}}$$

and hence

$$\max_{s\in[n-\sigma,n]} x_s \geq \max_{s\in[n-\sigma,n]} \left(-\frac{z_{s+\tau}}{p_{s+\tau}}\right). \tag{8}$$

Since the sequence (z_n) is eventually decreasing, then for sufficiently large n we have

$$\min_{s\in[n-\sigma,n]} (-z_{s+\tau}) = -z_{n+\tau-\sigma}. \tag{9}$$

Note that for every $s \in [n-\sigma, n]$

$$\left(-\frac{z_{s+\tau}}{p_{s+\tau}}\right) \geq \left(-\frac{z_{n+\tau-\sigma}}{p_{s+\tau}}\right) \geq -\frac{z_{n+\tau-\sigma}}{\max\limits_{s\in[n-\sigma,n]} p_{s+\tau}}.$$

Then, consequently, for every $s \in [n-\sigma, n]$ we have

$$x_s > -\frac{z_{n+\tau-\sigma}}{\max\limits_{s\in[n-\sigma,n]} p_{s+\tau}}$$

and

$$0 = \Delta z_n + q_n \max_{s\in[n-\sigma,n]} x_s \geq \Delta z_n - \frac{q_n}{\max\limits_{s\in[n-\sigma,n]} p_{s+\tau}} z_{n+\tau-\sigma}.$$

But from (7) and Lemma 5 it follows that the last inequality has no eventually negative solution which is a contradiction.
In the case where (x_n) is an eventually negative solution of (e) one arrives similarly to the inequality

$$0 = \Delta z_n + q_n \max_{s\in[n-\sigma,n]} x_s \leq \Delta z_n - \frac{q_n}{\max\limits_{s\in[n-\sigma,n]} p_{s+\tau}} z_{n+\tau-\sigma}.$$

From (7) and Lemma 5 it follows that the above inequality has no eventually positive solution which is a contradiction. Hence each solution of (e) is oscillatory. This completes the proof.

Acknowledgments

The authors thank the referee for valuable comments.

References

1. I. Gyori, G. Ladas, Oscillation theory of delay differential equations with applications, Clarendon, Oxford, New York, 1991.
2. D. Bainov, V. Petrov and V. Proicheva, Oscillation of neutral differential equations with "maxima", *Revista Math.*, 8(1)(1995), 171-180.

3. Y. Gao, G. Zhang, Oscillation of nonlinear first order neutral difference equations, *Appl. Math. E-notes* 1(2001), 5-10.
4. X. Luo, D. Bainov, Oscillatory and asymptotic behavior of second-order neutral difference equations with "maxima", *Int. Journal of Math. Analysis* 1(2007), 791-804.
5. X. Li, X. Zhou, Oscillation and nonoscillation of higher order neutral difference equations with maxima, *Int. Journal of Math. Analysis* 1(2007), 791-804.
6. M. Migda, J. Migda, A class of first-order nonlinear difference equations of neutral type, *Mathematical and Computer Modelling* 40 (2004) 297-306.
7. M. Migda, G. Zhang, On unstable neutral difference equations with "maxima", *Math. Slovaca*, 56 (2006), no. 4, 451-463.
8. E. P. Popov, Automatic regulation and control, Nauka, Moskwa, 1966.
9. B. G. Zhang, G. Zhang, Qualitative properties of functional differential equations with "maxima", *Rocky Mountain J. Math.* 29 (1999), 357-367.

Oscillatory Mixed Differential Difference Equations

Sandra Pinelas

Universidade dos Açores, Departamento de Matemática,
R. Mãe de Deus, 9500-321 Ponta Delgada, Portugal

In this work is studied the oscillatory behavior of the delay differential difference equation of mixed type

$$x'(t) = \sum_{i=1}^{\ell} p_i x(t - r_i) + \sum_{j=1}^{m} q_j x(t + \tau_j)$$

$$(r_i > 0,\ i = 1, ..., \ell;\ \tau_j > 0,\ j = 1, ..., m)$$

Some criteria are obtained in order to guarantee that all solutions of this equation are oscillatory. Some conditions for having nonoscillations are also given.

1. Introduction

The aim of this work is to study the oscillatory behavior of the differential difference equation of mixed type

$$x'(t) = \sum_{i=1}^{\ell} p_i x(t - r_i) + \sum_{j=1}^{m} q_j x(t + \tau_j) \tag{1}$$

where $x(t) \in \mathbb{R}$, $0 < r_1 < r_2 < ... < r_\ell$, $0 < \tau_1 < \tau_2 < ... < \tau_m$ and the coefficients p_i, q_j are real numbers.

As usual, we will say that a solution $x(t)$ of (1) oscillates if it has arbitrary large zeros. When all solutions oscillate (1) will be said oscillatory.

According to Krisztin [1, 2], the oscillatory behavior of (1) can be studied, as for delay equations, through the analysis of the zeros of the characteristic equation

$$\lambda = \sum_{i=1}^{\ell} p_i \exp(-\lambda r_i) + \sum_{j=1}^{m} q_j \exp(\lambda \tau_j). \tag{2}$$

In fact, assuming that $q_m \neq 0$, through [1, Corollary 5] one can conclude that the equation (1) is oscillatory if and only if the characteristic equation (2) has no real roots.

Therefore considering the function

$$N(\lambda) = \sum_{i=1}^{\ell} p_i \exp(-\lambda r_i) + \sum_{j=1}^{m} q_j \exp(\lambda \tau_j), \tag{3}$$

and noticing that $N(\lambda) \to \infty$, as $\lambda \to \infty$, faster then λ, one can say that equation (1) is oscillatory if, for every real λ, either

$$N(\lambda) > \lambda \tag{4}$$

or

$$N(\lambda) < \lambda. \tag{5}$$

2. Oscillations

With respect to condition (4) we obtain the following theorem.

Theorem 2.1. *If $p_i, q_j \geq 0$, for $i = 1, ..., \ell$ and $j = 1, ...m - 1$, $q_m > 0$ and*

$$e \sum_{j=1}^{m} \tau_j q_j > 1 \tag{6}$$

then the equation (1) is oscillatory.

Proof. For $\lambda \leq 0$ one has clearly

$$N(\lambda) = \sum_{i=1}^{\ell} p_i \exp(-\lambda r_i) + \sum_{j=1}^{m} q_j \exp(\lambda \tau_j) > 0 \tag{7}$$

and consequently $N(\lambda) > \lambda$.

If $\lambda > 0$ then

$$\begin{aligned} \frac{N(\lambda)}{\lambda} &= \sum_{i=1}^{\ell} \frac{\exp(-\lambda r_i)}{\lambda} p_i + \sum_{j=1}^{m} \frac{\exp(\lambda \tau_j)}{\lambda} q_j \\ &\geq \sum_{j=1}^{m} \frac{\exp(\lambda \tau_j)}{\lambda} q_j \\ &\geq e \sum_{j=1}^{m} \tau_j q_j > 1 \end{aligned}$$

and, in view of (6), $N(\lambda) > \lambda$.

Thus $N(\lambda) > \lambda$ for every real λ. □

Example 2.1. The equation

$$x'(t) = \sum_{i=1}^{\ell} p_i x(t - r_i) + x\left(t + \frac{1}{8}\right) + x\left(t + \frac{1}{7}\right) + \frac{1}{3} x\left(t + \frac{1}{3}\right)$$

where $p_i > 0$, for $i = 1, ..., \ell$ and $0 < r_1 < r_2 < ... < r_\ell$, is oscillatory since

$$e\left(\frac{1}{9} + \frac{1}{8} + \frac{1}{7}\right) \approx 1.0301.$$

A few simpler conditions that imply (6) can be used to obtain the oscillatory behavior (1). This is shown in the following corollary.

Corollary 2.1. *If $p_i, q_j \geqslant 0$, for $i = 1, ..., \ell$ and $j = 1, ...m - 1$, $q_m > 0$ and one of the following conditions holds*

i) $e\tau_1 \sum_{j=1}^{m} q_j > 1,$
ii) $eq \sum_{j=1}^{m} \tau_j > 1, \quad q = \min\{q_j : j = 1, ...m\},$
iii) $e\tau_1 qm > 1, \quad q = \min\{q_j : j = 1, ...m\},$

then the equation (1) is oscillatory.

Remark 2.1. Notice that the condition iii) of the Corollary 2.1 implies i) and ii) of the same corollary.

Remark 2.2. Notice that the Corollary 2.1 cannot to be applied to the equation of Example 2.1, since

$$e\tau_1 \sum_{j=1}^{m} q_j = e\frac{1}{8}\left(2 + \frac{1}{3}\right) \approx 0.79283,$$

$$eq \sum_{j=1}^{m} \tau_j = e\frac{1}{3}\left(\frac{1}{3} + \frac{1}{8} + \frac{1}{7}\right) \approx 0.54474,$$

$$eq\tau_1 m = 3e\frac{1}{3}\frac{1}{8} \approx 0.33979.$$

Example 2.2. Using the condition i) of the Corollary 2.1, the equation

$$x'(t) = \sum_{i=1}^{\ell} p_i x(t - r_i) + \frac{1}{2} x\left(t + \frac{1}{4}\right) + x\left(t + \frac{1}{2}\right) + \frac{1}{9} x\left(t + \frac{3}{4}\right)$$

where $p_i > 0$, for $i = 1, ..., \ell$ and $0 < r_1 < r_2 < ... < r_\ell$, is oscillatory since

$$e\tau_1 \sum_{j=1}^{m} q_j = e\frac{1}{4}\left(\frac{1}{2} + 1 + \frac{1}{9}\right) \approx 1.0949 > 1.$$

Notice that the condition ii) of the Corollary 2.1 cannot be used in this case since

$$eq\sum_{j=1}^{m}\tau_j = e\frac{1}{9}\left(\frac{1}{4}+\frac{1}{2}+\frac{3}{4}\right) \approx 0.45305.$$

Example 2.3. Using the condition ii) of the Corollary 2.1, the equation

$$x'(t) = \sum_{i=1}^{\ell} p_i x(t-r_i) + \frac{1}{2}x\left(t+\frac{1}{5}\right) + \frac{3}{4}x\left(t+\frac{1}{4}\right) + \frac{5}{9}x\left(t+\frac{1}{3}\right)$$

where $p_i > 0$, for $i = 1, ..., \ell$ and $0 < r_1 < r_2 < ... < r_\ell$, is oscillatory since

$$eq\sum_{j=1}^{m}\tau_j = \frac{e}{2}\left(\frac{1}{5}+\frac{1}{4}+\frac{1}{3}\right) = 1.0647 > 1.$$

Condition i) doesn't work for this equation since

$$e\tau\sum_{j=1}^{m}q_j = \frac{e}{5}\left(\frac{1}{2}+\frac{3}{4}+\frac{5}{9}\right) \approx 0.9816.$$

Example 2.4. Using the condition iii) of the Corollary 2.1, the equation

$$x'(t) = \sum_{i=1}^{\ell} p_i x(t-r_i) + x\left(t+\frac{1}{4}\right) + \frac{5}{4}x\left(t+\frac{1}{2}\right) + \frac{1}{2}x(t+1)$$

where $p_i > 0$, for $i = 1, ..., \ell$ and $0 < r_1 < r_2 < ... < r_\ell$, is oscillatory since

$$eq\tau_1 m = 3\frac{1}{2}\frac{1}{4}e = 1.0194 > 1.$$

The Theorem 2.1 expresses a larger relevance of the advanced part of the equation in order that (1) be oscillatory. A symmetric situation is obtained when the condition (5) is used to get oscillatory criteria. This is shown in the following theorem similar to [3, Theorem 2.2.1].

Theorem 2.2. *If $p_i, q_j \leqslant 0$, for $i = 1, ..., \ell$ and $j = 1, ...m-1$, $q_m < 0$ and*

$$e\sum_{i=1}^{\ell} r_i p_i < -1 \tag{8}$$

then the equation (1) is oscillatory.

Proof. If $\lambda \geqslant 0$ then

$$N(\lambda) = \sum_{i=1}^{\ell} p_i \exp(-\lambda r_i) + \sum_{j=1}^{m} q_j \exp(\lambda\tau_j) < 0 \tag{9}$$

and consequently $N(\lambda) < \lambda$.

For $\lambda < 0$ then

$$\frac{N(\lambda)}{\lambda} = \sum_{i=1}^{\ell} \frac{\exp(-\lambda r_i)}{\lambda} p_i + \sum_{j=1}^{m} \frac{\exp(\lambda \tau_j)}{\lambda} q_j$$
$$> -e \sum_{i=1}^{\ell} r_i p_i > 1$$

and consequently $N(\lambda) < \lambda$.

Thus for every real λ, one has $N(\lambda) < \lambda$. □

Example 2.5. The equation

$$x'(t) = -x\left(t - \frac{1}{7}\right) - \frac{1}{2}x\left(t - \frac{1}{4}\right) - \frac{1}{3}x\left(t - \frac{1}{3}\right) + \sum_{j=1}^{m} q_j x(t + \tau_j)$$

where $q_j < 0$, for $j = 1, ..., m$ and $0 < \tau_1 < \tau_2 < ... < \tau_m$, is oscillatory since

$$e \sum_{i=1}^{\ell} r_i p_i = e\left(-\frac{1}{9} - \frac{1}{8} - \frac{1}{7}\right) \approx -1.0301.$$

Analogously to Corollary 2.1 one can state the following corollary.

Corollary 2.2. *If $p_i, q_j \leqslant 0$, for $i = 1, ..., \ell$ and $j = 1, ...m-1$, $q_m < 0$ and*

iv) $er_1 \sum_{i=1}^{\ell} p_i < -1,$
v) $ep \sum_{i=1}^{\ell} r_i < -1, \quad p = \max\{p_i, i = 1, ..., \ell\},$
vi) $er_1 p\ell < -1, \quad p = \max\{p_i, i = 1, ..., \ell\},$

then the equation (1) is oscillatory.

Remark 2.3. In this case, as in Corollary 2.1, the condition vi) of Corollary 2.2 implies iv) or v) of the same corollary.

Remark 2.4. Notice that the Corollary 2.2 cannot to be applied to the equation of Example 2.5, since

$$er_1 \sum_{i=1}^{\ell} p_i = e\frac{1}{7}\left(-\frac{1}{3} - \frac{1}{2} - 1\right) \approx -0.7119,$$
$$ep \sum_{i=1}^{\ell} r_i = -e\frac{1}{3}\left(\frac{1}{3} + \frac{1}{4} + \frac{1}{7}\right) \approx -0.658,$$
$$er_1 p\ell = -3e\frac{1}{7}\frac{1}{3} \approx -0.38833.$$

Example 2.6. Using the condition iv) of the Corollary 2.2, one can conclude that the equation

$$x'(t) = -\frac{1}{4}x\left(t - \frac{1}{4}\right) - x\left(t - \frac{1}{3}\right) - x\left(t - \frac{1}{2}\right) + \sum_{j=1}^{m} q_j x(t + \tau_j)$$

is oscillatory for any $q_j < 0$, for $j = 1, ..., m$, and $0 < \tau_1 < \tau_2 < ... < \tau_m$, taking into account that

$$er_1 \sum_{i=1}^{\ell} p_i = e\frac{1}{4}\left(-2 - \frac{1}{4}\right) \approx -1.529 < -1.$$

Condition v) of the Corollary 2.2 doesn't work in this case

$$ep \sum_{i=1}^{\ell} r_i = -e\frac{1}{4}\left(\frac{1}{2} + \frac{1}{4} + \frac{1}{3}\right) \approx -0.7362.$$

Example 2.7. Through v) of the Corollary 2.2, the equation

$$x'(t) = -\frac{1}{2}x\left(t - \frac{1}{10}\right) - \frac{3}{4}x\left(t - \frac{1}{2}\right) - \frac{4}{5}x\left(t - \frac{2}{3}\right) + \sum_{j=1}^{m} q_j x(t + \tau_j)$$

is oscillatory for every $q_j < 0$, for $j = 1, ..., m$ and $0 < \tau_1 < \tau_2 < ... < \tau_m$, since

$$ep \sum_{i=1}^{\ell} r_i = -e\frac{1}{2}\left(\frac{1}{10} + \frac{1}{2} + \frac{2}{3}\right) \approx -1.7216.$$

In this case one cannot use condition iv) taking into account that

$$er_1 \sum_{i=1}^{\ell} p_i = e\frac{1}{10}\left(-\frac{1}{2} - \frac{3}{4} - \frac{4}{5}\right) \approx -0.55725.$$

Example 2.8. The condition vi) of the Corollary 2.2, is illustrated by the the equation

$$x'(t) = -\frac{1}{2}x\left(t - \frac{1}{2}\right) - x(t-1) - \frac{1}{4}x\left(t - \frac{3}{2}\right) + \sum_{j=1}^{m} q_j x(t + \tau_j)$$

where $q_j < 0$, for $j = 1, ..., m$ and $0 < \tau_1 < \tau_2 < ... < \tau_m$. As

$$er_1 p\ell = -3e\frac{1}{4}\frac{1}{2} \approx -1.0194,$$

one concludes that the equation is oscillatory.

3. Nonoscillations

With $0 < r_1 < r_2 < ... < r_\ell$, $0 < \tau_1 < \tau_2 < ... < \tau_m$ and p_i, q_j such that $q_m \neq 0$, we recall that (1) is nonoscillatory if and only if there exists a $\lambda_0 \in \mathbb{R}$ such that

$$N(\lambda_0) - \lambda_0 = 0.$$

In the regard of obtaining (1) oscillatory, all the results of the preceding section envolve specifically either delays or advances. In fact, the Theorem 2.1 and Corollary 2.1 state that (1) is oscillatory independently of the delays and the Theorem 2.2 and Corollary 2.2 state that (1) is oscillatory independently of the advances. Assumptions as the obtained in [4, Corollary 1] for delay equations, which imply the oscillatory behavior of (1) independently of the delays and advances are here no longer valid. That is, equation (1) cannot be oscillatory globally in the delays and advances simultaneously. However that is possible in view of the existence of nonoscillations as is shown in the following theorem.

Theorem 3.1. *If $p_\ell q_m < 0$ then the equation (1) is nonoscillatory independently of the delays and advances.*

Proof. Assume for example that $p_\ell > 0$ and $q_m < 0$. Then for every family of delays and advances one has, as $\lambda \to +\infty$,

$$N(\lambda) - \lambda = \sum_{i=1}^{\ell} p_i \exp(-\lambda r_i) + \sum_{j=1}^{m} q_j \exp(\lambda \tau_j) - \lambda \to -\infty$$

and as $\lambda \to -\infty$,

$$N(\lambda) - \lambda = \sum_{i=1}^{\ell} p_i \exp(-\lambda r_i) + \sum_{j=1}^{m} q_j \exp(\lambda \tau_j) - \lambda \to +\infty.$$

Then by continuity there exists at least a $\lambda_0 \in \mathbb{R}$ such that $N(\lambda_0) - \lambda_0 = 0$.

The case $p_\ell < 0$ and $q_m > 0$ can be obtained in a similar way. □

In the meantime (1) can be nonoscillatory in others situations, as the stated in the following theorem.

Theorem 3.2. *If p_i, $q_j > 0$, for $i = 1, ..., \ell$ and $j = 1, ...m$, and*

$$\sum_{i=1}^{\ell} p_i + \frac{1}{\tau_m} + \frac{1}{\tau_m} \ln \left(\tau_m \sum_{j=1}^{m} q_j \right) < 0 \tag{10}$$

then the equation (1) is nonoscillatory.

Proof. If $\lambda \leq 0$ then, by (7), $N(\lambda) - \lambda > 0$.

Let $\lambda > 0$. Notice that

$$N(\lambda) - \lambda = \sum_{i=1}^{\ell} p_i \exp(-\lambda r_i) + \sum_{j=1}^{m} q_j \exp(\lambda \tau_j) - \lambda$$
$$< \sum_{i=1}^{\ell} p_i + \exp(\lambda \tau_m) \sum_{j=1}^{m} q_j - \lambda.$$

The function

$$f(\lambda) = \sum_{i=1}^{\ell} p_i + \exp(\lambda \tau_m) \sum_{j=1}^{m} q_j - \lambda$$

has a minimum at

$$\lambda_1 = -\frac{1}{\tau_m} \ln \left(\tau_m \sum_{j=1}^{m} q_j \right)$$

and

$$f(\lambda_1) = \sum_{i=1}^{\ell} p_i + \frac{1}{\tau_m} + \frac{1}{\tau_m} \ln \left(\tau_m \sum_{j=1}^{m} q_j \right).$$

Consequently, by (10), we have $f(\lambda_1) = N(\lambda_1) - \lambda_1 < 0$.

Again by continuity we conclude the existence of a $\lambda_0 \in \mathbb{R}$ such that $N(\lambda_0) - \lambda_0 = 0$. □

Example 3.1. The equation

$$x'(t) = \frac{1}{9} x(t - r_1) + \frac{1}{4} x(t - r_2) + \frac{1}{5} x\left(t + \frac{1}{5}\right) + \frac{1}{10} x\left(t + \frac{1}{4}\right) + \frac{1}{2} x\left(t + \frac{1}{3}\right)$$

with $r_1 < r_2$, is nonoscillatory. In fact as

$$\left(\frac{1}{9} + \frac{1}{4}\right) + 3 + 3 \ln \left[\frac{1}{3}\left(\frac{1}{5} + \frac{1}{10} + \frac{1}{2}\right)\right] \approx -0.60416.$$

(10) is satisfied.

In the Theorem 3.2 one can see that nonoscillations occur independently of the delays, Analogously, one can obtain nonoscillations independently of the advances.

Theorem 3.3. *If $p_i, q_j < 0$, for $i = 1, ..., \ell$ and $j = 1, ...m$, and*

$$\sum_{j=1}^{m} q_j - \frac{1}{r_1} - \frac{1}{r_1} \ln \left(r_1 \sum_{i=1}^{\ell} |p_i| \right) > 0 \tag{11}$$

then the equation (1) is nonoscillatory.

Proof. If $\lambda \geq 0$ then by (9) we have $N(\lambda) - \lambda < 0$.

Letting $\lambda < 0$ we obtain

$$N(\lambda) - \lambda = \sum_{i=1}^{\ell} p_i \exp(-\lambda r_i) + \sum_{j=1}^{m} q_j \exp(\lambda \tau_j) - \lambda$$
$$> \exp(-\lambda r_1) \sum_{i=1}^{\ell} p_i + \sum_{j=1}^{m} q_j - \lambda$$

The function

$$g(\lambda) = \exp(-\lambda r_1) \sum_{i=1}^{\ell} p_i + \sum_{j=1}^{m} q_j - \lambda$$

has a maximum at

$$\lambda_2 = \frac{1}{r_1} \ln \left(-r_1 \sum_{i=1}^{\ell} p_i \right)$$

and

$$g(\lambda_2) = \sum_{j=1}^{m} q_j - \frac{1}{r_1} - \frac{1}{r_1} \ln \left(-r_1 \sum_{i=1}^{\ell} p_i \right).$$

Consequently, by (11), we have $g(\lambda_2) = N(\lambda_2) - \lambda_2 > 0$ and also by continuity one has $N(\lambda_0) - \lambda_0 = 0$, for some $\lambda_0 \in \mathbb{R}$. □

Example 3.2. The equation

$$x'(t) = -\frac{1}{8}x\left(t - \frac{1}{6}\right) - \frac{2}{5}x\left(t - \frac{1}{5}\right) - \frac{1}{6}x\left(t - \frac{1}{4}\right) - \frac{1}{4}x(t + \tau_1) - \frac{1}{10}x(t + \tau_2)$$

with $\tau_1 < \tau_2$, is nonoscillatory. Since

$$\left(-\frac{1}{4} - \frac{1}{10} \right) - 6 - 6 \ln \left[\frac{1}{6} \left(\frac{1}{8} + \frac{2}{5} + \frac{1}{6} \right) \right] \approx 6.6125,$$

condition (11) is verified.

References

1. T. Krisztin, Nonoscillations for functional differential equations of mixed type, Journal of Mathematical Analysis and Applications **254** (2000) 326-345.
2. T. Krisztin, Exponential boundedness and oscillation for solutions of linear autonomous functional differential systems, Dynamic Systems Appl. **4** (1995) 405-420.

3. I. Györi and G. Ladas, Oscillation Theory of Delay Differential Equations, Oxford University Press, 1991.
4. J. M. Ferreira and A. M. Pedro, Oscillations of delay difference systems, *J. Math. Anal. Appl.* **221** (1998) 364-383.

Dynamic Contact Problems in Linear Viscoelasticity

N. Pop*
North University of Baia Mare, Department of Mathematics and Computer Science, Baia Mare, Romania
E-mail: nicpop@ubm.ro, nicpop@gmail.com
www.ubm.ro

The dynamic contact problem in linear viscoelasticity will be discretized with finite element method and finite difference method. The concept of the a priori stability estimation for dynamic frictional contact will be analyzed by using the local split of the Coulomb model. The split will be implemented such that the global balance of energy will be preserved in the case of perfect stick, while in the case of slip an algorithmically consistent approximation will be produced pointwise on the contact interface.

Keywords: Variational inequalities, finite element method; dynamic contact problems with friction, Newmark algorithm, Newton-Raphson method.

1. Introduction

In this paper we will study a mathematical model of the dynamic process of a frictional contact problem between a deformable body and a foundation, under the consideration that the body is assumed to be viscoelastic with a linear elasticity operator, a nonlinear viscoelasticity operator.

This paper makes an approximation of this mathematical model with a sequence of variational inequalities that model the contact condition with the penalty method and the undifferentiable friction functional with a convex function.

Finite element methods, together with numerical schemes of finite differences for solving associated systems of nonlinear ordinary differential equations, are capable of modeling stick-slip motion, dynamic sliding, friction damping and related phenomena in a significant range of practical problems.

*Work partially supported by CEEX Grant No. 2-CEEX-06-11-96/19.09.2006 and partially supported by Grant No. 31/1.10.2007, PN-II-RU-MC-2007-1.

2. Variational formulation of the problem

Let us consider a linear viscoelastic body at a given time $t = 0$ situated in a domain $\Omega \subset \mathbb{R}^d$, where $d = 2$ or $d = 3$, with Lipschitz boundary $\partial\Omega = \bar{\Gamma}_U \cup \bar{\Gamma}_N \cup \bar{\Gamma}_C$ that is divided into three mutually disjoint measurable parts $\Gamma_U, \Gamma_N, \Gamma_C$ with $meas(\Gamma_U) > 0$, see Figure 1. We set $Q = \Omega \times (0, T)$ for $T > 0$. The body is fixed on Γ_U, so the normal displacement field vanishes there, and the friction between Γ_C and the foundation is given by Coulomb law. Volume forces of density f act in Ω and surface tractions of density h are applied on Γ_N.

We denote by $u = (u_1, \ldots, u_d)$ the displacement vector, by $\sigma = \{\sigma_{ij}\}$ the stress tensor, by $\varepsilon(u) = \{\varepsilon_{ij}(u)\}$ the linearized strain tensor, where $i, j = 1, \ldots, d$, by σ_N, σ_T and by u_N, u_T , the normal and tangential components of stress tensor and of the displacement vector, respectively. We assume the Kelvin-Voigt viscoelastic constitutive relation

$$\sigma_{ij} \equiv \sigma_{ij}(u, \dot{u}) = c^{(1)}_{ijkl}\varepsilon_{kl}(\dot{u}) + c^{(2)}_{ijkl}\varepsilon_{kl}(u)$$

where $c^{(1)}_{ijkl}$ and $c^{(2)}_{ijkl}$ with $i, j, k, l = 1, \ldots, d$ are the viscoelastic and elastic components tensors, respectively. Denoting by u_0 and u_1 the initial displacement and the initial velocity, respectively, we suppose that the mass density is constant, conveniently set equal to one. Then the classical formulation of the dynamic contact problem with friction in linear viscoelasticity, see [1] and [2], is: find a displacement field $u : Q \to \mathbb{R}^d$, such that:

$$\begin{array}{lll}
\ddot{u} - div\sigma = f & \text{in } Q & (1a) \\
\sigma_{ij}(u, \dot{u}) = c^{(1)}_{ijkl}\varepsilon_{kl}(\dot{u}) + c^{(2)}_{ijkl}\varepsilon_{kl}(u) & \text{in } Q & (1b) \\
u = 0 & \text{on } \Gamma_U \times (0, T) & (1c) \\
\sigma \cdot n = h & \text{on } \Gamma_N \times (0, T) & (1d) \\
\dot{u}_N \leq 0, \sigma_N \leq 0, \dot{u}_N \cdot \sigma_N = 0 & \text{on } \Gamma_C \times (0, T) & (1e) \\
\dot{u}_T = 0, \Rightarrow \ |\sigma_T| \leq F(0) \cdot |\sigma_N| & \text{on } \Gamma_C \times (0, T) & (1f) \\
\dot{u}_T \neq 0, \Rightarrow \ \sigma_T = -F(\dot{u}_T) \cdot |\sigma_N| \cdot \dfrac{\dot{u}_T}{|\dot{u}_T|} & \text{on } \Gamma_C \times (0, T) & (1g) \\
u(x, 0) = u_0, \quad \dot{u}(x, 0) = u_1 & \text{in } \Omega & (1h)
\end{array}$$

where n is the outward normal unit vector on $\partial\Omega$ and F is the coefficient of friction, f is body force, h is surface traction. The relation $(1a)$ is the equililibrium equation, $(1c)$ and $(1d)$ are the boundary conditions, $(1e)$ is the contact condition, $(1f)$ and $(1g)$ represent the Coulomb friction law and $(1h)$ are the initial conditions.

In order to obtain the variational form, (see [2]), of the problem (1), we need the following notations: $I \subset \mathbb{R}$ an interval, W a Banach space and

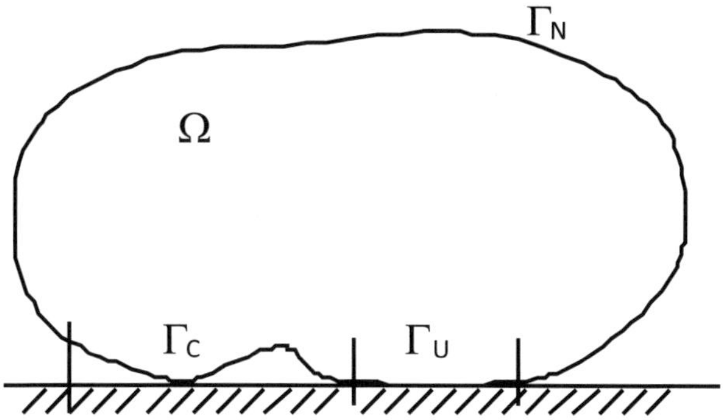

Fig. 1. The contact of a body with a foundation.

$B_0(\overline{I}, W)$ the set of the bounded functions, with the norm:

$$\|u\|^2_{L^2(I,W)} := \int_I \|u(t)\|^2_W dt$$

and the set of admissible functions:

$$K = \{v \in L^2(I, H^1(\Omega,^d))/v = 0 \; on \; \Gamma_U \times (0,T), \; v_N \leq 0 \; on \; \Gamma_C \times (0,T)\}$$

The variational form is the following variational inequality: find $\dot{u} \in K \cap B_0(\overline{I}, L^2(\Omega,^d))$ with $u(x,0) = u_0, \dot{u}(x,0) = u_1$ such that for every $v \in K$

$$\int_Q \{\ddot{u}(v - \dot{u}) + a(u, v - \dot{u})\}dxdt + \int_{S_C} F(\dot{u}_T)\,|\sigma_N(\dot{u}, u)|\,(|v_T| - |\dot{u}_T|)\,dsdt$$
$$\geq \int_Q f(v - \dot{u})dxdt \qquad (2)$$

where $S_C = \Gamma_C \times (0,T)$ and a bilinear form

$$a(u,v) = c^{(1)}_{ijkl}\varepsilon_{ij}(\dot{u})\varepsilon_{kl}(v) + c^{(2)}_{ijkl}\varepsilon_{ij}(u)\varepsilon_{kl}(v).$$

The first step in the sequence of the approximations is the penalty method for replacing the unilateral contact conditions by a nonlinear boundary condition dependent on the small parameter $\delta > 0$, $\sigma_N(\dot{u}, u) = 1/\delta\,[\dot{u}_N]_+$. The second step is the regularization method for the approximation of the module function $|\cdot|$, with a convex function $\Psi_\varepsilon(\cdot)$, that fulfils the following conditions: $|\Psi_\varepsilon(v) - |v|| \leq \varepsilon$ and $|grad\Psi_\varepsilon(v)| \leq 1$. Finally, we have to transform the variational inequality into a variational equality that will approximate the problem (1), using a test function $v = \dot{u} + \lambda w$ and after dividing by λ, when λ tends to zero, we obtain: find u with

$\ddot{u} \in (\overline{u}+V) \cap B_0(I, L^2(\Omega,^d))$ and $u(x,0) = u_0, \quad \dot{u}(x,0) = u_1$ s.t. for any $v \in V$

$$\int_{S_C} \{F(\dot{u}_T)1/\delta\,[\dot{u}_N]_+\, grad\Psi_\varepsilon(\dot{u}_T)v_T + 1/\delta\,[\dot{u}_N]_+\, v_N\}dsdt$$

$$+ \int_Q \{\ddot{u}v + a(u,v) - fv\} = 0 \qquad (3)$$

where the space is

$$V = \{v \in L^2(I, H^1(\Omega,^d))/v = 0 \ on\ \Gamma_U \times (0,T)\}$$

2.1. *Finite element approximations of the dynamic contact problems with friction*

Using standard finite element procedures, the approximation of the variational equation (3) can be constructed in finite-dimensional subspaces $V_h(\subset V \subset V')$. For certain (h) the approximations of displacements, velocities and accelerations at each time $t \in (0,T)$ are elements of V_h, $d^h\,(t)$, $v^h\,(t)$, $a^h\,(t) \in V_h$, where $d = u$, $v = \dot{u}$ and $a = \ddot{u}$. Within each element $\Omega^e_h(e = 1,\ldots,N_h)$ the components of the displacements, velocities and accelerations are expressed in the form:

$$d^h_k(t,x) = \sum_I^{N_e} d^I_k(t)N_I(x),$$

$$v^h_k(t,x) = \sum_I^{N_e} v^I_k(t)N_I(x)$$

$$a^h_k(t,x) = \sum_I^{N_e} a^I_k(t)N_I(x) \qquad (4)$$

where $k = 2$ or 3, N_e =the element's number of node, $d^I_k(t)$, $v^I_k(t)$, $a^I_k(t)$ are the nodal values of the displacements, velocities and accelerations, at the time t and N_I is the element shape function associated with the nodal point I. If is the number of the nodes of the finite element mesh from Ω, then this problem is equivalent to the following matrix problem:

Problem $P\varepsilon^h$. Find the function $d : [0,T] \to \mathbb{R}^{d\times N^\Omega_k}$, s.t.

$$Ma(t) + Kd(t) - P(d(t)) + J(d(t), v(t)) = F(t) \qquad (5)$$

with the initial conditions

$$d(0) = u_0, v(0) = u_1 \qquad (6)$$

Here we have introduced the following matrix notations: $d(t)$, $v(t)$, $a(t)$: the column vectors of nodal displacements, velocities and accelerations, respectively; M: standard mass matrix; K: standard stiffness matrix; $F(t)$: consistent nodal exterior forces vector; $P(d(t))$: vector of consistent nodal forces on Γ_C; $J\left(d(t), v(t)\right)$: vector of consistent nodal friction forces on Γ_C; u_0, u_1: initial nodal displacement, velocity.

The components of the element vector ${}^{(e)}P$ have the form: ${}^{(e)}P = -\int_{{}^{(e)}\Gamma_C} \sigma_N \cdot n \cdot N_I ds$ and the components of the element vector ${}^{(e)}J$ have the form: ${}^{(e)}J = -\int_{{}^{(e)}\Gamma_C} \sigma_T \cdot n \cdot N_I ds$. In order to obtain the components of the element vector P and J it is used a contact finite element, see [7], [8].

2.2. *Solution strategies for spatially discrete system and time steppining procedures*

The algorithms that we shall use for solving the discrete dynamical system involve variants of standard schemes used in nonlinear structural dynamics calculations: the Newmark-type algorithm or the central-difference scheme. Let us consider a partition of the time interval $I = \bigcup_{k=1}^{N} [t_{k-1}, t_k]$ with $0 = t_0 < t_1 < \ldots < t_N = T$, we denote $\Delta t = t_k - t_{k-1}$ for the length of the sub-interval $[t_{k-1}, t_k]$. In the dynamic case, the inertial terms are restored and issues associated with temporal accuracy and stability come to the fore (must analysis). If we denote in an ordinary differential equation (5) $K_N \equiv K - P + J$ (where the matrices P and J are nonlinear), and (5) becomes:

$$Ma(t) + K_N\left(d(t)\right) = F(t) \tag{7}$$

In dynamic case inertial terms cannot be neglected and the variable t in this case does have the interpretation of real time. We have to find approximations d_{k+1}, v_{k+1}, a_{k+1} at time t_{k+1} from equation (7), with given displacement vector d_k, velocity v_k and acceleration a_k at time t_k

$$Ma_{k+1} + K_N\left(d_{k+\alpha}\right) = F\left(t_{k+\alpha}\right) \tag{8}$$

$$d_{k+\alpha} = \alpha d_{k+1} + (1-\alpha) d_k$$

$$d_{k+1} = d_k + \Delta t v_k + \frac{\Delta t^2}{2}\left[(1-2\beta)\, a_k + 2\beta a_{k+1}\right]$$

$$v_{k+1} = v_k + \Delta t \left[(1-\gamma)\, a_k + \gamma a_{k+1}\right]$$

where α, β and γ are algorithmic parameters that define the stability and accuracy characteristics of the method. In particular, when $\alpha = 1$, the algorithm reduces to the classical Newmark algorithm. A wide range of algorithms exists corresponding to the different available choices of these parameters, we illustrate the implicit methods.

2.3. *Implicit methods*

To introduce the concept of an implicit method we examine the trapezoidal rule, which is simply a member of the Newmark family obtained by setting $\alpha = 1$, $\beta = 1/4$ and $\gamma = 1/2$. The substitution of these values into (8) yields:

$$\begin{cases} Ma_{k+1} + K_N(d_{k+1}) = F(t_{k+1}), \\ d_{k+1} = d_k + \Delta t v_k + \frac{\Delta t^2}{4}[a_k + a_{k+1}], \\ v_{k+1} = v_k + \frac{\Delta t}{2}[a_k + a_{k+1}]. \end{cases} \tag{9}$$

Combining the first two equations in (9) and solving it for d_{k+1}, gives us

$$\frac{4}{\Delta t^2}Md_{k+1} + K_N(d_{k+1}) = F(t_{k+1}) + M\left(a_k + \Delta t v_k + \frac{4}{\Delta t^2}d_k\right), \tag{10}$$

$$a_{k+1} = \frac{4}{\Delta t^2}(d_{k+1} - d_k) - \frac{4}{\Delta t}v_k - a_k$$

$$v_{k+1} = v_k + \frac{\Delta t}{2}[a_k + a_{k+1}]$$

This method is the most expensive procedure involved in updating the solution from t_k to t_{k+1}. First equation is not only fully coupled, but is also highly nonlinear, in general, due to interval force vector. We could write the first equation of (10) in terms of a dynamic incremental residual $R(d_{k+1})$ via

$$\begin{aligned} R(d_{k+1}) := F(t_{k+1}) - K_N(d_{k+1}) - \frac{4}{\Delta t^2}Md_{k+1} \\ +M\left(a_k + \Delta t v_k + \frac{4}{\Delta t^2}d_k\right) = 0 \end{aligned} \tag{11}$$

This system suggests that the same sorts of nonlinear solution are needed for implicit dynamic calculations as are needed in quasistatic.

2.4. *Stability solution for the implicit methods*

In this case, the integrator is second order accurate and unconditionally stable for linear problems, meaning that the spectral radii of the integrator remains less than 1 in modulus, for any time step Δt.

A criterion for stability (computation of critical points), in solving the contact problems, is that the second derivative of potential energy becomes zero (see [1]). In the iterative and incremental solving of the contact problems with penalty method the stability, consists in the kind of changes in the active set of contact nodes. For the gap or penetration $g \leq 0$, the contact nod becomes active, otherwise is inactive. The strategy for the choice of the active set may be a constant verification and reorganization of the active set after each iterative step, or a change of the active set only after convergence has been achieved. The iteration process might become more stabile by the second approach as changes in active set occur less frequently. The critical situations arise in transitions from sliding to adhesion of the contact nodes.

2.5. *Nonlinear equation solving with Newton-Raphson iterative method*

The implicit and explicit methods are valid only for linear or linearized problems. In this section we give a general framework for solving the nonlinear discrete equations associated with computation of an unknown state at step t_{k+1}, in either context of a dynamic contact problem formulation as in (11). In either case, the equation to be solved takes the form

$$R\left(d_{k+1}\right) = 0 \tag{12}$$

with R, a nonlinear function of the solution vector d_{k_1}, is considered.

The general concept of a Newton-Raphson iterative solution technique for (12) (identical with (10) and with (11)) is defined in iteration j by

$$R\left(d_{k+1}^{j}\right) + \left[\frac{\partial R}{\partial d}\right]_{d_{k+1}^{j}} \Delta d_k^j = 0 \tag{13}$$

following by the update

$$d_{k+1}^{j+1} = d_{k+1}^{j} + \Delta d_k^j \tag{14}$$

Iteration on j typically continue until the Euclidian norm $\left\|\Delta d_k^j\right\|$ is smaller than some tolerance.

The residual at iteration j , from (11) is of the form

$$R\left(d_{k+1}^{j}\right) \equiv F\left(t_{k+1}\right) - K_N\left(d_{k+1}^{j}\right) - \frac{4}{\Delta t^2} M d_{k+1}^{j} + M\left(a_k + \Delta t v_k + \frac{4}{\Delta t^2} d_k^{j}\right) = 0 \tag{15}$$

with (13), equation (15) to take the form

$$\left[\frac{4}{\Delta t} M + K_L\left(d_{k+1}^{j}\right)\right] \Delta d_j^k = R\left(d_{k+1}^{j}\right) \tag{16}$$

where the stiffness matrix $K_L\left(d_{k+1}^{j}\right)$ is given as

$$K_L\left(d_{k+1}^{j}\right) = \left(\frac{\partial K_N}{\partial d}\right)_{d_{k+!}^{j}} \tag{17}$$

We note that a variety of iterative procedures exist as alternatives to the Newton-Raphson nonlinear solution procedure (quasi-Newton, secant methods etc.). The scheme of solving the linearized dynamic contact problems is the following:

(1) initialization the set of the iterative count $t_k = 0$ $\Delta t = 0$, $k = 0$, $j = 0$, $d_j^k = 0$;
(2) compute the mass matrix M, the standard stiffness matrix K and a dynamic residual R;
(3) compute the contact nodal forces P and the contact friction forces J;
  (a) compute the normal gap g_N^j;
  (b) check for contact finite element status:
    IF $g_N^j > TOL$ then out of contact
      ELSE in contact. Check for frictional stick or slip contact status
    ENDIF.
  (c) compute total matrix K_L and residual R, this involves to compute the vectors $\left\{a_{k+1}^j, v_{k+1}^j, d_{k+1}^j\right\}$ from $k = 0$ to $N(k = N => t_{k+1} = T)$;
(4) check for convergence
    IF $\left|\Delta d_{k+1}^{j} - \Delta d_{k+1}^{j+1}\right| < TOL1$ then converge and exit
      ELSE go to step (6)
(5) update the displacement field $d_{k+1}^{j+1} = d_{k+1}^{j} + \Delta d_k^j$;
(6) set $j = j + 1$ and go back to (2).

3. Numerical example and concluding remarks

This example [11] is often studied for validation of computer codes, and has the advantage of being elementary and that of giving different contact areas and coefficient of friction, see Figure 2. Thus, we have an open contact area AB, a sliding area BC and a sticking area CD. The contact area was modeled with the contact finite element with three nodes, see [1].

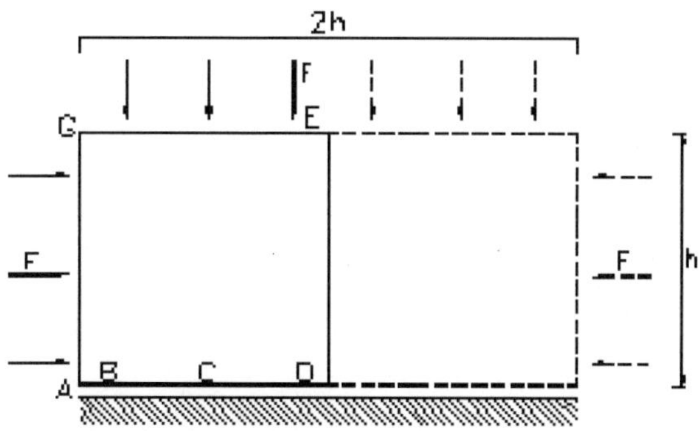

Fig. 2. Contact between a 2D elastic slab and a rigid plate.

(1) It is known that the matrix KN is wrong conditioned, if we split the normal and tangential stress from the contact boundary, in diagonal blocks matrices, these matrices blocks contain coefficients of the same order size, and with this procedure we will obtain a better conditioned matrix.
(2) The discontinuity of the Coulomb's friction law at zero sliding velocity is a major source of computational difficulties in friction problems. Even though, in the algorithms described in this section and the previous one, a regularized form of that law is used; those difficulties cannot be completely avoided. The situation which may arise when using the methods described here with a constant step is the following one: in unloading situations (passage from sliding to adhesion) the Newton-Raphson iterative techniques may fail to converge if ? is very small and the step too large. For small values of ? the radius of converge of the iterative scheme used is very small due to the step change in Ψ_ε on the interval $[-\varepsilon, \varepsilon]$.
(3) The critical situations arise in transitions from sliding to adhesion because it is then that the most important change in the solution occurs.

One simple remedy for these difficulties is to decrease the time step until two successive solutions are not too far apart.

References

1. P. Wriggers, *Computational Contact Mechanics*, (Springer-Verlag Berlin Heidelberg, 2006).
2. J. Jarusek and C. Eck, C. R. Acad. Sci. Paris t. 322, 497 (1996).
3. K. Bartosz, Nonlinear Analysis 65, 546 (2006).
4. W. Han and M. Sofonea, Applied Numerical Mathematics 57, 498 (2007).
5. T. A. Laursen and V. Chawla, Int. J. Num. for Meth. in Engng. 40, 863 (1997).
6. J. Jarusek, Czechoslovak Math. J. 46, 475 (1996).
7. N. Pop, American Journal of Applied Sciences 5(2), 152 (2008).
8. N. Pop, Carpathian Journal of Mathematics, 23, 177 (2007).
9. A. Amassad and C. Fabre, Int. J. Math. Sci. 32, 411 (2002).
10. A. Amasad, C. Fabre and M. Sofonea, IMA J. Numer. Anal. 38, 463 (2004).
11. M. Raous, P. Chabrand, F. Lebon, J. Mec. Theorique et App., Spec. issue 7,111 (1998).

Regularly Varying Decreasing Solutions of Half-Linear Dynamic Equations

Pavel Řehák*

Institute of Mathematics, Academy of Sciences of the Czech Republic, Žižkova 22, CZ-61662 Brno, Czech Republic
E-mail: rehak@math.muni.cz

Jiří Vítovec

Department of Mathematical Analysis, Faculty of Science, Masaryk University Brno, Janáčkovo náměstí 2a, CZ-60200 Brno, Czech Republic
E-mail: vitovec@math.muni.cz

Necessary and sufficient conditions in terms of the integral of the coefficient are derived under which positive decreasing solutions of a half-linear dynamic equation are (normalized) regularly varying of a known index.

Keywords: Regularly varying function, regularly varying sequence, time scale, half-linear dynamic equation.

1. Introduction

We consider the half-linear dynamic equation

$$(\Phi(y^{\Delta}))^{\Delta} - p(t)\Phi(y^{\sigma}) = 0 \tag{1}$$

on an unbounded time scale interval $\mathcal{I}_a = [a, \infty)$, $a > 0$, where p is a positive rd-continuous function, and $\Phi(u) = |u|^{\alpha-1} \operatorname{sgn} u$ with $\alpha > 1$. The aim of this contribution is to provide precise information on asymptotic behavior of positive decreasing solutions of (1), which always exist. We will show that all these solutions are normalized regularly varying if and only if the coefficient p satisfies certain integral condition. Moreover, the index of regular variation will be shown to be related to the limit behavior of the coefficient p. Our results unify and extend those from [7–9,11]. In addition

*Supported by the Grants KJB100190701 of the Grant Agency of ASCR and 201/07/0145 of the Czech Grant Agency, and by the Institutional Research Plan AV0Z010190503.

to the extension of linear results from [11], we also satisfactorily solve the problem of "normalized behavior" of solutions of (1).

In the next section we recall basic concepts which are useful for this work. The main results are contained in the third section.

2. Preliminaries

We assume that the reader is familiar with the notion of time scales. Thus note just that $\mathbb{T}$, σ, f^σ, μ, f^Δ, $\int_a^b f^\Delta(s)\,\Delta s$, and $e_f(t,a)$ stand for time scale, forward jump operator, $f\circ\sigma$, graininess, delta derivative of f, delta integral of f from a to b, and generalized exponential function, respectively. See [6], which is the initiating paper of the time scale theory, and the monograph [3] containing a lot of information on time scale calculus.

Basic properties of equation (1) can be found in [10. In particular, the solution space of (1) is homogeneous, the initial value problem involving (1) is uniquely solvable on the entire interval $\mathcal{I}_a$, and under the condition $p(t) > 0$, any (nontrivial) solution of (1) is eventually of one sign. Moreover, see [1], the class of positive decreasing solutions of (1) is always nonempty. Note that equation (1) covers half-linear differential equations, see e.g. [5], half-linear difference equations, see e.g. [?Chapter 8]book], and linear dynamic equations, see e.g. [3].

In [11], a *regularly varying function* $f : \mathbb{T} \to \mathbb{R}$ *of index* ϑ, $\vartheta \in \mathbb{R}$, is defined as a positive function on $\mathcal{I}_a$ satisfying $f(t) \sim Cg(t)$ where C is a positive constant and g is such that $\lim_{t\to\infty} tg^\Delta(t)/g(t) = \vartheta$. If $\vartheta = 0$, then f is said to be *slowly varying*. We always assume that the functions discussed here are of a sufficient smoothness. The totality of regularly varying functions of index ϑ is denoted by $\mathcal{RV}(\vartheta)$. The totality of slowly varying functions is denoted by $\mathcal{SV}$. Further, it is shown in [11] that $f \in \mathcal{RV}(\vartheta)$ iff it has the representation $f(t) = \varphi(t)e_\delta(t,a)$, where $\lim_{t\to\infty}\varphi(t) = \text{const} > 0$ and δ is a positively regressive function satisfying $\lim_{t\to\infty} t\delta(t) = \vartheta$. In [12], another representation is derived under the condition $\mu(t) = o(t)$ when $\vartheta \neq 0$ or $\mu(t) = O(t)$ when $\vartheta = 0$: $f \in \mathcal{RV}(\vartheta)$ iff $f(t) = t^\vartheta\varphi(t)e_\psi(t,a)$, where $\lim_{t\to\infty}\varphi(t) = \text{const} > 0$ and ψ satisfies $\lim_{t\to\infty} t\psi(t) = 0$. A positive function f is said to be *normalized regularly varying of index* ϑ, $\vartheta \in \mathbb{R}$, if it satisfies $\lim_{t\to\infty} tf^\Delta(t)/f(t) = \vartheta$. If $\vartheta = 0$, then f is said to be *normalized slowly varying*. The totality of normalized regularly varying functions of index ϑ is denoted by $\mathcal{NRV}(\vartheta)$. The totality of normalized slowly varying functions is denoted by $\mathcal{NSV}$. Normalized functions have the representations as above, where the function $\varphi(t)$ is replaced by a positive constant. It holds that $f \in (\mathcal{N})\mathcal{RV}(\vartheta)$ iff $f(t) = t^\vartheta L(t)$, where $L \in (\mathcal{N})\mathcal{SV}$. In [12], many other properties of regularly varying functions on time scales are derived. It is also shown there that $\mu(t) = O(t)$ when $\vartheta = 0$ and $\mu(t) = o(t)$ when $\vartheta \neq 0$ are somehow natural conditions in the theory and cannot be omitted. These conditions also occur in the main result of this paper which concerns regularly varying behavior of solutions of (1). For the theory of regular variation on a time scale with a "larger graininess", in particular on $\mathbb{T} = q^{\mathbb{N}_0}$, see [13]. Note that that theory is different in some important aspects.

Some basic references on the theory of regularly varying functions of real variable are [2,14], a reference on their relation to differential equations is [8], and a basic reference on the theory of regularly varying sequences is [4].

The following lemma is new and gives simple sufficient conditions for a regularly varying function to be normalized.

Lemma 2.1. *Let $f \in \mathcal{RV}(\vartheta)$, $\vartheta \in \mathbb{R}$. Assume that $\mu(t) = o(t)$ provided $\vartheta \neq 0$ and $\mu(t) = O(t)$ provided $\vartheta = 0$. If $f^\Delta(t) \leq 0$ and $f^\Delta(t)$ is nondecreasing for large t, then $f \in \mathcal{NRV}(\vartheta)$.*

Proof. Let $f \in \mathcal{RV}(\vartheta)$. Then

$$\lim_{t\to\infty} \frac{f(\tau(\lambda t))}{f(t)} = \lambda^\vartheta \quad \text{for all } \lambda > 0 \tag{2}$$

by [12], where $\tau : \mathbb{R} \to \mathbb{T}$ is defined as $\tau(t) = \max\{s \in \mathbb{T} : s \leq t\}$. Let $\lambda \in (0,1)$. Since $-f^\Delta$ is nonnegative and nonincreasing, we have $-f(t)+f(\tau(\lambda t)) = -\int_{\tau(\lambda t)}^t f^\Delta(s)\,\Delta s \geq -f^\Delta(t)[t-\tau(\lambda t)] \geq -f^\Delta(t)(1-\lambda)t$ for large t. This estimation and (2) imply $\limsup_{t\to\infty} -tf^\Delta(t)/f(t) \leq \limsup_{t\to\infty} (f(\tau(\lambda t))/f(t) - 1)\,(1-\lambda) = (\lambda^\vartheta - 1)/(1-\lambda)$. This estimation holds for every $\lambda \in (0,1)$. Taking now the limit as $\lambda \to 1-$, we obtain

$$\limsup_{t\to\infty} \frac{-tf^\Delta(t)}{f(t)} \leq \lim_{\lambda\to 1-} \frac{\lambda^\vartheta - 1}{1-\lambda} = -\vartheta. \tag{3}$$

In view of (3) for $f \in \mathcal{SV}$, we may now restrict ourselves to $\vartheta \neq 0$. We have $-f(t) + f(\tau(\lambda t)) = -\int_{\tau(\lambda t)}^t f^\Delta(s)\,\Delta s \leq -f^\Delta(\tau(\lambda t))(t - \tau(\lambda t))$ This estimation, (2), and $\lambda \in (0,1)$ imply

$$\begin{aligned}\lambda^\vartheta \liminf_{t\to\infty} \frac{-tf^\Delta(t)}{f(t)} &= \liminf_{t\to\infty} \frac{f(\tau(\lambda t))}{f(t)} \cdot \frac{-\tau(\lambda t) f^\Delta(\tau(\lambda t))}{f(\tau(\lambda t))} \\ &\leq \liminf_{t\to\infty} \frac{\tau(\lambda t)}{t-\tau(\lambda t)} \left(\frac{f(\tau(\lambda t))}{f(t)} - 1\right).\end{aligned} \tag{4}$$

Since for $x \in \mathbb{R}$, $\tau(x) \geq a$, $\tau(x) \leq x \leq \sigma(\tau(x)) = \tau(x) + \mu(\tau(x))$, we have $1 \leq x/\tau(x) \leq 1 + \mu(\tau(x))/\tau(x)$, and so $\lim_{x\to\infty} x/\tau(x) = 1$. Consequently, in view of (2) and (4),

$$\lambda^{\vartheta} \liminf_{t\to\infty} \frac{-tf^{\Delta}(t)}{f(t)} \geq \liminf_{t\to\infty} \frac{\lambda}{\lambda t/\tau(\lambda t) - \lambda} \left(\frac{f(\tau(\lambda t))}{f(t)} - 1 \right) = \frac{\lambda}{1-\lambda}(\lambda^{\vartheta} - 1)$$

for every $\lambda \in (0,1)$. Hence, $\liminf_{t\to\infty} -tf^{\Delta}(t)/f(t) \geq \lim_{\lambda\to 1-}(\lambda^{\vartheta} - 1)/(\lambda^{\vartheta-1} - \lambda^{\vartheta}) = -\vartheta$. From (3) and the latter estimate, we obtain $\lim_{t\to\infty} tf^{\Delta}(t)/f(t) = \vartheta$, $\vartheta \in \mathbb{R}$, which implies $f \in \mathcal{NRV}(\vartheta)$. □

3. Main Results

Along with (1), consider the generalized Riccati dynamic equation

$$w^{\Delta}(t) - p(t) + S(t) = 0, \tag{5}$$

where

$$S(t) = \lim_{\gamma(t)\to\mu(t)} \frac{w(t)}{\gamma(t)} \left(1 - \frac{1}{\Phi[1 + \gamma(t)\,\Phi^{-1}(w(t))]} \right), \tag{6}$$

Φ^{-1} being the inverse of Φ. The relation between (1) and (5), which will be utilized later, is following: $y(t)$ is a nonoscillatory solution of (1) having no generalized zero on $\mathcal{I}_a$, i.e., $y(t)y^{\sigma}(t) > 0$ for $t \in \mathcal{I}_a$ if and only if $w(t) = \Phi(y^{\Delta}(t)/y(t))$ satisfies (5) on $\mathcal{I}_a$ with $1 + \mu(t)\,\Phi^{-1}(w(t)) > 0$ on $\mathcal{I}_a$, see, for example, [10].

Theorem 3.1. *Let y be any positive decreasing solution of (1) on $\mathcal{I}_a$.*

(i) Assume $\mu(t) = O(t)$. Then $y \in \mathcal{SV}$ if and only if

$$\lim_{t\to\infty} t^{\alpha-1} \int_t^{\infty} p(s)\Delta s = 0. \tag{7}$$

Moreover, $y \in \mathcal{NSV}$.

(ii) Assume $\mu(t) = o(t)$. Then $y \in \mathcal{RV}(\Phi^{-1}(\lambda_0))$ if and only if

$$\lim_{t\to\infty} t^{\alpha-1} \int_t^{\infty} p(s)\Delta s = A > 0, \tag{8}$$

where λ_0 is the negative root of the algebraic equation

$$|\lambda|^{\beta} - \lambda - A = 0, \tag{9}$$

β is the conjugate number to α, i.e., $1/\alpha + 1/\beta = 1$. Moreover, $y \in \mathcal{NRV}(\Phi^{-1}(\lambda_0))$.

Proof. (i) "Only if": Let $y(t)$ be a slowly varying positive decreasing solution of (1) on $\mathcal{I}_a$. Then $y^\Delta(t)$ is negative and nondecreasing for large t. Hence, $y \in \mathcal{NSV}$ by Lemma 2.1. Let $w(t) = \Phi(y^\Delta(t)/y(t))$. Then $w(t) < 0$ and satisfies (5) with $1 + \mu(t)\,\Phi^{-1}(w(t)) > 0$ for $t \in \mathcal{I}_a$. Since $y \in \mathcal{NSV}$, we have $\lim_{t\to\infty} ty^\Delta(t)/y(t) = 0$, hence $\lim_{t\to\infty} t\,\Phi^{-1}(w(t)) = 0$, or $\lim_{t\to\infty} t^{\alpha-1}w(t) = 0$ (and also $\lim_{t\to\infty} w(t) = 0$). Thus, $\lim_{t\to\infty} Nt\,\Phi^{-1}(w(t)) = 0$ for any $N > 0$. Let N be so large that $\mu(t)/t \le N$ for $t \in \mathcal{I}_a$, thus $\mu(t) \le Nt$ for $t \in \mathcal{I}_a$, which is possible in view of $\mu(t) = O(t)$. Therefore we obtain $\lim_{t\to\infty} \mu(t)\,\Phi^{-1}(w(t)) = 0$. Note that $S(t)$ defined by (6) is positive for $t \in \mathcal{I}_a$, see [10]. Applying the Lagrange mean value theorem, $S(t)$ can be alternatively written as

$$S(t) = \frac{(\alpha-1)|w(t)|^\beta \xi^{\alpha-2}(t)}{[1+\mu(t)\,\Phi^{-1}(w(t))]^{\alpha-1}}, \tag{10}$$

where $0 < 1 + \mu(t)\,\Phi^{-1}(w(t)) \le \xi(t) \le 1$. We show that $\int_t^\infty S(s)\Delta s < \infty$. Since $\lim_{t\to\infty} \mu(t)\,\Phi^{-1}(w(t)) = 0$, $\xi(t) \to 1$ as $t \to \infty$, and we have $S(t) \le 2(\alpha-1)|w(t)|^\beta$ for large t. Further, since $\lim_{t\to\infty} t^{\alpha-1}w(t) = 0$, there exists $M > 0$ such that $|w(t)| \le Mt^{1-\alpha}$ for large t. Hence, for large t, with the use of the relation $\left(Mt^{1-\alpha}\right)^\beta = M^\beta t^{-\alpha}$,

$$\int_t^\infty S(s)\Delta s \le 2(\alpha-1)\int_t^\infty |w(s)|^\beta \Delta s \le 2(\alpha-1)M^\beta \int_t^\infty \frac{1}{s^\alpha}\Delta s < \infty. \tag{11}$$

Note that the integral $\int_t^\infty (1/s^\alpha)\Delta s$ is indeed convergent since $\mu(t) = O(t)$. Integration of (5) from t to ∞ and multiplication by $t^{\alpha-1}$ yield

$$-t^{\alpha-1}w(t) + t^{\alpha-1}\int_t^\infty S(s)\Delta s = t^{\alpha-1}\int_t^\infty p(s)\Delta s. \tag{12}$$

Equality (10) and the time scale L'Hospital rule give

$$\lim_{t\to\infty} t^{\alpha-1}\int_t^\infty S(s)\Delta s = \lim_{t\to\infty} \frac{-(\alpha-1)|w(t)|^\beta \xi^{\alpha-2}(t)}{[1+\mu(t)\,\Phi^{-1}(w(t))]^{\alpha-1}\left(t^{1-\alpha}\right)^\Delta}.$$

Now differentiating $t^{1-\alpha}$ and applying the Lagrange mean value theorem on this term with $t \le \eta(t) \le \sigma(t)$, we have, with the use of the relations $\lim_{t\to\infty} \xi(t) = 1$ and $\eta(t)/t \le 1 + \mu(t)/t \le 1 + N$,

$$\begin{aligned}\lim_{t\to\infty} t^{\alpha-1}\int_t^\infty S(s)\Delta s &= \lim_{t\to\infty} \frac{-(\alpha-1)|w(t)|^\beta \xi^{\alpha-2}(t)}{(1-\alpha)\eta^{-\alpha}(t)[1+\mu(t)\,\Phi^{-1}(w(t))]^{\alpha-1}}\\ &\le (1+N)^\alpha \lim_{t\to\infty} \left|t^{\alpha-1}w(t)\right|^\beta = 0.\end{aligned}$$

Hence, from (12), we get $\lim_{t\to\infty} t^{\alpha-1}\int_t^\infty p(s)\Delta s = 0$.

"If": Let $y > 0$ be a decreasing solution of (1) on $\mathcal{I}_a$, then $\lim_{t\to\infty} y^\Delta(t) = 0$. Indeed, if not, then there is $K > 0$ such that $y^\Delta(t) \le -K$ for $t \in \mathcal{I}_a$, and so $y(t) \le y(a) - (t-a)K$. Letting $t \to \infty$ we have $\lim_{t\to\infty} y(t) = -\infty$, a contradiction with $y > 0$. Therefore, integration of (1) from t to ∞ yields $\Phi(y^\Delta(t)) = -\int_t^\infty p(s)\Phi(y^\sigma(s))\Delta s$. Multiplying this equality by $-t^{\alpha-1}/\Phi(y(t))$ we obtain $-t^{\alpha-1}\Phi(y^\Delta(t))/\Phi(y(t)) = t^{\alpha-1}\int_t^\infty p(s)\Phi(y^\sigma(s))\Delta s/\Phi(y(t)) \le t^{\alpha-1}\int_t^\infty p(s)\Delta s$. Hence $0 < -t^{\alpha-1}\Phi(y^\Delta(t)/y(t)) \to 0$, or $0 < -ty^\Delta(t)/y(t) \to 0$ as $t\to\infty$, in view of (7). Thus $y \in \mathcal{NSV}$.

(ii) "Only if": Here we use the arguments similar to those in the proof of the part (i) "only if". Note just that, with $w = \Phi(y^\Delta/y)$, $\lim_{t\to\infty} t^{\alpha-1}w(t) = \lambda_0$ and $\lim_{t\to\infty}\mu(t)\Phi^{-1}(w(t)) = 0$ by $\mu(t) = o(t)$.

"If": Assume that (8) holds. Let y be a positive decreasing solution of (1) on $\mathcal{I}_a$. Let $v(t) = t^{\alpha-1}\Phi(y^\Delta(t)/y(t))$. Similarly as in the case "if" of part (i), we have $\lim_{t\to\infty} y^\Delta(t) = 0$ and $0 < -v(t) \le t^{\alpha-1}\int_t^\infty p(s)\Delta s$. Hence and due to (8), $v(t)$ is bounded. We will show that $\lim_{t\to\infty} v(t) = \lambda_0$, which implies $y \in \mathcal{NRV}(\Phi^{-1}(\lambda_0))$. First observe that $v(t)$ satisfies the modified Riccati equation

$$\left(\frac{v(t)}{t^{\alpha-1}}\right)^\Delta - p(t) + F(t) = 0, \tag{13}$$

where

$$F(t) = \lim_{\gamma(t)\to\mu(t)} \frac{v(t)}{t^{\alpha-1}\gamma(t)}\left(1 - \frac{1}{\Phi[1+\gamma(t)\,\Phi^{-1}(v(t)/t^{\alpha-1})]}\right),$$

with $1 + \mu(t)\,\Phi^{-1}(v(t)/t^{\alpha-1}) > 0$ for $t \in \mathcal{I}_a$. Since $\lim_{t\to\infty}(v(t)/t^{\alpha-1}) = 0$, integration of (13) from t to ∞ yields

$$-\frac{v(t)}{t^{\alpha-1}} = \int_t^\infty p(s)\Delta s - \int_t^\infty F(s)\Delta s. \tag{14}$$

If we write (8) as $t^{\alpha-1}\int_t^\infty p(s)\Delta s = A + \varepsilon_1(t) = |\lambda_0|^\beta - \lambda_0 + \varepsilon_1(t)$, where $\varepsilon_1(t)$ is some function satisfying $\lim_{t\to\infty}\varepsilon_1(t) = 0$, then multiplying (14) by $t^{\alpha-1}$ we obtain

$$-v(t) = |\lambda_0|^\beta - \lambda_0 - t^{\alpha-1}\int_t^\infty F(s)\Delta s + \varepsilon_1(t). \tag{15}$$

Applying the Lagrange mean value theorem, $F(t)$ can be written as $F(t) = (\alpha-1)|v(t)/t^{\alpha-1}|^\beta\zeta^{\alpha-2}(t)[1+\mu(t)\,\Phi^{-1}(v(t)/t^{\alpha-1})]^{1-\alpha}$, where $0 < 1 + \mu(t)\,\Phi^{-1}(v(t)/t^{\alpha-1}) \le \zeta(t) \le 1$. We show that we may write

$$t^{\alpha-1}\int_t^\infty F(s)\Delta s = t^{\alpha-1}\int_t^\infty \left[-\left(s^{1-\alpha}\right)^\Delta\right]|v(s)|^\beta\Delta s + \varepsilon_2(t), \tag{16}$$

$\lim_{t\to\infty} \varepsilon_2(t) = 0$. Denote $Q(t) = \zeta^{2-\alpha}(t)[1+\mu(t)\,\Phi^{-1}(v(t)/t^{\alpha-1})]^{1-\alpha}$. Since v is bounded and $\mu(t)=o(t)$, $\lim_{t\to\infty}\mu(t)\Phi^{-1}(v(t)/t^{\alpha-1})=0$, hence $\lim_{t\to\infty}\zeta(t)=1$, and so $\lim_{t\to\infty}Q(t)=1$. We have $t^{\alpha-1}\int_t^\infty F(s)\Delta s = t^{\alpha-1}\int_t^\infty \left[-\left(s^{1-\alpha}\right)^\Delta\right]|v(s)|^\beta\Delta s + t^{\alpha-1}\int_t^\infty H(s)\Delta s$, where $H(t)=F(t)-\left[-\left(t^{1-\alpha}\right)^\Delta\right]|v(t)|^\beta = (\alpha-1)|v(t)|^\beta Q(t)/(t^{\alpha-1})^\beta - (\alpha-1)|v(t)|^\beta/\iota^\alpha(t)$, with $t \le \iota(t) \le \sigma(t)$. Using the time scale L'Hospital rule, and with $t\le\omega(t)\le\sigma(t)$, we get

$$\lim_{t\to\infty} t^{\alpha-1}\int_t^\infty H(s)\Delta s = \lim_{t\to\infty}\frac{-(\alpha-1)|v(t)|^\beta\left(Q(t)/t^\alpha - 1/\iota^\alpha(t)\right)}{(1-\alpha)/\omega^\alpha(t)}$$
$$= \lim_{t\to\infty}|v(t)|^\beta\frac{\left(\iota(t)\,\omega(t)/t^2\right)^\alpha - \left(t\,\omega(t)/t^2\right)^\alpha}{(t\,\iota(t)/t^2)^\alpha} = 0,$$

where we use the fact that $\lim_{t\to\infty}\iota(t)/t=1$ and $\lim_{t\to\infty}\omega(t)/t=1$ following from $\mu(t)=o(t)$. Hence, $t^{\alpha-1}\int_t^\infty H(s)\Delta s=\varepsilon_2(t)$, with some $\varepsilon_2(t)$, where $\lim_{t\to\infty}\varepsilon_2(t)=0$, and so (16) holds. In view of (16), from (15) we get $-v(t)=|\lambda_0|^\beta-\lambda_0-t^{\alpha-1}\int_t^\infty\left[-\left(s^{1-\alpha}\right)^\Delta\right]|v(s)|^\beta\Delta s+\varepsilon(t)$, where $\varepsilon(t)=\varepsilon_1(t)-\varepsilon_2(t)$. Hence, $-v(t)=|\lambda_0|^\beta-\lambda_0-t^{\alpha-1}G(t)\int_t^\infty\left[-\left(s^{1-\alpha}\right)^\Delta\right]\Delta s+\varepsilon(t)$, where $m(t)\le G(t)\le M(t)$ with $m(t)=\inf_{s\ge t}|v(s)|^\beta$, $M(t)=\sup_{s\ge t}|v(s)|^\beta$, or

$$G(t)-v(t)=|\lambda_0|^\beta-\lambda_0+\varepsilon(t). \tag{17}$$

We show that $\lim_{t\to\infty}v(t)=\lambda_0$. Recall that $-v(t)>0$ is bounded from above. First assume that there exists $\lim_{t\to\infty}(-v(t))=L\ge 0$. Then from (17) we get $L^\beta+L=|\lambda_0|^\beta-\lambda_0$. If $L>-\lambda_0$, then $|\lambda_0|^\beta=L^\beta+L+\lambda_0>L^\beta$, contradiction. Similarly we get contradiction if $L<-\lambda_0$. Now assume that $\liminf_{t\to\infty}(-v(t))=L_*<L^*=\limsup_{t\to\infty}(-v(t))$. Let L_1 be defined by $\liminf_{t\to\infty}G(t)=L_1^\beta$ and L_2 be defined by $\limsup_{t\to\infty}G(t)=L_2^\beta$. In general, $L_*\le L_1\le L_2\le L^*$ are nonnegative reals. Assuming that at least one inequality is strict, which implies that at least on of the values is different from $-\lambda_0$, we come to a contradiction, arguing similarly as in the case when L existed. All these observations prove that the limit $\lim_{t\to\infty}v(t)$ exists and is equal to λ_0. □

Remark 3.1. (i) The statements (i) and (ii) in the main theorem could be unified, assuming $A\ge 0$ and $\lambda_0\le 0$.

(ii) The conditions $\mu(t)=O(t)$ and $\mu(t)=o(t)$, which are assumed in the theorem, are quite natural, and correspond with the general theory of regular variation, see also [12,13].

(iii) Observe that the condition y is decreasing does not need to be explicitly assumed. Moreover, note that we are actually dealing with all $\mathcal{SV}$ or $\mathcal{RV}(\Phi^{-1}(\lambda_0))$ solutions. Indeed, (1) possesses just eventually decreasing and increasing solutions which are convex. If a solution $y \in \mathcal{RV}(\vartheta)$, $\vartheta \leq 0$, were increasing, then there is $M > 0$ such that $y^{\Delta}(t) \geq M$ or there is $N > 0$ such that $y(t) \geq Nt$ for large t. This however contradicts to the fact that for any $y \in \mathcal{RV}(\vartheta)$ it holds $\lim_{t\to\infty} y(t)/t^{\nu} = 0$ provided $\vartheta < \nu$, see [12].

References

1. E. Akın-Bohner, Positive decreasing solutions of quasilinear dynamic equations, *Math. Comput. Modelling* **43** (2006), 283–293.
2. N.H. Bingham, C.M. Goldie, J.L. Teugels, *Regular Variation*, Encyclopedia of Mathematics and its Applications, Vol. 27, Cambridge Univ. Press, 1987.
3. M. Bohner, A. C. Peterson. *Dynamic Equations on Time Scales: An Introduction with Applications.* Birkhäuser, Boston, 2001.
4. R. Bojanić, E. Seneta, A unified theory of regularly varying sequences, *Math. Z.* **134** (1973), 91–106.
5. O. Došlý, P. Řehák, *Half-linear Differential Equations*, Elsevier, North Holland, 2005.
6. S. Hilger, Ein Maßkettenkalkül mit Anwendung auf Zentrumsmannigfaltigkeiten. Ph.D. dissertation, Universität of Würzburg, 1988.
7. J. Jaroš, T. Kusano, T. Tanigawa, Nonoscillation theory for second order half-linear differential equations in the framework of regular variation, *Result. Math.* **43** (2003), 129–149.
8. V. Marić, *Regular Variation and Differential Equations*, Lecture Notes in Mathematics 1726, Springer-Verlag, Berlin-Heidelberg-New York, 2000.
9. S. Matucci, P. Řehák, Regularly varying sequences and second-order difference equations, *J. Difference Equ. Appl.* **14** (2008), 17–30.
10. P. Řehák, Half-linear dynamic equations on time scales: IVP and oscillatory properties, *Nonl. Funct. Anal. Appl.* **7** (2002), 361–404.
11. P. Řehák, Regular variation on time scales and dynamic equations, Aust. J. Math. Anal. Appl, to appear.
12. P. Řehák, J. Vítovec, Regular variation on measure chains, submitted.
13. P. Řehák, J. Vítovec, q-regular variation and q-difference equations, submitted.
14. E. Seneta, *Regularly Varying Functions*, Lecture Notes in Mathematics 508, Springer-Verlag, Berlin-Heidelberg-New York, 1976.

Some Considerations on Fuzzy Dynamical Systems

R. Rodríguez-López

Departamento de Análisis Matemático,
Universidad de Santiago de Compostela,
Santiago de Compostela, 15782, Spain
E-mail: rosana.rodriguez.lopez@usc.es

We analyze the behavior of the solutions for initial value problems associated to certain fuzzy differential equations, paying special attention to its advantages and disadvantages in relation with the modelization of real phenomena subject to uncertainty and comparing the results obtained with the corresponding results in the ordinary case.

Keywords: Fuzzy numbers; Fuzzy differential equations; Diameter of level sets; Midpoint of level sets.

1. Introduction

We consider fuzzy differential equations, i.e., differential equations in the space of fuzzy numbers E^1, which contains the functions $u : \mathbb{R} \longrightarrow [0,1]$ satisfying that:

i) u is normal: there exists $x_0 \in \mathbb{R}$ with $u(x_0) = 1$.
ii) u is fuzzy convex: $\forall x, y \in \mathbb{R}$, $\lambda \in [0,1]$, $u(\lambda x + (1-\lambda)y) \geq \min\{u(x), u(y)\}$.
iii) u is upper-semicontinuous.
iv) The support of u, $[u]^0 = \overline{\{x \in \mathbb{R} : u(x) > 0\}}$ is a compact set.

E^1 is a complete metric space considering the distance d_∞ defined, for $u, v \in E^1$, by $d_\infty(u,v) = \sup_{\alpha\in[0,1]} d_H([u]^\alpha, [v]^\alpha)$, where d_H denotes the Hausdorff distance between nonempty compact convex subsets of $\mathbb{R}$ and

$$[u]^\alpha = \{x \in \mathbb{R} : u(x) \geq \alpha\}, \ \alpha \in (0,1], \qquad [u]^0 = \overline{\{x \in \mathbb{R} : u(x) > 0\}}$$

are the level sets of the fuzzy number u.

The addition and the multiplication by a non-zero scalar is extended to fuzzy numbers by the Zadeh's Extension Principle,[1] obtaining $[u+v]^a =$

$[u]^a + [v]^a$, $[\lambda u]^a = \lambda[u]^a$, for all $a \in [0,1]$, u, $v \in E^1$, and $\lambda \in \mathbb{R}$. On the other hand, if z_1, $z_2 \in E^1$, we call $w \in E^1$ the Hukuhara difference of z_1 and z_2, $w = z_1 -_H z_2$, if $z_1 = z_2 + w$. The Hukuhara difference of z_1 and z_2 is unique, provided it exists.

The integral of $f : [t_0, T] \longrightarrow E^1$ in $[t_0, T]$ $(T > t_0)$ is defined levelwise as $[\int_{[t_0,T]} f(t)\,dt]^\alpha = \{\int_{[t_0,T]} g(t)\,dt \mid g : [t_0, T] \longrightarrow \mathbb{R}$ measurable selection for $f_\alpha\}$, for $\alpha \in (0,1]$, where $f_\alpha(t) = [f(t)]^\alpha$. A fuzzy function f is integrable over $[t_0, T]$ if $\int_{[t_0,T]} f(t)\,dt \in E^1$. Obviously, every continuous function is integrable.

We consider differentiability of a fuzzy function in the sense of Hukuhara, that is, $f : [t_0, T] \to E^1$ is differentiable at $t \in [t_0, T]$ if, for some $\epsilon_0 > 0$, the Hukuhara differences $f(t+h) -_H f(t)$, $f(t) -_H f(t-h)$ exist in E^1, for $0 < h \leq \epsilon_0$ with $t \pm h \in [t_0, T]$ and there exists $f'(t) \in E^1$, the derivative in the sense of Hukuhara of f at t, such that $\lim_{h\to 0^+} \frac{f(t+h) -_H f(t)}{h}$, $\lim_{h\to 0^+} \frac{f(t) -_H f(t-h)}{h}$ exist and are equal to $f'(t)$.

In the ordinary case, if I is a real interval, $M \in \mathbb{R}$, and $\sigma : I \to \mathbb{R}$, equations $u'(t) + Mu(t) = \sigma(t)$, and $u'(t) = -Mu(t) + \sigma(t)$ are equivalent and have the same solution. However, in the fuzzy case, the effect produced on the level sets by the multiplication by a negative number $-1[u]^a = [-u_{ar}, -u_{al}]$, for $a \in [0,1]$, is the reason for the differences between the solutions to $u'(t) + 3u(t) = \sigma(t)$ and $u'(t) - 3u(t) = \sigma(t)$, provided both solutions exist. Solvability of the initial value problem for 'linear' fuzzy differential equations $u'(t) + Mu(t) = \sigma(t)$, $t \in I$, $u'(t) = -Mu(t) + \sigma(t)$, $t \in I$, $u'(t) = Mu(t) + \sigma(t)$, $t \in I$, $u'(t) - Mu(t) = \sigma(t)$, $t \in I$, where $M \in \mathbb{R}$, $M > 0$, as well as the expression of each solution[2] have been studied. Also in the cited reference, the expressions of the solutions to the previous problems are compared in terms of the midpoint[3] and the diameter of the level sets, $mp([u]^a) = \frac{1}{2}(u_{al} + u_{ar})$, $diam([u]^a) = u_{ar} - u_{al}$. For $\sigma(t) = \chi_{\{0\}}$, the solution was calculated[1,4,5] for a particular value of M. Initial value problems relative to the previous 'linear' equations with the derivative isolated have a unique solution by fuzzy Picard-Lipschitz Theorem.[1] However, the existence of solutions for the other equations depends on the validity of certain compatibility conditions[2] on data.

For modeling of ecological systems which are subject to imprecise factors[6,7] fuzzy differential equations are used, and a fuzzy analog of the logistic difference equation[8] has also been studied. Parametric LU-representation of fuzzy numbers[9] is used for simulation of fuzzy dynamical systems and continuous logistic model[10] has been studied from the fuzzy approach.

2. Comparison of the solutions to different linear fuzzy differential equations

In this section, we analyze the behavior of the solutions for initial value problems associated to 'linear' fuzzy differential equations, and compare the different expressions obtained for problems which has similar solutions in the classical case.

We recall some results[2] which provide the expressions of the solution for the i.v.p. relative to the different types of 'linear' f.d.e., specifying sufficient conditions for the existence of solution there where it is not guaranteed a priori.

We choose certain values for $M > 0$ and the independent term $\sigma(t) \in C(I, E^1)$. By means of the various examples, we analyze the behavior of the solutions of the different problems. We calculate the diameter and the mid-point of the level sets of the solution, this study allows a better comparison of the qualitative properties of the solutions, providing a useful information which can be taken into account in the process of defuzzification of a fuzzy solution in order to obtain a real solution for a problem influenced by uncertainty effects. The behavior of the solutions will suggest which is the best choice of 'linear' fuzzy problem to modelize a certain real process with uncertainty.

In the following, we consider $M > 0$, I a real interval $I = [0, T]$ with $T > 0$ or $I = [0, +\infty)$, $\sigma \in C(I, E^1)$, and $u_0 \in E^1$.

2.1. $u'(t) + Mu(t) = \sigma(t),\ t \in I,\ M > 0.$

Consider the 'linear' fuzzy initial value problem

$$u'(t) + Mu(t) = \sigma(t),\ t \in I, \qquad u(0) = u_0. \tag{1}$$

Theorem 2.1. *Problem (1) has a unique solution in I, given by*

$$u(t) = u_0 \chi_{\{e^{-Mt}\}} + \int_0^t \sigma(s) \chi_{\{e^{M(s-t)}\}}\, ds,\ t \in I, \tag{2}$$

if, for each $t \in I$, there exists $\beta > 0$ such that the Hukuhara differences $u(t+h) -_H u(t)$ and $u(t) -_H u(t-h)$ exist, for all $0 < h < \beta$.

Example 2.1. Take the particular case of problem (1)

$$u'(t) + 3u(t) = \chi_{[0,1]},\ t \in I, \qquad u(0) = u_0 = \chi_{\{1\}}. \tag{3}$$

Applying Theorem 2.1, we obtain the solution to (3) as $u(t) = \chi_{[e^{-3t}, \frac{2e^{-3t}+1}{3}]}$, $t \in I$. Note that the solution is given at each instant t by the characteristic function of a real interval, hence the endpoints of the level sets of the solution are independent of the level $a \in [0,1]$. The reason is that function σ and the initial condition u_0 are also given by characteristic functions. See Fig. 1.

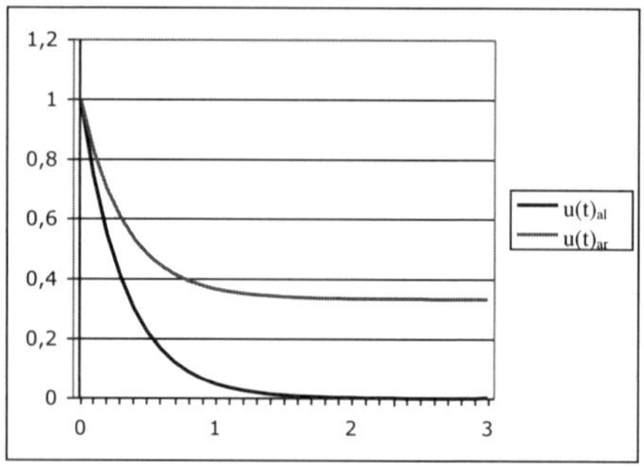

Fig. 1. Endpoints of the a-level set of the solution to problem (3).

On the other hand, the midpoint and the diameter of the level sets of the solution are $mp([u(t)]^a) = \frac{1}{6}(5e^{-3t}+1)$, $diam([u(t)]^a) = \frac{1}{3}(1-e^{-3t})$, $\forall a \in [0,1]$, $t \in I$. These values are again independent of the level $a \in [0,1]$, due to the choice of σ and u_0. See Fig. 2.

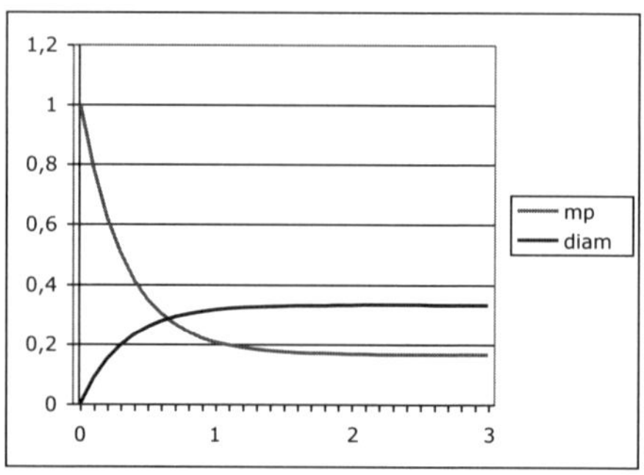

Fig. 2. Midpoint and diameter of the level sets of the solution to (3).

For a fuzzy function u differentiable in the sense of Hukuhara, the diameter of the a-level sets is a nondecreasing function in the variable t, for each a fixed. Thus, an interesting problem is the boundedness of the diameter of the level sets of the solutions of fuzzy differential equations (which are differentiable in the sense of Hukuhara) in order to control the fuzziness of the solutions by keeping the diameter of the level sets bounded by a certain degree of fuzziness. In this example, the solution has a very interesting behavior since the diameter of the a-level set of the solution stays bounded as t increases. On the other hand, note that, once the fuzzy solution has been obtained, one can choose the midpoint of the 1-level set of $u(t)$ as a real number which represents the fuzzy solution with a certain degree of accuracy, that is, in the process of defuzzification, we can select $v(t) = mp([u(t)]^1)$, for each t, and we get a real function which represents the fuzzy solution u. In the example presented, the midpoint of each level set tends to a fixed number ($\frac{1}{6}$) as t tends to $+\infty$, independently of $a \in [0,1]$.

2.2. $u'(t) = -Mu(t) + \sigma(t)$, $t \in I$, $M > 0$.

Next, consider the problem

$$u'(t) = -Mu(t) + \sigma(t), \ t \in I, \qquad u(0) = u_0. \tag{4}$$

Theorem 2.2. *Problem (4) has a unique solution on I, given by*

$$u(t)_{al} = -\frac{e^{Mt}}{2} U_1(t,a) + \frac{e^{-Mt}}{2} U_2(t,a), \ a \in [0,1], \ t \in I, \tag{5}$$

$$u(t)_{ar} = \frac{e^{Mt}}{2} U_1(t,a) + \frac{e^{-Mt}}{2} U_2(t,a), \ a \in [0,1], t \in I, \tag{6}$$

where

$$U_1(t,a) = diam([u_0]^a) + \int_0^t diam([\sigma(s)]^a) e^{-Ms}\, ds,$$

$$U_2(t,a) = (u_0)_{al} + (u_0)_{ar} + \int_0^t (\sigma(s)_{al} + \sigma(s)_{ar})\, e^{Ms}\, ds.$$

Example 2.2. Consider the particular case of (4)

$$u'(t) = -3u(t) + \chi_{[0,1]}, \ t \in I, \qquad u(0) = u_0 = \chi_{\{1\}}, \tag{7}$$

whose solution (see Fig. 3) is given, using Theorem 2.2, by the expression

$$[u(t)]^a = \left[\frac{2 - e^{3t} + 5e^{-3t}}{6}, \frac{e^{3t} + 5e^{-3t}}{6} \right], \ a \in [0,1], \ t \in I.$$

The midpoint and the diameter of the level sets of u are $mp([u(t)]^a) = \frac{1}{6}(5e^{-3t} + 1)$, $diam([u(t)]^a) = \frac{1}{3}(e^{3t} - 1)$, $a \in [0,1]$, $t \in I$ (see Fig. 4). Note that the diameter of the level sets tends to $+\infty$, situation not desirable from the point of view of applications, but the midpoint tends to $\frac{1}{6}$ as t tends to $+\infty$. In fact, the expression of the midpoint of $[u(t)]^a$ is the same of problem (3). As expected,[2] the expression of the diameter of the level sets of u can be derived from the expression of the diameter of the level sets of the solution to (3), replacing M by $-M$.

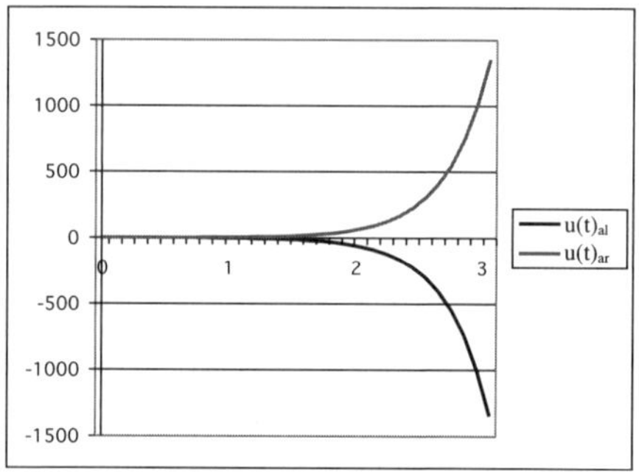

Fig. 3. Endpoints of the a-level set of the solution to problem (7).

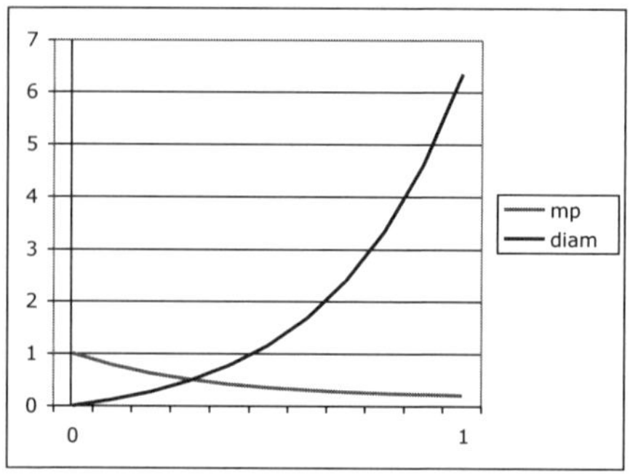

Fig. 4. Midpoint and diameter of the level sets of the solution to (7).

2.3. $u'(t) = Mu(t) + \sigma(t),\ t \in I,\ M > 0$

Let us study the problem

$$u'(t) = Mu(t) + \sigma(t),\ \ t \in I, \qquad u(0) = u_0. \tag{8}$$

Theorem 2.3. *Problem (8) has a unique solution on I, given by*

$$u(t) = u_0 \chi_{\{e^{Mt}\}} + \int_0^t \sigma(s) \chi_{\{e^{M(t-s)}\}}\, ds,\ t \in I. \tag{9}$$

Example 2.3. Next, we analyze the behavior of the solutions to problem

$$u'(t) = 3u(t) + \chi_{[0,1]},\ \ t \in I, \qquad u(0) = u_0 = \chi_{\{1\}}. \tag{10}$$

According to Theorem 2.3, the solution to (10) is given by $u(t) = \chi_{[e^{3t}, \frac{4e^{3t}-1}{3}]}$, $t \in I$, (see Fig. 5). The midpoint and the diameter of the level sets of the solution are $mp([u(t)]^a) = \frac{1}{6}(7e^{3t} - 1)$, $diam([u(t)]^a) = \frac{1}{3}(e^{3t} - 1)$, $a \in [0,1]$, $t \in I$ (see Fig. 6).

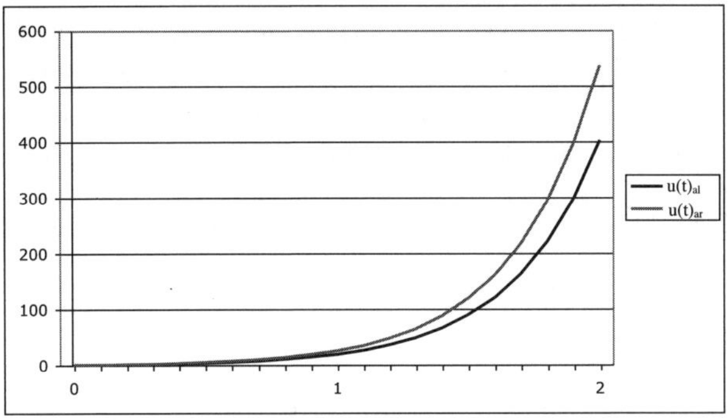

Fig. 5. Endpoints of the level sets of the solution to problem (10).

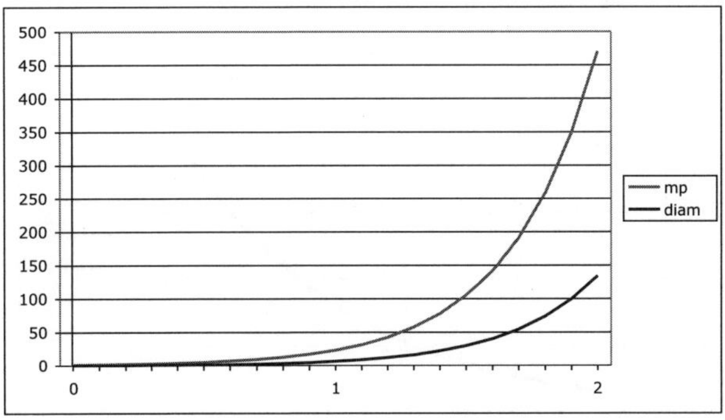

Fig. 6. Midpoint and diameter of the level sets of the solution to (10).

2.4. $u'(t) - Mu(t) = \sigma(t),\ t \in I,\ M > 0$

Finally, consider the problem

$$u'(t) - Mu(t) = \sigma(t),\ \ t \in I, \qquad u(0) = u_0. \tag{11}$$

Theorem 2.4. *Define*

$$W_1(t,a) = diam([u_0]^a) + \int_0^t diam([\sigma(s)]^a)e^{Ms}\,ds,$$
$$W_2(t,a) = (u_0)_{al} + (u_0)_{ar} + \int_0^t (\sigma(s)_{al} + \sigma(s)_{ar})\,e^{-Ms}\,ds.$$

Expressions

$$u(t)_{al} = -\frac{e^{-Mt}}{2}W_1(t,a) + \frac{e^{Mt}}{2}W_2(t,a), \tag{12}$$

$$u(t)_{ar} = \frac{e^{-Mt}}{2}W_1(t,a) + \frac{e^{Mt}}{2}W_2(t,a), \tag{13}$$

for $t \in I$, $a \in [0,1]$, represent the unique solution to problem (11) in I, if they define a fuzzy number, that is, if $u(t)_{al}$ is nondecreasing, $u(t)_{ar}$ is nonincreasing in a, and for every $t \in I$, there exists $\beta > 0$ such that the Hukuhara differences $u(t+h) -_H u(t)$, $u(t) -_H u(t-h)$ exist for $0 < h < \beta$.

Example 2.4. The level sets of the solution of the following problem

$$u'(t) - 3u(t) = \chi_{[0,1]}, \ t \in I, \qquad u(0) = u_0 = \chi_{\{1\}}, \tag{14}$$

whose endpoints are given by expressions (12) and (13), are

$$[u(t)]^a = \left[\frac{7e^{3t} + e^{-3t} - 2}{6}, \frac{7e^{3t} - e^{-3t}}{6}\right], \ a \in [0,1], \ t \in I.$$

See Fig. 7. The midpoint and the diameter of the level sets of the solution

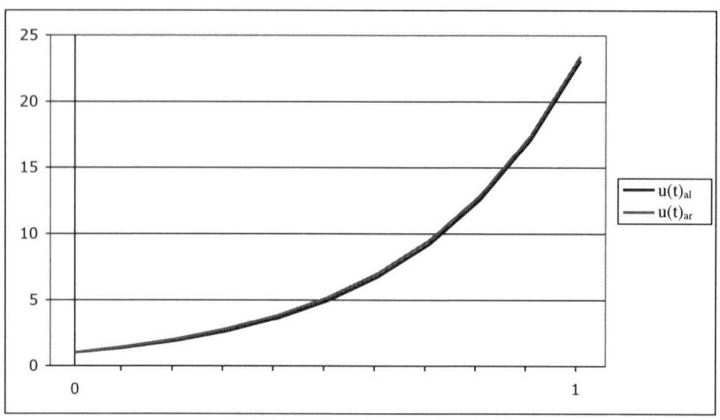

Fig. 7. Endpoints of the level sets of the solution to problem (14).

are

$$mp([u(t)]^a) = \frac{1}{6}(7e^{3t} - 1), \ diam([u(t)]^a) = \frac{1}{3}(1 - e^{-3t}), \ a \in [0,1], \ t \in I,$$

whose graphs are shown in Fig. 8. In this example, the diameter of the level sets is increasing and bounded by its limit as t tends to $+\infty$, which is $\frac{1}{3}$, thus the solution does not get very fuzzy as t increases.

In these examples we can observe the properties of invariance of midpoint (or diameter) of the level sets under certain changes in the equations. Note that[2] the change from $u'(t) + 3u(t) = \sigma(t)$ to $u'(t) = -3u(t) + \sigma(t)$, or from $u'(t) = 3u(t) + \sigma(t)$ to $u'(t) - 3u(t) = \sigma(t)$ preserves the midpoint of the level sets of the solution, while the diameter changes (replacing M by $-M$). On the other hand, passing from $u'(t) + 3u(t) = \sigma(t)$ to $u'(t) - 3u(t) = \sigma(t)$, or from $u'(t) = -3u(t) + \sigma(t)$ to $u'(t) = 3u(t) + \sigma(t)$, the diameter of the

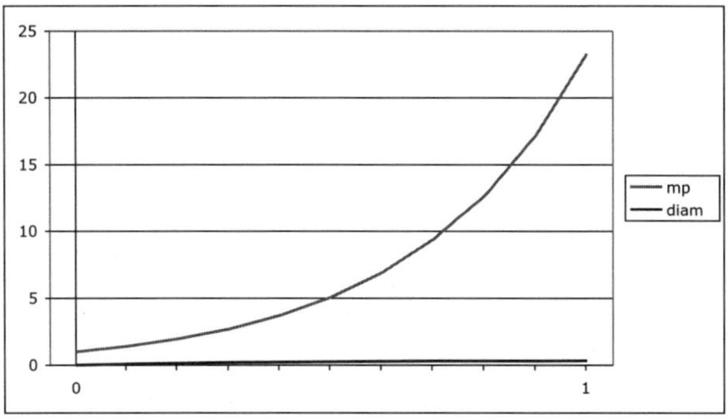

Fig. 8. Midpoint and diameter of the level sets of the solution to (14).

level sets remains invariant, while the expression of the midpoint changes (replacing M by $-M$). Both changes have to be made (in the midpoint and the diameter of the level sets) if we pass from $u'(t) + 3u(t) = \sigma(t)$ to $u'(t) = 3u(t) + \sigma(t)$, or from $u'(t) = -3u(t) + \sigma(t)$ to $u'(t) - 3u(t) = \sigma(t)$.

References

1. P. Diamond and P. E. Kloeden, *Metric spaces of fuzzy sets : theory and applications*, (World Scientific, Singapore, 1994).
2. J. J. Nieto, R. Rodríguez-López and D. Franco, Linear first-order fuzzy differential equations, *International Journal of Uncertainty, Fuzziness and Knowlegde-Based Systems* **14**, 687–709 (2006).
3. J. J. Nieto and A. Torres, Midpoints for fuzzy sets and their application in medicine, *Artificial Intelligence in Medicine* **27**, 81–101 (2003).
4. T. G. Bhaskar, V. Lakshmikantham and V. Devi, Revisiting fuzzy differential equations, *Nonlinear Anal.* **58**, 351–358 (2004).
5. P. Diamond, Brief note on the variation of constants formula for fuzzy differential equations, *Fuzzy Sets and Systems* **129**, 65–71 (2002).
6. L. C. de Barros, R. C. Bassanezi and P. A. Tonelli, Fuzzy modelling in population dynamics, *Ecological Modelling* **128**, 27–33 (2000).
7. R. C. Bassanezi, L. C. de Barros and P. A. Tonelli, Attractors and asymptotic stability for fuzzy dynamical systems, *Fuzzy Sets and Systems* **113**, 473–483 (2000).
8. E. Y. Deeba, A. de Korvin and E. L. Koh, On a fuzzy logistic difference equation, *Differential Equations Dynam. Systems* **4**, 149–156 (1996).
9. L. Stefanini, L. Sorini and M. L. Guerra, Simulation of fuzzy dynamical systems using the *LU*-representation of fuzzy numbers, *Chaos Solitons Fractals* **29**, 638–652 (2006).
10. J. J. Nieto and R. Rodríguez-López, Analysis of a logistic differential model with uncertainty, *International Journal of Dynamical Systems and Differential Equations* **1**, 164–176 (2008).

Asymptotic Behavior of Solutions of Third Order Difference Equations

Ewa Schmeidel, Małgorzata Zbąszyniak

Institute of Mathematics, Faculty of Electrical Engineering,
Poznań University of Technology
ul. Piotrowo 3a, 60-965 Poznań, Poland
e-mails: ewa.schmeidel@put.poznan.pl, mmielesz@math.put.poznan.pl

We consider a third order nonlinear difference equation

$$\Delta(r_n\Delta(p_n\Delta y_n)) = a_n y_{n+1} + f(n, y_n).$$

The sufficient conditions under which for any nonzero constant there exists a solution of the above equation which tends to this constants, are obtained. Also, the sufficient conditions for the existence a solution of the considered equation which can be written in the following form

$$y_n = \alpha_n \frac{u_n}{r_n} + \beta_n \frac{v_n}{r_n} + \gamma_n \frac{w_n}{r_n},$$

$$(\lim_{n\to\infty} \alpha_n < \infty \quad \text{and} \quad \lim_{n\to\infty} \beta_n < \infty \quad \lim_{n\to\infty} \gamma_n < \infty) \quad \text{are given.}$$

Here u, v and w are three linearly independent solutions of equation

$$\Delta(r_n\Delta(p_n\Delta y_n)) = a_n y_{n+1}.$$

Keywords: Nonlinear, difference equation, third order.

AMS Subject classification 39A10

1. Introduction

In this paper we consider the nonlinear difference equation of the form

$$\Delta(r_n\Delta(p_n\Delta y_n)) = a_n y_{n+1} + f(n, y_n), \quad n \in \mathbb{N} \tag{1}$$

where Δ denotes the forward difference operator $\Delta y_n = y_{n+1} - y_n$ for $y : \mathbb{N} \to \mathbb{R}$. Sequences (r_n), (p_n) are positive, and (a_n) is a real sequence. Function $f : \mathbb{N} \times \mathbb{R} \to \mathbb{R}$. Throughout this paper $\mathbb{N}$ denotes the set of positive integers, $\mathbb{R}$ the set of real numbers, $\mathbb{R}_+$ the set of positive real numbers. By a solution of (1), we mean a sequence (y_n) which satisfies equation (1) for sufficiently large n.

Third order difference equations with quasidifferences was considered, for example, by Andruch-Sobiło and Migda [3], Došla and Kobza [4-6], Graef and Thandapani [7], Kobza [10], Migda, Schmeidel and Drozdowicz [12], Thandapani and Mahalingam [13]. Background of difference equation theory can be found in [1-2], [8] and [9].

2. Main results

We begin with the following existence theorem.

Theorem 2.1. *Assume that*

$$\sum_{i=1}^{\infty} \frac{1}{p_i} \sum_{j=i}^{\infty} \frac{1}{r_j} \sum_{k=j}^{\infty} |a_k| < \infty, \tag{2}$$

and

$$\sum_{i=1}^{\infty} \frac{1}{p_i} \sum_{j=i}^{\infty} \frac{1}{r_j} \sum_{k=j}^{\infty} |f(k,\alpha)| < \infty, \tag{3}$$

where α is a nonzero real constant and the function f is continuous on the second argument. Then for every nonzero constant c, there exists a solution (y_n) of equation (1) *such that* $\lim_{n\to\infty} y_n = c$.

Proof. Assume that (2) and (3) hold with $c > 0$. (If $c < 0$ the proof is similar.) Let n_0 be so large that $\sum_{i=n_0}^{\infty} \frac{1}{p_i} \sum_{j=i}^{\infty} \frac{1}{r_j} \sum_{k=j}^{\infty} |a_k| \leq \frac{1}{4}$ and $\sum_{i=n_0}^{\infty} \frac{1}{p_i} \sum_{j=i}^{\infty} \frac{1}{r_j} \sum_{k=j}^{\infty} |f(k,c)| \leq \frac{c}{2}$. Set l^∞ be the Banach space of all real bounded sequences with "sup" norm. Let $S = \{y \in l^\infty : |y_n| \leq 2c \text{ for } n \geq n_0\}$. It is not difficult to see that S is nonempty convex and closed subset of l^∞. We define a mapping $T : S \to l^\infty$ as follows

$$Ty_n = c - \sum_{i=n}^{\infty} \frac{1}{p_i} \sum_{j=i}^{\infty} \frac{1}{r_j} \sum_{k=j}^{\infty} \left(a_k y_{k+1} + f(k, y_k)\right), \text{ for } n \in \mathbb{N}.$$

Hence $|Ty_n| \leq c + \frac{c}{2} + \frac{c}{2} = 2c$, for $n \geq n_0$. Thus, T maps S into itself. It is easy to see that T is continuous and $T(S)$ uniformly Cauchy. Therefore by Schauder's fixed point theorem there exists $y \in S$ such that $Ty_n = y_n$. We may verify that (y_n) is a solution of equation (1). Furthermore $\lim_{n\to\infty} y_n = c$. This completes the proof. $\square$

From now, assume that $p_n \equiv 1$ in the equation (1). Then this equation takes the following form

$$\Delta(r_n \Delta^2 y_n) = a_n y_{n+1} + f(n, y_n). \tag{4}$$

The following lemmas will be used.

Lemma 2.1. *The difference equation*

$$\Delta(r_n \Delta^2 z_n) = a_n z_{n+1}, \quad n \in \mathbb{N}, \tag{5}$$

where $a : \mathbb{N} \to \mathbb{R}$, $r : \mathbb{N} \to \mathbb{R}$ *is a nontrivial sequence which has linearly independent solutions* $u, v, w : \mathbb{N} \to \mathbb{R}$ *such that*

$$W_n = \begin{vmatrix} u_n & v_n & w_n \\ \Delta u_n & \Delta v_n & \Delta w_n \\ \Delta^2 u_n & \Delta^2 v_n & \Delta^2 w_n \end{vmatrix} = -1, \textit{ for all } n \in \mathbb{N}.$$

Proof. The proof is obvious and will be omitted. □

The next Lemma was proved by Migda in [11].

Lemma 2.2. *Let* (u_n), (v_n) *and* (w_n) *are linearly independent solutions of the difference equation* (5) *and let* $M_{i,k}(n)$ *are minors of the Casorati's matrix of these solutions. Then* $\Delta M_{2,k}(n) = M_{1,k}(n)$, *for* $k = 1, 2, 3, \quad n \in \mathbb{N}$ *and* $\Delta M_{3,k}(n) = M_{2,k}(n+1)$, *for* $k = 1, 2, 3, \quad n \in \mathbb{N}$.

We introduce the following useful definition.

Definition 2.1. The function $B : \mathbb{N} \times \mathbb{R} \to \mathbb{R}_+$ belongs to the set $B_F \quad (B \in B_F)$ when

- $B(n, x_1) \leq B(n, x_2)$ for $0 \leq x_1 \leq x_2$,
- $B(n, kx) \leq F(k) B(n, x)$ for $x > 0$ and every $k \in \mathbb{R}$,

where $F : \mathbb{R} \to \mathbb{R}$ is a continuous, nondecreasing function, such that $F(x) \neq 0$ for $x \neq 0$.

Theorem 2.2. *Assume that*

$$r_n : \mathbb{N} \to < \epsilon, \infty), \textit{ where } \epsilon > 0, \tag{6}$$

in equation (4), *and sequences* (u_n), (v_n) *and* (w_n) *denote three linearly independent solutions of this equation. Assume also that*

$$|f(n, x)| \leq B(n, |x|), \textit{ for all } x \in \mathbb{R} \textit{ and fixed } n \in \mathbb{N}, \tag{7}$$

where function $B \in B_F$*, and functions* F *fulfil the condition*

$$\int_{\epsilon}^{\infty} \frac{ds}{F(s)} = \infty, \tag{8}$$

and there exists a positive constant C *that*

$$|f(n,x)| > C|\Delta r_n|, \text{ for every } n \in \mathbb{N} \text{ and } x > \epsilon. \tag{9}$$

Let us denote

$$U_j = \max_{k=1,2,3} \{|u_j|, |v_j|, |w_j|, |M_{2,k}(j+1)|, |M_{3,k}(j+1)|\}, \tag{10}$$

for every $j \in \mathbb{N}$*. If*

$$\sum_{j=1}^{\infty} U_j B(j, U_j) < \infty, \tag{11}$$

and conditions (2) *and* (3) *hold then there exists a solution of equation* (4) *such that* $y_n = \alpha_n \frac{u_n}{r_n} + \beta_n \frac{v_n}{r_n} + \gamma_n \frac{w_n}{r_n}$*, where* $\lim_{n\to\infty} \alpha_n = \alpha$*,* $\lim_{n\to\infty} \beta_n = \beta$ *and* $\lim_{n\to\infty} \gamma_n = \gamma$ *(*α*,* β*,* γ*- constants).*

Proof. From Lemma 1 there exists three linearly independent solutions (u_n), (v_n) and (w_n) of equation (5) such that

$$W_n = \begin{vmatrix} u_n & v_n & w_n \\ \Delta u_n & \Delta v_n & \Delta w_n \\ \Delta^2 u_n & \Delta^2 v_n & \Delta^2 w_n \end{vmatrix} = -1. \tag{12}$$

First, we assume that (12) holds. Because assumptions of Theorem 1 hold then there exists a convergent solution of equation (4). Let us denote

$$\begin{aligned} A_n &= -M_{1,1}(n) r_n y_n + M_{2,1}(n) r_n \Delta y_n - M_{3,1}(n) r_n \Delta^2 y_n \\ B_n &= M_{1,2}(n) r_n y_n - M_{2,2}(n) r_n \Delta y_n + M_{3,2}(n) r_n \Delta^2 y_n \\ C_n &= -M_{1,3}(n) r_n y_n + M_{2,3}(n) r_n \Delta y_n - M_{3,3}(n) r_n \Delta^2 y_n, \end{aligned} \tag{13}$$

where y_n is a such convergent solution of equation (4). Then we obtain

$$\begin{aligned} &u_n A_n + v_n B_n + w_n C_n \\ &= [-u_n M_{1,1}(n) + v_n M_{1,2}(n) - w_n M_{1,3}(n)]\, r_n y_n + \\ &+ [u_n M_{2,1}(n) - v_n M_{2,2}(n) + w_n M_{2,3}(n)]\, r_n \Delta y_n + \\ &+ [-u_n M_{3,1}(n) + v_n M_{3,2}(n) - w_n M_{3,3}(n)]\, r_n \Delta^2 y_n. \end{aligned} \tag{14}$$

It is easy to check that $u_n M_{2,1}(n) - v_n M_{2,2}(n) + w_n M_{2,3}(n) = 0$ and $-u_n M_{3,1}(n) + v_n M_{3,2}(n) - w_n M_{3,3}(n) = 0$. From the above equalities, (14) and (12), we obtain $u_nA_n + v_nB_n + w_nC_n = [-u_nM_{1,1}(n) + v_nM_{1,2}(n) - w_nM_{1,3}(n)]r_ny_n = -W_nr_ny_n = r_ny_n$. Thus we get

$$y_n = \frac{u_n}{r_n}A_n + \frac{v_n}{r_n}B_n + \frac{w_n}{r_n}C_n. \tag{15}$$

Applying operator Δ for (13), and from Lemma 2 we obtain

$$\begin{aligned}\Delta A_n &= M_{2,1}(n+1)\Delta r_n \Delta y_{n+1} - M_{3,1}(n+1)f(n,y_n)\\ \Delta B_n &= -M_{2,2}(n+1)\Delta r_n \Delta y_{n+1} + M_{3,2}(n+1)f(n,y_n)\\ \Delta C_n &= M_{2,3}(n+1)\Delta r_n \Delta y_{n+1} - M_{3,3}(n+1)f(n,y_n).\end{aligned}$$

Summing both sides of the above equations over n, we get

$$\begin{cases} A_n = A_1 + \sum\limits_{j=1}^{n-1} M_{2,1}(j+1)\Delta r_j \Delta y_{j+1} - \sum\limits_{j=1}^{n-1} M_{3,1}(j+1)f(j,y_j)\\ B_n = B_1 + \sum\limits_{j=1}^{n-1} M_{2,2}(j+1)\Delta r_j \Delta y_{j+1} + \sum\limits_{j=1}^{n-1} M_{3,2}(j+1)f(j,y_j)\\ C_n = C_1 + \sum\limits_{j=1}^{n-1} M_{2,3}(j+1)\Delta r_j \Delta y_{j+1} - \sum\limits_{j=1}^{n-1} M_{3,3}(j+1)f(j,y_j).\end{cases} \tag{16}$$

Hence, we have

$$\begin{aligned}&|A_n| + |B_n| + |C_n| \le |A_1| + |B_1| + |C_1|\\ &+ \sum_{j=1}^{n-1} \left(|M_{2,1}(j+1)| + |M_{2,2}(j+1)| + |M_{2,3}(j+1)|\right)|\Delta r_j||\Delta y_{j+1}|\\ &+ \sum_{j=1}^{n-1} \left(|M_{3,1}(j+1)| + |M_{3,2}(j+1)| + |M_{3,3}(j+1)|\right)|f(j,y_j)|.\end{aligned} \tag{17}$$

Let us denote

$$h_n = |A_n| + |B_n| + |C_n|, \quad n \in \mathbb{N} \tag{18}$$

From (15), (6), (10) and (18), we obtain

$$y_j = \frac{1}{r_j}|u_jA_j + v_jB_j + w_jC_j| \le \frac{1}{\epsilon}U_j(|A_j| + |B_j| + |C_j|) = \frac{1}{\epsilon}U_jh_j. \tag{19}$$

From (17), (18) and (10), we get

$$h_n \le h_1 + 3\sum_{j=1}^{n-1} U_j|\Delta r_j||\Delta y_{j+1}| + 3\sum_{j=1}^{n-1} U_j|f(j,y_j)|. \tag{20}$$

From Theorem 1, there exists a constant $C_1 > 0$ that $|y_j| \le \frac{C_1}{2}$, for $j \in \mathbb{N}$. Then $|\Delta y_{j+1}| = |y_{j+2} - y_{j+1}| \le |y_{j+2}| + |y_{j+1}| \le C_1$, for $j \in \mathbb{N}$. From condition (9), we obtain

$$|\Delta r_j||\Delta y_{j+1}| \le \frac{C_1}{C}|f(j,y_j)|. \tag{21}$$

Hence, from (9), (18), (19) and (21), we get

$$\begin{aligned} h_n &\leq h_1 + 6C^* \sum_{j=1}^{n-1} U_j |f(j, y_j)| \\ &\leq h_1 + 6C^* \sum_{j=1}^{n-1} U_j |f(j, \tfrac{1}{r_j}(A_j u_j + B_j v_j + C_j w_j)| \\ &\leq h_1 + 6C^* \sum_{j=1}^{n-1} U_j (B_j, \tfrac{1}{\epsilon} U_j h_j), \end{aligned}$$

where $C^* = \max\{1, \frac{C_1}{C}\}$. We set $b_n = h_1 + 6C^* \sum_{j=1}^{n-1} U_j B(j, \frac{1}{\epsilon} U_j h_j)$. Then $b_n \geq h_n$, for $n \in \mathbb{N}$. Applying operator Δ for sequence (b_i), we obtain $\Delta b_i = 6C^* U_i B(i, \frac{1}{\epsilon} U_i h_i)$. From Definition 1, we get $\frac{\Delta b_i}{F(b_i)} \leq 6C^* F(\frac{1}{\epsilon}) U_i B(i, U_i)$. Since function F is nondecreasing then function $\frac{1}{F}$ is nonincreasing. Thus we obtain the following inequality $\frac{\Delta b_i}{F(b_i)} \geq \int_{b_i}^{b_{i+1}} \frac{dt}{F(t)}$. From above, we obtain

$$\int_{b_i}^{b_{i+1}} \frac{dt}{F(t)} \leq 6C^* F(\frac{1}{\epsilon}) U_i B(i, U_i), \quad \text{for all} \quad i \in \mathbb{N}. \tag{22}$$

By summation we get $\int_{b_1}^{b_n} \frac{dt}{F(t)} \leq 6C^* F(\frac{1}{\epsilon}) \sum_{i=1}^{n-1} U_i B(i, U_i)$. We will use the following notation $\int_{\epsilon}^{x} \frac{dt}{F(t)} = G(x)$. This implies that $\int_{b_1}^{b_n} \frac{dt}{F(t)} = G(b_n) - G(b_1)$. From (22), we obtain $G(b_n) \leq G(b_1) + 6C^* F(\frac{1}{\epsilon}) \sum_{i=1}^{n-1} U_i B(i, U_i)$. Function G is increasing then G^{-1} is increasing, also. From (8), we see that $G(b_1) + 6C^* F(\frac{1}{\epsilon}) \sum_{i=1}^{n-1} U_i B(i, U_i)$ belongs to the domain of function G^{-1}, for all $n \in \mathbb{N}$. Therefore there exists b_n such that

$$b_n \leq G^{-1}[G(b_1) + 6C^* F(\frac{1}{\epsilon}) \sum_{i=1}^{n-1} U_i B(i, U_i)].$$

We set $\sum_{j=1}^{\infty} U_j B(j, U_j) = S$. So, we obtain

$$h_n = |A_n| + |B_n| + |C_n| \leq G^{-1}(G(|A_1| + |B_1| + |C_1|) + 6C^* F(\frac{1}{\epsilon}) S) = \bar{C}.$$

From (7), (10), (19), (21) and properties of function f, we obtain

$$|M_{2,i}(j+1)||\Delta r_j||\Delta y_{j+1}| \leq U_j C^* |f(j, y_j)| \leq C^* F(\frac{1}{\epsilon} \bar{C}) U_j B(j, U_j)$$

and

$$|M_{3,i}(j+1)||f(j,y_j)| \le C^* F(\frac{1}{\epsilon}\bar{C})U_j B(j,U_j), i = 1,2,3, \quad j \in \mathbb{N}.$$

From (11) series

$$\sum_{j=1}^{\infty} M_{2,i}(j+1)\Delta r_j \Delta y_{j+1} \text{ and } \sum_{j=1}^{\infty} M_{3,i}(j+1) f(j,y_j), \quad i = 1,2,3$$

are absolute convergent. Then from (17) finite limits of A_n, B_n and C_n exist. Using (15), we get the thesis.
Now, we will prove this theorem for any three linearly independent solutions $(\tilde{u}_n)$, $(\tilde{v}_n)$ and $(\tilde{w}_n)$ of equation (5). Let (u_n), (v_n) and (w_n) be three linearly independent solutions of equation (5) fulfilling condition (12). Then for some constants c_i, $i = 1, 2, ..., 8, 9$ we have

$$u_n = c_1\tilde{u}_n + c_2\tilde{v}_n + c_3\tilde{w}_n, v_n = c_4\tilde{u}_n + c_5\tilde{v}_n + c_6\tilde{w}_n, w_n = c_7\tilde{u}_n + c_8\tilde{v}_n + c_9\tilde{w}_n.$$

Now,

$$\tilde{U}_j = \max_{k=1,2,3}\left\{|\tilde{u}_j|, |\tilde{v}_j|, |\tilde{w}_j|, |\tilde{M}_{2,k}(j+1)|, |\tilde{M}_{3,k}(j+1)|\right\}.$$

We will show that the condition (11) holds.
Set $\tilde{c} = \max_{k=1,\dots,9} |c_i|$. Hence $U_j \le \max\{3\tilde{c}, 9\tilde{c}^2\}\tilde{U}_j$. Let $c^* = \max\{3\tilde{c}, 9\tilde{c}^2\}$.
Therefore, we get inequalities

$$U_j B(j,U_j) \le c^* \tilde{U}_j B(j, c^* \tilde{U}_j) \le c^* \tilde{U}_j F(c^*) B(j, \tilde{U}_j),$$

and

$$\sum_{j=1}^{\infty} U_j B(j,U_j) < \infty.$$

We see that assumptions of Theorem 2 hold for solutions (u_n), (v_n) and (v_n) also. Then the thesis holds. This completes the proof of this Theorem. □

Now, we consider a special case of equation (4), where $a_n \equiv 0$. Hence this equation takes the following form

$$\Delta(r_n \Delta^2 y_n) = f(n, y_n). \tag{23}$$

Corollary 2.1. *Assume that conditions* (2), (3), (6)-(9) *and* (11) *hold. Let us denote*

$$U_j = \max\left\{ j, \sum_{i=1}^{j} \frac{j-i}{r_i}, \sum_{i=1}^{j} \frac{i+1}{r_i} \right\}.$$

Then there exists a solution of equation (23) *such that*

$$y_n = \frac{\alpha_n}{r_n} + \frac{\beta_n}{r_n} n + \frac{\gamma_n}{r_n} \sum_{i=1}^{n} \frac{n-i-1}{r_i},$$

where α_n, β_n, γ_n *have a finite limits.*

Proof. Equation $\Delta(r_n \Delta^2 z_n) = 0$ has three linearly independent solutions

$$u_n = 1, v_n = n \text{ and } w_n = \sum_{i=1}^{n-1} \sum_{j=1}^{i-1} \frac{1}{r_j} = \sum_{i=1}^{n} \frac{n-i-1}{r_i}.$$

Since assumptions of Theorem 2 hold, we get the thesis. □

References

1. R.P. Agarwal, Difference Equations and Inequalities. Theory, Methods and Applications, Marcel Dekker, Inc., New York 2000 (second edition).
2. R.P. Agarwal and P.J.Y. Wong, Advanced Topics in Difference Equations, Kluwer, Dordrecht, 1997.
3. A. Andruch-Sobiło, M.Migda, *Bounded solutions of third order nonlinear difference equations,* Rocky Mountain J. Math. 36 (1) (2006), 23-34.
4. Z. Došlá, A. Kobza, *Global asymptotic properties of third-order difference equations,* Comput. Math. Appl. 48 (1-2) (2004), 191-200.
5. Z. Došlá, A. Kobza, *Nonoscillatory solutions of third-order difference equations,* Proceedings of the Eighth International Conference on Difference Equations and Applications, 105–112, Chapman Hall/CRC, Boca Raton, FL, 2005.
6. Z. Došlá, A. Kobza, *On third-order linear difference equations involving quasi-differences,* Adv. Difference Equ. (2006), 1-13.
7. J.R. Graef, E. Thandapani, *Oscillatory and asymptotic behavior of solutions of third order delay difference equations,* Funkcial. Ekvac. 42 (3) (1999), 355-369.
8. S.N. Elaydi, An Introduction to Difference Equations, Springer, 1996.
9. W.G. Kelley, A.C. Peterson, Difference Equations, an Introduction with Applications, Academic press, 1991.
10. A. Kobza, *Property A for third order difference equations*, Studies of the University of Zilina, Math. Series 17 (2003), 109-114.
11. M. Migda, *Asymptotic properties of solutions of higher order difference equations,* Radovi Math. 5 (1989), 279-309.
12. M. Migda, E. Schmeidel, A. Drozdowicz, *Nonoscillation results for some third order nonlinear difference equation,* Folia Facultatis Scientiarium Naturalium Universitatis Masarykianae Brunensis Mathematica 13 (2003), 185-192.
13. E. Thandapani, K. Mahalingam, *Oscillatory properties of third order neutral delay difference equatons,* Demonstratio Math. 35 (2002), 325-337.

Difference Equations and Nonlinear Boundary Value Problems for Hyperbolic Systems

A. N. Sharkovsky
Institute of Mathematics, NASU,
Kiev, Ukraine

R. Severino and S. Vinagre
CIMA-UE, University of Évora,
Évora, Portugal

It is known that solutions of certain classes of linear hyperbolic systems with nonlinear boundary conditions and consistent initial conditions can be written via the iteration of a map of the interval. In this work we characterize the solutions of such problems, with a vortex as initial condition and the iteration of a bimodal map of the interval, using the bimodal topological invariants.

1. Introduction

There are nonlinear boundary value problems the investigation of which is reduced to the investigation of low-dimensional or even one-dimensional DS. The study of such problems was began still at the first half of 80th.[1] At that time, the theory of one-dimensional DS possessed certain tools allowing effectively and sufficiently deeply to study properties of solutions of such BVP.

Actually every achievement into the theory of low-dimensional DS can be transformed into new usefull results for these BVP. Prof. Sousa Ramos had initiated in Portugal investigations on application of symbolic dynamics to the study of various classes of low-dimensional and, first of all, one-dimensional DS. Later he offered to use results of these investigations in describing properties of BVP solutions. Our paper deals with just the such researches started still under direct participation of Prof. Sousa Ramos.

At the beginning of the paper we also consider one of the simplest n-dimensinal BVP with linear PDE, the investigation of which is reduced to one-dimensional DS. Generally, such reducing can be realized even for some classes of BVP that include piecewise linear PDE and essentially more complicated boundary conditions.

2. A Certain Class of Boundary Value Problems

Consider the following linear hyperbolic system

$$\begin{cases} \dfrac{\partial u}{\partial t} = \displaystyle\sum_{i=1}^{m} a_i \frac{\partial u}{\partial x_i} + a\frac{\partial u}{\partial y} \\ \dfrac{\partial v}{\partial t} = \displaystyle\sum_{i=1}^{m} b_i \frac{\partial v}{\partial x_i} + b\frac{\partial u}{\partial y}, \end{cases} \tag{1}$$

with $u = u(t, x_1, x_2, \ldots, x_m, y)$ and $v = v(t, x_1, x_2, \ldots, x_m, y)$ being functions of the arguments $t \in \mathbb{R}^+$, $x_1, \ldots, x_m \in \mathbb{R}$, and $y \in [0, 1]$. Assume both u and v are subject to the nonlinear boundary conditions

$$\begin{cases} u|_{y=0} = v|_{y=0} \\ u|_{y=1} = f(v|_{y=1}), \end{cases} \tag{2}$$

with $f : \mathbb{R} \to \mathbb{R}$, and to some initial conditions

$$\begin{cases} u|_{t=0} = u_0(x_1, x_2, \ldots, x_m, y) \\ v|_{t=0} = v_0(x_1, x_2, \ldots, x_m, y). \end{cases} \tag{3}$$

A similar problem was considered by Maistrenko[2] where the solution representation was given and the continuous dependence on perturbations was investigated.

It is convenient for us to represent the general solution of (1) in the form

$$\begin{aligned} u\,(t, x_1, ..., x_m, y) &= \phi(x_1 - \tfrac{a_1}{a}y, \ \ldots, \ x_m - \tfrac{a_m}{a}y, \ t + \tfrac{1}{a}y), \\ v\,(t, x_1, ..., x_m, y) &= \psi(x_1 - \tfrac{b_1}{b}y, \ \ldots, \ x_m - \tfrac{b_m}{b}y, \ t + \tfrac{1}{b}y), \end{aligned}$$

where ϕ, ψ are arbitrary (differentiable) functions. Then, the equality $\phi = \psi$ follows from the first boundary condition in (2). Using the second condition, we obtain that, for $a < 0 < b$, the solution of the boundary value problem (1), (2) can be written as

$$\begin{aligned} u\,(t, x_1, ..., x_m) &= w(x_1 - \tfrac{a_1}{a}y, \ \ldots, \ x_m - \tfrac{a_m}{a}y, \ t + \tfrac{1}{a}y), \\ v\,(t, x_1, ..., x_m) &= w(x_1 - \tfrac{b_1}{b}y, \ \ldots, \ x_m - \tfrac{b_m}{b}y, \ t + \tfrac{1}{b}y), \end{aligned} \tag{4}$$

where $w(x_1, \ldots, x_m, y)$, with $x_1, \ldots, x_m \in \mathbb{R}$, $y \geq 1/a$, is the solution of the difference equation

$$w(x_1 + c_1, \ldots, x_m + c_m, \ y + d) = f\big(w(x_1, \ldots, x_m, \ y)\big), \tag{5}$$

$c_i = a_i/a - b_i/b$, $i = 1, ..., m$, $d = 1/b - 1/a$. For this difference equation, the initial condition $w_0 = w_0(x_1, \ldots, x_m, y)$, $x_1, \ldots, x_m \in \mathbb{R}$ and $y \in$

$[y_0, y_0 + d)$ with $y_0 = 1/a$, follows from the initial conditions (3) for the problem (1),(2):

$$w_0 = \begin{cases} u_0(x_1 + a_1 y, \dots, x_m + a_m y,\ ay) & if\ \ y \in [1/a, 0], \\ v_0(x_1 + b_1 y, \dots, x_m + b_m y,\ by) & if\ \ y \in (0, 1/b). \end{cases} \tag{6}$$

Hence, the solution of the difference equation (5) with the initial condition $w = w_0$ if $x_1, \dots, x_m \in \mathbb{R},\ y \in [1/a, 1/b)$ can be represented for $y \in [nd + 1/a, nd + 1/b)$ in the form

$$w(x_1, \dots, x_m, y) = f^n\big(w_0(x_1 - nc_1, \dots, x_m - nc_m,\ y - nd)\big). \tag{7}$$

This means that, with (4) and (7), the solution of the problem (1), (2), (3) also is computed via the iterations of the map f. The latter allows us to realize sufficiently deep study of solution properties of the boundary value problems in the form (1), (2).

Consider the (simplest) case $m = 1$, namely, the linear hyperbolic system

$$\begin{cases} \dfrac{\partial u}{\partial t} = -\dfrac{\partial u}{\partial x} - \dfrac{\partial u}{\partial y} \\ \dfrac{\partial v}{\partial t} = \dfrac{\partial v}{\partial x} + \dfrac{\partial v}{\partial y}, & x \in \mathbb{R},\ y \in [0, 1],\ t > 0, \end{cases} \tag{8}$$

under the nonlinear boundary conditions

$$\begin{cases} u|_{y=0} = v|_{y=0} \\ v|_{y=1} = f\left(u|_{y=1}\right), \end{cases} \tag{9}$$

with f a bimodal map of the interval $I = [-1, 1]$, and subject to some initial conditions

$$\begin{cases} u|_{t=0} = u_0(x, y) \\ v|_{t=0} = v_0(x, y). \end{cases} \tag{10}$$

In this case, the solution of (8)-(10), at times $t = 2n$, is given by

$$\begin{cases} u(2n,\, x,\, y) = f^n\big(u_0(x, y)\big) \\ v(2n,\, x,\, y) = f^n\big(v_0(x, y)\big), & n \in \mathbb{N}_0. \end{cases}$$

Suppose now that we assume a vortex as initial condition,

$$\begin{cases} u_0(x, y) = 2\,y - 1 \\ v_0(x, y) = -2\,x, & (x, y) \in \mathfrak{R}_\alpha, \end{cases}$$

defined on a region $\mathfrak{R}_\alpha = \{x^2 + (y - 1/2)^2 = \alpha^2/4, \alpha < 1\}$ of the plane, see Figure 1, and choose $f(x) = -0.21 - 2.4x + 0.21x^2 + 3.4x^3$. The

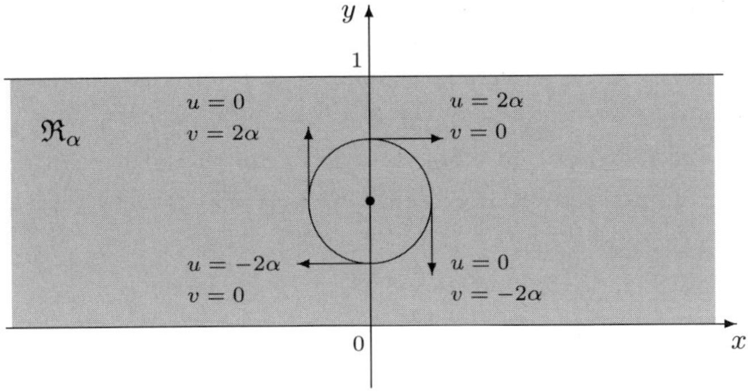

Fig. 1. The inital vortex defined on $\mathfrak{R}_\alpha$.

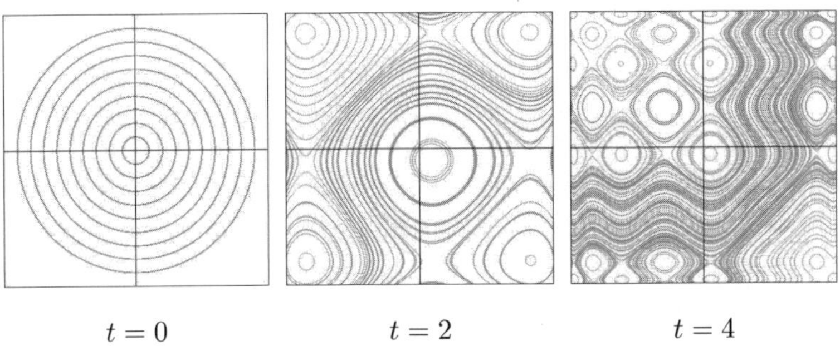

following sequence of pictures gives us snapshots, at times $t = 0, 2, 4$, for the evolution of the initial vortex under the map f. We claim that there is a striking resemblance with turbulence phenomena, e.g., the growth with time of the number of vortices, worth exploring. The object of this paper is to use symbolic dynamics techniques to describe both the growth number of vortices and its spatial distribution of the BVP solution for the family of bimodal maps of the interval $[-1, 1]$,

$$f(x) = f_{ab}(x) = ax^3 + bx^2 + (1 - a)x - b\,, \quad (a, b) \in \Omega,$$

with Ω a suitable region of the plane, see Skjolding et al.[3]

3. A Bimodal Boundary Condition

In this section we consider the class of BVPs described by (8), with a vortex as initial condition, and subject to a boundary condition characterized by a bimodal map of the interval f. The first observation is that the vortices center, at a time $t = 2n$, are related with the zeros of the map f^n.

Let $\zeta_1 < \cdots < \zeta_{2N+1}$ denote the zeros of f^n, for some integer $N > 0$, and assume that $(f^n)'(\zeta_k) \neq 0$, for $k = 1, \ldots, 2N+1$. Then, it is an easy exercise to see that the centers of the vortices of the BVP solution, at time $t = 2n$, occur precisely at the points $P(\eta_{2i+1}, \xi_{2j+1})$ and $P(\eta_{2i}, \xi_{2j})$, for $i, j = 1, \ldots, N$, with

$$\begin{aligned} \eta_k &= -\zeta_k/2, \qquad k = 1, \ldots, 2N+1 \\ \xi_k &= (\zeta_k + 1)/2. \end{aligned}$$

Next, we define the growth number of vortices of a bimodal BVP as

$$\rho(f) = \lim_{k \to +\infty} (N_k)^{1/k},$$

in which N_k denotes its number of vortices, at time $t = 2k$. The following theorem relates $\rho(f)$ with an important dynamical property of the map f.

Theorem 3.1. *The growth number of vortices of a BVP characterized by a bimodal map f of the interval I is equal to its growth number of laps.*

Proof. Since we are going to use symbolic dynamics techniques, consider the following bimodal alphabet: let A and B be the addresses of the critical points of the map, and L, M, R the addresses of every point belonging, respectively, to the subinterval on the left of the first critical point, to the subinterval defined by both critical points, and to the subinterval on the right of the second critical point.

Given the itinerary $\text{it}(x) = S_0 S_1 \cdots$ of a point $x \in I$, we define its n-itinerary as the subsequence of the first $n+1$ symbols, $\text{it}_n(x) = S_0 S_1 \cdots S_n$. Consider now the following symbolic classification of n-itineraries: we say that a n-itinerary $\text{it}_n(x)$ is of Type 1, if none of its symbols are equal to A or B; of Type 2, if S_n is the only of its symbols equal to A or B; and of Type 3, if S_k is equal to A or B, for some $0 \leqslant k \leqslant n-1$.

Let us denote by $\#1(n)$, $\#2(n)$, and $\#3(n)$ the number of n-itineraries of type one, two, and three, respectively. From its definition, we have $\#3(n+1) = \#3(n) + \#2(n)$. Furthermore, there is a precise relation between the number of laps of f^n and Type 3 n-itineraries.

Lemma 3.1. *The number of laps of f^n can be computed from the number of Type 3 n-itineraries, $\ell(f^n) = \#3(n) + 1$.*

For latter use, it is convenient to distinguish the sequences of Type 2 ending with a symbol $S_n = A$, from those ending with $S_n = B$. Let $\#2_A(n)$ denote the number of Type 2 n-itineraries with $S_n = A$, and $\#2_B(n)$ the number of those such that $S_n = B$.

Since not all laps of f^n correspond to zero crossings, the problem is to count the number of positive and negative extrema of f^n. The following two lemmas state that some of that information can be obtain from a subsequence of the n-itineraries.

Lemma 3.2. *Given a Type 3 n-itinerary $\text{it}_n(x) = S_0 \cdots S_n$, let k be the largest integer such that $S_k = A$. Then, x is a maximum of f^n if the parity of the subsequence $S_{k+1} \cdots S_n$ is equal to $+1$, and a minimum otherwise.*

Proof. From the expression for the second derivative of $f^n(x)$,

$$\begin{aligned}(f^n)''(x) = {} & f''(x)\, f'(x^{(1)})\, f'(x^{(2)}) \cdots f'(x^{(n-2)})\, f'(x^{(n-1)}) + \\ & + f'(x)^2 f''(x^{(1)})\, f'(x^{(2)}) \cdots f'(x^{(n-2)})\, f'(x^{(n-1)}) + \\ & + f'(x)^2 f'(x^{(1)})^2 f''(x^{(2)}) \cdots f'(x^{(n-2)})\, f'(x^{(n-1)}) + \ldots \\ & \cdots + f'(x)^2 f'(x^{(1)})^2 f'(x^{(2)}) \cdots f'(x^{(n-2)})^2 f''(x^{(n-1)})\end{aligned}$$

with $x^{(k)} = f^k(x)$, $k = 1, 2, \ldots$, we observe that, if $S_0 \cdots S_n$ has only one symbol $S_k = A$, all terms are null except one, in which case

$$(f^n)''(x) = f'(x)^2 \cdots f'(x^{(k-1)})^2\, f''(x^{(k)})\, f'(x^{(k+1)}) \cdots f'(x^{(n)})\,.$$

Therefore, the signal of $(f^n)''(x)$ is equal to the signal of $f'(x^{(k+1)}) \cdots f'(x^{(n)})$, i.e., the parity of the subsequence $S_{k+1} \cdots S_n$, and the required statement follows easily.

Now observe that, if $S_0 \cdots S_n$ has two symbols $S_{k-m} = S_k = A$, then both the second and the third derivatives of f^n are zero, and that the only term different from zero of its fourth derivative is given by

$$\begin{aligned}(f^n)^{(4)}(x) = 3 f'(x)^4 \cdots f'(x^{(k-m-1)})^4\, f''(x^{(k-m)})^2\, f'(x^{(k-m+1)})^2 \cdots \\ \cdots f'(x^{(k-1)})^2\, f''(x^{(k)})\, f'(x^{(k+1)}) \cdots f'(x^{(n-1)}),\end{aligned}$$

from which we get the same conclusion. By direct computation it is further possible to analyze the situation for any number of symbols A and prove the stated result. □

Similarly, there is an analogous result for n-itineraries with a symbol B.

Lemma 3.3. *Given a Type 3 n-itinerary $\text{it}_n(x) = S_0 \cdots S_n$, let k be the largest integer such that $S_k = B$. Then, x is a minimum of f^n if the parity of the subsequence $S_{k+1} \cdots S_n$ is equal to $+1$, and a maximum otherwise.*

Since after any symbol A or B of an itinerary comes the corresponding kneading sequence of the map f, the lemmas above guarantee that we can get the number of positive and negative extrema, i.e., the number of zero crossings of f^n, from a detailed analysis of the kneading data $\mathcal{K}_f$. However, to avoid some cumbersome technicalities, we will assume now a very simple form for the kneading sequences of f. The generalization for any kneading data, being not easy, follows closely.

Suppose $\mathcal{K}_f = (R^{m-1}A, L^{k-1}B)$, for some integers $m, k > 2$. Then, every Type 3 n-itinerary with $S_n = A$ corresponds to a negative maximum, otherwise it corresponds to a positive maximum. Likewise, every Type 3 n-itinerary with $S_n = B$ corresponds to a positive minimum, otherwise it corresponds to negative minimum. Therefore, the number of zero crossings of f^n, $\#z(n)$, is given by

$$\#z(n) = \#3(n) + 1 - \sum_{n-mj \geqslant 0} \#2_A(n - mj) - \sum_{n-kj \geqslant 0} \#2_B(n - kj).$$

Thus, the following inequality holds

$$\#z(n) > \#3(n) + 1 - \sum_{n-mj \geqslant 0} \#2(n - mj) - \sum_{n-kj \geqslant 0} \#2(n - kj),$$

with $n - kj \neq n - mj$, from which we have

$$\#z(n) > \#3(n) + 1 - \sum_{n-mj \geqslant 0} (\#3(n - mj + 1) - \#3(n - mj)) - \\ - \sum_{n-kj \geqslant 0} (\#3(n - kj + 1) - \#3(n - kj)).$$

Then, since $\#z(n) < \ell(f^n)$, we conclude that the growth number of zero crossings of the map f is equal to the growth number of Type 3 n-itineraries, and thus to the growth number of its laps, as required. □

This theorem gives a generalization to bimodal maps of a previous result by Severino et al [4], stated for unimodal maps of the interval. However, there is a significant difference, since the vortices spatial distribution for a bimodal BVP is no longer necessarily symmetric, as it was for the unimodal class.

4. The Spatial Distribution of Vortices for a Bimodal BVP

Although, at the first sight, it might look like a straightforward generalization from the unimodal situation, the bimodal class of BVPs exhibits spatial patterns of vortices which never occur for the former family of BVPs. Thus, we now turn our attention to the characterization of the vortices' spatial distribution of a bimodal BVP.

Given a BVP associated with a bimodal map f, with critical points c_1 and c_2, let

$$\delta(f) = 1 - \lambda(c_2) - \lambda(c_1),$$

where $\lambda : [0,1] \to [0,1]$ is the map introduced by Milnor and Thurston[5],

$$\lambda(x) = \lim_{t \to 1/s} \frac{L([-1,x],t)}{L([-1,1],t)},$$

with s the growth number of laps of f, and $L(J,t)$ is given by

$$L(J,t) = \sum_{n=1}^{\infty} \ell(f^n|J)\, t^{n-1}, \quad J \subset [-1,1].$$

Now, it is important to make a few observations about this quantity: first, it is obvious that, for any antisymmetric bimodal map f, we have $\delta(f) = 0$. Therefore, $\delta(f)$ measures the deviation of the map f from the antisymmetric situation. Second, it uses the values of λ at c_1 and c_2 because our symbolic methods are best suitable to deal with data taken from the critical points of the map. Finally, introducing $I_1 = [-1, c_1]$ and $I_2 = [c_2, 1]$, we find that $\delta(f) = \Delta_f(I_2) - \Delta_f(I_1)$ is the difference between

$$\Delta_f(I_2) = \lim_{t \to 1/s} \frac{L([c_2,1],t)}{L([-1,1],t)}$$

and

$$\Delta_f(I_1) = \lim_{t \to 1/s} \frac{L([-1,c_1],t)}{L([-1,1],t)}.$$

As pointed out by Milnor and Thurston,[5] $\Delta_f(I_2)$ and $\Delta_f(I_1)$ can be regarded as the probability of a randomly chosen lap of the map f^n, for a large value of n, is totally contained in I_2 and I_1, respectively. But, since the arguments that allow us to proof Theorem 3.1 can be use to state that the growth number of zeros of f in I_2 and I_1 are equal to the growth number of laps of f contained in these subintervals, we can conclude that the proposed quantity $\delta(f)$ measures the difference of probabilities of a randomly chosen zero of f^n, for large n, to be in I_2 and in I_1. Therefore, $\delta(f)$ give us a asymptotical measure of the difference of the percentages of vortices of the BVP in the two regions $\mathcal{R}_1 = I_1 \times I_1$ and $\mathcal{R}_2 = I_2 \times I_2$. In the following figure we pictured these two regions and, schematically, identified the position of the vortices for the bimodal map considered and for a time $t = 4$ (compare it with the picture given from the numerical solution of the BVP). The next theorem gives an expression for $\delta(f)$ in terms of the growth

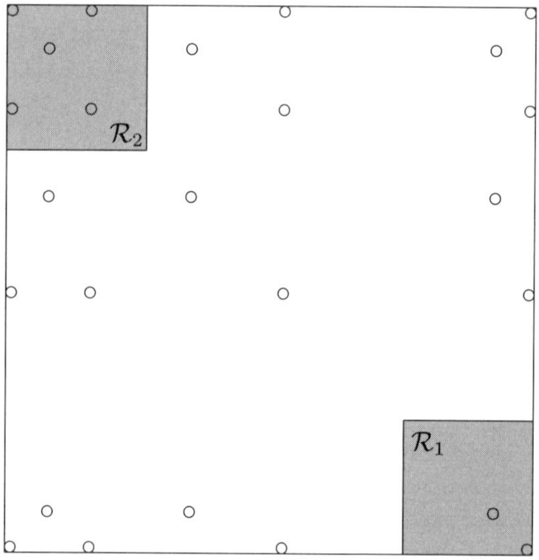

number and a second topological invariant introduced by Sousa Ramos, see Almeida et al[6], for bimodal maps.

Theorem 4.1. *For a BVP associated with a bimodal map f, we have* $\delta(f) = (3s + 2r - 3)/(2s)$.

Proof. It is an easy consequence of the definition of $\delta(f)$ and the bimodal second topological invariant $r = r(f)$. □

5. Conclusions

From the results shown above, Theorem 3.1 and Theorem 4.1, we conclude that the topological classification of the family of bimodal maps establishes a classification of the space-time evolution of the bimodal class of BVPs with an initial vortex.

References

1. A. N. Sharkovsky, Yu. L. Maistrenko, E. Yu. Romanenko, *Difference Equations and Their Applications* (Naukova Dumka, 1986 (Russian); Kluwer Academic Publishers, 1993).
2. V. L. Maistrenko, *On continuous dependence of solutions of n-dimensional boundary value problems on singular perturbations of nonlinear boundary conditions*, Tech. Rep. 86.22, Inst. Math., Acad. Sci. of Ukr.SSR, Kiev (1986).

3. H. Skjolding, B. Branner-Jorgensen, P. L. Christiansen and H. Jensen, *SIAM Journal of Applied Mathematics* **43**, 520 (1985).
4. R. Severino, A. Sharkovsky, J. Sousa Ramos and S. Vinagre, *Grazer Mathematische Berichte* **350**, 210 (2006).
5. J. Milnor and W. Thurston, *On iterated maps of the interval*, in *Dynamical Systems*, ed. J. C. Alexander, Lecture Notes in Math., Vol. 1342 (Springer,1988), pp. 465–563.
6. P. Almeida, J. P. Lampreia and J. Sousa Ramos, *Topological invariants for bimodal maps*, in *Iteration Theory (Batschuns, 1992)*, (World Scientific, 1996), pp. 1–8.

Conservation Laws of Several (2+1)-Dimensional Differential-Difference Hierarchies

Zuo-Nong Zhu

Department of Mathematics, Shanghai Jiao Tong University,
800 Dongchuan Road, Shanghai, 200240, P R China

In this paper, by solving the (2+1)-dimensional discrete spectral equations, we demonstrate the existence of infinitely many conservation laws for several (2+1)-dimensional differential-difference hierarchies and obtain the corresponding conserved densities and associated fluxes.

1. Introduction

In recent years there has been wide interest in the study of nonlinear integrable differential-difference systems. It is well known that integrable differential-difference systems not only possess rich mathematical structures, such as the Lax pairs, the Hamiltonian structures, infinitely many conservation laws, the Bäcklund transformation, and soliton solutions, but also have applications in many fields, such as mathematical physics, numerical analysis, statistical physics, quantum physics, etc. It is known that demonstrating the existence of infinitely many conservation laws for a discrete system and deriving the corresponding conserved densities and the associated fluxes are very interesting and important. There are several motives to find conservation laws of a discrete system. First, the existence of infinitely many conservation laws is an important indicator of integrability of the discrete system. The second, from physical view, it is important to know whether there exist conservation laws for a discrete system, since the first few conservation laws usually have physical meaning, such as the conserved momentum and energy. The third, the conservation laws are also very useful to numerical analysis for a discrete system. This is due to the fact that we hope the solution to a difference model satisfies the corresponding conservation laws.

By solving the corresponding discrete spectral problem, infinitely many conservation laws for a number of (1+1)-dimensoional differential-difference hierarchies have been given[1−3]. In this paper, our aim is to demonstrate the existence of infinitely many conservation laws for several (2+1)-dimensional differential-difference hierarchies and to obtain the corresponding conserved densities and associated fluxes by solving the (2+1)-dimensional discrete spectral equations.

2. Conservation laws for several (2+1)-dimensional differential-difference hierarchies

For a (2+1)-dimensional differential-difference system

$$F\left(\frac{\partial q_n}{\partial t}, \frac{\partial^2 q_n}{\partial t^2}, \frac{\partial q_n}{\partial y}, \frac{\partial^2 q_n}{\partial y^2}, \ldots, q_{n-1}, q_n, q_{n+1}, \ldots\right) = 0, \tag{1}$$

where $q_n = q(n,t,y)$ (with n a discrete variable, and t and y continuous variable), if there exist functions ρ_n, η_n and J_n such that

$$\left(\frac{\partial \rho_n}{\partial t} + \frac{\partial \eta_n}{\partial y}\right)|_{F=0} = (E-1)J_n, \tag{2}$$

then equation (2) is called the conservation law of equation (1), with ρ_n, η_n being the conserved density and J_n the associated flux. In this section, by solving the (2+1)-dimensional discrete spectral equation, we will demonstrate the existence of infinitely many conservation laws for (2+1)-dimensional Benjamin-Ono (BO) lattice hierarchy, (2+1)- dimensional Blaszak-Szum (BS) three-field lattice hierarchy, and (2+1)-dimensional BS four-field lattice hierarchy. The corresponding conserved quantities, conserved densities and the associated fluxes will be given.

(1) The discrete spectral problems for several (2+1)-dimensional differential-difference hierarchies

In order to derive the conservation laws, let us first give the discrete spectral problems for (2+1)-dimensional BO lattice hierarchy, (2+1)- dimensional BS three-field lattice hierarchy, and (2+1)-dimensional BS four-field lattice hierarchy. The three differential-difference equation hierarchies admit the following (2+1)-dimensional discrete isospectral problem:

$$E\psi(n,t,y,\lambda) = U(u(n,t,y), \lambda + \partial_y)\psi(n,t,y,\lambda), \tag{3}$$

$$\frac{d\psi(n,t,y,\lambda)}{dt} = V(u(n,t,y), \lambda + \partial_y)\psi(n,t,y,\lambda), \tag{4}$$

where E is a shift operator defined by $Ef_n = f_{n+1}$. For the (2+1)-dimensional BO lattice hierarchy, U and V take the form

$$U = \lambda + \partial_y - u_n, \tag{5}$$

$$V = \sum_{l=0}^{m} a_{l,m}(\lambda + \partial_y)^{m-l}; \tag{6}$$

for the (2+1)-dimensional BS three-field lattice hierarchy, the matrix U and $V = (V_{i,j})_{3\times 3}$ are given by

$$U = \begin{pmatrix} 0 & 1 & 0 \\ \lambda - v_{n-1} + \partial_y & -p_{n-1} & -u_{n-1} \\ 1 & 0 & 0 \end{pmatrix} \tag{7}$$

$$V_{31} = a = \sum_{j=0}^{m} a_{j,m}(\lambda + \partial_y)^{m-j}, \qquad V_{32} = b = \sum_{j=0}^{m} b_{j,m}(\lambda + \partial_y)^{m-j},$$

$$V_{33} = c = \sum_{j=0}^{m} c_{j,m}(\lambda + \partial_y)^{m-j},$$

$$V_{11} = (Eb)(\lambda + \partial_y - v_{n-1}) + Ec, V_{12} = Ea - (Eb)p_{n-1}, V_{13} = -(Eb)u_{n-1},$$

$$V_{21} = \Big(E^2a - (E^2b)p_n\Big)(\lambda + \partial_y - v_{n-1}) - (E^2b)u_n,$$

$$V_{22} = (E^2b)(\lambda + \partial_y - v_n + p_{n-1}p_n) + E^2c - (E^2a)p_{n-1},$$

$$V_{23} = \Big((E^2b)p_n - E^2a\Big)u_{n-1};$$

for the (2+1)-dimensional BS four-field lattice hierarchy, the matrix U and $V = (V_{i,j})_{4\times 4}$ are written as

$$U = \begin{pmatrix} 0 & 1 & 0 & 0 \\ 0 & 0 & 1 & 0 \\ \lambda - u_{n-2} + \partial_y & -v_{n-2} & -p_{n-2} & -q_{n-2} \\ 1 & 0 & 0 & 0 \end{pmatrix}, \tag{8}$$

and

$$
\begin{aligned}
V_{41} &= a, V_{42} = b, V_{43} = c, V_{44} = d,\\
V_{11} &= (Ec)(\lambda + \partial_y - u_{n-2}) + Ed, V_{12} = Ea - (Ec)v_{n-2}, V_{13} = Eb - (Ec)p_{n-2},\\
V_{14} &= -(Ec)q_{n-2}, V_{21} = \left(E^2 b - (E^2 c)p_{n-1}\right)(\lambda + \partial_y - u_{n-2}) - (E^2 c)q_{n-1},\\
V_{22} &= (E^2 c)(\lambda + \partial_y - u_{n-1} + p_{n-1}v_{n-2}) + E^2 d - (E^2 b)v_{n-2},\\
V_{23} &= E^2 a - (E^2 c)(v_{n-1} - p_{n-1}p_{n-2}) - (E^2 b)p_{n-2},\\
V_{24} &= (E^2 c)(p_{n-1}q_{n-2}) - (E^2 b)q_{n-2},\\
V_{31} &= \left[E^3 a - (E^3 c)(v_n - p_{n-1}p_n) - (E^3 b)p_{n-1}\right](\lambda + \partial_y - u_{n-2})\\
&\quad -(E^3 b)q_{n-1} + (E^3 c)(p_n q_{n-1}),\\
V_{32} &= \left(E^3 b - (E^3 c)p_n\right)(\lambda + \partial_y - u_{n-1})\\
&\quad -(E^3 c)(q_n - v_n v_{n-2} + p_n p_{n-1} v_{n-2}) - (E^3 a)v_{n-2} + (E^3 b)(p_{n-1}v_{n-2}),\\
V_{33} &= (E^3 c)(\lambda + \partial_y - u_n + p_n v_{n-1} + v_n p_{n-2} - p_n p_{n-1} p_{n-2})\\
&\quad + E^3 d - (E^3 a)p_{n-2} + (E^3 b)(p_{n-1}p_{n-2} - v_{n-1}),\\
V_{34} &= -(E^3 a)q_{n-2} + (E^3 c)(v_n q_{n-2} - p_n p_{n-1} q_{n-2}) + (E^3 b)(p_{n-1}q_{n-2}).
\end{aligned}
$$

The integrability condition of the spectral problem (3-4) leads to the following discrete operator zero curvature equation:

$$\frac{\partial U}{\partial t}\phi = [(EV)U - UV]\phi, \tag{9}$$

where ϕ is an arbitrary function. By using the discrete operator zero curvature equation and taking the above U and V, the (2+1)-dimensional BO lattice hierarchy, the (2+1)-dimensional BS three-field lattice hierarchy, and the (2+1)-dimensional BS four-field lattice hierarchy are derived respectively[4,5]. Here we write down the first equations for the three (2+1)-dimensional differential-difference hierarchies. The first equations for the (2+1)-dimensional BO hierarchy are, respectively,

$$u_{n,t_2} = u_{n,y} + 2u_n u_{n,y} + H u_{n,yy}, \tag{10}$$

$$
\begin{aligned}
u_{n,t_3} &= u_{n,yyy} + u_{n,yy} + u_{n,y}(3u_n^2 + 2u_n + 3(E-1)^{-1}u_{n+1,y})\\
&\quad + 3u_n(E-1)^{-1}u_{n+1,yy} + (E-1)^{-1}(3(u_n u_{n,y})_y + 2u_{n,yy})\\
&\quad + 3(E-1)^{-2}u_{n+1,yyy}
\end{aligned} \tag{11}
$$

where the operator $H = (E+1)(E-1)^{-1}$. The term V associated with the two equations is, respectively,

$$V_2 = (\lambda + \partial_y)^2 + \lambda + \partial_y + 2(E-1)^{-1}u_{n,y}, \tag{12}$$

$$\begin{aligned} V_3 = {} & (\lambda + \partial_y)^3 + (\lambda + \partial_y)^2 + 3(E-1)^{-1}u_{n,y}(\lambda + \partial_y) \\ & + 3(E-1)^{-2}u_{n+1,yy} + (E-1)^{-1}(3u_n u_{n,y} + 2u_{n,y}) \end{aligned} \tag{13}$$

It is interesting to note that equation (10) yields the discrete KP equation[6–8]

$$(E-1)(q_{n,t} + q_{n,y} - 2q_n q_{n,y}) = (E+1)q_{n,yy}, \tag{14}$$

where we have set $q_n = u_n + 1$. The first equations for the (2+1)-dimensional BS three-field hierarchies are given by the following equations:

$$\begin{pmatrix} u_n \\ v_n \\ p_n \end{pmatrix}_{t_0} = \begin{pmatrix} u_n H^{-1} p_{n-1} \\ u_{n+1} - u_n + (E+1)^{-1} p_{n,y} \\ v_{n+1} - v_n - p_n H^{-1} p_n \end{pmatrix}, \tag{15}$$

$$\begin{pmatrix} u_n \\ v_n \\ p_n \end{pmatrix}_{t_0} = \begin{pmatrix} u_n(v_n - v_{n-1}) \\ p_n u_{n+1} - p_{n-1}u_n + v_{n,y} \\ u_{n+2} - u_n + p_{n,y} \end{pmatrix}. \tag{16}$$

The corresponding matrix V of the two equations is, respectively,

$$V = \begin{pmatrix} (E+1)^{-1}p_{n-1} & 1 & 0 \\ \lambda - v_{n-1} + \partial_y & -(E+1)^{-1}p_{n-1} & -u_{n-1} \\ 1 & 0 & (E+1)^{-1}p_{n-2} \end{pmatrix} \tag{17}$$

$$V = \begin{pmatrix} \lambda + \partial_y & 0 & -u_{n-1} \\ -u_n & \lambda + \partial_y & 0 \\ p_{n-2} & 1 & v_{n-2} \end{pmatrix} \tag{18}$$

The first equations for the (2+1)-dimensional BS four-field hierarchies are, respectively,

$$\begin{pmatrix} u_n \\ v_n \\ p_n \\ q_n \end{pmatrix}_{t_0} = \begin{pmatrix} q_{n+1} - q_n + Ap_{n,y} \\ u_{n+1} - u_n + v_n(1-E)Ap_n \\ v_{n+1} - v_n - p_n(E^2 - 1)Ap_n \\ q_n(E-1)Ap_{n-1} \end{pmatrix}, \tag{19}$$

$$\begin{pmatrix} u_n \\ v_n \\ p_n \\ q_n \end{pmatrix}_{t_0} = \begin{pmatrix} (E-1)[q_n A(p_n + p_{n-1})] + \partial_y[A(v_n + v_{n+1} - p_n A p_{n+1})] \\ q_{n+2} - q_n + (u_{n+1} - u_n) A (E+1) p_n \\ -v_n (E-1) A [v_{n+1} + v_n - p_n A p_{n+1}] + A(p_n + p_{n+1})_y \\ u_{n+2} - u_n - v_{n+1} A p_{n+2} + v_n A p_n \\ -p_n A (v_{n+2} - v_{n+1}) + p_n (E^2 - 1) A (p_n A p_{n+1}) \\ q_n (E-1) A (v_n + v_{n-1} - p_{n-1} A p_n) \end{pmatrix} \tag{20}$$

$$\begin{pmatrix} u_n \\ v_n \\ p_n \\ q_n \end{pmatrix}_{t_0} = \begin{pmatrix} v_n q_{n+1} - v_{n-1} q_n + u_{n,y} \\ p_n q_{n+2} - p_{n-1} q_n + v_{n,y} \\ q_{n+3} - q_n + p_{n,y} \\ q_n (u_n - u_{n-1}) \end{pmatrix} \tag{21}$$

Here the operator A is defined by $A = (E^2+E+1)^{-1} = \sum_{k=0}^{\infty}(E^{3k} - E^{3k+1})$. The matrix V associated with the three equations are given, respectively,

$$V = \begin{pmatrix} A p_{n-2} & 1 & 0 & 0 \\ 0 & A p_{n-1} & 1 & 0 \\ \lambda + \partial_y - u_{n-2} & -v_{n-2} & -A(p_{n-1} + p_{n-2}) & -q_{n-2} \\ 1 & 0 & 0 & A p_{n-3} \end{pmatrix} \tag{22}$$

$$V = \begin{pmatrix} A(v_{n-2} + v_{n-1} - p_{n-2} A p_{n-1}) & A(p_{n-1} + p_{n-2}) & 1 & 0 \\ \lambda + \partial_y - u_{n-2} & -A(v_{n-2} + p_{n-1} A p_n) & -A p_{n-2} & -q_{n-2} \\ V_{31} & V_{32} & V_{33} & V_{34} \\ A(p_{n-2} + p_{n-3}) & 1 & 0 & A(v_{n-3} + v_{n-2} - p_{n-3} A p_{n-2}) \end{pmatrix} \tag{23}$$

where

$$V_{31} = (A p_{n-1})(u_{n-2} - \lambda - \partial_y) - q_{n-1}, V_{32} = \lambda + \partial_y - u_{n-1} + v_{n-2} A p_{n-1}$$

$$V_{33} = p_{n-2} A p_{n-1} - A(v_{n-1} + p_n A p_{n+1}), \qquad V_{34} = q_{n-2} A p_{n-1}$$

$$V = \begin{pmatrix} \lambda + \partial_y & 0 & 0 & -q_{n-2} \\ -q_{n-1} & \lambda + \partial_y & 0 & 0 \\ 0 & -q_n & \lambda + \partial_y & 0 \\ v_{n-3} & p_{n-3} & 1 & u_{n-3} \end{pmatrix} \tag{24}$$

We remark here that the (2+1)-dimensional BS differential-difference equations are the generalizations of the (1+1)-dimensional differential-difference equations constructed by Blaszak and Marciniak[9]. Equation (15) is also viewed as a two-dimensional generalization of a differential-difference equation[10].

(2) Conservation laws for the (2+1)-dimensional BO lattice hierarchy

Here we first suppose the eigenfunctions ψ_n is an analytical function of the arguments. The (2+1)-dimensional BO lattice equations correspond to the discrete spectral problem

$$\frac{\psi_{n+1}}{\psi_n} = \lambda - u_n + \partial_y ln\psi_n. \tag{25}$$

Let $\Gamma_n = \frac{\psi_{n+1}}{\psi_n}$, the spectral problem leads to the following discrete Riccati equation:

$$\Gamma_n\Gamma_{n+1} - \Gamma_n^2 + (u_{n+1} - u_n)\Gamma_n = \frac{\partial \Gamma_n}{\partial y} \tag{26}$$

We can give a series solution to the equation. Expand Γ_n with respect to λ by the Laurent series

$$\Gamma_n = \sum_{j=1}^{\infty} \lambda^{-j} w_n^{(j)}, \tag{27}$$

and substitute it into equation (26), we obtain

$$\begin{aligned}
w_n^{(1)} &= exp(\int_0^y (u_{n+1} - u_n)dy),\\
w_n^{(2)} &= w_n^{(1)}[1 + \int_0^y (E-1)exp(\int_0^y (u_{n+1} - u_n)dy)dy],\\
w_n^{(i)} &= w_n^{(1)}[1 + \int_0^y exp(\int_0^y (u_n - u_{n+1})dy) \sum_{l+s=i} w_n^{(l)}(w_{n+1}^{(s)} - w_n^{(s)})dy], i \geq 3.
\end{aligned}$$

On the other hand, we have

$$\frac{\partial}{\partial t}\ln\Gamma_n = (E-1)\frac{\partial}{\partial t}\ln\psi_n. \tag{28}$$

For equations (10) and (11), $\frac{\partial}{\partial t} ln\psi_n$ is described by the following equations, respectively,

$$\frac{\partial}{\partial t}\ln\psi_n = u_n + u_n^2 + Hu_{n,y} + (2u_n + 1)\Gamma_n + \Gamma_n^2 + \frac{\partial \Gamma_n}{\partial y} \tag{29}$$

and

$$\frac{\partial}{\partial t}\ln\psi_n = u_n^3 + u_n^2 + Hu_{n,y} + u_{n,yy} + 3u_n(E-1)^{-1}u_{n+1,y}$$
$$+ 3(E-1)^{-2}u_{n+1,yy} + 3(E-1)^{-1}(u_n u_{n,y})$$
$$+ \Gamma_n^3 + (3u_n+1)\Gamma_n^2 + (3u_n^2 + 2u_n + 3(E-1)^{-1}u_{n,y})\Gamma_n$$
$$+ \frac{\partial}{\partial y}\left[\Gamma_n\Gamma_{n+1} + \frac{1}{2}\Gamma_n^2 + (u_{n+1} + 2u_n + 1)\Gamma_n\right] \tag{30}$$

It follows from equation (28) that

$$\frac{\partial}{\partial t}ln\Gamma_n + \frac{\partial}{\partial y}(1-E)\Gamma_n = (E-1)(u_n + u_n^2 + Hu_{n,y} + (2u_n+1)\Gamma_n + \Gamma_n^2) \tag{31}$$

and

$$\frac{\partial}{\partial t}ln\Gamma_n + \frac{\partial}{\partial y}(1-E)[\Gamma_n\Gamma_{n+1} + \frac{1}{2}\Gamma_n^2 + (u_{n+1} + 2u_n + 1)\Gamma_n]$$
$$= (E-1)[u_n^3 + u_n^2 + Hu_{n,y} + u_{n,yy} + 3u_n(E-1)^{-1}u_{n+1,y}$$
$$+ 3(E-1)^{-2}u_{n+1,yy} + 3(E-1)^{-1}(u_n u_{n,y})$$
$$+ \Gamma_n^3 + (3u_n+1)\Gamma_n^2 + (3u_n^2 + 2u_n + 3(E-1)^{-1}u_{n,y})\Gamma_n]. \tag{32}$$

Noting that

$$\frac{\partial}{\partial t}ln\Gamma_n = \frac{\partial}{\partial t}\ln w_n^{(1)} + \frac{\partial}{\partial t}(\sum_{k=1}^{\infty}\frac{(-1)^{k+1}\lambda^{-k}}{k}\Phi^k), \tag{33}$$

where

$$\Phi = \sum_{j=0}^{\infty}\lambda^{-j}s_j, \qquad s_j = \bar{w}_n^{(j+2)} = \frac{w_n^{(j+2)}}{w_n^{(1)}}, \tag{34}$$

and making a comparison of the powers of λ on both sides of equations (31) and (32), we obtain infinitely many conservation laws of the 2+1 dimension BO lattice equations (10) and (11),

$$\frac{\partial}{\partial t}\rho_n^{(j)} + \frac{\partial}{\partial y}\beta_n^{(j)} = (E-1)J_n^{(j)}, \qquad j = 0, 1, 2, ... \tag{35}$$

For equation (10), the conserved densities and the associated fluxes are

$$\rho_n^{(0)} = \int_0^y (u_{n+1} - u_n)dy, \qquad \beta_n^{(0)} = 0, \qquad J_n^{(0)} = u_n + u_n^2 + Hu_{n,y},$$
$$\rho_n^{(1)} = \bar{w}_n^{(2)}, \qquad \beta_n^{(1)} = (1-E)w_n^{(1)}, \qquad J_n^{(1)} = (2u_n+1)w_n^{(1)} \tag{36}$$

and

$$\rho_n^{(j)} = s_{j-1} - \frac{1}{2}\sum_{l_1+l_2=j-2} s_{l_1}s_{l_2} + \frac{1}{3}\sum_{l_1+l_2+l_3=j-3} s_{l_1}s_{l_2}s_{l_3} - \ldots\ldots$$
$$+\frac{(-1)^{j-1}}{j-2}\sum_{l_1+l_2+\ldots+l_{j-2}=2} s_{l_1}s_{l_2}\ldots s_{l_{j-2}} + (-1)^j s_0^{j-2}s_1 + \frac{(-1)^{j+1}}{j}s_0^j, \tag{37}$$

$$\beta_n^{(j)} = (1-E)w_n^{(j)}, \quad J_n^{(j)} = (2u_n+1)w_n^{(j)} + \sum_{l+s=j} w_n^{(l)}w_n^{(s)}, \quad j \geq 2 \tag{38}$$

For equation (11), the conserved densities and the associated fluxes can be written as

$$\rho_n^{(0)} = \int_0^y (u_{n+1}-u_n)dy, \qquad \beta_n^{(0)} = 0,$$
$$J_n^{(0)} = u_n^3 + u_n^2 + Hu_{n,y} + u_{n,yy} + 3u_n(E-1)^{-1}u_{n+1,y}$$
$$+3(E-1)^{-2}u_{n+1,yy} + 3(E-1)^{-1}(u_nu_{n,y})$$
$$\rho_n^{(1)} = \bar{w}_n^{(2)}, \qquad \beta_n^{(1)} = (1-E)[(1+2u_n+u_{n+1})w_n^{(1)}],$$
$$J_n^{(1)} = (3u_n^2 + 2u_n + 3(E-1)^{-1}u_{n,y})w_n^{(1)}, \tag{39}$$

and $\rho_n^{(j)}$, $\beta_n^{(j)}$ and $J_n^{(j)}$ $(j \geq 2)$ are presented by equation (37) and the following equations respectively,

$$\beta_n^{(j)} = (1-E)[(1+2u_n+u_{n+1})w_n^{(j)} + \sum_{l+s=j} w_n^{(l)}(\frac{w_n^{(s)}}{2} + w_{n+1}^{(s)})]$$
$$J_n^{(j)} = (3u_n^2 + 2u_n + 3(E-1)^{-1}u_{n,y})w_n^{(j)}$$
$$+ (3u_n+1)\sum_{l+s=j} w_n^{(l)}w_n^{(s)} + \sum_{l+s+m=j} w_n^{(l)}w_n^{(s)}w_n^{(m)}, \quad j \geq 2 \tag{40}$$

It is interesting to note that equations (35)-(38) with $q_n = u_n + 1$ yield infinitely many conservation laws of the discrete KP equation.

(3) Conservation laws for the (2+1)-dimensional BS three lattice-field equations

The (2+1)-dimensional BS three lattice-field equations (15) and (16) relate to the discrete spectral problem

$$\psi_{n+1} + p_{n-1}\psi_n + u_{n-1}\psi_{n-2} = (\lambda - v_{n-1} + \partial_y)\psi_{n-1}, \tag{41}$$

which leads to the following discrete Riccati-type equation:

$$\Gamma_{n-1}\Gamma_n\Gamma_{n+1}(\Gamma_{n+2}-\Gamma_n)+\Gamma_{n-1}\Gamma_n(p_{n+1}\Gamma_{n+1}-p_n\Gamma_n)$$
$$+\Gamma_{n-1}\Gamma_n(v_{n+1}-v_n)+u_{n+1}\Gamma_{n-1}-u_n\Gamma_n=\Gamma_{n-1}\frac{\partial\Gamma_n}{\partial y}, \quad (42)$$

The discrete Riccati-type equation also has the solution given by the Laurent series (27) with

$$w_n^{(1)}=u_{n+1}, \qquad w_n^{(2)}=u_{n+1}[v_{n+1}-(E-1)^{-1}\partial_y\ln u_{n+2}],$$
$$w_n^{(i)}=-u_{n+1}(E-1)^{-1}[\frac{C(n)}{u_{n+1}u_{n+2}}]$$

where

$$C(n)=\sum_{l+s=i}w_n^{(l)}w_{n+1,y}^{(s)}+(v_{n+1}-v_{n+2})\sum_{l+s=i}w_n^{(l)}w_{n+1}^{(s)}$$
$$+p_{n+1}\sum_{l+s+m=i}w_n^{(l)}w_{n+1}^{(s)}w_{n+1}^{(m)}-p_{n+2}\sum_{l+s+m=i}w_n^{(l)}w_{n+1}^{(s)}w_{n+2}^{(m)}$$
$$-\sum_{l+s+m+r=i}w_n^{(l)}w_{n+1}^{(s)}w_{n+2}^{(m)}(w_{n+3}^{(r)}-w_{n+1}^{(r)}), i=3,4,......$$

Note that $\frac{\partial}{\partial t}\ln\psi_n$ for equations (15) and (16) are written as, respectively,

$$\frac{\partial}{\partial t}\ln\psi_n=\Gamma_n+(E+1)^{-1}p_n, \quad (43)$$

$$\frac{\partial}{\partial t}\ln\psi_n=\Gamma_n\Gamma_{n+1}+p_n\Gamma_n+v_n-\lambda, \quad (44)$$

we thus obtain the following two discrete conserved equations:

$$\frac{\partial}{\partial t}ln\Gamma_n=(E-1)(\Gamma_n+(E+1)^{-1}p_n), \quad (45)$$

$$\frac{\partial}{\partial t}ln\Gamma_n=(E-1)(\Gamma_n\Gamma_{n+1}+p_n\Gamma_n+v_n). \quad (46)$$

From the above two equations, we obtain infinitely many conservation laws of the (2+1)- dimensional BS three lattice-field equations (15) and (16),

$$\frac{\partial}{\partial t}\rho_n^{(j)}=(E-1)J_n^{(j)}, \qquad j=0,1,2,..... \quad (47)$$

where the conserved densities have the form

$$\rho_n^{(0)}=lnu_{n+1}, \qquad \rho_n^{(1)}=v_{n+1}-(E-1)^{-1}\partial_y lnu_{n+2},$$

and $\rho_n^{(j)}, j \geq 2$ is described by equation (37). The associated fluxes are given, respectively,

$$J_n^{(0)} = (E+1)^{-1} p_n, \quad J_n^{(1)} = u_{n+1}, \quad J_n^{(j)} = w_n^{(j)}, \quad j \geq 2 \tag{48}$$

and

$$J_n^{(0)} = v_n, \quad J_n^{(1)} = p_n u_{n+1}, \quad J_n^{(j)} = p_n w_n^{(j)} + \sum_{l+s=j} w_n^{(l)} w_{n+1}^{(s)}. \tag{49}$$

(4) Conservation laws for the (2+1)-dimensional BS four lattice-field equations

The discrete spectral problem corresponding to the (2+1)- dimensional BS four lattice-field equations (19)-(21) reads

$$\psi_{n+1} + p_{n-2}\psi_n + v_{n-2}\psi_{n-1} + q_{n-2}\psi_{n-3} = (\lambda - u_{n-2} + \partial_y)\psi_{n-2}, \tag{50}$$

which leads to the discrete Riccati-type equation,

$$\begin{aligned} &\Gamma_{n-1}\Gamma_n\Gamma_{n+1}\Gamma_{n+2}(\Gamma_{n+3} - \Gamma_n + p_{n+1}) + \Gamma_{n-1}\Gamma_n\Gamma_{n+1}(v_{n+1} - p_n\Gamma_n) \\ &\quad + \Gamma_{n-1}\Gamma_n(u_{n+1} - u_n - v_n\Gamma_n) + q_{n+1}\Gamma_{n-1} - q_n\Gamma_n = \Gamma_{n-1}\frac{\partial \Gamma_n}{\partial y}. \end{aligned} \tag{51}$$

The Laurent series (27) solves the last equation, where

$$\begin{aligned} &w_n^{(1)} = q_{n+1}, \qquad w_n^{(2)} = q_{n+1}[u_{n+1} - (E-1)^{-1}\partial_y \ln q_{n+2}], \\ &w_n^{(i)} = -q_{n+1}(E-1)^{-1}\left[\frac{D(n)}{q_{n+1}q_{n+2}}\right], \qquad i \geq 3 \end{aligned} \tag{52}$$

with

$$\begin{aligned} D(n) = &\sum_{l+s=i} w_n^{(l)} w_{n+1,y}^{(s)} + (u_{n+1} - u_{n+2}) \sum_{l+s=i} w_n^{(l)} w_{n+1}^{(s)} \\ &+ (v_{n+1} - v_{n+2}) \sum_{l+s+m=i} w_n^{(l)} w_{n+1}^{(s)} (w_{n+1}^{(m)} + w_{n+2}^{(m)}) \\ &+ (p_{n+1} - p_{n+2}) \sum_{l+s+m+\mu=i} w_n^{(l)} w_{n+1}^{(s)} w_{n+2}^{(m)} (w_{n+1}^{(\mu)} + w_{n+3}^{(\mu)}) \\ &+ \sum_{l+s+m+\mu+\gamma=i} w_n^{(l)} w_{n+1}^{(s)} w_{n+2}^{(m)} w_{n+3}^{(\mu)} (w_{n+1}^{(\gamma)} - w_{n+4}^{(\gamma)}). \end{aligned}$$

Note that $\frac{\partial}{\partial t}\ln\psi_n$ for equations (19)-(21) are written as, respectively,

$$\frac{\partial}{\partial t}\ln\psi_n = \Gamma_n + (E^2+E+1)^{-1}p_n, \qquad (53)$$

$$\frac{\partial}{\partial t}\ln\psi_n = \Gamma_n\Gamma_{n+1} + \Gamma_n(E^2+E+1)^{-1}(p_{n+1}+p_n) + (E^2+E+1)^{-1}[v_{n+1}+v_n - p_n(E^2+E+1)^{-1}p_{n+1}], \qquad (54)$$

$$\frac{\partial}{\partial t}\ln\psi_n = u_n + v_n\Gamma_n + p_n\Gamma_n\Gamma_{n+1} + \Gamma_n\Gamma_{n+1}\Gamma_{n+2}, \qquad (55)$$

we have the following three discrete conserved equations:

$$\frac{\partial}{\partial t}ln\Gamma_n = (E-1)\left[\Gamma_n + (E^2+E+1)^{-1}p_n\right] \qquad (56)$$

$$\frac{\partial}{\partial t}ln\Gamma_n = (E-1)[\Gamma_n\Gamma_{n+1} + \Gamma_n(E^2+E+1)^{-1}(p_{n+1}+p_n) + (E^2+E+1)^{-1}(v_{n+1}+v_n - p_n(E^2+E+1)^{-1}p_{n+1})], \qquad (57)$$

$$\frac{\partial}{\partial t}ln\Gamma_n = (E-1)\left(u_n + v_n\Gamma_n + p_n\Gamma_n\Gamma_{n+1} + \Gamma_n\Gamma_{n+1}\Gamma_{n+2}\right). \qquad (58)$$

The above equations yield infinitely many conservation laws of the (2+1)-dimensional BS four lattice-field equations (19)-(21), respectively,

$$\frac{\partial}{\partial t}\rho_n^{(j)} = (E-1)J_n^{(j)}. \qquad j = 0,1,2,..... \qquad (59)$$

where

$$\rho_n^{(0)} = lnq_{n+1}, \qquad \rho_n^{(1)} = u_{n+1} - (E-1)^{-1}\partial_y \ln q_{n+2}, \qquad (60)$$

and $\rho_n^{(j)}, j \geq 2$ is given by equation (37). The corresponding fluxes are, respectively,

$$J_n^{(0)} = (E^2+E+1)^{-1}p_n, \quad J_n^{(1)} = q_{n+1}, \quad J_n^{(j)} = w_n^{(j)}, \quad j \geq 2, \qquad (61)$$

$$J_n^{(0)} = (E^2+E+1)^{-1}[v_{n+1}+v_n - p_n(E^2+E+1)^{-1}p_{n+1}],$$
$$J_n^{(1)} = q_{n+1}(E^2+E+1)^{-1}(p_{n+1}+p_n),$$
$$J_n^{(j)} = w_n^{(j)}(E^2+E+1)^{-1}(p_{n+1}+p_n) + \sum_{l+s=j} w_n^{(l)}w_{n+1}^{(s)}, \quad j \geq 2 \qquad (62)$$

and

$$J_n^{(0)} = u_n, \quad J_n^{(1)} = v_n q_{n+1},$$
$$J_n^{(j)} = v_n w_n^{(j)} + p_n \sum_{l+s=j} w_n^{(l)}w_{n+1}^{(s)} + \sum_{l+s+m=j} w_n^{(l)}w_{n+1}^{(s)}w_{n+2}^{(m)}, \quad j \geq 2 \qquad (63)$$

(5) Conservation laws for several (1+1)-dimensional lattice-field equations

In this sub-section, we derive infinitely many conservation laws for several (1+1)-dimensional lattice-field equations which are the (1+1)-dimensional reductions of the (2+1)- dimensional BS lattice-field equations. Infinitely many conservation laws for (1+1)-dimensional reductions of the (2+1)-dimensional BS lattice-field equations (16) and (21) have been constructed[3]. Here we give the conservation laws for the (1+1)-dimensional reductions of the (2+1)- dimensional BS lattice-field equations (15), (19) and (20). In the (1+1)-dimensional case, the discrete Riccati-type equations (42) and (51) reduce to the following equations, respectively,

$$\Gamma_{n-1}\Gamma_n\Gamma_{n+1} + p_n\Gamma_{n-1}\Gamma_n + (v_n - \lambda)\Gamma_{n-1} + u_n = 0, \tag{64}$$

$$\Gamma_{n-1}\Gamma_n\Gamma_{n+1}\Gamma_{n+2} + p_n\Gamma_{n-1}\Gamma_n\Gamma_{n+1} + v_n\Gamma_{n-1}\Gamma_n + (u_n - \lambda)\Gamma_{n-1} + q_n = 0 \tag{65}$$

The solutions to the two equations can be given by the Laurent series (27), where $w_n^{(j)}\,(j \geq 1)$ are, respectively,

$$\begin{aligned} &w_n^{(1)} = u_{n+1}, \qquad w_n^{(2)} = v_{n+1}u_{n+1}, \qquad w_n^{(3)} = u_{n+1}(v_{n+1}^2 + p_{n+1}u_{n+2}), \\ &w_n^{(i)} = v_{n+1}w_n^{(i-1)} + p_{n+1}\sum_{l+s=i-1} w_n^{(l)}w_{n+1}^{(s)} + \sum_{l+s+m=i-1} w_n^{(l)}w_{n+1}^{(s)}w_{n+2}^{(m)}, \end{aligned} \tag{66}$$

and

$$\begin{aligned} &w_n^{(1)} = q_{n+1}, \qquad w_n^{(2)} = q_{n+1}u_{n+1}, \qquad w_n^{(3)} = q_{n+1}(u_{n+1}^2 + v_{n+1}q_{n+2}) \\ &w_n^{(i)} = u_{n+1}w_n^{(i-1)} + v_{n+1}\sum_{l+s=i-1} w_n^{(l)}w_{n+1}^{(s)} + p_{n+1}\sum_{l+s+m=i-1} w_n^{(l)}w_{n+1}^{(s)}w_{n+2}^{(m)} \\ &\qquad + \sum_{l+s+m+\mu=i-1} w_n^{(l)}w_{n+1}^{(s)}w_{n+2}^{(m)}w_{n+3}^{(\mu)}, \qquad i \geq 4 \end{aligned} \tag{67}$$

We thus obtain infinitely many conservation laws described by equation (47) for the (1+1)-dimensional reductions of the (2+1)- dimensional BS lattice-field equations (15), (19) and (20). The conserved densities $\rho_n^{(j)}$ is given by equation (37). Here we write down the following conserved densities and associated fluxes for the three (1+1)-dimensional differential-difference equations, respectively,

$$\rho_n^{(0)} = ln u_{n+1}, \quad \rho_n^{(1)} = v_{n+1}, \quad \rho_n^{(2)} = p_{n+1}u_{n+2} + \frac{1}{2}v_{n+1}^2, \cdots, \tag{68}$$

and

$$\rho_n^{(0)} = ln q_{n+1}, \quad \rho_n^{(1)} = u_{n+1}, \quad \rho_n^{(2)} = v_{n+1}q_{n+2} + \frac{1}{2}u_{n+1}^2, \cdots, \tag{69}$$

$$J_n^{(0)} = (E+1)^{-1}p_n, \quad J_n^{(1)} = u_{n+1}, \quad J_n^{(2)} = v_{n+1}u_{n+1}, \quad J_n^{(j)} = w_n^{(j)}, \tag{70}$$

$$J_n^{(0)} = (E^2+E+1)^{-1}p_n, \quad J_n^{(1)} = q_{n+1}, \quad J_n^{(2)} = u_{n+1}q_{n+1}, \quad J_n^{(j)} = w_n^{(j)}, \tag{71}$$

and

$$\begin{aligned} J_n^{(0)} &= (E^2+E+1)^{-1}[v_{n+1} + v_n - p_n(E^2+E+1)^{-1}p_{n+1}], \\ J_n^{(1)} &= q_{n+1}(E^2+E+1)^{-1}(p_{n+1}+p_n), \\ J_n^{(j)} &= w_n^{(j)}(E^2+E+1)^{-1}(p_{n+1}+p_n) + \sum_{l+s=j} w_n^{(l)}w_{n+1}^{(s)}, \quad j \geq 2 \end{aligned} \tag{72}$$

3. Conclusions

As is well known, infinitely many conservation laws is an important integrable property for a differential-difference system. Specially, there is less work on the infinitely many conservation laws for the (2+1)- dimensional differential-difference hierarchy. In this paper, by solving the corresponding (2+1)-dimensional discrete spectral equations, we have demonstrated the existence of infinitely many conservation laws for several (2+1)- dimensional lattice hierarchies and have derived the corresponding conserved densities and associated fluxes. To the best of our knowledge, the explicit constructions of infinitely many conserved quantities associated with the (2+1)- dimensional differential-difference hierarchies discussed in the paper are new. It should be marked that some integrable properties on (2+1)-dimensional BO equations (10) and (2+1)-dimensional BS equation (15), such as lump solution, the Lie symmetries, the Bäcklund transformation have been derived[8].

Acknowledgments

The work was supported by the National Natural Science Foundation of China under Grant No. 10671125, and by the Shanghai Pujiang Program, and also in part by the Ministry of Education and Science of Spain under contract no MTM2006-14603. I am very grateful to the referees for their useful remarks and suggestions on the original manuscript.

References

1. T. Tsuchida, H. Ujino, M. Wadati, Integrable semi-discretization of the coupled modified KdV equations, J. Math. Phys., **39** (1998) 4785.

2. T. Tsuchida, H. Ujino, M. Wadati, Integrable semi-discretization of the coupled nonlinear Schrödinger equations, J. Phys. A, **32** (1999) 2239.
3. Z. N. Zhu, et al., Infinitely many conservation laws for the Blaszak-Marciniak four-field integrable lattice hierarchy, Phys. Lett. A, **296** (2002) 280; Infinitely many conservation laws and integrable discretizations for some lattice soliton equations, J. Phys. A: Math. Gen., **35** (2002) 5079; New matrix Lax representation for a Blaszak-Marciniak four-field lattice hierarchy and its infinitely many conservation laws, J. Phys. Soc. Jpn., **71** (2002) 1864.
4. M. Blaszak and Szum, Lie algebraic approach to the construction of (2+1)-dimensional lattice-field and field integrable Hamiltonian equations, J. Math. Phys., **42** (2001) 225.
5. Z. N. Zhu, 2+1 dimensional lattice hierarchies derived from discrete operator zero curvature equation, *Proceedings of the international conference on difference equations, special functions and orthogonal polynomials, Munich, Germany, 2005. Edited by S. Elaydi, J. Cushing, R. Lasser, A. Ruffing, V. Papageorgiou and W. Van Assche, World Scientific* (2007) 762.
6. E. Date, M. Jimbo, and T. Miwa, Method for Generating Discrete Soliton Equations. I-II, J. Phys. Soc. Jpn., **51** (1982) 4116-4131; Method for Generating Discrete Soliton Equation. III, **52** (1983) 388; Method for Generating Discrete Soliton Equation. IV-V, ibid, **52** (1983) 761-771.
7. T. Tamizhmani, S. Kanaga Vel, and K. M. Tamizhmani, Wronskian and rational solutions of the differential-difference KP equation, J. Phys. A, **31** (1998) 7627.
8. H. W. Tam, X. B. Hu, and X. M. Qian, Remarks on several 2+1 dimensional lattices, J. Math. Phys., **43** (2002) 1008.
9. M. Blaszak and K. Marciniak, R-matrix approach to lattice integrable systems, J. Math. Phys., **35** (1994) 4661.
10. Y. T. Wu and X. B. Hu, A new integrable differential-difference system and its explicit solutions, J. Phys. A, **32** (1999) 1515.